Modern Methods in
Analytical Morphology

Modern Methods in Analytical Morphology

Edited by

Jiang Gu

Deborah Research Institute
Browns Mills, New Jersey

and

Gerhard W. Hacker

Institute of Pathological Anatomy
Salzburg, Austria

Springer Science+Business Media, LLC

Library of Congress Cataloging-in-Publication Data

On file

Proceedings of an International Workshop on Modern Analytical Methods in Histochemistry,
held September 14–18, 1992, in Salzburg, Austria

ISBN 978-0-306-44838-6   ISBN 978-1-4615-2532-5 (eBook)
DOI 10.1007/978-1-4615-2532-5

© 1994 Springer Science+Business Media New York
Originally published by Plenum Press New York in 1994

PREFACE

While advances in modern medicine largely parallel our understanding of morphology, discoveries in morphology are propelled by developments of new tools and means to visualize and measure tissue elements. The invention of dissecting, light, fluorescence and electron microscopes together with advances in labeling and staining techniques are among the stepping stones of morphological progress. Today, we are in an exciting new era when classical morphology is being combined with developments from other disciplines. The combination of morphology and immunology resulted in immunocytochemistry; morphology and molecular biology led to *in situ* hybridization and *in situ* PCR. Adding computer science to morphology gave birth to image analysis. Combining laser technology and the microsope evolved into confocal microscope. For more than a decade, modern morphology has continued to develop by merging with other disciplines at a rate that is still gathering momentum, providing exciting and dynamic new frontiers for other biological fields.

"Modern Methods in Analytical Morphology," based largely on the "First International Workshop on Modern Methods in Analytical Histochemistry," is an updated review of the current trends in the field. It covers an extensive array of new technical developments in major disciplines of modern morphology. The authors are not only leaders in their fields but also have extensive "hands on" experience with "bench work." Their chapters are written in a comprehensive manner including discussion of both theoretical considerations and practical applications to give the readers a broad view of the topics covered. The book is also intended to be a practical guide for those who wish to develop these techniques. Most of the recommended protocols are based on the experience of the contributing authors and have not been previously published. We hope that this book will serve as a comprehensive review and a practical guide to understanding and performing related techniques.

Jiang Gu, M.D., Ph.D.
Gerhard W. Hacker, Ph.D.

ACKNOWLEDGMENTS

We would like to thank the Deborah Research Institute, Browns Mills, NJ, USA, and the Institute for Pathological Anatomie, Salzburg, Austria, for their support in completing this book. We would like to thank the many contributing authors for sharing their knowledge and expertise in this volume.

In particular, our gratitude is extended to Dr. David R. Kersten, Michele Forte, Nancyleigh Carson, Dr. Cristina Xenachis, Dr. Robyn Rufner, Dr. Yao Tang, Denise Brunelle and those whose names are not printed in the book but who selflessly donated their time and knowledge in commenting on and correcting the content and the English language. Last, but not least, we wish to thank Gayle Englund for her tireless effort in formatting the book contributed by multilingual authors, each of whom had a different format and style and bringing the book into its current form.

CONTENTS

Chapter 1

Chapter 2

Chapter 3

Chapter 4

Chapter 5

Chapter 13

Chapter 14

Chapter 15

Chapter 16

Chapter 17

Chapter 18

Chapter 19

CHAPTER 1

THE USE OF SILVER STAINS IN THE IDENTIFICATION OF NEUROENDOCRINE CELL TYPES

Lars Grimelius[1], Huici Su[1,2] and Gerhard W. Hacker[3]

[1] Institute of Pathology,   University of Uppsala, Uppsala, Sweden
[2] Department of Histology and Embryology, Fourth Military Medical University, Xian, P.R. China
[3] Institute of Pathological Anatomy, Head, Immunohistochemistry and Biochemistry Unit, General Hospital, Salzburg, Austria

SUMMARY

Silver stains have been widely applied to identify cell types and tissue components in routinely processed tissue sections. They have declined somewhat in importance since the introduction of immunohistochemical techniques. However, some silver techniques are still useful for both histology and histopathology. In this chapter, three silver staining methods — Masson, Grimelius, and Sevier-Munger, all useful for identification of peripheral neuroendocrine cells — are discussed and commented on, and their chemical background, when known, is briefly reviewed. For Masson and Grimelius techniques, microwave irradiation protocols are outlined that allow completion of the stains in only a few minutes.

INTRODUCTION

Silver staining techniques can visualize several different types of cells and tissue components and have therefore been widely applied in both histology and histopathology. Although their importance has been somewhat on the decline since the introduction of immunohistochemical methods, some silver staining techniques are still useful in both

---

[1]Address for Correspondence: Dr. Lars Grimelius, M.D., Professor, Uppsala University Hospital, Institute of Pathology, S-751 85 Uppsala, Sweden. Phone: 46-018-66 30 00; FAX: 46-018-55 27 39.

research and routine histopathology to demonstrate neuroendocrine cells, nerve tissue, reticulum, melanin, fungi, spirochetes,[1] mitotic figures,[2] among others.

Almost all available silver staining techniques have been empirically developed and the chemical background is known for only a few of them. In the following, interest is concentrated on three silver stains frequently used on histological sections for the identification of peripheral neuroendocrine cells and their related tumors. These techniques (Masson,[3] Grimelius,[4,5] and Sevier-Munger[6]) are based on commercial readily available and well-known chemical substances, which give reproducible results in formalin-fixed tissue sections routinely processed to paraffin.

A distinction is made between argentaffin and argyrophilic reactions. Cells showing an argentaffin reaction contain one or more chemical substances which retain silver ions (from an ammoniacal silver solution) and also reduce them to metallic silver. Cells displaying an argyrophilic reaction retain silver ions from the silver solution, but visible metallic silver appears only after a subsequent reducing process brought about by an external agent or agents. The Masson stain belongs to the former category; the other two stains, Grimelius, and Sevier-Munger, belong to the latter.

In all these three silver methods, the staining is caused by metallic silver deposits in secretory granules.[7-9] Frequency and distribution of the silver grains on these granules vary to some extent in the different endocrine cell types. The staining procedures of the silver methods take from 1-3 hours. By using microwave irradiation technique, the duration of staining can be shortened to a few minutes without affecting the staining quality. In the present chapter, we comment on these three silver stains and describe their processing, including the application of microwave irradiation techniques to the Masson and Grimelius methods.

THE MASSON TECHNIQUE

The Masson stain has been modified many times, the best-known variants being those described by Hamperl in 1952,[10] and Singh in 1964.[11] In 1986, Portela-Gomes and Grimelius[12] described a simple modification where the sections were stained in a preheated (60°C) 5% ammonium silver solution for 10-20 min, followed by rinsing and mounting. By using microwave irradiation, the same staining results can be obtained by using a 1% silver ammonium solution (for details, see *staining protocols*).

On an experimental basis, Barter and Pearse[13,14] found firm evidence that the reaction product of serotonin with paraformaldehyde caused the argentaffin reaction. Lundqvist *et al.*[15] demonstrated by a dot-blot technique that dopa, dopamine, noradrenalin, adrenalin, and 5-hydroxytryptamine (serotonin) also gave rise to this silver reaction.

The Masson stain visualizes the enterochromaffin (EC) cells in gastrointestinal mucosa (*Fig. 1*), as well as their related tumors (so-called classic carcinoids) located mainly in mid-gut, in melanin deposits in the skin (*Fig. 2*), and in malignant melanomas.

A more specific way to demonstrate the presence serotonin is, of course, to use immunohistochemical staining with serotonin antibodies. However, a comparison of the frequency and distribution of argentaffin and serotonin immunoreactive cells in mid-gut carcinoids revealed more silver-positive cells than serotonin immunoreactive cells.[16] An explanation for this discrepancy could be that the argentaffin serotonin-negative cells contain chemical substances other than serotonin, thus causing the silver reaction. Serotonin immunoreactivity has been demonstrated in other neuroendocrine cells besides the enterochromaffin cells, e.g., secretin[17] and PYY cells.[18] Also, it is very difficult to raise specifically binding antibodies to serotonin which is a very small molecule usually not immunogenic by itself. Only a very few antibodies are available which can be used reliably for immunocytochemistry. Hence, the serotonin immunostaining does not demonstrate the enterochromaffin cells as specifically as does the Masson stain.

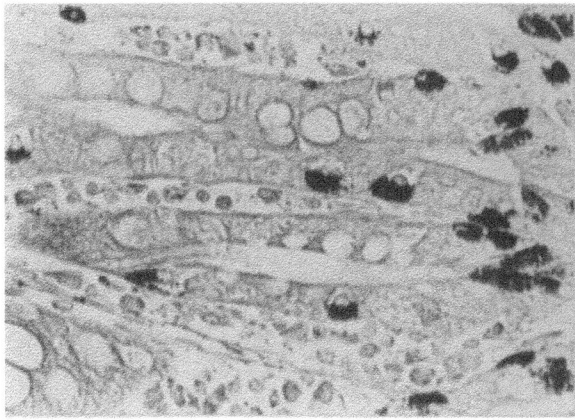

Figure 1. Human duodenal mucosa showing argentaffin (enterochromaffin) cells. At the bottom of the crypts Paneth's cells are seen containing yellow-brown granules, which contrast distinctly against the black stained enterochromaffin cells. Masson stain (microwave procedure). Original magnification X330.

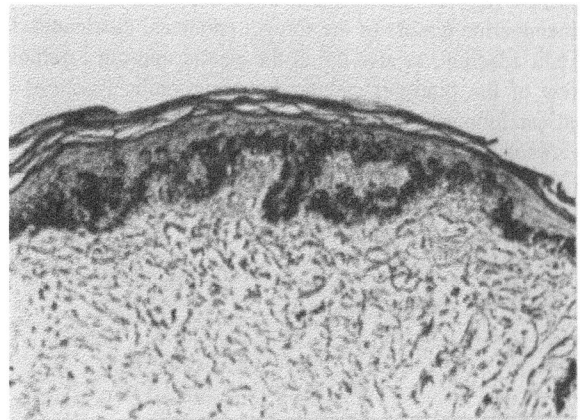

Figure 2. Human skin showing melanin in the basal cells. Masson stain (microwave procedure). Original magnification X160.

GRIMELIUS SILVER NITRATE STAIN

The Grimelius technique[4] (1968) was initially devised to demonstrate a non-insulin cell fraction of the human pancreatic islets. In the first staining step, a 0.03% silver nitrate solution (pH 5.6) is used and the 'impregnation' process takes place at 37°C for 24 hours or at 60°C for 3 hours. The reducing solution consists of aqueous hydroquinone-sodium sulphite solution, and the reducing process takes place at 45°C for 1 min. When the staining technique was applied to gastrointestinal mucosa to demonstrate neuroendocrine cells, a more intense argyrophil reaction was obtained than when the silver nitrate concentration in the 3 hour "impregnation" variant was increased from 0.03% to 0.07%. A further intensification of staining was obtained when the temperature of the reducing solution was increased from 45°C to 55-58°C.[19]

A modified Grimelius staining method has been developed where microwave irradiation is applied in both staining steps.[20] By using this modification, the staining procedure can be shortened to 3 min (for staining details, see *staining protocols*).

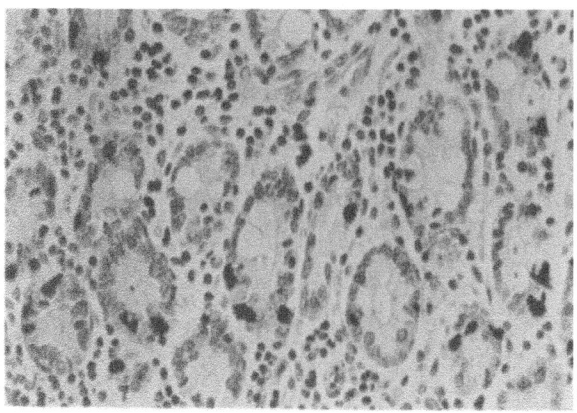

Figure 3. Human duodenal mucosa showing argyrophil cells. Grimelius stain. Original magnification X230.

The argyrophilic reaction occurs in most neuroendocrine cell types of the gastrointestinal mucosa (*Fig. 3*) except in CCK and somatostatin cells. It is positive e.g. in glucagon and PP cells of the endocrine pancreas, in thyroid C cells, and in adrenal medulla (noradrenalin). Argyrophilic cells also occur in pituitary gland and in paraganglia. Most of the neuroendocrine tumors of the foregut (stomach, duodenum, lung, pancreas) show an argyrophilic reaction, as also do all the classic mid-gut carcinoid tumors (Fig. 4), but only a few of the hind-gut carcinoids. Argyrophilic reactions also appear in medullary thyroid carcinomas, pheochromocytomas, and paragangliomas. The staining intensity varies between the different neuroendocrine cell types and also between different neuroendocrine tumors[19]. Insulin cells of the pancreatic islets do not show the argyrophilic reaction as mentioned above, but this reaction does occur in some insulinomas.[21]

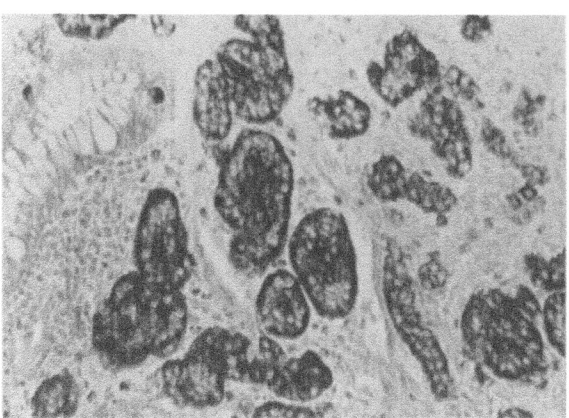

Figure 4. Human mid-gut carcinoid tumor showing argyrophil reaction in the tumor nests. Grimelius stain. Original magnification X160.

Chemical Background

Rindi *et al.*[22] showed by a dot-blot technique that chromogranin A, a granular protein, gives rise to the Grimelius argyrophilic reaction. This result has been confirmed by Lundqvist *et al.*[15] who also showed by using the same technique that dopamine, noradrenalin, and serotonin cause the Grimelius silver-positive reaction. By sequential staining of pancreatic islets and intestinal mucosa with immunohistochemical technique and silver staining on the same tissue section, it was shown that chromogranin A immunoreactivity and the Grimelius argyrophilic reaction occur in the same cells.

THE SEVIER-MUNGER TECHNIQUE

The argyrophil Sevier-Munger technique[6] was initially developed to demonstrate neural tissue, but the staining method also visualizes the enterochromaffin-like (ECL) and D_1 cells of the human gastric mucosa (*Fig. 5*) and the enterochromaffin cells of the gastrointestinal mucosa, and gastric inhibitory peptide (GIP) cells of the intestinal mucosa. The initial "impregnation" takes place in a 20% aqueous silver nitrate solution, while the subsequent staining procedure occurs in a"physical developer", i.e., a solution containing both silver ions and a reducing agent (for details, see *staining protocols*. See also the chapter by Danscher, in this book.)

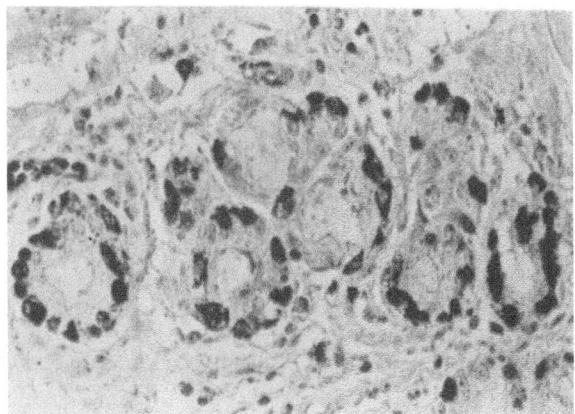

Figure 5. Human gastric oxyntic mucosa showing linear hyperplasia of the enterochromaffin-like (argyrophil) cells. The patient suffered from pernicious anemia with hypergastrinemia. Sevier-Munger stain. Original magnification X330.

The ECL cells and tumors derived for this cell type (ECLoma) contain histamine, a biogenic amine which can be demonstrated in tissue fixed in formalin-carbutamide fixative or in freeze-dried tissue vapor-fixed in diethylpyrocarbonate (DEPC).[23,24] In formalin-fixed, paraffin embedded tissue, ECL cells can be demonstrated indirectly by combining the staining results of the Sevier-Munger and Masson methods on adjacent sections; the ECL cells are those cells stained with the Sevier-Munger stain but failing to show argentaffin (Masson) reaction. Weak argyrophilic reaction has also been reported to occur in pancreatic PP cells. The chemical background to the Sevier-Munger stain is still not known.

Modified Masson stain (Portela-Gomes and Grimelius, 1986)[12]

Ammonium hydroxide silver solution is prepared by dissolving 5 g silver nitrate in 100 ml double-distilled water and then adding ammonium hydroxide drop by drop until the brown-black precipitation disappears. Then, add another few drops of 5% silver nitrate solution until the solution is slightly cloudy.

1. Preheat the ammoniacal silver solution to 60°C and transfer the sections into this solution for 5-20 min (check the staining in the light microscope and interrupt at an optimal staining).
2. The sections are then rinsed in water and mounted in Canada balsam, DPX (BDH Biochemicals, UK) or Cytoseal Mounting Medium (Curtin Matheson Scientific, Inc., Wayne, NJ, USA). If necessary, counterstain the sections with nuclear fast red or methyl green.

Microwave Procedure for Masson Staining (modified)

Prepare the ammoniacal silver solution as above, but use 1 g silver instead of 5 g. The solution should be slightly cloudy. Filter the solution into a plastic Coplin jar.

1. Place the deparaffinized sections in the silver solution and irradiate in a microwave oven (e.g., Miele Electronic M696) at 450 W for 1 min. Do not put more than 8-10 slides into 100 ml silver solution.
2. Allow the sections to remain in the hot silver solution for a few minutes until they attain a light golden color.
3. Rinse well, dehydrate clear in xylene and mount in Canada balsam, DPX, or Cytoseal, as above.

Grimelius Staining (modified)[19]

Silver solution: Dissolve 70 mg silver nitrate in 100 ml of Walpole's acetate buffer solution, pH 5.6 diluted 1:10 (buffer preparation: 48 ml of 0.2 M acetic acid + 452 ml of 0.2 M sodium acetate in 500 ml glass double-distilled water). The buffer alone can be stored for months in the fridge (4-8°C).
Reducing Solution: Dissolve 1 g hydroquinone and 5 g sodium sulphite (anhydr.) in 100 ml of distilled water.
The silver and reducing solutions should both be freshly prepared. Formaldehyde and Bouin's fluid are the best fixatives for the Grimelius stain. Glutaraldehyde, ethanol, or fixative containing ethanol cause a weaker argyrophil reaction, or it may fail to arise altogether. Heavy metals in the fixative cause silver precipitation on the sections.

1. Place the sections in the silver solution at room temperature and transfer them later to an oven maintaining a temperature of 60°C for 3 h. Do not stain more than 6-8 glass slides per 100 ml silver solution.
2. Wipe dry the glass slides **around** the sections (the sections themselves should not dry out!) and transfer them to a prewarmed (57-58°C) reducing solution for a minimum of 1 minute. The reducing time is not critical, i.e., the sections can be left for more than 1 min in this solution.
3. Rinse the sections, dehydrate, and mount in Canada balsam, Cytoseal or DPX (avoid eukitt, entellan, glycerine/gelatine as mounting medium as they may remove the silver grains).

If the argyrophilic reaction appears weak, it can be enhanced in the following way: transfer the well-rinsed sections to a freshly prepared silver solution (see above) at room temperature for 10 min. Wipe the glass slides **around** the sections as above and immerse them in a freshly prepared prewarmed (57-58°C) reducing solution (see above) for 1 min. Rinse the sections, dehydrate, clear in xylene, and mount. Counterstain, if necessary, with methyl green.

Microwave Procedure for Grimelius Staining[20]

Use the silver solution as above, but increase the silver nitrate concentration to 1g/90 ml diluted buffer solution. The composition of the reducing solution is as mentioned earlier.

1. Immerse the sections in 90 ml silver solution and irradiate at 450 W for 100 seconds until the solution starts to boil. Leave the slides in the hot solution for a further 1 min.
2. Rinse the sections in distilled water for a few seconds and then transfer them to 40 ml of the reducing solution preheated (450W for 90s) until it starts to boil. Leave the sections in the solution until they assume a golden-brown color. If sections are not yet successfully stained, proceed with paragraphs 3 and 4.
3. After careful rinsing, return the slides to the silver solution (filtered before re-use) and irradiate at 450 W for 45 s until it starts to boil.
4. Reheat the reducing solution (filtered before re-use) by irradiation at 450 W for 45 s until it starts to boil, and transfer the slides to the hot solution until the sections assume a golden-brown color.
5. Rinse carefully and, if necessary, counterstain with 0.2% nuclear fast red for 3 min. Dehydrate, clear in xylene and mount in Canada balsam, DPX, or Cytoseal.

Sevier-Munger Staining (modified)[6,19]

Silver solution: Dissolve 10 g silver nitrate in 50 ml glass double-distilled water and filter it into a Coplin jar.

Ammonium hydroxide-silver solution: Add 28-30% (conc.) ammonium hydroxide drop by drop to 50 ml of a 10% silver nitrate solution until the dark brown precipitate has **almost** disappeared. Add 0.5 ml of a sodium carbonate solution (prepared by dissolving 8 g $Na_2CO_3.(10H_2O)$ in 30 ml glass double-distilled water) and shake well. Add further 25 drops of conc. ammonium hydroxide during vigorous shaking. The solution should be clear. Filter the solution into a Coplin jar. Add 10 drops of 0.8% formaldehyde (2 ml conc. formalin plus 98 ml distilled water) just before use.
The best fixatives for this staining are formaldehyde and Bouin's fluid.

1. Preheat the 20% silver solution to 60°C and leave the deparaffinized sections at that temperature for 15 min.
2. Rinse the sections thoroughly in tap water and then in double-distilled water and transfer them to the Coplin jar containing the ammoniacal silver-formalin solution. The Coplin jar should be placed in a hot oven (60°C) for 15 to 30 min. Check the staining under the light microscope and interrupt the process when the cells appear black.
3. Rinse the sections well in distilled water. Counterstaining is usually not necessary. Dehydrate, clear in xylene and mount in Canada balsam, DPX, or Cytoseal.

REFERENCES

1. J.D. Bancroft and A. Stevens. Theory and Practice of Histological Techniques. Churchill Livingstone, New York (1990).
2. C. Busch and J. Vasko, Differential staining of mitoses in tissue sections and cultured cells by a modified methenamine-silver method, *Lab Invest* **59**:876 (1988).
3. P. Masson, La glande endocrine de l'intestin chez l'hormone, *CR Acad Sci Paris* **15**:59 (1914).

4. L. Grimelius, A silver nitrate stain for A2 cells of human pancreatic islets, *Acta Soc Med Upsal* **73**:243 (1968).

5. L. Grimelius, The argyrophil reaction in islet cells of adult human pancreatic islet cells, with reference to a new silver nitrate procedure, *Acta Soc Med Upsal* **73**:271 (1986).

6. A.C. Sevier and B.L. Munger, A silver method for paraffin sections of neural tissue, *J Neuropath Exp Neurol* **24**:130 (1965).

7. L. Grimelius, An electron microscopic study of silver stained adult human pancreatic islet cells, with reference to a new silver nitrate procedure, *Acta Soc Med Upsal* **74**:28 (1969).

8. G. Vassallo, C. Capella, and E. Solcia, Endocrine cells of the human gastric mucosa, *Z. Zellforsch* **118**:49 (1971).

9. G. Vassallo, C. Capella, and E. Solcia, Grimelius silver stain for endocrine cell granules, as shown by electron microscopy, *Stain Techn* **46**:7 (1971).

10. H. Hamperl, Über argyrophile Zellen, *Virchows Arch Path Anat* **321**:482 (1952).

11. I. Singh, A modification of the Masson-Hamperl method for staining argentaffin cells, *Anat. Anz.* **115**:81 (1964).

12. G.M. Portela-Gomes and L. Grimelius, Identification and characterization of enterochromaffin cells with different staining techniques, *Acta Histochem* **79**:161 (1986).

13. R. Barter and A.G.E. Pearse, Detection of 5-hydroxy-tryptamine in mammalian enterochromaffin cells, *Nature* **171**:810 (1953).

14. R. Barter and A.G.E. Pearse, Mammalian enterochromaffin cells as the source of serotonin (5-hydroxytryptamine), *J Path Bact* **69**:25 (1955).

15. M. Lundqvist, H. Arnberg, J. Candell, M. Malmgren, E. Wilander, L. Grimelius, and K. Öberg, Silver stains for identification of neuroendocrine cells. A study of the chemical background, *Histochem J* **22**:615 (1990).

16. M. Lundqvist and E. Wilander, Small intestinal chromaffin cells and carcinoid tumours: a study with silver stains, formalin-induced fluorescence and monoclonal antibodies to serotonin, *Histochem J* **16**:1247 (1984).

17. Y. Cetin, Secretin cells of the mammalian intestine contain serotonin, *Histochemistry* **93**:601 (1990).

18. A.I.C. Lukinius, J.L.E. Ericsson, M.K. Lundqvist, and E.M.O. Wilander, Ultrastructural localization of serotonin and polypeptide YY (PYY) in endocrine cells of the human rectum, *J Histochem Cytochem* **34**:719 (1986).

19. L. Grimelius and E. Wilander, Silver stains in the study of endocrine cells of the gut and pancreas, *Invest Cell Pathol* **3**:3 (1980).

20. M.E. Boon and L.P. Kok, "Microwave Cookbook of Pathology; The Art of Microscopic Visualization," Coulomb Press Leyden, Leiden (1987).

21. W. Creutzteldt, Pancreatic endocrine tumors - the riddle of their origin and hormone secretion, Israel *J Med Sci* **11**:762 (1975).

22. G. Rindi, R. Buffa, F. Sessa O. Tortora, and E. Solcia, Chromogranin A, B and C immunoreactivities of mammalian endocrine cells. Distribution, distinction from costored hormones/prohormones and relationship with the argyrophil component of secretory granules, *Histochemistry* **85**:19 (1986).

23. E. Solcia, C. Capella, G. Vassallo, and R. Buffa, Endocrine cells of the gastric mucosa, *Int Rev Cytol* **42**:223 (1975).

24. R. Håkansson, G. Böttcher, E. Ekblad, P. Panula, M Simonsson, M. Dahlsten, T. Hallberg, and F. Sundler, Histamine in endocrine cells in the stomach. A survey of several species using a panel of histamine antibodies, *Histochemistry* **86**:5 (1986).

CHAPTER 2

QUANTITATIVE IMMUNOHISTOCHEMISTRY FOR THE INVESTIGATION OF REGULATORY PEPTIDES IN HEALTH AND DISEASE

Giorgio Terenghi and Julia M. Polak[1]

Department of Histochemistry
Royal Postgraduate Medical School
Hammersmith Hospital, London W12, UK

SUMMARY

Immunohistochemical studies often present the problem of quantifying the immunoreactive structures in an objective and reliable way, particularly when examining samples which may present possible changes. The advent of computerized image analysis has made possible the quantitative evaluation of immunoreactive structures identified with chromogenic or immunofluorescent methods. However, there are a number of potential problems which can be encountered during image analysis quantification. Size changes due to fixation shrinkage have been observed during specimen preparation, while tissue cutting can cause compression of the sections. Non-specific tissue autofluorescence can be reduced by the use of counterstaining or filters. Also, during imaging it is important to stabilize the power supply to provide uniform illumination. Video cameras tend to be more light-sensitive at the center of the field than at the edges, resulting in uneven illumination. For analysis, the average image input can be modified by defining a specific area or deleting unwanted signal. Before the image is measured, grey levels need to be defined to discriminate the image intensity, and analysis parameters should be selected.

Using quantitative immunohistochemistry, it is possible to provide an accurate assessment of nerve density or of immunostained surface area from images containing a large number of immunoreactive structures. These measurements can be made rapidly.

[1] Address for correspondence: Julia M. Polak, D.Sc., M.D, FRCPath, Professor, Dept. Histochemistry, Royal Postgraduate Medical School, Hammersmith Hospital, London, UK; Tel:081-740-3231; Fax:081-743-5362

Modern Methods in Analytical Morphology, Edited by
J. Gu and G.W. Hacker, Plenum Press, New York, 1994

Statistically significant differences may be demonstrated when comparing control, experimental, or pathological tissues.

INTRODUCTION

Immunohistochemistry has been used widely to study in detail the mapping of regulatory peptides of the diffuse neuroendocrine system (DNES). In morphological studies, it is often necessary to obtain a precise and objective quantitative assessment of the identified structures, particularly in experimental and pathological studies where tissue changes may be present. In some situations, manual counts can be carried out, but this method is time-consuming and not always accurate. The introduction of computerized image analysis has opened up the possibilities of this methodology. Quantification procedures have become more automated, fast, and precise. Furthermore, fluorescent-labelled antibodies[1-3] or chromogenic reporter molecules[4,5] can be used for a detailed assessment of the results. In this chapter, we will give an overview of the quantification technique, and examine some of the problems that may be encountered.

METHODOLOGY

The preparation of tissue samples is pivotal to obtain good results from the imaging process. Hence, it is important to be aware of the potential problems to be avoided during preparation, staining, and quantification of the tissue. Also, it must be remembered that an accurate calibration of the measuring procedures is essential to achieve reliable and reproducible results.[6-8]

During fixation, the tissue samples might undergo a certain amount of shrinkage. This problem is particularly evident in soft or elastic tissues as demonstrated by the comparison of tissue dimensions before and after fixation. Hence, standardization of the fixation technique is essential to minimize possible differences between the samples under investigation. Tissue cutting can also cause compression of the sections. In tissues where this problem is particularly evident, it is recommended that a correction factor be determined in order to express the measurement data in terms of the original dimensions.[9,10] When comparative quantification is carried out, for example between control and experimental tissues, all samples must be processed at the same time to minimize differences of procedure due to sectioning and immunostaining.

The use of immunofluorescent preparations can be limiting because of the fluorescence fading due to exposure during screening or photography. It has been claimed that the rate of fading can be decreased using specific solutions for mounting specimens,[11] but these are not always effective. Storage of the preparations at room temperature is not recommended. However, it has been found that quantification measurements were not significantly changed after storage of the immunofluorescent staining in a freezer up to three weeks,[12] although storage at 4°C has proved adequate in our experience.

During quantification, interference problems can be encountered because of background autofluorescence. In this case, image contrast and resolution can be improved by using a cut-off filter of appropriate wavelength or a suitable counterstain, such as Evans Blue[13] or Pontamine Sky Blue.[14] Also, the treatment of the sections with the serum of the species of the second layer antibody may prove useful in decreasing the background signal.[15] Summing several images captured sequentially and inputting an average of them into the frame store for measurement can be an effective way to reduce random noise. With this approach, it is possible to increase the signal/noise ratio of the image, as the random noise present in one frame is unlikely to be repeated in successive frames.

Uneven illumination of the section is generally caused by the microscope optical system. This may give a brighter area in the center of the field and a darker surrounding area. This problem can bias the segmentation and thresholding of the image. The use of coated lenses and filters has proved beneficial in solving this problem, but a correction procedure can be inserted into the computer program. This operates by subtracting the averaged image of the background illumination from the section images, hence eliminating, to a large extent, the problems caused by microscope lighting and optics.[16] Fluctuations in the main current supply due to other electrical equipment are common, but they can be controlled by using a stabilizer transformer which isolates the image analyzer.

Image editing is often required to separate the areas of interest from the background noise or from irrelevant structures present in the field. The most common procedure is to define the area interactively using measuring frames drawn with a digitizer. Sophisticated programs can be prepared whereby specific cell shapes or structures are identified automatically. Following editing, it is advisable to enhance the image before segmentation to further increase the signal/noise contrast. This is particularly important when measuring weakly stained structures. The segmentation of the image is carried out by substraction of the background, using the maximum grey value of background as the threshold at which to separate the image to be measured from the background. It is possible to set an automatic threshold value, but the results might be biased by the intrinsic variations of immunostaining and background in each sample if the threshold value is set at the same level for all specimens. Routinely, the image segmentation is carried out interactively. Subjective factors due to the observer selecting the threshold levels may influence the measurements substantially. Attempts have been made to establish the influence of the observer or of the image characteristics on the threshold setting. A large variation was found between observers in the selection of threshold values of a given image, although an increase of viewing experience contributed to a decrease in variability. Also, a better discriminatory power to detect true differences between samples can be obtained by employing several observers measuring a limited number of fields rather than one observer measuring numerous fields. This indicates that the subjectivity related to the interaction between observer and image analyzer can bias the results heavily.[6]

The binary image obtained by the thresholding process is then measured with the chosen parameters which are selected according to the system under investigation. The measurements are generally made on the whole field. The automatic counting includes the total number of thresholded pixels within a defined selected area. In our studies on changes of immunoreactive nerves in different pathological conditions, the immunostained structures were measured mainly by using three different parameters of quantification: number of fibers per field (count/field), total immunostained area of the field (μm^2/field), and number of intercept counts (nerve/mm). The latter was related to the length and number of the nerve fibers. Fine nerve terminals mainly contribute to counts/field while large fibers and nerve bundles have a determinant role in defining the immunostained area/field. The tissue under study is generally divided into definite areas. For example, in the skin the immunoreactivity in epidermis/ papillary dermis is measured separately from that around the sweat glands. For each immunostained peptide or marker, a predetermined number of sections are analyzed. In each section, several fields (objective magnification X20) are measured for every defined area. Adjacent non-overlapping fields of epidermis/ papillary dermis are selected by scanning along the superficial layer of the skin and measurements are carried out on alternate fields. Sweat glands are identified within the dermis and measurements made in all the fields necessary to include all the glands present in the section.

Point counts and area measurements of immunoreactive structures are the most common parameters of quantification. However, it is also possible to quantify the levels of antigens using either the immunohistochemical supra-optimal dilutions (S.O.D.) method[17] or the microdensitometric measurement of the immunostaining intensity.[18] The S.O.D. method is based on the reduction of sensitivity of immunostaining with increasing dilution of primary antisera. In effect, this procedure renders the lower levels of antigen undetectable and, thus, demonstrates differences in the amount of antigen not detected with the optimal dilution of antisera saturating all available antigenic sites. Using this technique, it was possible to recognize an increase of calcitonin gene-related peptide (CGRP)-containing pulmonary endocrine cells in hypoxic lung.[19] Microdensitometric analysis of immunostaining intensity has the advantage over the above methods of providing a more precise and accurate assessment of the magnitude of changes. However, the method is technically demanding and it requires a very accurate standardization of the staining procedure.[20,21] By applying microdensitometry, it was possible to establish precisely the amount of CGRP antigen increase in the endocrine cells of hypoxic animals.[18] Intensity of staining can also be calculated from immunofluorescence preparations,[16] although a different approach is needed from that used on chromogenic preparations. This includes more strict controls to take into account the possibility of fluorescence fading and the variability of immunostaining between samples. Radiolabelled secondary antibodies have been used as an alternative. The method is based on the use of radiolabelled antibodies in conjunction with isotopic calibration standards simultaneously exposed to autoradiography film. The optical density of the resulting film images is then measured by image analysis and referred to the standards for calculation of relative amounts of antigen present.[22]

APPLICATIONS

The involvement of neuropeptides in human and experimental diseases is now very evident. Morphologically, this can be seen as an alteration of peptide immunoreactivity. Often, the extent of these changes needs to be defined more objectively by image analysis because of the difficulty in assessing them by visual microscopical examination.

Alterations of peripheral neuropeptide-containing nerves have been studied in cold-related diseases, particularly in primary Raynaud's phenomenon. The finding that intravenous injections of CGRP cause an improvement of cutaneous blood circulation in these patients[23] is consistent with the decrease of CGRP-immunoreactive nerves around microvessels in the papillary dermis found by manual counting.[24] However, it was only by the use of image analysis quantification that it was also possible to detect a decrease of protein gene product 9.5 (PGP), a pan neuronal marker and vasoactive intestinal peptide (VIP)-immunoreactive nerves around cutaneous microvessels and sweat glands, respectively[3] (Figs. 1-3). This clearly demonstrates the advantage of image analysis quantification in precisely identifying small changes in large populations of immunostained structures which can not be easily assessed by subjective examination.

Peripheral sensory and autonomic neuropathy is commonly found in diabetic patients. In semiquantitative studies of human skin biopsies of severely neuropathic patients, PGP, CGRP and VIP-immunoreactive fibers showed a marked decrease.[25] Further quantitative studies on long term diabetic patients showed that the reduction of immunoreactive fibers was progressive, and more evident in neuropathic patients (i.e. with abnormal small fiber functional tests) than in non-neuropathic patients (i.e. with normal tests). Also, the decrease of VIP-immunoreactive fibers showed a statistically significant correlation with the variation of functional tests in sweat glands (acetylcholine-stimulated sweat output and sympathetic skin response).[26]

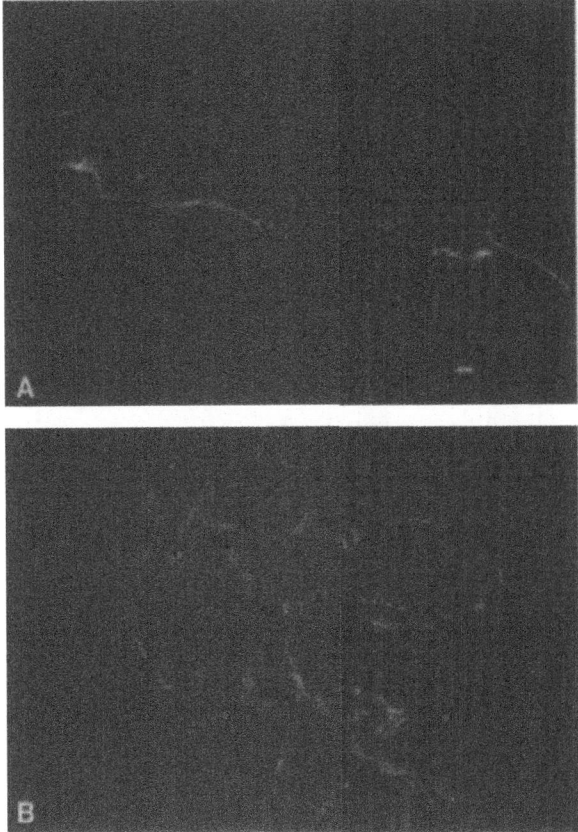

Figure 1. Control skin showing staining for **(A)** PGP-immunoreactive nerve fibers in epidermis and subepidermis and **(B)** VIP-immunoreactive nerve fibers around sweat glands. Indirect immunofluorescence method.

When patients with diabetes of different lengths were examined, it was possible to define an increase of PGP- and VIP-immunoreactive nerves around the sweat glands of patients with diabetes of less than 3 years, in contrast to the sharp decrease of these nerves with the progression of the pathology (*Fig. 4*). Similarly, a time related change in PGP- and CGRP-immunoreactive nerves was found around the capillaries of the papillary dermis[27] (*Fig. 5*). It is noteworthy that these morphological changes were observed prior to the demonstration of abnormal neurophysiological tests. These were used in diagnostic clinics to assess the extent of neuropathy. Endothelin-1 (ET-1) is a vasoconstrictor peptide which is produced by and released from endothelial cells of all blood vessels, including microvessels of the skin.[28] In diabetic patients, quantification of ET-1 immunostaining in blood vessels showed that in early diabetic patients there was an increase of immunoreactivity when compared to control subjects followed by a sharp decrease in patients with longer duration of diabetes (*Fig. 6*). Also, there was a significant correlation between the increase of ET-1 immunoreactivity and the early stages of retinopathy as assessed by clinical tests[29] (*Fig. 7*). These results indicate the potential of quantitative immunohistochemistry on skin biopsies as a non-invasive and repeatable test for studying the evolution of diabetes.

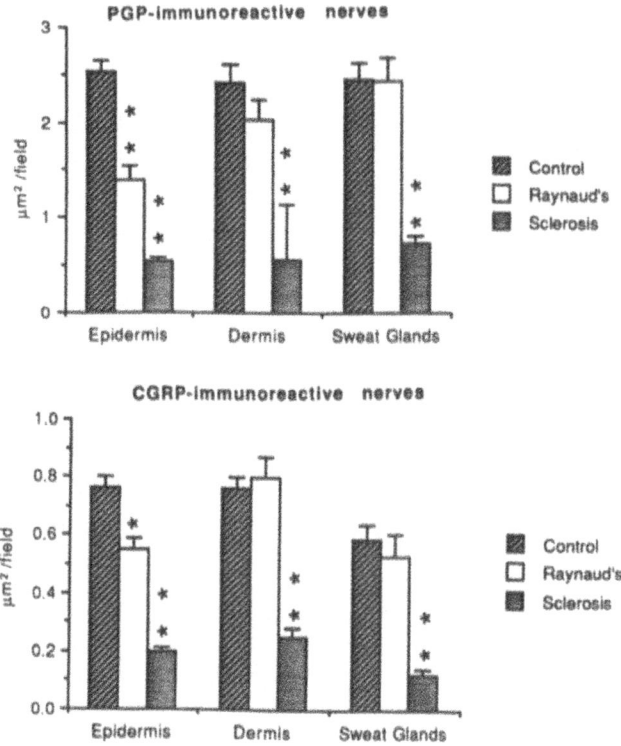

Figure 2. Histograms showing the total field fluorescent area (μm²) of PGP- and CGRP-immunoreactive nerves in different areas of the skin of controls and patients with primary Raynaud's phenomenon and with Raynaud's phenomenon secondary to systemic sclerosis. The values are expressed as mean, with the SEM represented by error bars. *P=0.005, **P<0.001.

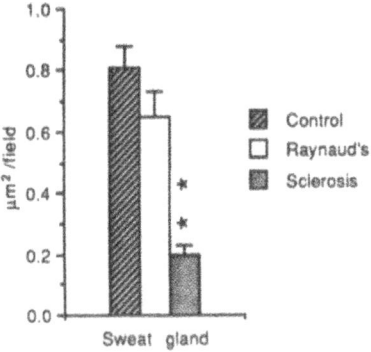

Figure 3. Histogram showing the total field fluorescent area (μm²) of VIP-immunoreactive nerves around the sweat glands of controls and patients with Raynaud's phenomenon, either primary or secondary to systemic sclerosis. The values are expressed as mean, with the SEM represented by error bars. **P<0.001.

14

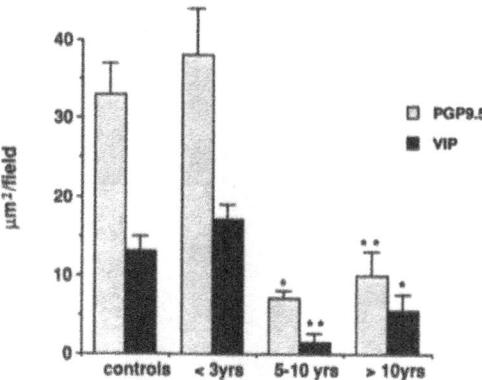

Figure 4. Histogram showing the total field fluorescent area (μm²) of PGP- and VIP-immunoreactive nerves around the sweat glands of controls and diabetic patients divided according to the duration of the disease. The values are expressed as mean, with the SEM represented by error bars. *P<0.02, **P<0.001.

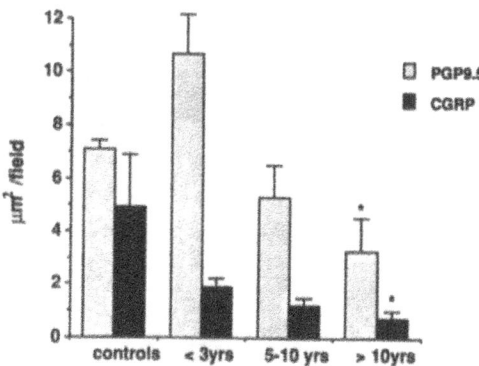

Figure 5. Histogram showing the total field fluorescent area (μm²) of PGP- and CGRP-immunoreactive nerves in the epidermis and subepidermis of controls and diabetic patients, divided according to the duration of the disease. The values are expressed as mean, with the SEM represented by error bars. *P<0.05.

The pathological changes associated with Alzheimer's disease have been well documented in the central nervous system.[30] Several studies using manual counts have correlated the number of plaques seen in post mortem brain samples with the severity of the dementia observed in patients.[31,32] Recently, immunohistochemical investigations have shown that ßA4 protein is a useful marker for detecting the extent of plaque deposits.[33] By combining image analysis quantification and ßA4 immunostaining, it has been possible to identify the precise changes of this protein between different brain areas of Alzheimer's disease patients.[34] Furthermore, a significant correlation was found between the amount of ßA4 plaques in the temporal cortex and the severity of Alzheimer's-type dementia.[35]

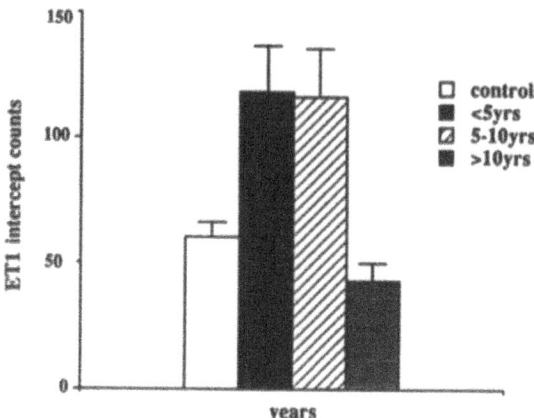

Figure 6. Histogram showing intercept counts for endothelin-1 (ET-1) immunostaining in control and diabetic patients. An increase of measurable ET-1 immunoreactivity was found in patients with diabetes duration for less than 10 years (P=0.06), with a decrease in the patients with longer term diabetes. The values are expressed as mean values, with the SD represented by error bars.

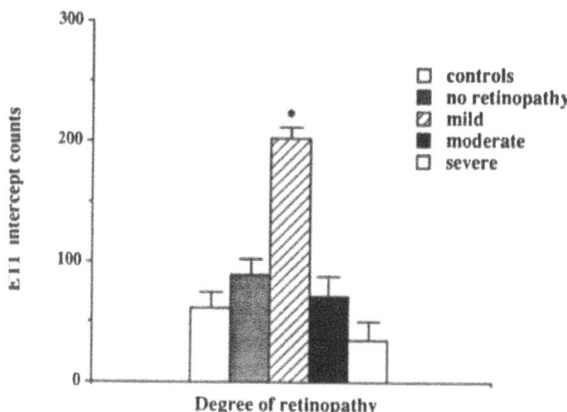

Figure 7. Histogram showing the correlation between the severity of retinopathy and endothelin-1 (ET-1) immunoreactivity measured with intercept counts. A significant increase (*P=0.005) was observed in patients with mild retinopathy compared to controls, while a decrease was found in patients with severe retinopathy. The values are expressed as mean values, with the SD represented by error bars.

CONCLUSIONS

From these examples, it is evident that image analysis quantification can be a useful tool for research into normal and pathological conditions. It may provide important information which is not always obtainable with subjective or semiquantitative microscopical examination of the tissue. The use of image analysis quantification, in all its different forms and in combination with other techniques, can improve the morphological assessment of different tissues and expand the current understanding of different patho-physiological events.[14]

16

REFERENCES

1. T. Cowen, C. Alafaci, H.A. Crockard, and Burnstock, Origin and postnatal development of nerves showing 5-hydroxytryptamine-like immunoreactivity supplying major cerebral arteries of the rat, *Neurosci Lett* **78**:121 (1987).

2. J.D. Gale, J.C.J. Alberts, and T. Cowen, A quantitative study of changes in old age of 5-hydroxytryptamine-like immunoreactivity in perivascular nerves of the rabbit. *J Auton Nerv Syst* **28**:51 (1989).

3. G. Terenghi, C.B. Bunker, Y-F. Liu, D.R. Springall, T. Cowen, P.M. Dowd, and J.M. Polak, Image analysis quantification of peptide-immunoreactive nerves in skin of patients with Raynaud's phenomenon and systemic sclerosis, *J Path* **164**:245 (1991).

4. L.F. Agnati, K. Fuxe, A.M. Janson, M. Zoli, and A. Haustrand, Quantitative analysis, computer assisted morphometry and microdensitometry applied to immunostained neurones, *in*: "Immunocytochemistry: modern methods and applications", Polak JM, Van Noorden S (eds), John Wright & Sons Ltd, Bristol, pp 205 (1986).

5. R.J.I. Hitchcock, M.J. Pemble, A.E. Bishop, L. Spitz, and J.M. Polak, The ontogeny and distribution of neuropeptides in the human fetal and infant oesophagus, *Gastroenterology* **102**:840 (1992).

6. R. Jagoe, J.H. Steel, V. Vucicevic, N. Alexander, S. Van Noorden, R. Wooton, and J.M. Polak, Observer variation in quantification of immunocytochemistry by image analysis. *Histochemistry* **23**:541 (1991).

7. R.E. Coggeshall, A consideration of neural counting methods. *TINS* **15**:9(1992).

8. T.J. Stephenson, Image analysis - Critical summaries. *J Pathol* **166**:83 (1992).

9. T. Cowen and G. Burnstock, Image analysis of catecholamine fluorescence. *Brain Res Bull* **9**:81 (1982).

10. T. Cowen, C. Alafaci, H.A. Crockard, and G. Burnstock, 5HT containing nerves to cerebral arteries of the gerbil originate in the superior cervical ganglion. *Brain Res* **384**:51 (1986).

11. G.D. Johnson and G.M. Araujo, A simple method of reducing the fading of immunofluorescence during microscopy, *J Imm Meth* **43**:349 (1981).

12. T. Cowen, Image analysis of FITC-immunofluorescence histochemistry in perivascular substance P-positive nerves, *Histochemistry* **81**:609 (1984).

13. I.S. De la Lande and J.G. Waterson, Modification of autofluorescence in the formaldehyde-treated rabbit ear artery by Evans Blue, *J Histochem Cytochem* **16**:281 (1968).

14. T. Cowen, A.J. Haven, and G. Burnstock, Pontamine Sky Blue: a counterstain for background autofluorescence in fluorescence and immunofluorescence histochemistry, *Histochemistry* **82**:205 (1985).

15. J.M. Polak and S. Van Noorden S (eds), "Immunocytochemistry - Modern methods and applications", John Wright & Sons, Bristol (1986).

16. T. Cowen and C. Thrasivoulou, Cerebrovascular nerves in old rats show reduced accumulation of 5-hydroxytryptamine and loss of nerve fibres, *Brain Res* **513**:237 (1990).

17. L. Vacca-Galloway, Differential immunostaining for substance P in Huntington's disease and normal spinal cord: significance of serial (optimal, supraoptimal and end-point) dilutions of primary antiserum in comparing biological specimens, *Histochemistry* **83**:561 (1985).

18. J.T. McBride, D.R. Springall, R.I.D. Winter, and J.M. Polak, Quantitative immunocytochemistry shows calcitonin gene-related peptide-like immunoreactivity in lung endocrine cells is increased by chronic hypoxia in the rat, *Am J Resp Cell Mol Biol* **3**:587 (1990).

19. D.R. Springall, G. Collina, G. Barer, A.J. Suggett, D. Bee, and J.M. Polak, Increased intracellular levels of calcitonin gene-related peptide-like immunoreactivity in pulmonary endocrine cells of hypoxic rats, *J Pathol* **155**:259 (1988).

20. R.H. Benno, L.W. Tucker, T.H. Joh and D.J. Reis, Quantitative immunocytochemistry of tyrosine hydroxylase in rat brain. I. Development of a computer assisted method using the peroxidase-antiperoxidase technique, Brain Res **246**:225 (1982).

21. R.R. Mize, R.N. Holdefer, and L.B. Nabors, Quantitative immunocytochemistry using an image analyzer. I. Hardware evaluation, image processing, and data analysis, *J Neurosci Method* **26**:1 (1988).

22. A.P. Davenport, S.J. Augood, D.E. Lawson and P.C. Emson, The use of quantitative immunocytochemistry (QICC) to measure calbindin D28K-like immunoreactivity in the rat brain, *Cell Mol Biol* **36**:1 (1990).

23. C.B. Bunker, J.C. Foreman, D. O'Shaughnessy, C. Reavly, and P.M. Dowd, Calcitonin gene-related peptide in the treatment of severe Raynaud's phenomenon, *Br J Dermatol* **121**:43 (1989).

24. C.B. Bunker, G. Terenghi, D.R. Springall, J.M. Polak, and P.M. Dowd, Deficiency of calcitonin gene-related peptide in Raynaud's phenomenon, *The Lancet* **336**:1530 (1990).

25. D.M. Levy, S.S. Karanth, D.R. Springall, and J.M. Polak, Depletion of cutaneous nerves and neuropeptides in diabetes mellitus: an immunocytochemical study, *Diabetologia* **32**:427 (1989).

26. D.M. Levy, G. Terenghi, X-H. Gu, R.R. Abraham, D.R. Springall, and J.M. Polak, Quantitative immunohistochemistry of nerves and neuropeptides in diabetic skin: relationship to tests of neurological function, *Diabetologia* **35**:889 (1992).

27. G. Properzi, S. Francavilla, G. Poccia, P. Aloisio, X-H. Gu, G. Terenghi, and J.M. Polak, Early increase precedes a depletion of VIP and PGP 9.5 in the skin of insulin dependent diabetics, *J Pathol* in press.

28. G. Terenghi, H.A. Bull, C.B. Bunker, D.R. Springall, Y. Zhao, J. Wharton, P.M. Dowd and J.M. Polak, Endothelin-1 in human skin: immunohistochemical, receptor binding and functional studies, *J Card Pharmacol* **17 (Suppl** 7):S467 (1991).

29. G. Properzi, G. Terenghi, X-H. Gu, G. Poccia, R. Pasqua, S. Francavilla and J.M. Polak, An increase of endothelin-1 immunoreactivity in cutaneous microvessels of early diabetic patients is associated with developing microangiopathy, *Lab Invest* (in press).

30. R. Katzman, Alzheimer's disease, *N Eng J Med* **314**:964 (1986).

31. G. Blessed, B.E. Tomlinson, and M. Roth, The association between quantitative measures of dementia and of senile changes in the cerebral grey matter of elderly subjects, *Br J Psychiatry* **114**:797 (1968).

32. G.K. Wilcock and M.M. Esiri, Plaques, tangles and dementia - a quantitative study, *J Neurol Sci* **56**:343 (1981).

33. S.M. Gentleman, C. Bruton, D. Allsop, S.J. Lewis, J.M. Polak, and G.W. Roberts, A demonstration of the advantages of immunostaining in the quantification of amyloid plaque deposits, *Histochem* **92**:355 (1989).

34. S.M. Gentleman, D. Allsop, C.J. Bruton, R. Jagoe, J.M. Polak, and G.W. Roberts, Quantitative differences in the deposition of ßA4 protein in the sulci and gyri of frontal and temporal isocortex in Alzheimer's disease, *Neurosci Lett* **136**:27 (1992).

35. S.M. Gentleman, R.M. Perry and G.W. Roberts, Correlation between ß-amyloid protein (ßAP) load and mental test scores in Alzheimer's disease, *Neuropath Appl Neurobiol* **17**:531 (1991).

CHAPTER 3

IMMUNOGOLD-SILVER STAINING (IGSS) FOR DETECTION OF ANTIGENIC SITES AND DNA SEQUENCES

Gerhard W. Hacker,[1] Cornelia Hauser-Kronberger,[1] Anton-Helmut Graf,[2] Gorm Danscher,[3] Jiang Gu,[4] and Lars Grimelius[5]

[1]Institute of Pathological Anatomy
Immunohistochemistry and Biochemistry Unit
General Hospital, Salzburg, Austria
[2]Department of Gynecology and Obstetrics
General Hospital, Salzburg
[3]Institute of Anatomy, Department of Neurobiology
University, Aarhus, Denmark
[4]Deborah Research Institute, Browns Mills, NJ, USA
[5]Institute of Pathology, University Hospital
Uppsala, Sweden

SUMMARY

It is often difficult to obtain intense and consistent immunostaining in paraffin sections of routinely formalin-fixed tissues. However, it is possible to overcome some of these problems by applying immunogold-silver staining (IGSS) which is highly sensitive, provided that an appropriate silver intensification method (autometallography) is chosen. The silver acetate autometallographic technique that we developed,[1] is very efficient in combination with its low sensitivity to light exposure during the intensifying process. Hence this method can be used in daylight, facilitating optimization of staining intensity by direct visual control under a light microscope. IGSS with silver acetate

[1] Address for correspondence: Univ.-Doz. Dr. Gerhard W. Hacker, Ph.D., Associate Professor, Salzburg General Hospital, Institute of Pathological Anatomy, Immunohistochemistry and Biochemistry Unit, Muellner Hauptstr. 48,A5020 Salzburg, Austria. Tel:++43-662-4492-4730; Fax: ++43-662-4482-882

Modern Methods in Analytical Morphology, Edited by
J. Gu and G.W. Hacker, Plenum Press, New York, 1994

autometallography can sometimes detect traces of antigen where other immunocytochemical methods have failed. Another advantage of this method is its strong staining intensity, permitting counterstaining with hematoxylin and eosin for better morphological evaluation. As the primary antibodies in the method can often be more diluted than the conventional method, IGSS is also less costly than most other immunostaining techniques.

The chapter presents a detailed explanation of IGSS methods for immunocytochemistry and *in situ* hybridization, together with extensive literature references for guidance in the application of the techniques.

INTRODUCTION

One achievement of the progress in histochemistry is the development of highly efficient techniques for specific detection of chemical substances by light microscopy. Among these are immunocytochemical techniques for detecting peptides, proteins and amines (and other substances against which antibodies can be raised), lectin histochemistry for tracing carbohydrates, and *in situ* hybridization techniques for the demonstration of specific nucleic acid sequences. Immunogold- silver staining (IGSS) methods first described by Holgate *et al.* in 1983,[2,3] are based on an autometallographic method described earlier by Danscher.[4-8] During the past decade, IGSS techniques have proved highly reliable, especially for the efficient and highly sensitive detection of substances in tissues prepared by routine formalin or Bouin's fluid fixation followed by paraffin embedding.

HISTORY

Colloidal gold as a marker for immunoelectron microscopy was first introduced by Faulk & Taylor in 1971.[9] Since then, the immunogold-staining (IGS) technique has gained widespread application by virtue of its various advantages over other, non-particulate immunostaining techniques. Geoghegan *et al.* were the first[10] to utilize the red color produced by colloidal gold sols for the purpose of light microscopy. Several applications and modifications of this staining technique have been described.[11-14] However, when applied to light microscopy, all of these IGS methods were less sensitive than most other immunocytochemical techniques. The next major advance was the introduction of silver enhancement (autometallography) of colloidal gold particles by Danscher[4-8] and Holgate *et al.*[2,3] This led to the so-called immunogold-silver staining (IGSS) techniques, which have demonstrably improved sensitivity and detection efficiency.

Originally, IGSS was an indirect IGS method combined with silver intensification by autometallography (earlier called "physical development").[2-8] Silver lactate, used as the ion source, forms a shell of metallic silver around each gold particle. This reducing process is catalyzed by hydroquinone in a citrate buffer of low pH. Gold particles enlarged in size and conglomerate if sufficiently near each other.[2,5,15] Under the light microscope, the precipitate is clearly visualized as grayish-black, standing out distinctly against the unreacting background.

The original IGSS method was an indirect immunostaining technique by which specific primary antibodies against the substance to be detected were reacted with a second layer of antibodies to immunoglobulin of the primary antibody species. The second layer antibodies were adsorbed to colloidal gold particles.[2,3] This method was subsequently improved in its sensitivity and specificity for the demonstration of various chemical substances.[16-18] IGSS has also been used for *in situ* hybridization[19-21] and lectin histochemistry.[22,23] In addition to the indirect method, direct methods,[21] bridging methods,[24] streptavidin-biotin methods,[25] protein A-gold-silver staining,[26] and various other combinations[27,28] have been described. IGSS has also been used in multiple immunostaining[2,3,11,29] and with epi-illumination.[11] Color modifications of the black autometallographic product have been discussed.[30] Indirect IGSS has also been called by other names, such as the "silver-intensified gold (SIG)" technique. However, we feel that Holgate's original definition of "immunogold-silver staining" method is the most appropriate. Multiple terms should be avoided.

IMMUNOGOLD-SILVER STAINING FOR IMMUNOCYTOCHEMISTRY

Originally, Holgate and his colleagues utilized antisera to immunoglobulins as the primary layer, while the second antiserum was adsorbed to 20 nm diameter gold particles. The technique was applied to reactive human tonsil specimens fixed in formol sublimate and embedded in paraffin.[2] They clearly demonstrated the superior sensitivity of the IGSS method over standard immunoperoxidase methods. These findings were confirmed by the same group when applying the technique to detect membrane surface antigens in non-Hodgkin lymphomas.[3]

Judging from Holgate's publications, there appears to be great potential for the IGSS method in demonstrating substances present in only small quantities or susceptible to formalin fixation and paraffin embedding, especially for regulatory peptides. However, upon testing of the original method, incredibly high background staining was observed. Therefore, we attempted to modify the technique to facilitate a highly sensitive demonstration of various substances in routinely formalin-fixed and paraffin-embedded tissue sections.[16] Our modification has been applied for diagnostic immunohistopathology,[17] cytology,[31-33] and immunoelectron microscopy.[15] IGSS methods can be applied to sections prepared in paraffin,[2,3,11,16,17] resin,[15,24,35] cryostat,[34,36] to cell cultures (Hellman, Hacker & Grimelius, unpublished), blotting techniques.[37] It may even be used the silver-enhanced enzyme linked immuno-adsorbent assay (SELISA).[38] Recently, various applications of IGSS for immunostaining ultrathin sections have been discussed.[13,14,39,40]

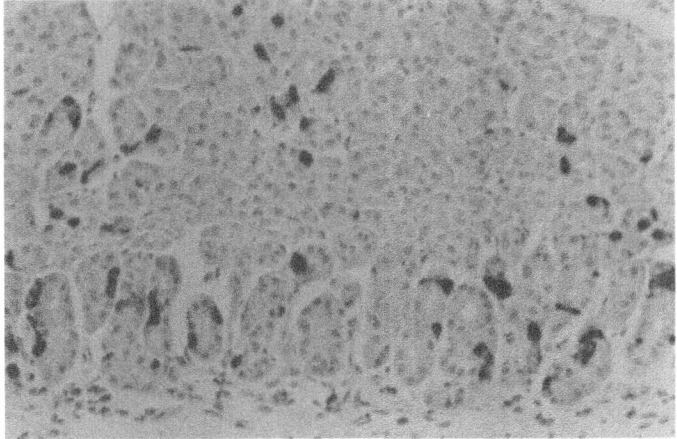

Figure 1. Detection of histamine in rat stomach using mouse monoclonal antibodies. A strong reaction is seen in the enterochromaffin-like (ECL) cells. At present, demonstration of histamine is only possible when specially fixed tissue is available. In this case, diethyl-pyrocarbonate (DEPC) vapor fixation has been used. 4 μm thick paraffin section, indirect IGSS method (5 nm gold particle diameter) with silver acetate autometallography, light hematoxylin and eosin counterstaining. Original magnification x 300.

Advantages of IGSS Methods

IGSS techniques have a number of advantages over other immunocytochemical methods. For instance, they have considerable value in immunohisto-pathology, sometimes giving positive immunostaining where other methods have failed. This greatly facilitates the demonstration of substances present in only minute quantities.[11,13,16,17,27] The indirect IGSS procedure outlined in the *Appendix (Protocol 1)* is also very quick, taking only 3 hours. When using microwave irradiation to incubate the primary and secondary antibody layers, the entire procedure takes only about 30 minutes (Gu & Hacker, unpublished).

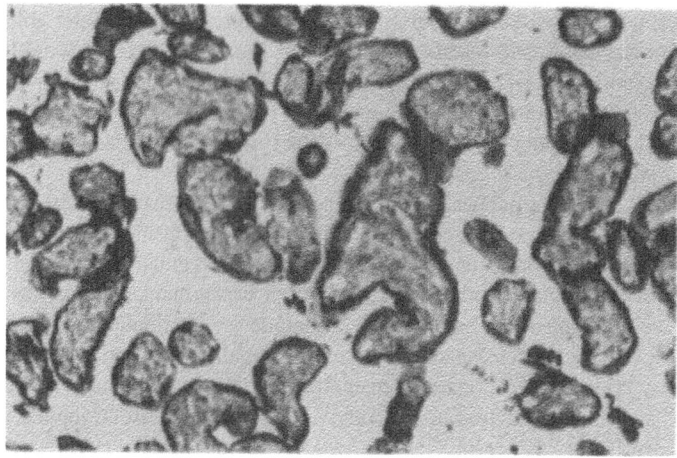

Figure 2. Immunostaining for human placental lactogen using mouse monoclonal antibodies in human placenta. Strong expression is seen in the lining trophoblasts. Formalin-fixed, 4 μm thick paraffin section, indirect IGSS method (5 nm gold particle diameter) with silver acetate autometallography, light hematoxylin and eosin counterstaining. Original magnification x 300.

It is also an important fact that in IGSS methods, hazardous reagents such as diamino-benzidine tetrahydrochloride (DAB) are avoided. Positive reactions can be readily identified, thanks to the very intense grey-black signal. That helps in the screening of sections at even low microscopic magnification, and, therefore, may facilitate a more rapid diagnosis in immunohistopathology. It also allows the use of conventional counterstaining (hematoxylin and eosin, and/or nuclear fast red) to aid morphological assessment.[11,16,17] The finding that many primary antibodies could be diluted considerably more when using IGSS methods than in immunoenzyme techniques[2,11,16,17] might help to reduce costs. IGSS methods have a high detection efficiency which is closely related to the type of silver amplification used and also to the quality of the immunogold reagents used.[1,24,26,27]

The Indirect Method

IGSS, as originally described by Holgate[2] and modified by our group,[16] is an indirect method. We have found the indirect method superior to direct methods, streptavidin gold-silver staining, and protein A-gold-silver staining as it takes the least time and gives the best detection efficiency. The indirect IGSS procedure consists an unlabeled primary antibody to react with tissue antigen, and a gold-adsorbed secondary antibody directed against immunoglobulins of the species in which the primary antibody has been raised. The minute gold particles that accumulate at antigenic sites can subsequently be visualized by autometallography.[1,4-8,22] The procedure performed in our laboratories is detailed in *Protocol 1*. The essential steps of the technique are discussed in detail in the following sections.

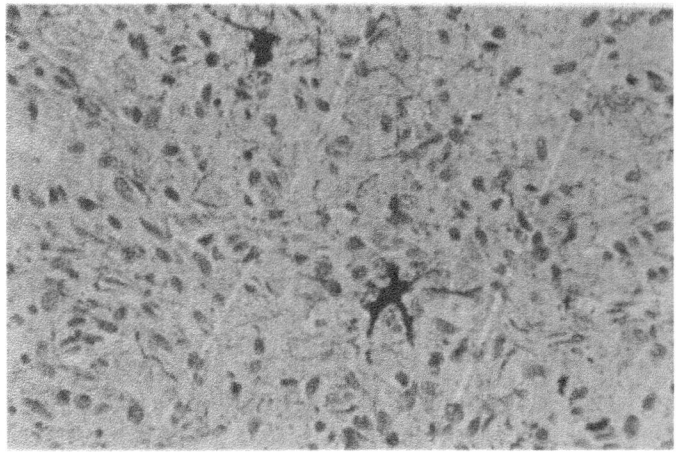

Figure 3. Detection of differentiated astrocyte-like cells in human brain of a case with glioblastoma, using mouse monoclonal antibodies to glial fibrillary acidic protein (GFAP). 4 μm thick paraffin section, indirect IGSS method (5 nm gold particle diameter) with silver acetate autometallography, light hematoxylin and eosin counterstaining. Original magnification x 430.

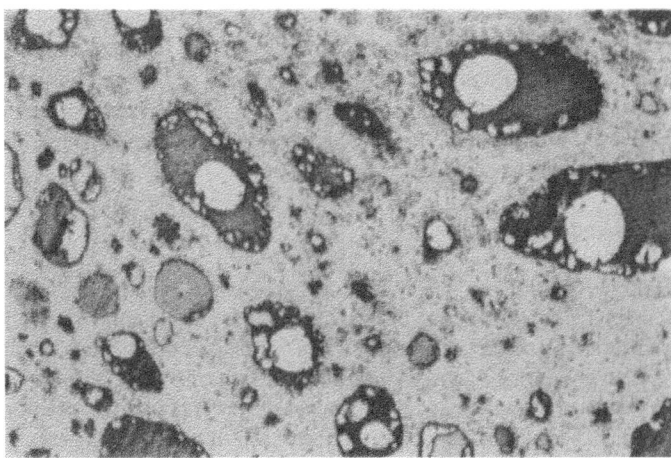

Figure 4. Demonstration of thyroglobulin in a case of a benign tumor of human thyroid gland, using rabbit polyclonal antibodies. 4 μm thick paraffin section, indirect IGSS method (5 nm gold particle diameter) with silver acetate autometallography, light eosin counterstaining. Original magnification x 180.

Tissue Preparation

We have found that IGSS methods are particularly suitable for use with paraffin and resin sections. Cryosections may be problematical due to high background levels, though this can be avoided by modifying the staining technique. Postfixation of cryosections with cross-linking fixatives such as neutral-buffered formalin may improve the staining result.

Various fixatives are suitable for fixation of the fresh tissue by immersion or perfusion. Phosphate-buffered formaldehyde solution (4%) is preferable, so are Bouin's fluid or Stefanini's/Zamboni's solution.[41,42] Parabenzoquinone, sometimes used as a special fixative for regulatory peptides,[42] should be avoided for IGSS as it has a reducing effect on the silver ions and causes non-specific silver precipitation. We have found Stefanini's/Zamboni's solution to give excellent results for peptide antigens.

To achieve the best results, the choice of fixation should be optimized for each antigen to be detected. As a general rule, one should use a fixative at low concentrations and for a short fixation period. Also, the embedding process should be mild (we have tested paraffin, LR-White, Epon and Araldite). However, if the tissue sample is left too long in the fixative, in alcohols (isopropanol or ethanol are preferable), or in xylene, as well as in the melted paraffin, the staining results will often be suboptimal. Higher temperatures (>60°C) of the fixation and embedding steps can also cause a poorer staining result.

Treatment with Lugol's Iodine

The need to pretreat sections with Lugol's iodine was first reported by Holgate et al.[1] Although protocols without Lugol's solution have been published by some authors.[15,27,43] Others - including our group - have found that iodine treatment with subsequent destaining using sodium thiosulfate improves the staining efficiency, for almost all antigens in paraffin sections tested so far.[1-3, 16-18,21,31] If the iodine treatment is omitted, the silver intensification procedure must be prolonged. We, therefore, suggest retaining the iodine/sodium thiosulfate steps in the staining protocol. If staining problems occur, then it might be worth trying to stain without Lugol's. For instance, we found that the demonstration of amplified DNA or mRNA in the in situ polymerase chain reaction (in situ PCR) or by in situ nucleic acid sequence based amplification (in situ NASBA) with IGSS is best achieved without Lugol's iodine treatment.[45] (see also the chapter by Zehbe et al. in this book). In recent experiments, we found that alternative oxidizing treatments (e.g. with hydrogen peroxide or potassium permanganate) may also help to yield better IGSS results, but Lugol's iodine appears to be the best. Nevertheless, this detection method is still undergoing development and no definitive conclusions can be drawn at this time.

Washing Buffers and Detergents

Most protocols in literature recommend Tris-buffered saline or phosphate buffered saline with a pH of about 7.2 as the washing buffer. In one series of experiments, we found that a high salt concentration and the addition of Triton X-100 or Tween 80 or 20 to the buffer system before applying primary antibody improved the staining especially for intracellular or intranuclear antigens. The increased NaCl concentration in the buffer system appears to have a considerable background lowering effect on the unspecific staining. The buffer system applied before the immunogold incubation may be adjusted to pH 8.2 to stabilize the gold reagent and to achieve optimum results.[16] In order to make the staining protocol easier to handle, it may also be feasible to use for both buffers a pH of about 7.6, for all washing steps before the autometallographic procedure. To avoid unspecific reactions, adding 0.04-0.1% cold water fish gelatin to all buffers may be helpful.[46] We recommend trying the different buffer combinations in the system used.

Primary Antibodies

A variety of antigens have been successfully demonstrated in routinely processed sections by applying IGSS methods. Polyclonal rabbit, guinea pig, or sheep antisera and monoclonal mouse or rat antibodies have been used to detect regulatory peptides, intermediate filament proteins, enzymes, cell surface antigens, tumor markers, plant and fungus antigens, receptors, bacteria, and viruses.[1-3,13-18,24,26,28,29,31-36] A primary antibody incubation for 60-90 min at room temperature is both convenient and effective, especially for routine use; this incubation time can even be significantly reduced by microwave incubation (see also the chapter by Gu *et al.* in this book). Prolonged incubation, e.g. overnight at 4°C, as well as repeated application of the same antibody, may increase the detection efficiency even further. This may also allow the use of higher antibody dilutions. As a guideline, antibody dilutions for a one hour incubation in IGSS methods are usually equal to, or even higher than those used in PAP or APAAP methods for overnight incubation. As in other immunocytochemical methods, the optimal concentration of primary (and secondary) antibodies should be chosen after testing a range of dilutions. Normally, with such a sensitive method, unspecific staining due to bad quality of primary antibodies (especially when using polyclonal antisera) may present problems.

Secondary Antibodies

Affinity-purified secondary antibodies adsorbed to gold particles (often referred to as immunogold-reagents) are available from several companies: Amersham (Bucks, England, and Arlington Heights, IL, USA), Aurion (Wageningen, NL), Dakopatts (Glostrup, Denmark, and Santa Barbara, CA, USA), Cambridge Research Biochemicals (Cambridge, England), BioClin and BioCell (both: Cardiff, Wales), Sigma (Deisenhofen, FRG), Nanoprobes (Browns Mills, NY, USA), and others. It should be mentioned that quality and prices differ greatly. Every laboratory should, therefore, test different immunogold-reagents at its own establishment.

Dilution of the immunogold antibodies should also be optimized by titration tests. Usually a dilution between 1:25 and 1:200 is optimal. Tris-buffered saline, pH 7.6-8.2, is used as diluent and should contain 0.8 to 1% bovine serum albumin (BSA) to prevent an aggregation of gold particles, and to lower background staining.[16] Once diluted, the immunogold reagent may be kept at 4°C for several days without significant loss of light microscopical labeling properties.

As demonstrated by Lackie[15] and Gu,[40] immunolabelling is most dense when using small gold particles. Such immunogold reagents can penetrate sections better and achieve greater particle density at the antigenic sites than can larger ones. However, electron microscopy studies by Gu *et al.*[40] have shown that this effect on labeling density is seen only with gold particles measuring down to 5 nm diameter; i.e. 1 nm particles did not appear to give denser labeling than 5 nm gold particles. We, too, found that 5 nm gold particles were the best for light and electron microscopic applications, giving high labeling density and good penetration of the section. In our view, "electron microscopical grade" immunogold reagents are better for light microscopy than are "LM-grade" reagents. In our laboratories, we often use a mixture of several optimally diluted immunogold batches from different manufacturers, each mixed in equal proportions. This may sometimes help to avoid non-labelling of primary antibody subgroups of immunoglobulins, caused by over-purification of immunogold by affinity chromatography.

In a recently developed reagent, "nanogold", 1.4 nm gold particles are encased in an organic shell and covalently attached to the secondary antibody or streptavidin.[47] The authors claim that the new reagent gives improved resolution, stability, uniformity and sensitivity of staining with complete absence of aggregation. In our hands, staining obtained with those new probes was comparable to that obtained with "ordinary" immunogold reagents and did not show any advantages.

Glutaraldehyde Postfixation

Postfixation in 2% glutaraldehyde, after rewashing in buffer to remove unbound immunogold reagent, may prevent the release of gold reagent from its binding sites in the low-pH environment of the silver enhancement process. However, the need for postfixation should be tested, as this step is only necessary for certain batches of immunogold. Glutaraldehyde should be diluted in phosphate-buffered saline (PBS, pH 7.2) and can be kept in this buffer at 4°C up to 2 weeks. Meanwhile, it can be re-used several times. Diluting glutaraldehyde in Tris-buffer systems (e.g., TBS) is also possible, however, then it has to be freshly prepared for each use.

Figure 5. Detection of very fine nerve fibers in rat cerebellum by application of mouse monoclonal antibodies to neurofilament proteins 150 and 200 kD. In this case, the new "nanogold" secondary layer has been used, in which 1.2 nm gold particles are surrounded by an organic sheath which is covalently bound to the immunoglobulin molecule. 4 μm thick paraffin section, indirect IGSS method with silver acetate autometallography, nuclear fast red counterstaining. Original magnification x 430.

Autometallographic (AMG) Silver Amplification

In the 19th century, a process was discovered by which gold particles of less than one tenth of a nanometer in diameter become encapsulated in silver when treated in a solution of silver ions and reducing molecules.[46,48] Liesegang introduced this process in histology to intensify silver in tissue sections previously exposed to a silver salt,[48] using silver nitrate as the silver ion donor in his AMG developer. Liesegang's physical developer was that used by photographers. Reduction of the developers autocatalytic activity improved the specificity of the technique.[4] However, the modifications of Liesegang's original developer were quite modest. In essence, silver nitrate was replaced by silver lactate and the level of hydroquinone was reduced. This improved the technique significantly and made possible the demonstration of nanometer-sized crystal lattices of sulphides and selenides of silver, mercury and zinc, as well as that of metallic gold.[4-8] A detailed historical overview, as well as an exact description of AMG process, are given in the chapter by G. Danscher, in this book.

Danscher's AMG technique expands colloidal gold particles many fold within a few minutes of silver lactate development.[15,46,49,50] The metallic silver shells encasing the original gold particles ultimately cause conglomeration of gold and silver when gold particles are sufficiently close together.[15] This is seen in the bright field light microscope as brownish-grayish or even black areas. The speed at which the colloidal gold particles expand is greatly dependent on the type of AMG developer used and the temperature at which the process takes place.[15,50] Commercial silver enhancement kits proved to be less effective than self-made AMG solutions.[50] (See also the chapter by Krenács on EM pre-embedding IGSS, in this book.)

The silver lactate procedure should preferably be executed under a lightproof cover on the laboratory bench. If not, the silver lactate developer tends to darken due to autocatalysis. Some of the minute silver particles that cause the developer to darken will adhere to the surfaces of the tissue sections if they have not been covered with an ultrathin membrane of gelatin prior to the development procedure.[5,46] Such unspecific silver grains will affect the quality of the staining. However, if the sections have been dipped in a 0.1-0.5% gelatin solution prior to AMG development, the surplus of silver grains can be flushed with 45°C warm water.[4,5,46] Alternatively, 0.1% gelatin can be added to the working buffers used in IGSS.

The introduction of the less light-sensitive silver acetate as AMG silver ion donator[1,22,46] made it possible to intensify gold particles in daylight conditions, i.e. without cover. By using silver acetate AMG, the development procedure can proceed in light up to 20 min without observable darkening of the developer. This also makes possible observation of the silver amplification process in the bright field light microscope.[1] The silver acetate AMG procedure, as outlined in *Protocol 1*, is therefore recommended for IGSS at the light microscopic level. For electron microscopic application, both silver lactate and silver acetate AMG proved to be very useful; addition of a colloid (gum arabic) in the setup can improve the overall performance of the developer.[1,4,46,50] To achieve optimal staining, water purity and purity of the chemicals used are crucial. The glass slides may be washed in 10% Farmer's solution before attachment of the tissue sections. The pH of the AMG solution is critical. It should be adjusted for pH 3.8, neutral conditions may cause unspecific argyrophilic-type reactions, e.g. in pancreatic A-cells or in collagen fibers. High quality glass-distilled water is recommended for all washes before and during the autometallographic process. Even traces of impurity (e.g. chloride or remnants of old silver developer) will cause a precipitate and make the intensification solutions ineffective. Therefore, only scrupulously cleansed glassware should be used. Contact between metallic objects and the silver solutions should be avoided. For the same reason, use only plastic tweezers.

IMMUNOGOLD-SILVER STAINING FOR *IN SITU* DNA HYBRIDIZATION

In diagnostic histopathology, DNA or RNA hybridization probes can be used to identify viral genomes for the detection of infected cells. Biotin or digoxigenin are now used with great success, because they satisfy the requirement of most pathology laboratories in that the probes can be stored and handled without the hazard of radioactivity. Non-radioactive labels are cheaper, easier to handle, and give a higher resolution than radioactive probes. Recently, several combinations of optimized protocols for *in situ* DNA and RNA hybridization, using biotin-labeled probes and IGSS techniques, have been described by our group.[21] An optimum working scheme for DNA detection by conventional *in situ* hybridization combined with IGSS is outlined in *Protocol 2*.

Our test system included human papillomavirus (HPV) subtypes, cytomegalovirus (CMV), adenovirus 5, hepatitis-B virus (HBV), Epstein-Barr virus (EBV) and herpes simplex virus (HSV) DNA which were detected in infected tissue sections and cytological preparations. Prehybridization and hybridization protocols were optimized for the biotin-labeled DNA probes used. As a predigestion protease, we recommend the use of proteinase K, which gave the highest signal to noise ratio and sensitivity. Of course, both the duration of the predigestion and the concentration of proteinase K need to be optimized according to the strength of tissue fixation and the type of tissue being investigated. Also, batch variations of proteinase K have been observed. In our system, the use of acetic anhydride in triethanolamine, followed by incubation with formamide and dextran sulfate in buffer, did not affect the staining appearance and was, therefore, excluded. To detect the hybrids formed, treatment with Lugol's iodine was not absolutely necessary to obtain the hybridization signal, though the signal to noise ratio was best when Lugol's treatment was used after the hybridization step, not before. Applications of IGSS for *in situ* hybridization PCR for the most sensitive detection of DNA sequences are described in this book by Zehbe *et al.* For this particular application, oxidizing treatment such as Lugol's iodine, as well as detergents, have to be strictly avoided. We have now also been successful in using IGSS techniques in combination with a thermocycler-independent *in situ* NASBA method for single copy mRNA and DNA detection, and for the first time applied *in situ PCR* techniques for preembedding electron microscopy (Zehbe, Muss, Hacker, *et al.* unpublished).

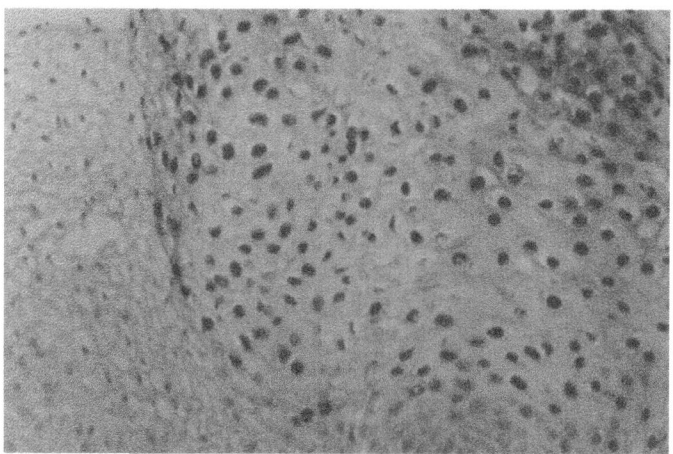

Figure 6. *In situ* hybridization using a biotinylated cDNA probe to human papillomavirus types 6 and 11 (HPV 6/11) and subsequent direct IGSS method with 1 nm gold adsorbed antibodies to biotin. Strong staining with very high resolution has been obtained in numerous nuclei of koilocytes in this case of condylomata acuminata. 4 μm thick paraffin section, silver acetate autometallography, light hematoxylin and eosin counterstaining. Original magnification x 300.

PROSPECTS

Due to the numerous and varied advantages of IGSS techniques when compared with other immunocytochemical methods, we hope that IGSS will be much more widely applied in future. One hinderance is that many pharmaceutical companies assert various advantages of other techniques, e.g. the ABC system and the APAAP techniques, far more vigorously than those of IGSS. One possible reason for this is that most companies appear to be unable to produce their own high quality immunogold reagents. Presently, only a few reagents (Table 1) seem to fulfill all the quality requirements for light and electron microscopical IGSS of the highest standard. Another hinderance is that first-time users of IGSS techniques sometimes get too much "dirty" visible background. This is usually due to poor optimization of each individual step in this highly sensitive technique; often higher antibody dilutions should be used; antisera may be of too low quality or titer and insufficiently specific. Commercially produced developers may be suboptimal; glassware may be insufficiently cleaned, or the water quality inadequate. Staining protocols used may not be optimized. With time, many more researchers and routine histopathologists will be convinced by the literature reports that IGSS is a most sensitive and efficient group of techniques. It gives a better contrast against background and permits counterstaining with hematoxylin and eosin to gain a better assessment of the morphology. It also eliminates the necessity of blocking the intrinsic peroxidase in the enzyme activity, as necessary for immunoperoxidase methods to negate the background. All these advantages should ultimately make IGSS the technique most preferred for many applications.

SUGGESTED PROCEDURES

PROTOCOL 1: <u>INDIRECT IMMUNOGOLD-SILVER STAINING</u>

A modification of the Holgate IGSS-method is recommended.[1,2,16,24] All steps are performed at room temperature (20°C).

Poly-L-lysine (hydrobromide, PLL, MW > 150.000; Sigma, FRG) is used as the adhesive for the sections or cells.[51] A small drop of 0.1% PLL in distilled water is applied with a Pasteur pipette to a clean glass slide and carefully smeared onto the slide with a second slide. PLL-coated slides may be stored for not longer than 1 week at room temperature before use, although it is preferable to use freshly prepared slides to take up tissue sections after cutting. This PLL coating may be applied for paraffin, resin, or cryostat tissue sections, as well as for cytological preparations. Alternatively, APES (see *Protocol 2*) may be used as a section or cell adhesive.[52]

Immunocytochemistry:

1. Paraffin sections: Dewax in xylene and rehydrate through graded alcohols.
   Cryostat sections: Air-dry for 1 hour, postfix in 4% phosphate-buffered formaldehyde solution for 2 min.
   Resin sections: Treat with saturated sodium ethoxide for about 10-20 min under microscopical control, wash in ethanol for 3x2 min.
2. Wash in tap water (3 min) followed by a short wash in distilled water (some sec).
3. Immerse in Lugol's iodine (1% iodine in 2% potassium iodide; ready made up from Merck (no. 9261, Darmstadt, FRG) (5 min).
4. Rinse briefly in tap water, followed by a short wash in distilled water (some sec).
5. Treat with 2.5% aqueous sodium thiosulfate until sections become colorless (usually less than 30 sec).
6. Wash in tap water (2 min), followed by a short wash in distilled water (some sec).
7. Immerse in TBS pH 7.6 containing 0.1% cold water fish gelatin. If necessary, add 0.1% Triton X-100 or 0.5% Tween 80 into this buffer (10 min). Higher salt concentrations (e.g., 2.5% NaCl) may help to reduce background.
8. Apply normal serum of the secondary antibody species (1:10 in TBS pH 7.6).
9. Incubate with primary antibody (60 min or overnight). The dilution should be tested carefully. Antibody diluent is 0.1 M phosphate- or Tris-buffered saline (TBS or PBS, pH 7.2-7.6) containing 0.1% bovine serum albumin and 0.1 % sodium azide.
10. Wash in TBS pH 7.6 containing 0.1% cold water fish gelatin (3 min).
11. Apply normal serum 1:10 as in step 2.
12. Incubate with gold-adsorbed second layer antibodies for 60 min. Optimum dilution is between 1:25 and 1:200 and should be determined by titration. Diluent is TBS pH 7.6-7.2 containing 0.8% bovine serum albumin (BSA).
13. Wash in TBS pH 7.6 containing 0.1% cold water fish gelatin (2x3 min).
14. Immerse in 0.1 M PBS (pH 7.2) (2 min).
15. Postfix in 2% glutaraldehyde in PBS (2 min).
16. Perform silver acetate autometallography.

SILVER ACETATE AUTOMETALLOGRAPHY.[1]

a. Rinse the sections carefully about 5 times in glass-distilled water (about 30 sec each) followed by 3 washes (each 3 min) in the same.

b. Solution A: Dissolve 100 mg silver acetate in 50 ml of deionized and, preferably, glass-distilled water. This solution should be made fresh for every run. Silver acetate crystals can be readily dissolved by continuous stirring within about 15 min, provided that powdered crystals are used (Table 4). Larger crystals may be dissolved using an ultrasonic bath. It is advisable to filter the solution shortly before use, at least for EM applications.

c. Citrate buffer: Dissolve 23.5 g trisodium citrate dihydrate and 25.5 g citric acid monohydrate in 850 ml of deionized or distilled water. This buffer can be kept at 4°C for at least 2-3 weeks. Before use, adjust to pH 3.8 with citric acid.

d. Solution B: Dissolve 250 mg hydroquinone in 50 ml citrate buffer.

e. Just before use, mix solution A with solution B.

f. Silver enhancement: Place the slides vertically in a glass container (suggested 100 ml volume, holding up to 19 slides) and cover them with the mixture of solutions A and B. Normally, optimal IGSS development takes about 3-15 min at room temperature. When no Lugol's iodine or other oxidation treatments are used, development duration is greatly prolonged. In this case, some shielding from daylight is recommended (dark box). Staining intensity may be checked in light microscope during enhancement process which is stopped by immersion in photographic fixer.

g. Photographic fixer (e.g., Ilfospeed or Agefix, Table 2) diluted 1:20 may be used to stop enhancement process immediately. (This solution can be re-used for many stainings). Leave the slides in this solution for about 1 min. Alternatively, a 2.5% aqueous solution of sodium thiosulfate can be used.

h. Rinse the slides carefully in tap water for at least 3 min. After silver enhancement, sections can be counterstained with hematoxylin and eosin and/or nuclear fast red, dehydrated and mounted in Canada balsam or DPX (BDH Chemicals, Poole, England).

PROTOCOL 2: *IN SITU* DNA HYBRIDIZATION.[21]

This is an optimized DNA hybridization protocol using a direct method and gold-labeled anti-biotin antibodies with silver acetate autometallography following the hybridization with a biotinylated DNA probe.[21] Recommended section or cell adhesive is aminopropyl-triethoxysilane (APES = 3-(triethoxysilyl)-propylamine, Merck, FRG):[52] Clean the glass slides with acetone, dry, incubate for 5 min with 2% APES dissolved in acetone, then wash in acetone, distilled water, followed by drying. Glass slides prepared in this way may be stored at room temperature for several months.

Prehybridization Treatments

1. Deparaffinize the sections using xylene and rehydrate them through graded alcohols (e.g. isopropanol). Cell monolayers (imprints, smears) may also be used, but have to be fixed in 4% phosphate-buffered formalin (10 min) after having dried for at least one hour.

2. Immerse the sections in 20 mM PBS (5 min).

3. Soak in 0.3% Triton X-100 (10-15 min) to permeabilize sections.

4.   Proteolytic treatment: Digest the tissue sections slightly with 0.1% proteinase K in PBS for about 20 min (time and concentration depend on strength of fixation of tissue, size of section, and enzyme batch) at 37°C.
5.   Wash in PBS or 2 x SSC (standard sodium citrate buffer) (3 x 2 min).
6.   Wash in distilled water (2 min), immerse in 50 %, 70 % and 98% isopropanol (1 min each) and air dry the sections at room temperature.
7.   Prehybridization (optional): Incubate with 50 % deionized formamide and 10% dextran sulfate in 2 x SSC, at 50°C for 5 min. Drain off excess.

TABLE 1. Antibodies and other immunogold reagents used for immunogold-silver staining and for detection of the *in situ* hybridization products. The reagents and dilutions reported showed optimal visualization of the antigenetic or hybridization sites with none to low levels of unspecific background staining.

Probe	Dilution	Code No.	Source
AuroProbe One anti-biotin, (1nm gold-adsorbed anti-biotin antibodies)	1:25	RPN 473	Amersham, Bucks, England
AuroProbe EM GAR G5 (goat anti-rabbit immunoglobulins, 5 nm gold-adsorbed)	1:50	RPN 420	Amersham
AuroProbe EM GAM IgG G5 (goat anti-mouse IgG, 5 nm gold-adsorbed)	1:50	RPN 424	Amersham
Goat anti-rabbit EM, 6 nm gold-adsorbed	1:40	106.011	Aurion, Wageningen, NL
Goat anti-rabbit GP ultrasmall, 0.8 nm gold-adsorbed	1:40	100.011	Aurion
Goat anti-mouse IgG/M EM 6 nm gold-adsorbed	1:40	106.044	Aurion
Goat anti-mouse IgG/M ultrasmall 0.8 nm gold-adsorbed	1:40	100.044	Aurion
Nanogold anti-mouse (from goat) 1.4 nm gold-attached	1:40	2001	Nanoprobes, Stony Brook, NY
Nanogold anti-rabbit (from goat) 1.4 nm gold-attached	1:40	2003	Nanoprobes

TABLE 2. Commercial Sources of Chemicals Used.

Product	Code	Source
Silver acetate	85140	Fluka, Buchs, Switzerland
Hydroquinone	53960	Fluka
Trisodiumcitrate-dihydrate	6448	Merck, Darmstadt, Germany
Citric acid	244	Merck
Tween 80	56023	BDH Chemicals, Poole, UK
Cold water fish gelatin, 40%	900.033	Aurion, Wageningen, NL

Hybridization

Particular care must be taken to insure that sections will never dry from the hybridization step on.

8. Apply a small drop (about 20 µl of ready-to-use biotinylated cDNA probe, or 20 ng of nick-translated probe) of probe mixture onto the section, which is then covered with a rectangular (22 x 22 mm, or circular 19 mm in diameter) coverslip.
9. Place slides on a 92°C heating block and incubate for 5-10 min.
10. Transfer slides into a 37°C oven and incubate for further 30-120 min or overnight.
11. Remove coverslips by soaking with 4 x SSC.
12. Wash under stringent conditions in 4 x SSC, 2 x SSC, 0.1 x SSC, 0.05 x SSC (see below), and distilled water (each > 5 min) at room temperature or (preferably) at 37°C.

Steps 13-15 are optional and may not apply for certain DNA detection setups such as *in situ* PCR.

13. Immerse sections in Lugol's iodine (1% iodine in 2% potassium iodide) (5 min) and briefly rinse them in tap water and distilled water.
14. Treat with 2.5% aqueous sodium thiosulfate until sections become colorless, then wash in tap water (2 min).
15. Wash in distilled water (2 min).

Detection of Hybridization Sites:

16. Immerse in TBS pH 7.6 containing 0.1% cold water fish gelatin (2 x 3 min).
17. Incubate with a mixture of equal amounts of optimally diluted gold-adsorbed anti-biotin antibodies (Table 3) and streptavidin adsorbed to colloidal gold (1 and or 5 nm) for 60 min at room temperature. Optimum dilution is usually between 1:25 and 1:50. Antibody diluent is TBS pH 7.6 containing 0.8 % bovine serum albumin (BSA).
18. Wash in TBS pH 7.6 containing 0.1% cold water fish gelatin (2 x 3 min).
19. Immerse in 0.1 M PBS pH 7.2 (2 min).
20. Post-fix in 2% glutaraldehyde in PBS (2 min).
21. Apply silver acetate autometallography as in Protocol 1, counterstain and coverslip.

Buffer Systems Used in These Protocols:

TBS: For facilitating the IGSS protocols, 0.05 Tris-buffered saline (TBS) pH 7.6 is used in all washing. Background staining may be reduced by addition of 2.5% NaCl into the wash before primary antibody incubation. Penetration of antibodies through cell membranes and into nuclei may be improved by addition of 0.1% Triton X-100 or 0.5% Tween 80 to the buffer wash and before primary antibody incubation. For *in situ* hybridization or *in situ* PCR applications, detergents and iodine treatments should be avoided under certain circumstances.

PBS: 0.1 M phosphate-buffered saline pH 7.2

SSC: Standard sodium citrate buffer.
[Preparation of 20 x SSC: dissolve 175.32 g NaCl and 88.23 g Na-citrate, ad 1 liter H_2O; adjust pH to 7.0 with HCl or citric acid; premixed concentrate is available from Sigma (no. S-6639; FRG)].

ACKNOWLEDGEMENTS

For the joint collaboration in IGSS-related projects, we sincerely thank doctors J.M. Polak and D.R. Springall (London, England), P. Lackie (Southampton, England), H.C. Su (Xian, PR China), T. Krenács (Szeged, Hungary), S. Dechet, A. Schiechl, U. Sonnleitner-Wittauer, J. Thurner, H. Adam, and O. Dietze (Salzburg, Austria).

REFERENCES

1. G.W. Hacker, L. Grimelius, G. Danscher, G. Bernatzky, W. Muss, H. Adam, and J. Thurner, Silver acetate autometallography: an alternative enhancement technique for immunogold-silver staining (IGSS) and silver amplification of gold, silver, mercury and zinc in tissues, *J. Histotechnol.* 11:213 (1988).
2. C.S. Holgate, P. Jackson, P.N. Cowen, and C. Bird, Immunogold-silver staining: new method of immunostaining with enhanced sensitivity, *J Histochem Cytochem* 31:938 (1983).
3. C.S. Holgate, P. Jackson, I. Lauder, P. Cowen, and C. Bird, Surface membrane staining of immunoglobulins in paraffin sections of non-Hodgkin's lymphomas using immunogold-silver staining technique, *J Clin Path* 36:742 (1983).
4. G. Danscher, Histochemical demonstration of heavy metals. A revised version of the sulphide silver method suitable for both light and electron microscopy, *Histochemistry* 71:1 (1981).
5. G. Danscher, Localization of gold in biological tissue. A photochemical method for light and electron microscopy, *Histochemistry* 71:81 (1981).
6. G. Danscher, Light and electron microscopical localisation of silver in biological tissue, *Histochemistry* 71:77 (1981).
7. G. Danscher and J.O.R. Norgaard, Light microscopic visualisation of colloidal gold on resin-embedded tissue, *J Histochem Cytochem* 31:394 (1983).
8. G. Danscher, Autometallography. A new technique for light and electron microscopical visualization of metals in biological tissue (gold, silver, metal sulphides and metal selenides), *Histochemistry* 81:331 (1984).
9. W.P. Faulk and G.M. Taylor, An immunocolloid method for the electron microscope, *Immunochemistry* 8:1081 (1971).
10. W.D. Geoghegan, J.J. Scillian, and G.A. Ackerman, The detection of human B-lymphocytes by both light and electron microscopy utilizing colloidal gold-labelled anti-immunoglobulin, *Immunol Commun* 7:1 (1978).
11. J. De Mey, G.W. Hacker, M. De Waele, and D. Springall, Gold probes in light microscopy, in: "Immunocytochemistry - Modern Methods and Applications," J.M. Polak and S. Van Noorden, eds., Wright, Bristol, 1971.
12. J. Gu, J. De Mey, M. Moeremans, and J.M. Polak, Sequential use of the PAP and immunogold-staining methods for the light microscopical double staining of tissue antigens. Its application to the study of regulatory peptides in the gut, *Reg Peptides* 1:465 (1981)

13. J. Roth, Applications of imunocolloids in light microscopy. Preparation of protein A-silver and protein A-gold complexes and their application for the localisation of single and multiple antigens in paraffin sections, *J. Histochem Cytochem* **30**:691 (1982).

14. J. Roth, The colloidal gold marker system for light and electron microscopic cytochemistry, *in*: "Immunocytochemistry, Vol. 2," G.R. Bullok and P. Petrusz, eds., Academic Press, London, 217 (1983).

15. P.M. Lackie, R.J. Hennessy, G.W. Hacker, and J.M. Polak, Investigation of immunogold-silver staining by electron microscopy, *Histochemistry* **83**:545 (1985).

16. D.R. Springall, G.W. Hacker, L. Grimelius, and J.M. Polak, The potential of the immunogold-silver staining method for paraffin sections, *Histochemistry* **81**:603 (1984).

17. G.W. Hacker, D.R. Springall, S. Van Noorden, A.E. Bishop, L. Grimelius, and J.M. Polak, The immunogold-silver staining method. A powerful tool in histopathology, *Virch Arch A* **406**:449 (1985).

18. G.W. Hacker, J.M. Polak, D.R. Springall, J. Ballesta, A. Cadieux, J. Gu, J.Q. Trojanowski, D. Dahl, and P. Marangos, Antibodies to neurofilament proteins and other brain proteins reveal the innervation of peripheral organs, *Histochemistry* **82**:581 (1985).

19. I.M. Varndell, J.M. Polak, K.L. Sikri, C.D. Minth, S.R. Bloom, and J.E. Dixon, Visualisation of messenger RNA directing peptide synthesis by *in situ* hybridization using a novel single-stranded cDNA probe, *Histochemistry* **81**:597 (1984).

20. P. Liesi, J.P. Julien, P. Vilja, F. Grosveld, and L. Rechardt, Specific detection of neuronal cell bodies: *in situ* hybridisation with a biotin-labeled neurofilament cDNA probe, *J Histochem Cytochem* **34**:923 (1986).

21. G.W. Hacker, A.H. Graf, C. Hauser-Kronberger, G. Wirnsberger, A. Schiechl, G. Bernatzky, U. Wittauer, H. Su., H. Adam, J. Thurner, G. Danscher, and L. Grimelius, Application of silver acetate autometallography and gold-silver staining methods for *in situ* DNA hybridization, *Chinese Med J* **106**:83 (1993).

22. E. Skutelsky, V. Goyal, and J. Alroy, The use of avidin-gold complex for light microscopic localization of lectin receptors, *Histochemistry* **86**:291 (1987).

23. J. Schmidt and W. Peters, Localization of glycoconjugates at the tegument of the tapeworms Hymenolepis nana and H. microstoma with gold labelled lectins, *Parasitol Res* **73**:80 (1987).

24. G. W. Hacker, Silver-enhanced colloidal gold for light microscopy, *in*: "Colloidal gold - principles, methods, and applications." Vol. 1, M.A. Hayat, ed., Academic Press Inc., San Diego, USA, 297 (1989).

25. G. Coggi, P. Dell'Orto, and G. Viale, Avidin-Biotin methods, *in*: "Immunocyto-chemistry - Modern Methods and Applications," J.M. Polak and S. Van Noorden, eds., Wright, Bristol, U.K, 54 (1986).

26. O. Fujimori and M. Nakamura, Protein A gold-silver staining method for light microscopic immunohistochemistry, *Arch Histol Jap* **48**:449 (1985).

27. L. Scopsi and L.-I. Larsson, Increased sensitivity in immunocytochemistry. Effects of double application of antibodies and of silver intensification on immunogold and peroxidase-antiperoxidase staining techniques, *Histochemistry* **82**:321 (1985).

28. T. Krenács, E. Dobo, and Z. Lslik, Characteristics of endocrine pancreas in chronic pancreatitis, as revealed by simultaneous immunocyto-chemical demonstration of hormone production, *J.Histotechol.* **13**:213 (1990).

29. T. Krenács, L. Krenács, B. Bozky, and B. Ivnyi, Double and triple immunocyto-chemical labelling at the light microscopical level in histopathology, *Histochem J* **22**:530 (1990).

30. P. Fritz, J. Hoenes, J. Schenk, A. Mischlinski, A. Grau, J.G. Saal, H.V. Tuczek, H. Multhaupt, and G. Pfleiderer, Color development of immunogold- labelled antibodies for light microscopy, *Histochemistry* **85**:209 (1986).

31. D.R. Springall, Immunocytochemistry in diagnostic cytology, *in*: "Immunocytochemistry - Modern Methods and Applications," J.M. Polak and S. Van Noorden, eds., Wright, Bristol. 547 (1986).

32. M. De Waele, J. De Mey, M. Moeremans, M. De Brabander, and B. Van Camp, Immunogold staining method for the light microscopic detection of leucocyte cell surface antigens with monoclonal antibiodies: its application to the enumeration of lymphocyte subpopulations, *J Histochem Cytochem* **31**:376 (1983).

33. M. De Waele, J. De Mey, W. Renmans, C. Labeur, Ph. Reynaert, and B. Van Camp, An immunogold-silver staining method for the detection of cell-surface antigens in light microscopy, *J Histochem Cytochem* **34**:935 (1986).

34. L. Bastholm, L. Scopsi, and M.H. Nielsen, Silver-enhanced imunogold staining of semithin and ultrathin cryosections, *J Electr Misrosc Tech* **4**:175 (1986).

35. A.N. Van Den Pol, Silver-intensified gold and peroxidase as dual immunolabels for pre- and postsynaptic neurotransmitters, Science **228**:332 (1985).

36. K. Westermark, M. Lundqvist, G.W. Hacker, A. Karlsson, and B. Westermark, Growth factor receptors in thyroid follicle cells, *Acta Endocrinol (Copenh) Suppl* **281**:252 (1987).

37. M. Moeremans, G. Daneels, A. Van Dijk, G. Langanger, and J. De Mey, Sensitive visualisation of antigen antibody reaction in dot and blot immune-overlay assays with immunogold and immunogold-silver staining, *J Immunol Meth* **74**:353 (1984).

38. B.F. Rocks, N. Patel, M.P. Bailey, Use of a silver-enhanced gold-labelled immunoassay for detection of antibodies to the human immunodeficiency virus in whole blood samples, *Ann Clin Biochem* **28**:155 (1991).

39. T. Krenács, B. Ivnyi, B. Bozky, Z. Lszik, L. Krancs, Z. Rzga, and J. Ormos, Postembedding immunoelectron microscopy with immunogold-silver staining (IGSS) in Epon 812, Durcupan ACM and LR-White resin embedded tissues, *J Histotechnol* **14**:75 (1991).

40. J. Gu, M. D'Andrea, C.Z. Yu, M. Forte, L.B. McGrath, Quantitative evaluation of indirect immunogold-silver electron microscopy, *J Histotechnology* **16**:19 (1993).

41. M. Stefanini, C. De Martino, and L. Zamboni, Fixation of ejaculated spermatozoa for electron microscopy, *Nature* **216**:173(1967).

42. L. Zamboni, and C. De Martino, Buffered picric acid-formaldehyde: a new rapid fixative for electron microscopy, *J Cell Biol* **35**:148A (1967).

43. A.E. Bishop, J.M. Polak, S.R. Bloom, and A.G.E. Pearse, A new universal technique for the immunocytochemical localisation of peptidergic innervation, *J Endocrinol* **77**:25 (1978).

44. O. Van Laere, L. De Wael, and J. De Mey, Immuno gold staining (IGS) and immuno gold silver staining (IGSS) for the identification of the plant pathogenetic bacterium Erwinia amylovora (Burrill) Winslow *et al.*, *Histochemistry* **83**:397 (1985).

45. I. Zehbe, G.W. Hacker, J. Sllstrm, E. Rylander, and E. Wilander, Detection of single copies of HPV in SiHa cells *in situ* polymerase chain reaction (*in situ* PCR) combined with immunoperoxidase staining and immunogold-silver staining (IGSS) techniques, *Anticancer Res* **12**:2165 (1992).

46. G. Danscher, G.W. Hacker, L. Grimelius, and J.O.R. Norgaard, Autometallo-graphic silver amplification of colloidal gold, *J Histotechnol* (in press).

47. J.F. Hainfeld, and F.R. Furuya, A 1.4-nm gold cluster covalently attached to antibodies improves immunolabeling *J Histochem Cytochem* **40**: 177 (1992).

48. R. Liesegang (ed.): Die Kolloidchemie der histologischen Silber-Färbungen, *Kolloid Beihefte* **3**:1 (1911).

49. G. Danscher, and J.O.R. Norgaard, Ultrastructural autometallography: a method for silver amplification of catalytic metals, *J Histochem Cytochem* **33**:706 (1985).

50. Y.-D. Stierhof, B.M. Humbel, R. Hermann, M.T. Otten, and H. Schwarf, Direct visualization of silver enhancement of ultra-small antibody-bound gold particles on immunolabeled ultrathin resin sections, *Scanning Microsc* **6**:1003 (1992).

51. W.-M. Huang, S.J.Gibson, P. Facer, J. Gu, and J.M. Polak, Improved section adhesion for immunocytochemistry using high molecular weight polymers of L-lysine as a slide coating, *Histochemistry* **77**:275 (1983).

52. P.H. Maddox, and D. Jenkins, 3-aminopropyltriethoxysilane (APES): a new advance in section adhesion, *J Clin Pathol* **40**:1256 (1987).

CHAPTER 4

SINGLE AND MULTIPLE IMMUNOFLUORESCENCE - AN OVERVIEW

Cornelia Hauser-Kronberger[1] and Gerhard W. Hacker

Institute of Pathological Anatomy
Immunohistochemistry and Biochemistry Unit
Salzburg General Hospital
A-5020 Salzburg, Austria

SUMMARY

The first attempts to conjugate immunoglobulins with labels were made by using ordinary dyes, but these were unsatisfactory because they were not visible under the microscope. Coons devised the idea of localizing substances in tissues by means of specific antibodies labeled with a fluorescent dye. One of the first and still most widely used labels was fluorescein isothiocyanate (FITC). Originally, the antibody itself was labeled (direct method), and later the more sensitive and versatile indirect method was introduced, utilizing an unlabeled primary antibody followed by a second layer of labelled antibody directed against immunoglobulin of the species in which the primary antibody had been raised. Bridging techniques using a third layer of antibody, or streptavidin-biotin-complex methods are even more sensitive and are included today in most double immunofluorescent staining techniques. The disadvantages of immunofluorescent methods include the necessity of an expensive fluorescence microscope, relative difficulty in appreciating background details, and fading of the staining with prolonged exposure to ultraviolet light. Nevertheless, because of the speed, the simplicity (especially of double staining) and the high specificity, the use immunofluorescent methods has remained popular in basic research and pathological diagnosis. It also appears true that these

[1] Address for correspondence to: Dr. Mag. Cornelia Hauser-Kronberger, Ph.D., Institute of Pathological Anatomy, Immunohistochemistry and Biochemistry Unit, Salzburg General Hospital, Muellner Hauptstr. 48, A-5020 Salzburg, Austria, Phone:++43/662/4482/4730; Fax:++43/662/4482/88

methods are the best techniques for selectively demonstrating regulatory peptides in nerve fibers.

INTRODUCTION

Historical Review

Fluorescent histochemical methods have been widely used over the last thirty years for the localization of biogenic amines.[1,2] Immunocytochemical techniques, especially immunofluorescent (IF) methods, although already invented in the early forties (Albert H. Coons[3,4]), only came into broad use since the sixties. Coons' ingenious idea was the localization of antigen in tissues sections by means of specific antibodies, labeled by a fluorescent dye. Since then, the most widely used label was fluorescein-isothiocyanate (FITC), giving a bright green color under the fluorescent microscope. The development of these techniques has also led to a series of further modifications and expansions in the use of fluorescent histochemical methods.

THEORETICAL AND PRACTICAL CONSIDERATIONS

Direct Method

In direct methods, only one layer of antibodies labeled with a fluorescent dye is applied to the tissue section. The main disadvantages of these methods are a low sensitivity and the need to label every single antiserum by fluorescent dyes. Such labeled antibodies can hardly be transferred to other detection systems.

Indirect Method

In indirect methods, tissue sections are first incubated with unlabeled primary antibodies. The sites of the primary antibody are then indirectly identified by labeled secondary antibodies raised against immunoglobulin of the species providing the primary antibody. At least two secondary antibody molecules can bind to each primary antibody molecule; therefore, indirect methods are more sensitive than direct methods. Smaller amounts of antigen can be detected, and a stronger signal is given for the same number of bound primary antibodies. An additional advantage is that the labeled secondary antibody system can be applied to identify any primary antibody raised in that species.

Bridging Methods

An unlabeled primary antibody is reacted with the tissue antigen. As a second step, an unlabeled or labeled bridging antibody raised against immunoglobulins of the species providing the primary antibody is bound to the primary antibody. In a third step, labeled or unlabeled antibodies from the same species as the one in which the primary antibody has been raised are attached to secondary antibodies. Therefore, the secondary antibody acts as a bridging antibody. Such methods are even more sensitive than ordinary indirect techniques.[5] A further modification is the use of biotinylated second layer antibodies, followed by their detection utilizing labeled streptavidin.[6]

Fixation

Essential conditions for accurate localization of antigen by antibodies are a good preservation of the tissue morphology and also of the antigenic sites, the production of specific antibodies, and a very efficient labeling and binding reaction of the antibodies.

It is necessary and conventional to treat fresh tissue with various chemical fixatives, such as formalin, picric acid, para-benzoquinone and others, in order to preserve structure and prevent tissue components from elution during histological staining procedures. Soluble antigens must be fixed with a suitable fixative before being immunostained. Different antigen-groups often require different fixation methods and solutions for optimal preservation.[7] The standard fixation with formalin can sometimes induce too strong cross-linking and, therefore, is not always the best choice. A compromise between good tissue preservation and antigen availability has to be reached for each antigen. The method of fixation depends not only on the antigen to be localized, but also on the tissue structure in which it is present.[8]

Permeabilization

Thick cryostat sections, and especially whole-mount preparations, need to be permeable to the applied antibodies. The use of detergents like Triton-X 100 or Tween, in concentrations from 0.1 % to 1%, is the most conventional way of achieving permeabilization. However, permeabilization with detergents should be avoided if the detection of membrane antigens is desired.

For the detection of antigens contained in fine nerve fibers in tissue sections, the use of fairly thick sections (>10 μm) is recommended, obtained in a cryostat or vibratome from prefixed tissue specimens. Immunofluorescent techniques are sometimes more effective than other immunocytochemical methods because the thickness of the section does not always allow sufficient penetration of transmitted light to give a clear signal and background.[9]

Adhesion to Sections

To prevent sections from becoming detached from the glass slide during immuno-staining, it is useful to coat the slide with poly-L-lysine[10] or aminopropyl- triethoxysilane (APES).[11] Cryostat sections should be melted onto the coated slide and dried for at least 1 hour at room temperature or for 30 minutes at 37°C before being stained. Dried cryostat sections may also be stored at -20°C for at least 2 weeks before use.

Working dilution of antibodies

Every antibody should be tested on known positive controls to find the optimal working dilution for the particular stain applied. This dilution will vary also according to the incubation period and temperature. For immunofluorescent techniques, the incubation periods for the primary antibody vary from a few hours to 48 hours at room temperature or from 24 hours up to 72 hours at 4°C. In most cases, polyclonal primary antibodies are applied at dilutions of 1/50 to 1/1000. Monoclonal antibodies need to be tested from concentrated to dilutions of 1/1000 in positive reference tissue sections. The working dilution may have to be altered for another tissue or another fixation method. An increased concentration does not necessarily mean better staining, because an optimally diluted antibody provides less non-specific binding and less background staining and gives a better contrast. Examples of dilutions for secondary antibodies used in our laboratory are given in *Table 1*.

Table 1. Examples of Secondary Fluorescent Antibodies and Labeled Streptavidin, and Their Proposed Optimal Dilutions for Single and Multiple Immunofluorescence.

Antibody	Donor	Code	Dilution	Source
anti-mouse IgG, FITC-conjugated	rabbit	4221-1	1/25	BioMakor, Israel
anti-rabbit IgG, FITC-conjugated	goat	4271-1	1/25	BioMakor, Israel
anti-mouse IgG, RITC-conjugated	nk	AP181R	1/25	Chemicon, USA
anti-rabbit IgG, RITC-conjugated	nk	AP182R	1/25	Chemicon, USA
anti-sheep IgG, FITC-conjugated	donkey	nk	1/40	Sigma, FRG
anti-mouse IgG, FITC-conjugated	goat	nk	1/80	Sigma, FRG
streptavidin, Texas Red-conjugated	-	RPN1233	1/50	Amersham, UK
streptavidin, phycoerythrin-conjugated	-	RPN1235	1/50	Amersham, UK
streptavidin, AMCA™-conjugated	-	RPN1234	1/50	Amersham, UK
biotinylated anti-rabbit IgG	donkey	1004	1/50	Amersham, UK
biotinylated anti-mouse IgG	sheep	1001	1/50	Amersham, UK

nk = not known

Storage

Primary antibodies as well as FITC- and RITC-conjugated reagents may be kept at 4°C for short periods. For longer periods, they should be stored in small undiluted aliquots at -20°C. Generally, irradiation even of daylight to the fluorescent antibodies or to the stained sections should be avoided to prevent fading.

Labels

Some of the most frequently used fluorescent dyes are summarized in *Table 2*. Fluorescein isothiocyanate (FITC) and rhodamine isothiocyanate (RITC) are well-known labels, the former fluorescing green at an excitation wavelength of 450-490 nm, the latter, red at 510-560 nm. Newer fluorescent labels are Texas Red and phycoerythrin. Texas Red (sulphonylchloride) (*Fig. 1*) fluoresces red at the same wavelength as rhodamine, but is more stable under the irradiation. Phycoerythrin, which also gives red fluorescence at a wavelength of 510-560 nm, has the advantage in that its excitation optimum for fluorescence is near to that of FITC; but in this case, it shows orange fluorescence. Thus

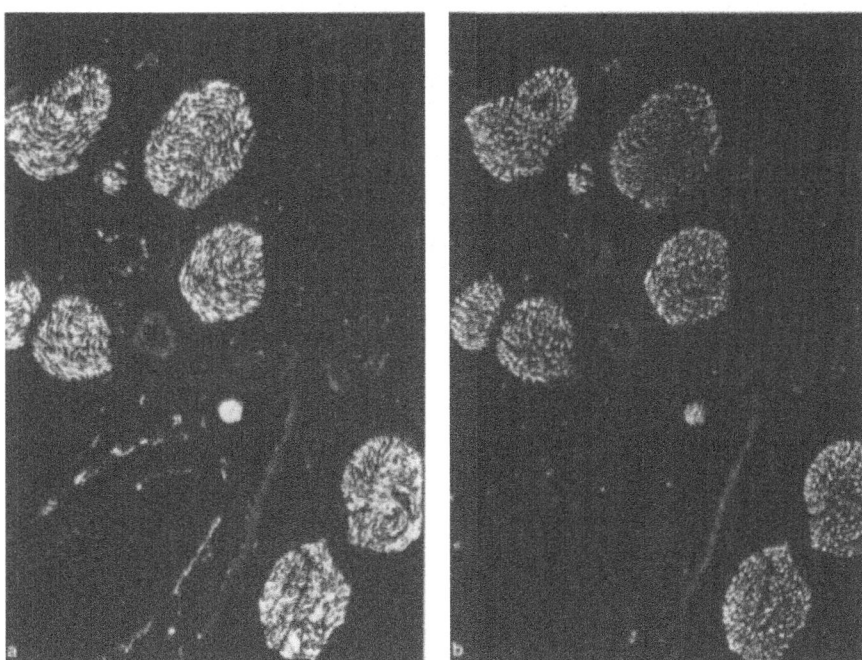

Figure. 1. Double immunofluorescence micrographs.

Figure 1a. A rabbit polyclonal antibody to protein gene-product (PGP) 9.5 was applied in a section from the human penis. Secondary antibodies were FITC-labeled, giving a green fluorescence.
Figure 1b: The same section was stained simultaneously with a mouse monoclonal antibody against neurofilament protein (NFP, 200 kD) and a streptavidin-Texas Red system, giving a red fluorescence. 14 μm thick, Stefanini's/Zamboni's fixed cryostat section. Original magnification x 120.

in a double fluorescent immunostain, both labels can be seen together without the need for switching filters (*Fig. 2*). Another label similar to RITC but more stable under irradiation with UV-light, is the tetramethylrhodamin-isothiocyanate (TRITC),[13] which fluoresces red at a 510-590 nm excitation wavelength. The recently introduced label AMCA shows blue fluorescence at an excitation wavelength of 330-380nm and can be used for multiple immunofluorescence studies (*Fig. 3*).

Table 2. Examples of Labels Used in Immunofluorescence.

Label	Abbr.	Color	Excitation Optimum	Filter for Leica-Reichert Polyvar Microscope
Fluorescein-isothiocyanate	FITC	green	450-490 nm	B1, B2
Rhodamine-isothio cyanate	RITC	red	510-560 nm	G2
Tetramethylrhodamin-is othiocyanate	TRITC	red	510-590 nm	G2
Sulphonylchloride	Texas Red	red	510-590 nm	G2
Phycoerytrin		orange	450-490 nm	B1, B2
		red	510-560 nm	G2
AMCA™		blue	330-380 nm	U1, V2

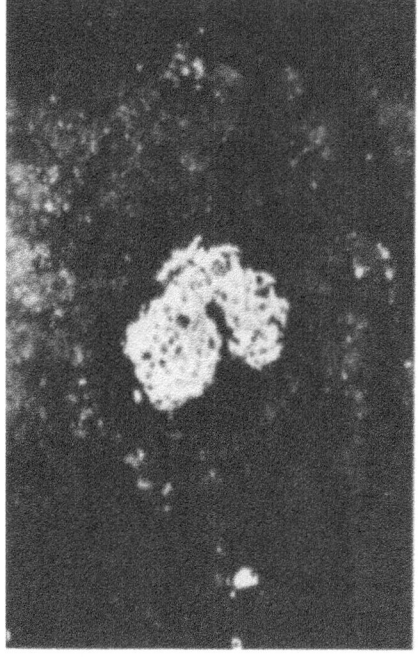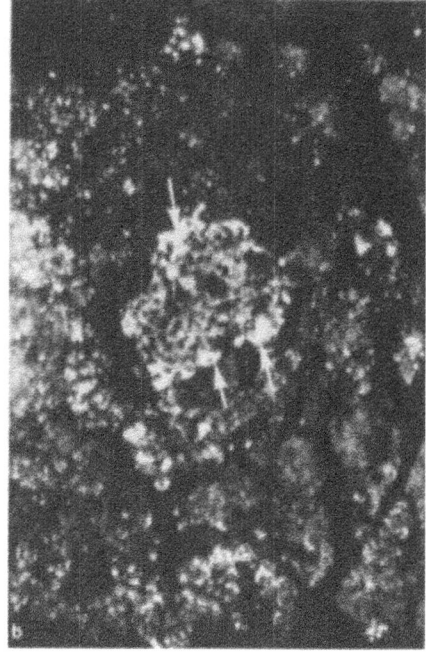

Figure 2. Double immunofluorescence micrographs with a mouse monoclonal antibody to insulin (Fig. 2a) and a rabbit polyclonal antibody to glucagon (Fig. 2b). For detection, a combination of FITC in an indirect method (Fig. 2a) with a biotin-streptavidin-phycoerythrin system (Fig. 2b) was used.
Figure 2a: With green rhodamine filters, the insulin-immunoreactive cells show a red color, FITC is not visible here.
Figure 2b: The originally green picture given with blue FITC-filters in the microscope. It simultaneously shows insulin-immunoreactive cells in grey (small arrows; original color: orange) and glucagon-immunoreactive cells in white (big arrows; original color: green). Lipid granules show autofluorescence with both filter systems. 14 µm thick, Stefanini's/Zamboni's-fixed cryostat section. Original magnification x 300

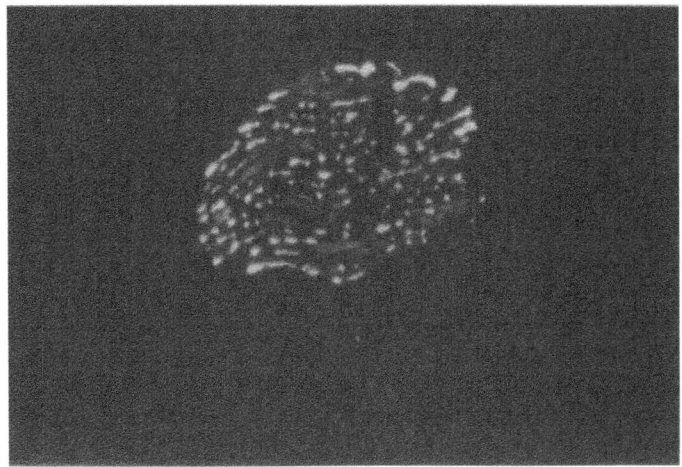

Figure 3. Single indirect immunofluorescence staining with a mouse monoclonal antibody to neurofilament protein (NFP, 200 kD) in the human penis, using a blue fluorescing streptavidin-AMCA™ system. 14 μm thick, Stefanini's/Zamboni's- fixed cryostat section. Original magnification x 300

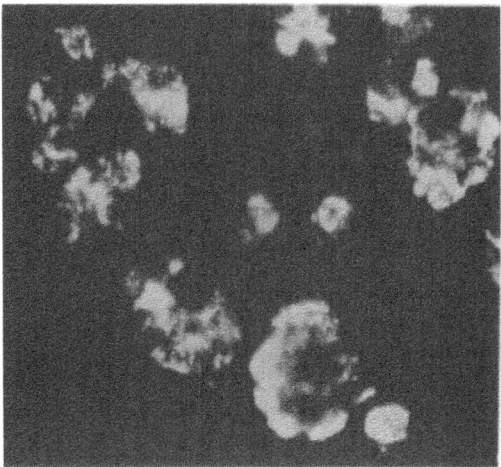

Figure 4. Single indirect immunofluorescence staining with an mouse monoclonal antibody to a receptor protein for vasoactive intestinal polypeptide (VIP) located in glandular acini in the human nasal mucosa. 14 μm thick, Stefanini's/Zamboni's-fixed cryostat section. Original magnification x 120.

Autofluorescence

A major problem seen with immunofluorescent techniques is the presence of variable amounts of autofluorescence which may obscure the specific fluorescent label. This difficulty is particularly marked in tissue sections from human material, especially from aged patients or also from older animals due to the higher levels of connective tissue, a major source of autofluorescence. Endogenous tissue components showing auto-fluorescence include collagen, elastin, cartilage, keratin, lipids, porphyrins, and lime (*Fig. 2*). Elastic fibers may show autofluorescence induced by formaldehyde (catecholamines), but this usually is no problem because it is removed by water by the end of the immunostaining sequence.[14] Auto-fluorescence may sometimes be altered or reduced with a counterstain such as Evans Blue[15] or Pontamine sky blue 5BX (0.05% Pontamine sky blue in PBS containing 1% dimethyl-sulfoxide)[16] which fluoresces red when using a blue filter for green FITC-fluorescence.

OPTIMAL PROCEDURES

The optimal procedure can be divided into three major steps:

1. Tissue Preparation and Fixation
2. Staining Procedure
3. Interpretation and Storage

In the following text, some procedures are explained, and protocols are given which we found to perform well for the demonstration of regulatory peptides and various other chemical substances in tissue sections.

Preparation and Fixation

Freshly dissected specimens should be fixed by immersion as soon as possible after dissection. Whenever applicable, instillation or perfusion fixation should be applied.

The most versatile fixation solution in our hands proved to be Stefanini's/ Zamboni's-fixative at 4°C.[17-19] It contains picric acid and paraformaldehyde dissolved in phosphate buffered saline (PBS) pH 7.2. Often, 4-10% neutral buffered formalin or 4% paraformaldehyde solution at a temperature of 4°C are suitable immersion fixatives for immunofluorescent techniques. Great attention should be given to optimize the fixation-period, depending on the size and type of the samples. After fixation, specimens should be thoroughly washed in PBS containing 10-20% sucrose for at least 24 h at 4°C. For the production of cryostat sections, the samples are embedded in OCT-compound (Tissue Tec, USA) and shock frozen.

Preparation of Stefanini's/Zamboni's Fixative[17]

Dissolve 20 g p-formaldehyde in 150 ml saturated picric acid and heat up for 2 hours at 60°C. Neutralize by adding, dropwise, 2.5% NaOH until the solution becomes clear. Filtrate and add PBS to get 1000 ml. The fixative should have a final pH of about <7.4.

Double Immunofluorescence Staining

For many purposes, detection of a single antigen in a tissue section or on a protein blot is adequate, but the ability to detect two antigens within the same section or blot proved to be very useful and can be optimally achieved using biotin-streptavidin systems.[20]

The availability of streptavidin-biotin complexes labeled with fluorescent dyes and antibodies labeled with other fluorescent labels facilitates dual immuno-fluorescent staining. This is possible because the emission maximum of FITC is clearly distinct from the maximum of RITC, Texas Red, or AMCA, so that the signal from FITC can be distinguished from either of the other fluorochromes. Generally, double labels cannot be viewed simultaneously, and different filters of high quality with exactly defined wavelengths must be used. However, it is possible to discriminate between phycoerythrin and FITC using a standard filter for FITC. Even in the latter system, the best photographic results will be achieved by making composite photographs by exposing the film twice using different filters in turn, since phycoerythrin is usually brighter than fluorescein and requires shorter exposure times (*Fig. 5-8*).

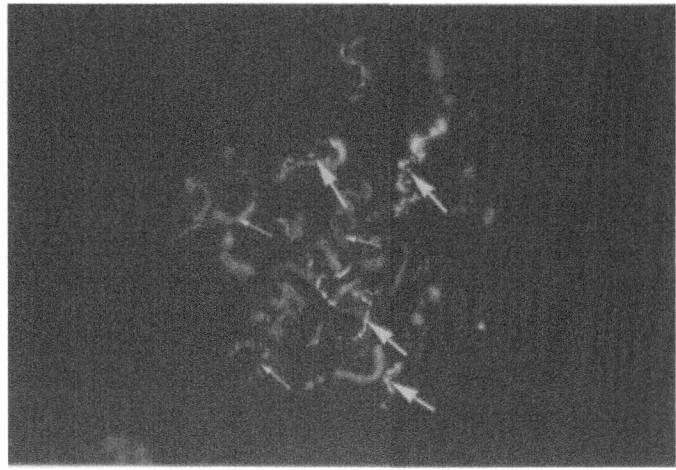

Figure 5. Double immunofluorescence with a mouse monoclonal antibody to neurofilament protein (NFP 200 kD, grayish nerve fibers, small arrows) and a rabbit polyclonal antibody to vasoactive intestinal polypeptide (VIP, white nerve fibers, big arrows) in the human epiglottis. Double exposure from the same section, one time with green fluorescence filter used for red emission, and one time with blue fluorescence filter used for green emission. 14 µm thick, Stefanini's/Zamboni's-fixed cryostat section. Original magnification x 300.

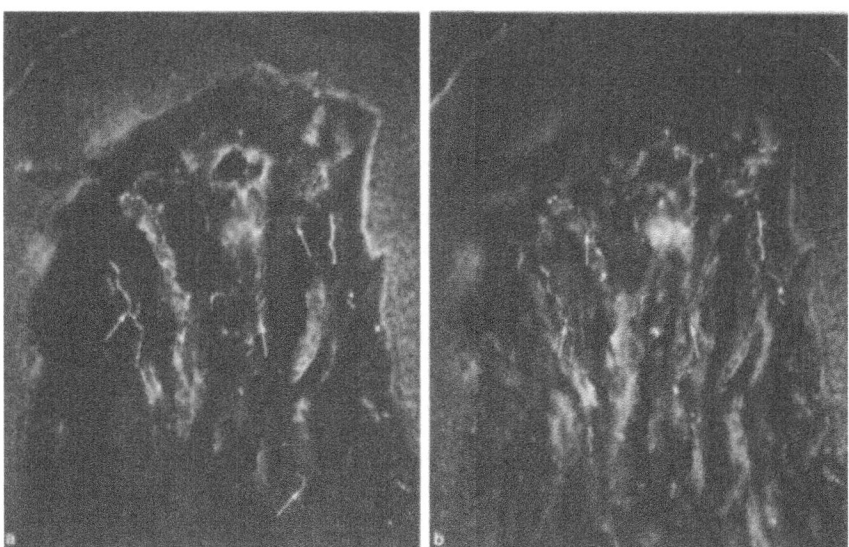

Figure 6. Double immunofluorescence with a rat monoclonal antibody to substance P (Fig. 6a) and a rabbit polyclonal antibody to calcitonin gene-related peptide (CGRP, Fig. 6b) in the human clitoris. Examples for the detection of regulatory peptides co-localized in fine varicose nerve fibers (arrows). 14 µm thick, Stefanini's/Zamboni's-fixed cryostat section. Original magnification x 240.

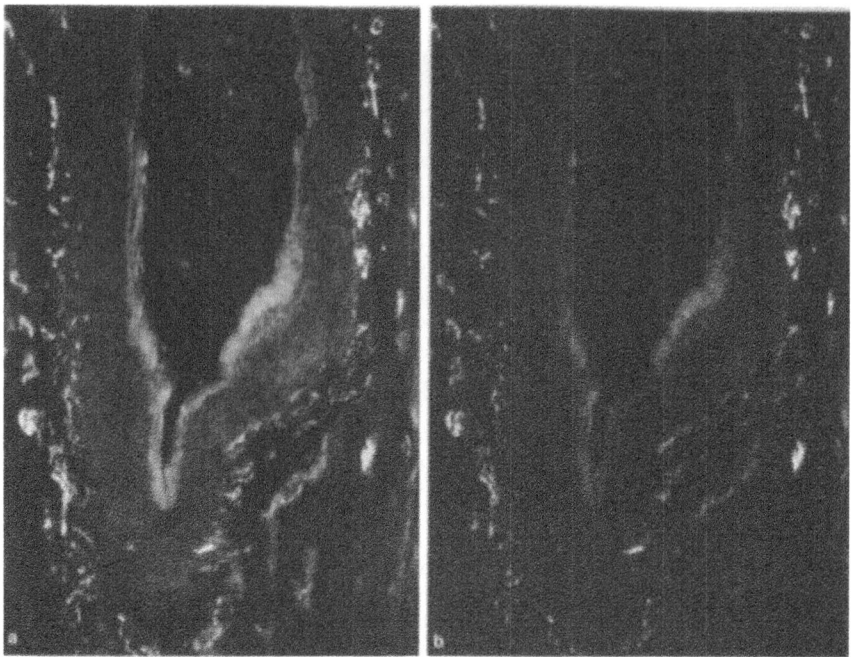

Figure 7. Double immunofluorescence with a mouse monoclonal antibody to vasoactive intestinal polypeptide (VIP, Fig. 7a) and a polyclonal rabbit antibody to peptide histidine methionine (PHM, Fig. 7b) showing delicate nerve fibers around an artery in the human nasal mucosa. 14 μm thick, Stefanini's/Zamboni's-fixed cryostat section. Original magnification x 300.

Triple Staining

The availability of the blue fluorescent label AMCA opens the possibility of a simple triple staining using a combination of immunofluorescent techniques, even if the antigens are localized within the same structures. AMCA may be conjugated to streptavidin and may be used in a streptavidin-biotin-complex system. Basically for multiple staining, antibodies raised in three different species are necessary.

Mounting-Media

The medium we use to mount coverslips on immunofluorescence stained preparations is *Glycergel* (Dakopatts, Glostrup, DK, or Carpinteria, CA, USA), which is relatively firm at room temperature. Before use, heat it in warm tap water or in a microwave oven, until it is liquified. Glycergel has the advantage that it gets relatively firm a few minutes after mounting.

The traditional mountant is buffered glycerine. Solutions of one part or nine parts glycerine and one part PBS are often recommended for mounting immuno-fluorescence-stained sections, but this medium does not get firm.[21] Some fluorescent labels fade rapidly under ultraviolet irradiation. Modified mounting media have been developed to retard this fading. Media with photostabilizers, which do not get hard either, may improve the photographic reproduction of fine details.

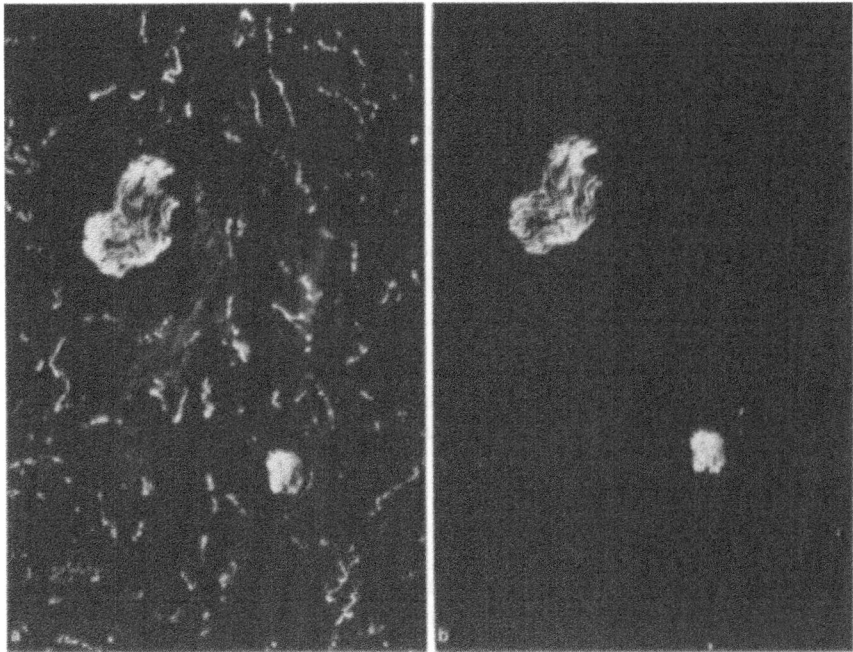

Figure 8. Double immunofluorescence with a polyclonal rabbit antibody to protein gene-product (PGP) 9.5 (Fig. 8a) and a mouse monoclonal antibody to neurofilament protein (NFP, 200 kD) (Fig. 8b) in the human penis. As clearly seen in these micrographs, antibodies to PGP 9.5 also reveal most fine varicose nerve fibers, whereas antibodies to NFP only give reaction in axons of thicker nerve bundles. 14 μm thick, Stefanini's/Zamboni's-fixed cryostat section. Original magnification x 240.

Storage of Immunofluorescence Stained Sections

Sections stained with immunofluorescent techniques can be easily stored in dark at 4°C for at least one year without noticeable fading. Fading can be even further retarded if the sections are stored at -70°C or by adding 0.1% p-phenylene-diamine (PPD) to the mounting media.[22]

RECOMMENDED STAINING PROTOCOLS
Protocol 1. Indirect Immunofluorescence (Single Staining) (*Fig. 4*)
1. Prefixed cryostat sections (> 10 μm) are mounted on poly-L-lysine or APES coated slides and dried for 1 hour at room temperature.
2. Background blocking: Apply 10 % normal serum from the species of the secondary antibody, diluted in PBS (pH 7.2-7.4) containing 0.1 % bovine serum albumin (BSA) and 0.1 % Triton X-100 for 15 min.
3. First layer antibody is diluted in PBS at the established dilution and incubated for 48-72 hours at 4°C or for 24 hours at room temperature.
4. Rinse in PBS (3x5 min).
5. Apply second layer (raised against immunoglobulin of the primary antibody species, FITC- or RITC-conjugated) at the appropriate dilution in PBS, for 1 hour at room temperature.
6. Rinse in PBS (3x5 min).
7. Rinse in distilled water (3 min).
8. Mount in Glycergel (Dakopatts, Glostrup, DK, or Carpinteria, CA, USA) or in a mixture of PBS and glycerine (1:1).

Protocol 2. Double Immunofluorescence Staining[19,23]

1. Prefixed cryostat sections (> 10 μm thick) are mounted on poly-L-lysine- or APES-coated slides and dry for 1 hour at room temperature.
2. Background blocking: Apply 10% normal serum from the species of the secondary antibodies in PBS (pH 7.2-7.4), containing 0.1% BSA and 0.1% Triton X-100, for 15 min.
3. Two unlabeled primary antibodies raised in different species are applied simultaneously in a 1:1 mixture of the established dilutions (the diluted antibodies have to be mixed together before application) for 24 hours at room temperature or for 48-72 hours at 4°C.
4. Rinse in PBS three times.
5. Apply second layers together (as 1:1 mixture) directed against immuno-globulins of both species providing the primary antibodies, diluted in PBS, (FITC-, and RITC-, TRITC-, or Texas Red-conjugated) for 1 hour at room temperature.
6. Rinse in PBS (3x5 min).
7. Rinse in distilled water (3 min).
8. Mount in Glycergel or in a mixture of PBS and glycerine (1:1), as above.

Protocol 3. Double- and Triple-Immunofluorescence Staining with a Biotin-Streptavidin-Complex System

1. Prefixed cryostat sections (> 10 μm) are mounted on poly-L-lysine- or APES-coated slides and dried at room temperature for 1 hour.
2. Background blocking: Apply 10% normal serum from the species of the secondary antibodies in PBS (pH 7.2-7.4) containing 0.1% BSA and 0.1% Triton X-100, for 15 min.
3. Three primary antibodies raised in different species are applied simultaneously in equal amounts of the established dilutions in PBS, which are mixed together before application, for 24 hours at room temperature or for 48-72 hours at 4°C.
4. Rinse in PBS (3x5 min).
5. Apply second layers (a 1:1:1 mixture of diluted antisera directed against immunoglobulins of two species providing primary antibodies; e.g. FITC- and RITC-conjugated, and a biotinylated antibody against immunoglobulins of the third primary antibody species) simultaneously in the established dilutions for 1 hour at room temperature.
6. Rinse in PBS (3x5 min).
7. Apply third layer (streptavidin conjugated with AMCA™) in the established dilution for 1 hour at room temperature.
8. Rinse in PBS (3x5 min).
9. Rinse in distilled water (3 min).
10. Mount in Glycergel or in a mixture of PBS and glycerine (1:1), as above.

Appendix: 0.1 M phosphate-buffered-saline (PBS) 8.7 g NaCl + 0.272 g KH_2PO_4 + 1.136g Na_2HPO_4 in 1000 ml distilled water pH 7.4

DISCUSSION

Application of immunofluorescent methods requires relatively little laboratory time, which is a distinct advantage over immunoenzyme methods. In order to obtain more morphologic details, hematoxylin may be used as a counterstain without quenching the fluorescence. This can be visualized by switching the microscope from fluorescence to bright field mode.[24] Another counterstain we generally apply for FITC-labelled single immunofluorescence is pontamine-sky blue which can be watched simultaneously with FITC green-fluorescence and gives a red counterstain with this filter system.[16] Other advantages of immunofluorescence are a high sensitivity and the simplicity of performing single and multiple immunostaining. Immunofluorescent techniques show a high contrast, especially on black and white photomicrographs.

Disadvantages of immunofluorescent methods include that they do not work properly on paraffin sections,[25] that it is difficult to judge background details even when using counterstains, and sometimes, a higher concentration of the primary antibody than in other methods may be necessary. Fading of the stained sections can be decreased or avoided by adequate storage or by adding certain chemicals to the mounting media. Another disadvantage is the necessity of an expensive fluorescent microscope with an appropriate filter set.

Immunofluorescent techniques should be used if the antigen is sensitive to alcohol, temperature, and paraffin embedding. Alcohol dehydration should be avoided. Such techniques are recommended for the detection of chemical substances in fine structures (e.g. regulatory peptides in delicate nerve fibers) or if multiple staining is needed in the same tissue section and the antigens are located within the same morphological structures. The histology of tissue must be well known, and no direct further interpretation should be needed. Therefore, the results from immunofluorescent stainings should not be solely for obtaining information, but also to get further information (e.g. in histopathology).[22]

Immunofluorescent techniques should not be used if the antigen is stable, easily detectable, and rich amounts of antigen are present, other methods of choice give better morphologic resolution. ABC, peroxidase, or immunogold-silver staining (IGSS) are methods of choice (see also the chapters by Hacker and by Krenács, in this book).

As pointed out previously, immunofluorescent and other immunohistochemical methods are complementary rather than competetive techniques. Methods using enzymes as labels are basic in immunohistochemical laboratories but require the complementary power of immunofluorescence for two main reasons: immuno-fluorescence is an elegantly rapid screening and detecting method; it is the most scientifically acceptable method for investigating double- and triple-labeled preparations with sufficient clarity. This is especially noteworthy when the substances of interest are localized within the same structures and are therefore simultaneously labeled by two or three different antibodies. In this case, the viewing of different labels by an easy switching of filters is a great advantage. Comparable intermediate colors of double enzymatic reactions would be much more difficult to judge.

It is unlikely that the final practical limits of the technique have been reached. Combinations applying different fluorochromes or antibodies of different isotypes can still be improved.

ACKNOWLEDGEMENTS

We gratefully acknowledge the dedicated technical assistance of Sylvia Dechet, Walter Kaufmann and Angelika Schiechl (all Salzburg, Austria).

REFERENCES

1. B. Falck, N-A. Hillarp, G. Thieme, and A. Torp. Fluorescence of catecholamines and related compounds with formaldehyde. *J Histochem Cytochem* **10**:348 (1962).
2. O. Lindvall and A. Björklund. The glyoxylic acid fluorescence histochemical method: a detailed account of the methodology for the visualization of central catecholamine neurons. *Histochemistry* **39**:97 (1974).

3. A.H. Coons, H.J. Creech, and R.N. Jones. Immunological properties of an antibody containing a fluorescent group. *Proc Soc Exp Biol Med* **47**:200 (1941).

4. A.H. Coons, E.H. Leduc, and J.M. Connolly. Studies on antibody production. I a method for the histochemical demonstration of specific antibody and its application to a study of the hyperimmune rabbit. *J Exp Med* **102**:49 (1955).

5. L.A. Sternberger, P.H. Hardy, J.J. Cuentis, and H.G. Meyer. The unlabelled antibody enzyme method of immunocytochemistry preparations and properties of soluble antigen-antibody complex (horseradish peroxidase-antihorseradish peroxidase) and its use in identification of spirochetes. *J Histochem Cytochem* **18**:315 (1970).

6. G. Coggi, P. Dell 'Orto, and G. Viale. Avidin-biotin methods, *in* "Immunocytochemistry - modern methods and applications," J.M. Polak and S. Van Noorden, *eds.*, Bristol, U.K., 54 (1986).

7. W.W. Hancock, G.J. Becker, and R.C. Atkins. A comparison of fixatives and immunohistochemical techniques for use with monoclonal antibodies to cell surface antigens. *Am J Clin Pathol* **78**:825 (1982).

8. G.D. Johnson, E.J. Holboro. Immunofluorescence. *in*: "Handbook of Experimental Immunology," D.M. Weir and I.A. Herzenberg eds., Blackwell Scientific Publications, Oxford (1986).

9. L.W. Poulter, M. Chilosi, G.J. Seymour, S. Hobbs, and G. Janossy. Immunofluorescence membrane staining and cytochemistry, applied in combination for analysing cell interaction *in situ*. *in*: "Immunocytochemistry, Practical Applications in Pathology and Biology," J.M. Polak, S. Van Norden eds., Wright PSG, Bristol (1983).

10. W.M. Huang, S.J. Gibson, P. Facer, J. Gu, and J.M. Polak. Improved section adheasion for immunocytochemistry using high molecular weight polymers of L-lysine as a slide coating. *Histochemistry* **77**:275 (1983).

11. P.H. Maddox and D. Jenkins. 3-aminopropyltriethoxysilane (APES): a new advance in section adhesion. *J Clin Pathol* **40**:1256 (1987).

12. J.A. Titus, R. Haugland, S.O. Sharrow, and D.M. Segal. Texas Red, a hydrophilic, red-emitting fluorophore for use with fluorescein in dual parameter flow microfluorimetric and fluorescence microscopic studies. *J Immunol Methods* **50**:193 (1982).

13. R. Hiramoto, K. Engel, and D. Pressman. Tetramethylrhodamine as immuno-histochemical fluorescent label in the study of chronic thyroiditis. *Proc Soc Exp Biol* **97**:611 (1985).

14. I.S. de la Lande and J. G. Waterson. Modification of autofluorescence in the formaldehyde-treated rabbit ear artery by Evans Blue. *J Histochem Cytochem* **16**:281 (1968).

15. M. Goldman. Fluorescent Antibody Methods. Academic Press, New York (1968).

16. T. Cowen, A.J. Haven, and G. Burnstock. Pontamine sky blue: a counterstain for background autofluorescence in fluorescence and immunofluorescence histochemistry. *Histochemistry* **82**:205 (1985).

17. M. Stefanini, C. De Maritino, and L. Zamboni. Fixation of ejaculated spermatova for electron microscopy. *Nature* **216**:173 (1967),

18. L. Zamboni, and C. de Martino. Buffered picric acid formaldehyde: a new rapid fixative for electron microscopy. *J Cell Biol* **35**:148A (1967).

19. C. Hauser-Kronberger, G.W. Hacker, F. Sundler, J. Thurner, and K. Albegger. Distribution and co-localization of immunoreactive helospectin with vasoactive intestinal polypeptide and peptide histidine methionine in human nasal mucosa, soft palate and larynx. *Eur Arch Otorhinolaryngol* **249**:201 (1992).

20. B.L. Wang, and L.I. Larsson. Simultaneous demonstration of multiple antigens by indirect immunofluorescence or immunogold staining. Histochemistry **83**:47 (1985).

21. A.J. Balaton, A.M. Dalix, and R. Otriol. An improved mounting medium for immunofluorescence microscopy. *Arch Pathol Lab Med* **109**:108 (1985).

22. G. Wick, K.N. Traill, and K. Schauenstein. Immunofluorescence Technology, Selected and Theroetical and Clinical Aspects. Elsvier Biomedical Press, Amsterdam (1982).

23. W. Kummer. Simultanous immunohistochemical demonstration of vasoactive intestinal polypeptide and its receptor in human colon, *Histochem J* **22**:249 (1990).

24. M. Chilosi, G. Pizzolo, and C. Vincenzi. Haematoxylin counterstaining of immunofluorescence preparations. *J Clin Pathol* **36**:114 (1983).

25. S. Huang, H. Minassian, and J.D. More. Application of immunofluorescent staining in paraffin sections by trypsin digestion. *Lab Invest* **35**:383 (1976).

CHAPTER 5

IMMUNOCYTOCHEMISTRY OF WHOLE-MOUNT PREPARATIONS

Gian-Luca Ferri[1] and Rosa Maria Gaudio

Department of Cytomorphology, University of Cagliari
Cagliari, Italy

SUMMARY

Virtually any organ and tissue can be studied by means of immunostained "whole-mount preparations". These are unsectioned portions, or microdissected fragments of an organ largely retaining the original three-dimensional morphology and tissue context. Immunocytochemistry of whole-mounts may be relatively complex and time-consuming, but is highly rewarding. On such preparations, the enormous selectivity of immunostaining provides unique opportunities to visually follow and "dissect out" specific tissue components in their original three-dimensional geometry and organ/tissue context.

INTRODUCTION

Living organisms and their components are obviously three-dimensional in shape and cellular organization. However, what we are accustomed to stain, or immunostain, and examine under the microscope are two-dimensional histological sections. When using these, one attempts to reconstruct the complexity of biological tissues from a multitude of adjacent, "flat" sections, comprising little more than thin cell slices. Thus, it is a great step forward when one can prepare and immunostain organ and tissue fragments in such a way, as to show under the microscope both the feature/s of interest and the actual tissue organization in their true three-dimensional geometry.

According to different organ structure and research requirements, whole-mounts can be

---

[1] Address for correspondence: Gian-Luca Ferri, M.D., Department of Cytomorphology, University of Cagliari, via Porcell 2, 09124 Cagliari, Italy, Telephone: +39 70 651979; Fax: +39 70 657972

prepared for immunostaining and visualization in very flexible ways. Whole flat organs (such as iris, or diaphragm),[1,2] tissue laminae,[3,4] oriented samples of highly "modular" tissues,[5,6] and whole brains (e.g. from insects)[7] for instance, all can fall within the wide range of "whole-mounts." A direct, "hands-on" interaction of the investigator with the tissue/s of interest is encouraged, so that the high resolving power of immunostaining itself may become a virtual microdissection tool.

Understandably, a successful application of immunocytochemical methods to whole-mount preparations is more critical, complex, and time-consuming than on-section work.[3,6,8] We hope to be able to convince the reader that the sort of morphology and information produced by this approach can be so comprehensive, as to be both scientifically rewarding and economical.

Due to the practical scope of chapter 5, we shall restrict ourselves to the delineation of methodology and techniques. A few applicative examples will be added concerning gut endocrine cells and visceral nerves. Hints and suggestions found in the literature but not extensively tested in our laboratory will be given as such. On this basis, the various approaches to the production and immunostaining of whole-mount preparations can certainly be extended to many different systems and tissues.

HISTORICAL BACKGROUND

"Whole-mount" preparations for microscopy probably date back to the early days of microscopic anatomy itself, when dissociated fragments of tissue were used to follow (for example, the three-dimensional pattern of a nerve, a sensory organ, a blood vessel, etc.). Thus, the "dissociation method" was listed among the established approaches for the study of nervous tissues by Cajal.[9] Essentially, the method consisted of the microdissection of nerve tissues, carried out in such a way as to follow the three-dimensional morphology of nerve bundles and cell groups and to visualize them "whole." For instance, using a similar method, Ehrenberg discovered myelinated nerves in 1833.[9] A direct descendant of the same approach is today's neuropathological technique of nerve teasing, described as "extremely useful... because it provides a three-dimensional image" of nerves and nerve fibers.[10]

At about the same time, it became clear that the combination of a selective staining with the use of such "thick" tissue fragments was very advantageous. The observer was thus in the position of microscopically detecting and following the few features of interest, both in three-dimension and within the complexity of tissue context. In this light, the use of selective staining of whole-mount preparations is probably about a century old.[8] Interestingly, such techniques were pushed by Golgi and others as far as to stain whole-mount "samples" of considerable size, such as the whole human brain.[11]

Some 35 years ago, the development of immunocytochemistry made available staining methods that intrinsically displayed an unprecedented degree of selectivity and specificity. Various methods of pre-embedding immunostaining involving the use of some techniques for the improvement of antibody penetration into the tissue were introduced later (see the chapter of Kummer et al. in this book).

In 1980, a reliable and widely applicable method for the production of whole-mount immunocytochemistry was published,[3] which reported examples of peptide- immuno-staining of whole-mount plexuses from guinea pigs' gut. Thus, a most remarkable exploitation of the potentials of whole-mount immunocytochemistry by Furness, Costa and associates followed.[4] Shortly thereafter, while working with Dr. Julia M. Polak in London, one of us developed a similar methodology applied to the human gut which permitted visualization of whole mucosal endocrine cells and gut nerves.[5,12,13]

During the last ten years or so, a wealth of applications followed especially in the field of visceral innervation of virtually all organ systems. On the whole, it is now clear that whole-mount immunocytochemistry has far-reaching implications in many research areas in which microimaging technology is of use. It has been applied to many different questions in biological and medical research and to specialized areas of pathology. Although even a short account of such wealth of reports is beyond the scope of this chapter, a few examples relevant to preparative and immunostaining methodology will be mentioned in the following sections.

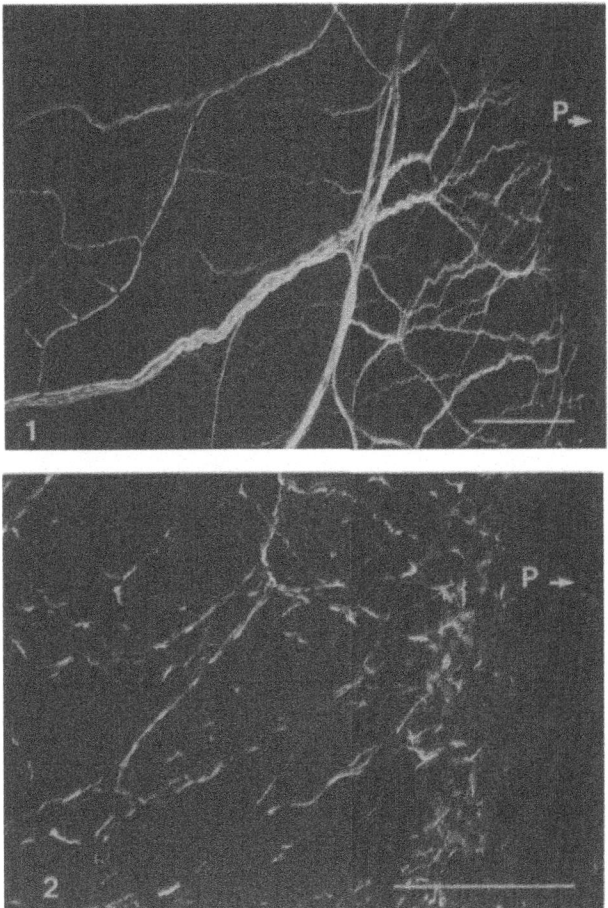

Figures 1 and 2. Whole-mount rat iris, stained by immunofluorescence with an antiserum against neurofilament proteins,[26] here revealing sensory nerve fibers.[1] The whole iris was delicately stretched on a teflon-masked slide, air-dried, permeabilized with 1% triton X-100 and immunostained (without further triton; immunostaining time: 3 days). In a normal iris (*Fig. 1*), numerous thin, unmyelinated fibers are shown, while a few beaded, myelinated fibers are conspicuous (thin arrows: Ranvier's nodes). After 2,5-hexanedione, a multitude of segmental dilations are revealed along thin fibers (*Fig. 2*). Scale bars: 0.5 μm; P + short arrow: pupillary margin.

WHOLE-MOUNT PREPARATIONS: A WIDE RANGE OF POSSIBILITIES

When we think of a part of the body which is a thin, transparent lamina, such as the cornea, the idea of putting it "whole" under the microscope sounds almost natural. In fact, several organs and structures lend themselves almost directly to whole-mount immunostaining. At the other extreme, visualizing the microscopic morphology of a parenchyma in three-dimensions may seem an impervious task.

Whole-Organ Mounts

Let us take two naturally "flat and thin" organs, a rat iris and the diaphragm of the same animal, and immunostain them for neurofilament proteins (*Figs. 1-4*). Under the microscope, the whole-mount iris will show the entire distribution of a multitude of unmyelinated thin sensory fibers, among which a few thinly-myelinated ones are conspicuous (*Fig. 1*).[1,2] In the diaphragm, the full extent of phrenic nerve bundle distribution and branching can be followed (*Fig. 3*).[2,14] Motor end-plates are hardly seen since neurofilaments are degraded as soon as they reach this area.[15]

In the same way, many structures can certainly be considered for whole-organ immunostaining, though they are not as "flat" as the two examples provided. Whole small organs, such as ganglia lying on the organs' surface, can be revealed in their entirety (*Fig. 6*).[14] Understandably, the smaller the organism or its component parts, the easier it may become to immunostain major portions of it as whole-mounts. Although immunostaining of a whole brain, as has been done for insects[7] may represent a somewhat extreme example, it is conceivable that the limits of whole-mount immunostaining will soon broaden significantly. Recently introduced microimaging technology, such as confocal microscopy, will be instrumental in making it possible to analyze complex structures which can hardly be visualized with conventional microscopy.[16]

Layer Separation

Several organs, such as the gut and other hollow viscera, are formed by associating various concentric components, each of which is endowed with specific functions. Since each component in itself is formed as a relatively thin lamina, one can split the viscus wall into the composing layers, which will then be used for immunostaining. Depending on whether the thin viscera of small rodents or the gut of human adults are being studied,[3,6] watchmakers' forceps or stronger instruments including a scalpel are advisable. The wall is stripped or otherwise split into the main component layers. Separation takes place at the organ's natural cleavage planes: along muscularis mucosae, between submucosa and muscularis externa, between circular and longitudinal gut muscle, etc. In general, separation can be carried out on fresh tissues but preservation is best in tissues separated after fixation (or air-drying and rehydration: see below). The separated layers e.g. mucosa, submucosa, muscularis, longitudinal muscle with myenteric plexus, or even the muscularis mucosae alone) are then immunostained as whole-mounts, with or without further dissection to reduce their thickness.[3,5,8,13]

Epithelial Suspensions

The intestinal epithelial layer can be very effectively separated from the lamina propria and remainder of the wall by incubation with EDTA.[17] The resulting preparations contain villi-crypts clusters, or neat crypts (if produced from the colon) with intact endocrine cells, and can be immunostained in suspension.[13] Cells or cell sheets intact in the three-dimensional geometry and ultrastructure can be produced from fixed material such as liver by simple mechanical dissociation or by ultrasound treatments.[18]

Microdissected Slices

A further, different way of producing whole-mount preparations can be applied to many organs, which are in themselves too thick to go under the microscope but have a modular construction, i.e. are composed of large numbers of similar, repetitive units. Thus, one can take advantage of the orderly three-dimensional arrangement of such units and dissociate or slice the organ either parallel or perpendicular to the repetitive units, along a plane of symmetry, or along specific structures present inside the organ. Fragments will be obtained which are small enough to be immunostained whole but still display the three-dimensional organization and features of the organ. Thus, the original cell populations can be visualized in their whole geometry, or the original vascular and nerve supply may be selectively revealed by immunostaining.[5,19]

The gut mucosa is a good example to illustrate the technique. It is formed of thousands

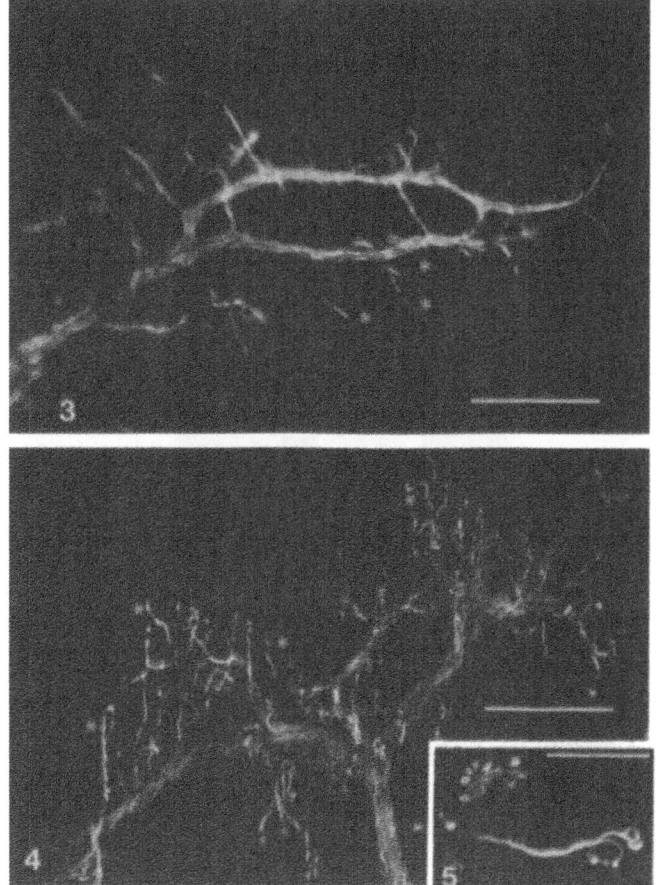

Figures 3-5. Whole-mounts of rat diaphragm (*Figs. 3,4*) and detail of end-plates from oesophageal muscularis (*Fig. 5*), stained by immunofluorescence for neuronal intermediate filament proteins: neurofilament proteins[26] (*Figs. 3,4*), and peripherin[36] (*Fig. 5*). Rat diaphragms and oesophageal segments (opened longitudinally) were stretched on slides progressively while drying (oesophagus: mucosal side up), then dried out and rehydrated. For the diaphragm, roughly 3 x 6 mm pieces were taken across a phrenic nerve branch. The muscularis layer was separated from the oesophageal wall with watchmakers' forceps, under a dissection microscope. All whole-mounts were immersion immunostained with 1% triton throughout (immunostaining time: 5 days). In a normal rat (*Fig. 3*), the entire distribution of phrenic nerve fascicles is revealed, while motor end-plates (*) are virtually devoid of immunoreactivity. Axonal changes, with distal accumulations of immunoreactive material and deformation of end-plates (*), are seen in 2,5-hexanedione (*Fig. 4*), or acrylamide intoxication (*Fig. 5*). Scale bars of Figures 3,4: 250 µm; Fig. 5: 50 µm.

of villi, crypts or glands joined to the lamina propria component with its neuro-vascular supply, immune component, etc. In this instance, whole-mounts are obtained by slicing parallel to the luminal surface (*Fig. 7*), or between adjacent rows of villi or crypts, thus producing a piece of tissue with a single row of villi and a few rows of crypts (Fig. 8). For example, such preparations can be immunostained for a neuropeptide and nerves may be observed from the luminal surface of the mucosa (*Fig. 7*) or from the lateral aspect of crypts and villi (*Fig. 8*). If we add in immunostaining an antibody for a hormonal peptide, populations of endocrine cells are revealed in their three-dimensional shape and distribution (*Fig. 8*).

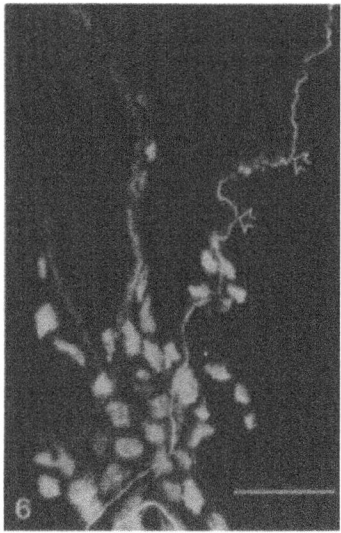

Figure 6. Rat trachea (membranous part) with adjacent tracheal ganglion, stained by immunofluorescence for neurofilament proteins.[26] The trachea was opened anteriorly, placed flat on a slide (mucosal side up) and the membranous part was stretched while drying. After drying-out and rehydration, the membranous part was dissected free of cartilages and immersion-immunostained with 1% triton throughout (immuno-staining time: 4-5 days). A number of immunostained perikarya are seen, together with fiber bundles branching into the organ. A larger-diameter fiber is also apparent (arrows). Scale bar: 100 µm.

Pathology

As an example of pathological changes as seen on whole-mounts, preparations comparable to those presented above (*Figs. 1 and 3*) are taken from rats treated with the neurotoxicants 2,5-hexanedione (*Fig. 2,4*), or acrylamide (*Fig. 5*) which largely affect neurofilaments in distal axons.[20,21] Alterations in thin sensory fibers of the iris show very clearly on whole-mounts, with a multitude of segmental dilations[14] (*Fig. 2*). In the diaphragm and oesophagus, accumulations of intermediate filaments (s-i), immunoreactive material in motor end-plates and preterminal axons are obvious (*Figs. 4,5*).[2,14]

The technique of nerve teasing has been mentioned earlier. Provided compatible fixatives are used, teased nerve preparations are well suitable for immunostaining, thus further extending the value of such "three-dimensional, approach to neuropathology.[10]

PRACTICAL CONSIDERATIONS

"Fixation" of Tissue Samples

Choice of fixation, or other methods for tissue pre-treatment, largely depends on the types of antigens to be demonstrated. Routine buffered formalin can certainly be used as long as the antigens of interest are demonstrable on the corresponding sections. Fixation must be comparatively short to avoid both excessive tissue hardening and antigen damage. Paraformaldehyde (freshly made 4% paraformaldehyde solution in phosphate buffer) and paraformaldehyde-picric acid mixture (Stefanini's/Zamboni's fixative[22]) are widely used. Periodate-lysine-paraformaldehyde (PLP)[23] and p-benzoquinone solution[24] are useful alternatives in view of their less strong cross-linking activity, making such fixatives more respectful of many immunoreactive epitopes.

Another approach to tissue pre-treatment which cannot in itself be regarded as a fixation method,[25] consists in the removal of water content from the tissue sample either by simple air-drying[1,2] or by various freeze-substitution or freeze-drying methods.[25] Desiccation is very effective in preserving certain cellular components, namely, the polymeric protein T8 (protein f)forming intermediate filaments (at least the neuroglial elements).[1,2,14] It is conceivable that dehydration blocks enzymes degrading such proteins,[15] since triton permeabilization of fresh tissues disrupts neurofilaments[15] and nerve morphology (G.L. Ferri and S. Zareh, unpublished observations).

In the present context, the degree of tissue hardening induced by fixative should be considered, in view of its relevance to both microdissection and final mounting of preparations. Aldehydes, depending on concentration and fixation time, can make tissues rather rigid. Preparations tend to retain the shape taken during fixation, so that flattening (by pinning, etc.) is essential. Layer separation generally works well, and all components including the mucosa are hardened enough to withstand the manipulation. On the other hand, samples fixed in benzoquinone solution remain almost as pliable as in the fresh state, especially if fixation was relatively short. Many tissue components, especially the mucosa, remain soft and deformable and can thus be flattened effectively after coverslipping provided that the preparations are not dehydrated, but mounted in PBS-glycerol.

Flat Preparations

The latter possibility is of interest since preparations can be thinned by simply applying pressure onto the coverslip. When using a conventional microscope (for transmitted light, or fluorescence), such artefactual "flattening" of the immunostained preparations is of considerable help. In fact, a good deal of focussing may be required to follow the features of interest through the depth of the specimen. Some of the pictures enclosed (*Figs. 3, 4 and 8*) were especially selected in order to provide the reader with some visual impression of this problem.

Non-Flattened Preparations

In other cases, the original thickness of the preparations must be respected, e.g. for confocal microscopy and three-dimensional quantitative study. In the latter case, various precautions, such as non-stretched, somewhat longer fixation, and the insertion of spacers, between slide and coverslip for mounting are advisable.[16] Where air-drying would have been the method of choice, freeze-substitution (see below) or freeze-drying[25] is most appropriate. Such techniques can be expected to preserve at least their original proportions, if not true tissue dimensions.[25]

Permeabilization

Penetration and diffusion into the tissue of antibodies and other reagents are major limiting factors in immunocytochemistry of whole-mount preparations. Thus, some method of permeabilization is required in order to permit an effective and even immunostaining. Dehydration-clearing-rehydration[3] is very satisfactory for a wide variety of tested antigens including many hormonal and neuronal peptides,[4,8,13] soluble cytoplasmatic proteins (such as S-100[12]), enzyme markers (dopamine-beta-hydroxylase, DBH, and neuron-specific: enolase, NSE).[6]

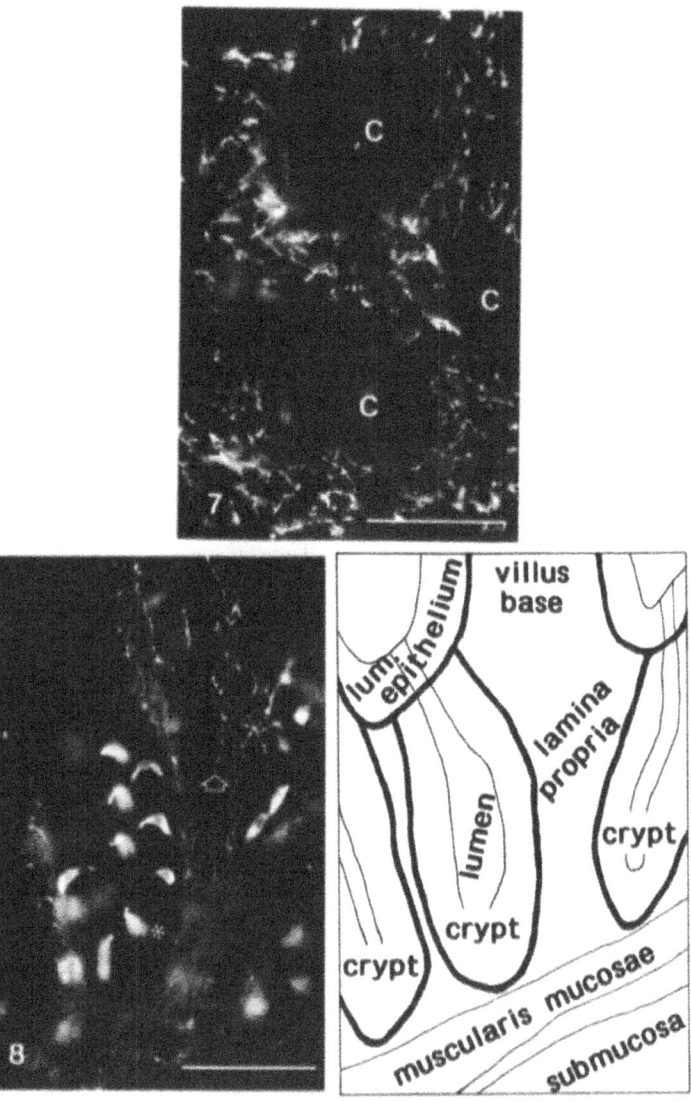

Figures 7, 8. Microdissected preparations of human gut mucosa, permeabilized by dehydration-clearing-rehydration and stained by immunofluorescence with antibodies[13] to the neuropeptide vasoactive intestinal polypeptide (VIP: both), and the hormonal peptide enteroglucagon (Fig. 8 only). Samples of full-thickness human bowel were pinned on cork (mucosal side up) and immersion-fixed in benzoquinone solution. Under a dissecting microscope, the colonic mucosa was sliced through parallel to the luminal surface, and the luminal epithelium was delicately scraped. Thus, the subepithelial lamina propria and its rich, typically varicose VIP-immunoreactive nerve supply was fully revealed (*Fig. 7*; C: crypts). The ileal mucosa (*Fig. 8*) was micro-sliced by hand between adjacent rows of villi, so that preparations containing a single row of villi (with multiple crypts) were obtained (compare with explanatory scheme, on right). Note VIP nerves (arrow), and the typical three-dimensional morphology of enteric endocrine cells, with basal accumulation of immunoreactive material (granules) and the apical prolongation, reaching the crypt lumen. Of these, many lie perpendicular to the plane of focus (e.g.*), so that only the apical prolongation is sharp. Scale bar:100 μm.

On the whole, one can probably expect all antigens immunostainable on wax sections to behave well with this method. In some cases, especially with neurofilament proteins, immunoreactivity was somewhat reduced by the above procedure as opposed to permeabilization carried out with the detergent triton X-100 (G.-L. Ferri and S. Zareh, unpublished observations). Permeabilization and immunostaining with 1% triton X-100 was used routinely with polymeric proteins, such as neurofilament proteins and peripherin, which are not triton-soluble.[15,26] Such treatment is most effective after air-drying or freeze-substitution in acetone (see below), but works well also with benzoquinone-fixed tissues. For other purposes, 1% triton X-100 is usually too extractive, producing loss of neuropeptide immunoreactivity and nerve fibre fragmentation. Lower concentrations, or short permeabilization steps may be worth trying in specific cases. For very thin whole-mounts, the inclusion of 0.3% triton X-100 in all antisera was reported to produce very satisfactory immunofluorescence.[1] Another permeabilizing agent which has been used for neuropeptide immunostaining, is dimethylsulfoxide (DMSO: 3 x 10 min, after 3 x 30 min in 80% alcohol for the rat iris).[27]

A peculiar problem presents itself when immunostaining whole-mounts which have been labelled with lipophilic dyes, e.g. the carbocyanine retrograde tracers DiI and DiO.[28] In fact, the removal of membrane lipids produced by either of the above permeabilization methods also washes out the lipophilic label. So far, some immuno-fluorescence labelling can be obtained without permeabilization after very prolonged incubations of vibratome sections.[29] Alternatively, DiI labelling was photoconverted to an insoluble product with diaminobenzidine.[30] Further efforts are warranted towards methods for permeabilization (or vehiculation) which would let antibodies penetrate into cells without too much membrane disruption. In the present case, for instance, whole-mount immunocytochemistry would make an exceedingly useful combination with carbocyanine retrograde tracing since DiI and the like can travel along the membrane of fixed cells and are retained for extended periods in living tissues. This method should work well in fixed whole-mount preparations, provided cell membranes are not interrupted by sectioning.[28-30]

Electron Microscopy

Whole-mounts can be advantageously used for electron microscopy by using light microscopical immunostaining to select and sample tissue. Ida Llewellyn-Smith[31] fixed tissues with paraformaldehyde-picric acid mixture containing 0.05% glutaraldehyde,[31] followed by 50% alcohol and 0.1% NaCNBH$_3$ (sodium cyanoboro-hydride) to rapidly remove fixatives. This allows sufficient penetration of antibodies while preserving ultrastructure.

Choice of Immunostaining Method

Immunoperoxidase methods, especially the peroxidase-antiperoxidase technique,[32] can be employed quite successfully even with comparatively thick tissues, such as the adult human gut[13] or whole insect brain.[7] In our laboratory, however, we prefer the quicker immunofluorescence methods,[33] which have the additional advantage of easily permitting double, or multiple immunostaining. In our hands, it has not always been easy to avoid an unwanted precipitation of diaminobenzidine on the surface of peroxidase- immunostained preparations. Some post-staining microdissection is required in order to expose the clean underlying staining.[13] Of course, various other immunoenzymatic methods could be worth trying.

Background Reduction

Some autofluorescence is commonly seen on the relatively thick whole-mounts. The degree depends on tissue composition and thickness, fixation used, excitation wavelength, and spectral discrimination of the filter system. For instance, we could not effectively visualize the retrograde tracer fluorogold on whole-mounts of paraformaldehyde-fixed rat rectum (muscularis only), using the Leitz filter system "A" (UV excitation). We had to resort to sections (G.-L. Ferri and W. Neuhuber, unpublished observations). Using appropriate, narrow band excitation and emission wavelengths it may be possible to overcome many such problems.[34] Treatment with 1 mg/ml borohydride in PBS has been used to decrease tissue autofluorescence.[16]

Routinely, our antibodies are diluted in PBS containing normal sera from two animal species: the one from the donor of the secondary antibody and the one from the donor of the tissues which are being immunostained. Defatted dried milk could also be tried. For biotin-avidin amplified immuno-fluorescence, blocking treatments with avidin and biotin are applied (see relevant protocol).

Antisera Concentrations

Empirically, it is easy to determine that antisera, antibodies, and enzyme-antibody complexes are to be used at higher concentrations on whole-mounts as compared to sections of identical tissues. Since the tissue volume is no longer insignificant (as it would be with sections), concentration gradients form from surface to center of the sample and reagents are "diluted" by the volume of the tissue itself (mainly composed of water). In general, primary antisera/antibodies work well on whole-mounts when they are two to five times more concentrated than optimal on sections (under identical conditions). The same applies to secondary antibodies, enzyme complexes, avidin/ streptavidin, etc., all of which are routinely titrated on whole-mounts before extensive use.

Free-Floating and On-Slide Immunostaining

In general, immersion-immunostaining is preferable since antibodies will be free to effectively penetrate free-floating preparations from all sides. To prevent drying out of preparations, containers for immersion-immunostaining must not be too large and must be kept in a humid atmosphere. We routinely use multiwell microtiter plates, of 150-1000 μl well capacity. To improve tissue visibility, the plate's back can be painted in a contrasting colour. To increase washing efficiency, tissues are moved to larger containers at each such step.

Delicate tissues (e.g. iris) are immunostained on slides, either adhering on the same slides on which they were air-dried or, if liquid-fixed (and appropriately permeabilized), layered flat on slides in thick "drops" of incubation media. A preliminary permeabilization step (either dehydration-clearing-rehydration or 1% triton, for 4-8 h) is often enough and then immunostaining can be carried out without triton.

OPTIMAL PROCEDURES

I. TISSUE PRE-TREATMENT - Liquid fixation

1. Pin tissue flat on cork sheet, under moderate tension. For fragile tissues (e.g. iris), stretch delicately on flat surface, cover with filter paper and press gently, lift paper with adhering tissue (Do not allow to dry out!)
2. Immerse in fixative, ensuring that fixative penetrates between tissue and cork sheet.
3. Fix at temperature indicated, for appropriate time, mixing occasionally.
4. Decant fixative and wash with PBS (2-5 changes, in 2-3 hours).
5. Store in PBS containing 0.02% NaN$_3$, at +4°C.

Table 1. Guideline of Fixation Times (Hours)

Tissue	Para	BQS
rat iris	2.0	0.5
needle biopsy	2-3	1.0
rat gut, trachea, bladder (open)	2-3	1-1.5
adult human intestine	3.0	2.0
parenchyma: 1 mm slice	3.0	2.0
parenchyma (2-4 mm slice)	4-5	2-3

Para = 4% paraformaldehyde; OR: paraformaldehyde-picric acid mixture;[22] OR: periodate-lysine-paraformaldehyde (PLP)[23]
BQS = p-benzoquinone solution:[24] 0.4% in PBS.

BQS must be yellow at start of fixation, while later darkening is normal. Times appropriate for normal tissues, change proportionally for abnormal thickness; mix at start and repeatedly during fixation; viscera to be cut open and pinned flat, ensuring fixative penetrates from all sides. For routine buffered formalin, keep fixation short, as close as possible to times indicated for "para".

Air-Drying

1. Layer tissue on glass slide.
2. Delicately stretch from edges, with thin forceps or blunt needles.
    This will be easier when tissue starts drying out.
3. Dry out completely (several hours to overnight).
4. Store in box in dry place.

Freeze-substitution in acetone (from [35] modified)

1. Layer tissue onto metal spatula (preferably teflon-sprayed).
2. Freeze by quick immersion into melting freon, cooled with liquid nitrogen.
3. Transfer to acetone pre-cooled to -70°C.
4. Leave to freeze-substitute for several weeks, at -70°C.
5. Bring to room temperature; take out tissues, let acetone evaporate.
6. Rehydrate in distilled water, then transfer to PBS.

II. MICRODISSECTION

Possible on fresh tissues, for best preservation carry out after tissue pre-treatment (any of above protocols).

Under dissecting microscope, with scalpel blade, cut a slice of tissue containing a few rows of repetitive units (e.g. intestinal villi, glands, etc.).

- With blade, scrape off surface cell layer/s to expose underlying connective tissue, nerves, blood vessels, etc.
- Strip off fragment of muscle to expose nerves, end-plates, etc.
- Dissect out surface component, e.g. choroid plexus
- layer separately.

Subdivide samples to size appropriate for immunostaining.

III. PERMEABILIZATION

Dehydration-clearing-rehydration

1. Dehydrate through ascending alcohols: 70% and 90% alcohol (30 min each), absolute alcohol (3 x 30 min).
2. Clear in xylene (**at least** 2 x 45 min).
3. Rehydrate: absolute alcohol (3 x 20 min), 90% and 70% alcohol (20 min each), and distilled water (3 x 10 min).
4. Rinse in PBS (3 x 10 min); immunostain, WITHOUT Triton.

With detergent: 1% Triton X-100

1. Immerse slides with adhering tissues or liquid-fixed tissues in PBS containing 1% Triton X-100 (1 h).
2. Delicately remove from slides; microdissect as appropriate.
3. Immunostain with 1% Triton X-100 throughout (antibodies and washings).

Or, for very thin tissues, to be stained on slides

1. Immerse slides with adhering tissues, or liquid-fixed tissues, in PBS containing 1% Triton X-100 (4-8 h).
2. Wash well in distilled water (about 5 x 15 min), then PBS (about 2 x 15 min).
3. Incubate on slide WITHOUT Triton, wash by immersion.

IV. IMMUNOSTAINING

Indirect Immunofluorescence

1. Transfer permeabilized tissues to incubation containers, or leave on slides.
2. Drain PBS; cover with appropriate "antibody diluent" (with/without Triton); incubate 4 h.
3. Drain; cover with primary antibody, or antibodies (brought to final dilution with the same diluent as above).
4. Incubate 2-4 overnights.
5. Wash in PBS (with/without Triton): 5-8 changes in 4-8 h.
6. Drain; cover with fluorochrome-labelled secondary antibody, or antibodies (in diluent as above); incubate overnight in the dark.
7. Wash as in point 5, preferably overnight, in the dark.

8. Layer tissues on glass slides, and/or coverslip with glycerine-PBS (3:2, with 0.02% NaN$_3$ as preservative); seal with nail polish.

Biotin-Avidin Amplification for Immunofluorescence

1. Transfer permeabilized tissues to incubation containers, or leave on slides.
2. Drain PBS; cover with avidin solution (in appropriate "antibody diluent", with/without Triton); incubate 3-4 h.
3. Wash in PBS, about 3 x 15 min.
4. Drain; cover with biotin "blocking" solution (with/without Triton); incubate 1-2 h.
5. Wash as in point 3.
6. Drain; cover with primary antibody (of antibodies brought to final dilution in the same diluent as above).
7. Incubate 2-4 overnights.
8. Wash in PBS (with/without Triton): 5-8 changes in 4-8 h.
9. Drain; cover with biotinylated secondary antibody (in same diluent, as above: for double/multiple staining, fluorochrome labelled secondary antibody/bodies are included); incubate overnight (if fluorochrome/s included, in the dark).
10. Wash as in point 8 (where appropriate, in the dark).
11. Drain; cover with fluorochrome-labelled avidin, or streptavidin (brought to final dilution with PBS); incubate overnight, in the dark.
12. Wash as in point 8, in the dark.
13. Mount and/or coverslip, as for indirect immunofluorescence.

Peroxidase-Antiperoxidase

1. Cover permeabilized tissues with 3 μl/ml H$_2$O$_2$ (30 volumes; in distilled water, or PBS); incubate for 1 h.
2. Wash in five changes of PBS, 10 min each.
3. Transfer to incubation containers, or leave on slides.
4. Drain PBS; cover with appropriate "antibody diluent" (with/without Triton); incubate 4 h.
5. Drain; cover with primary antibody (brought to final dilution with the same diluent as above).
6. Incubate 2-4 days.
7. Wash in PBS (with/without Triton): 5-8 changes in 4-8 h.
8. Drain; cover with unlabelled secondary antibody (in same diluent, as above); incubate overnight.
9. Wash as in point 7.
10. Drain; cover with peroxidase-antiperoxidase complex (in antibody diluent as above, but **with merthiolate**); incubate overnight.
11. Wash as in point 7.
12. Layer free-floating tissues flat in small dishes (slide-adhering tissues will be immersion-treated).
13. Cover tissues with filtered diaminobenzidine (0.25 mg/ml, in PBS: ensure free-floating tissues are exposed on both sides).
14. 5-10 min later add H$_2$O$_2$ (0.3 μl/ml, 30 volumes), mix; develop for 5-20 min (check microscopically for background).
15. Wash thoroughly in water, then distilled water; where appropriate, remove surface tissue with DAB precipitate.

16.   Dehydrate through ascending alcohols, clear in xylene.
17.   Coverslip with Canada balm, or other non-shrinking medium.

STEP-BY-STEP HINTS AND COMMENTS

General
- PBS: 0.05 mol/l phosphate buffer in 0.15 mol/l NaCl saline
- Incubate and wash routinely at room temperature.
- Wash free-floating samples in about 20 times as much PBS. Mix repeatedly by pipetting some PBS out, then back onto tissues.
- For prolonged washes in the same change of PBS (e.g. overnight), place at +4°C, or add preservative (NaN$_3$, 0.02%).
- When using fluorochromes, incubate and wash in the dark.
- Keep covered to prevent dirt from reaching (and sticking to) whole-mounts. Also, wear lab coat at all times.
- Do not allow tissues to dry out at any stage (including alcohol and xylene steps). Incubate plates and slides in a humid chamber.
- Immunostaining can be stopped (overnight) at the following stages: alcohols (any), xylene, washings between antibodies (with/without Triton, as appropriate: place at 4°C), antibody diluent.

Incubation
- Multiple free-floating samples can be pooled together in the same incubation container, with incubation mixture to cover.
- Before adding incubation mixture, drain previous incubation medium or PBS (from washes) to avoid unwanted dilution.
- For free-floating samples, drain from bottom of incubation container with thin plastic pipette.
- Mix free-floating samples thoroughly when starting (2-3 times) and during each incubation (3-4 times per day).

On-Slide Immunostaining
- Wash carefully to remove Triton (antibodies would spread away)
- Teflon-masked slides (e.g. multispot slides from Hendley-Essex, Essex, UK) help retain a thick layer of incubation media onto tissues: clean and dry thoroughly the treated area at each step to fully retain hydrorepellent properties.

Antibodies and Reagents

Antisera diluents are: PBS + 3% normal serum (pool) of the same species as the donor of secondary antibody + 3% normal serum (pool) of the same species as the animal from which tissue is immunostained + **either**: 0.02% NaN$_3$, **or** 0.05% Na merthiolate. With peroxidase methods, avoid NaN$_3$ which inhibits the enzyme.

Avidin-biotin blocking steps: an appropriate avidin to biotin concentration ratio must be respected (e.g. 100 ug/ml avidin, 20 ug/ml biotin, as in blocking kit by Vector Laboratories). More concentrated solutions are often required for whole-mounts.

Labelled avidin/streptavidin: dilute in PBS only since serum may contain biotin-like substances.

Biotin-avidin immunofluorescence: an additional, final incubation step with free biotin may improve fluorescence (use 2.5 ug/ml, or more).[34]

Triton does not dissolve easily, thus a pre-diluted stock is very convenient for quick use (1:4 in PBS; keep at room temperature).

ACKNOWLEDGEMENTS

W. Barr and Vector Laboratories are thanked for advice on avidin-biotin blocking; M.-M. Portier, D. Dahl and J.M. Polak for antibodies; A. Cadau for photographic work; E. Solcia, V. Eusebi and E. Santoro for unceasing encouragement. Partly supported by the Italian CNR and Ministry of University and Research.

REFERENCES

1. A. Seiger, D. Dahl, C. Ayer-LeLievre, and H. Bjorklund, Appearance and distribution of neurofilament immunoreactivity in iris nerves, *J Comp Neurol* **223**:457 (1984).
2. G.L. Ferri, S. Zareh, M. Sbraccia, L. Abelli, N. Frontali, D. Dahl, Heterogeneous visceral nerve changes in acrylamide intoxication, *Exp Brain Res* **87**:363 (1991).
3. M. Costa, R. Buffa, J.B. Furness, and E. Solcia, Immunohistochemical localization of polypeptides in peripheral autonomic nerves using whole mount preparations, *Histochemistry* **65**:157 (1980).
4. J.B. Furness and M. Costa. "The Enteric Nervous System," Churchill Livingstone, Edinburgh (1987).
5. G.L. Ferri, P.L. Botti, P. Vezzadini, G. Biliotti, S.R. Bloom, and J.M. Polak, Peptide-containing-innervation of the human intestinal mucosa: an immunocytochemical study on whole-mount preparations, *Histochemistry* **76**:413 (1982).
6. G.L. Ferri, Human gut neuroanatomy: methodology for a quantitative analysis of nerve elements and neurotransmitter diversity in the human "enteric nervous system," *Basic Appl Histochem* **32**:117 (1988).
7. A.R. Aitken, R.E. Sandemann, and D.C. Sandeman, Preparation of immuno-peroxidase-labelled whole mounts of invertebrate brains, *J Neurosci Meth* **21**:1 (1987).
8. M. Costa and J.B. Furness, Immunohistochemistry on whole mount preparations, *in*: "Immunohistochemistry," A.C. Cuello, ed., Wiley, Chichester (1983).
9. S.R. Cajal. "Textura del Sistema Nervioso del Hombre y de los Vertebrados," Nicolas Moya, Madrid (1899).
10. P.S. Spencer, M.C. Bischoff, and H.H. Schaumburg, Neuropathological methods for the detection of neurotoxic disease, *in*: "Experimental and Clinical Neurotoxicology," P.S. Spencer and H.H. Schaumburg, ed., Williams and Wilkins, Baltimore (1980).
11. P.L. Williams and R. Warwick, eds. "Gray's Anatomy, 35th ed.", Churchill Livingstone, Edinburgh (1980).
12. G.L. Ferri, L. Probert, D. Cocchia, F. Michetti, P.J. Marangos, and J.M. Polak, Evidence for the presence of S-100 protein in the glial component of the enteric nervous system, *Nature* **297**:409 (1982).
13. G.L. Ferri, T.E. Adrian, M.A. Ghatei, D.J. O'Shaughnessy, L. Probert, Y.C. Lee, A.M.J. Buchan, J.M. Polak, and S.R. Bloom, Tissue localization and relative distribution of regulatory peptides in separated layers from the human bowel. *Gastroenterology* **84**:777 (1983).
14. G.L. Ferri, S. Zareh, A. Amadori, A. Bastone, M. Sbraccia, D. Dahl, N. Frontali, 2,5-hexanedione-induced accumulations of neurofilament-immunoreactive material throughout the rat autonomic nervous system, *Brain Res* **444**:383 (1988).
15. W.W. Schlaepfer, Neurofilaments: structure, metabolism and implications in disease, *J Neuropathol Exp Neurol* **46**:117 1987).
16. R. Bacallao and E.H.K. Stelzer, Preservation of biological specimens for observation in a confocal fluorescence microscope and operational principles of confocal fluorescence microscopy. *Meth Cell Biol* **31**:437 (1989).
17. T.S. Gaginella, H.S. Makhjian, and T.M. O'Dorisio, Vasoactive intestinal peptide: quantification by radioimmunoassay in isolated cells, mucosa, and muscle of the hamster intestine, *Gastroenterology* **74**:718 (1978).
18. K. Yamashita, K. Moriaki, and M. Sano, A simple and rapid method of dissociating hepatocytes from fixed liver of the mouse, *Stain Technol* **56**:29 (1981).
19. G.L. Ferri, P. Botti, G. Biliotti, L. Rebecchi, S.R. Bloom, L. Tonelli, G. Labo, and J.M. Polak, VIP-, substance P- and met-enkephalin-immunoreactive innervation of the human gastroduodenal mucosa and Brunner's glands, *Gut* **25**:948 (1984).

20. P.S. Spencer, H.H. Schaumburg, M.I. Sabri, and B. Veronesi, The enlarging view of hexacarbon neurotoxicity, *CRC Crit Rev Toxicol* 7:279 (1980).

21. H.H. Schaumburg, H.M. Wisniewski, P.S. Spencer, Ultrastructural studies of the dying-back process. 1. Peripheral nerve terminal and axon degeneration in systemic acrylamide intoxication. *J Neuropathol Exp Neurol* 33:260 (1974).

22. M. Stefanini, C. De Martino, and C. Zamboni, Fixation of ejaculated spermatozoa for electron microscopy, *Nature* 216:173 (1967).

23. I.W. McLean and P.K. Nakane, Periodate-lysine-paraformaldehyde fixative. A new fixative for immunoelectron microscopy, *J Histochem Cytochem* 22:1077 (1974).

24. A.E. Bishop, J.M. Polak, S.R. Bloom, and A.G.E. Pearse, A new universal technique for the immunocytochemical localization of peptidergic innervation, *J Endocrinol* 77:25 (1978).

25. A.G.E. Pearse. "Histochemistry, Theoretical and Applied, IV edition", Churchill Livingstone, Edinburgh (1980).

26. D. Dahl and A. Bignami, Preparation of antisera to neurofilament protein from chicken brain and human sciatic nerve, *J Comp Neurol* 176:645 (1977).

27. I.L. Gibbins and J.L. Morris, Co-existence of neuropeptides in sympathetic, cranial autonomic and sensory neurons innervating the iris of the guinea-pig, *J Autonom Nerv System* 21:67 (1987.

28. M.G. Honig and R.I.Hume, DiI and DiO: versatile fluorescent dyes for neuronal labelling and pathway tracing, *TINS* 12:333 (1989).

29. A.J. Elberger and M.G. Honig, Double-labeling of tissue containing the carbocyanine dye DiI for immunocytochemistry, *J Histochem Cytochem* 38:735 (1990).

30. C.S. von Bartheld, D.E. Cunningham, and E.W. Rubel, Neuronal tracing with DiI: decalcification, cryosectioning, and photoconversion for light and electron microscopic analysis, *J Histochem Cytochem* 38:725 (1990).

31. I.J. Llewellyn-Smith, M. Costa, and J.B. Furness, Light and electron microscopic immunocytochemistry of the same nerves from whole mount preparations, *J Histochem Cytochem* 33:857 (1985).

32. L.A. Sternberger. "Immunocytochemistry, II ed.", Wiley & Sons, New York (1979).

33. A.H. Coons, E.H. Leduc, and J. Connolly, Studies of antibody production. I. A method for the histochemical demonstration of specific antibody and its application to a study of the hyperimmune rabbit, *J Exp Med* 102:49 (1955).

34. R.P. Haugland. "Handbook of Fluorescent Probes and Research Chemicals, 5th edition: 1992-1994", K.D. Larison, ed., Molecular Probes Inc, Eugene (1992).

35. S. Ochs and R.A. Jersild, Cytoskeletal organelles and myelin structure of beaded nerve fibers, *Neuroscience* 22:1041 (1987).

36. M. Escurat, M. Gumpel, F. Lachapelle, F. Gros, and M.M. Portier, Comparative study of the expression of two intermediate filament proteins: peripherin and the 68 kDa neurofilament protein during development of the rat embryo, *C R Acad Sci Paris, Ser III-vie* 306:447 (1988).

CHAPTER 6

MICROWAVE IMMUNOCYTOCHEMISTRY

Jiang Gu, M.D., Ph.D.[1]
Deborah Research Institute
Browns Mills, New Jersey, U.S.A.

SUMMARY

Microwave technology is changing the way morphological tissue samples are prepared. It has been employed in tissue fixation, processing, drying, antigen retrieval and various methods of staining. The biggest advantage of this technique is that it speeds up routine procedures tremendously while retaining the desired quality. In some instances, the result obtained with microwave treatment are superior to those obtained with routine procedures. Microwave techniques are fairly straight forward but require much more precision in execution when compared to conventional procedures. In this chapter, development and application of microwave technology in relation to fixation, antigen retrieval and immunocytochemistry are reviewed and discussed with reference to our own experiences. Some protocols tested in the author's laboratories are recommended for general purposes. This chapter is intended to give morphologists a comprehensive guide to developing and to applying microwave procedures in their practice.

INTRODUCTION

Microwave is a form of electromagnetic radiation energy that is transmitted through space as waves with wave lengths ranging from 1 mm to 1 m. In a vacuum, microwave travels at the speed of light. It also reflects and accents shadows like light. It is capable of interacting with dipolar molecules by imparting kinetic energy and altering electric fields. This form of radiation makes dipolar molecules rotate 180 degrees rapidly thereby increasing intermolecular and intramolecular motion and the temperature. It was recently

---

[1]Address for correspondence: Jiang Gu, M.D., Ph.D., Chairman, Scientific Affairs, Deborah Research Institute, Trenton/Pine Mill Road, Browns Mills, NJ 08015-1799, U.S.A. Phone: 609-893-1016; Fax: 609-893-2441

recognized that molecules with an uneven distribution of electrical charge can also be oscillated by microwave irradiation. Such energy is not strong enough to alter covalent bonds but interacts readily with steric bonds, essential for many biochemical reactions.[1]

Microwaves penetrate materials so that heat is generated inside the object. It penetrates a few centimeters deep into a typical tissue sample. The depth of penetration is positively related to the energy output of the microwave oven and inversely proportional to the rate of energy absorption of the sample, the less microwave energy an object absorbs, the deeper the microwave travels. Whether or not and how much heat is generated within the target is determined by many factors, including irradiation duration, frequency, power output, and the specific dielectric properties of the irradiated object. In most materials, including morphological tissue samples and food, microwave power absorption is proportional to the water content, i.e. the more liquid contained in the materials, the more easily and quickly the materials will be heated by microwave irradiation. Therefore, the water content of a tissue sample, among other factors, should be taken into consideration when adapting a microwave treatment procedure. However, pure ice absorbs much less microwave energy than water.

The generation of heat by increased internal molecular motion, generated by microwave energy is a much faster process than is the conventional method of heating which relies on thermoconduction. Microwave heating is also a much more evenly distributed process within the heated object. Because of these properties, microwave has become widely used in many areas other than cooking, such as freeze drying, curing of glues, vulcanization of rubber, synthesis of superconducting cirimex, etc. Its expanded applications are gradually being embraced for biomedical research and clinical diagnosis. The increasing number of reports utilizing microwave technology in biomedicine, particularly morphology, signifies an important advancement in microwave technology.

In a microwave oven, microwaves are generated by a magnetron through interaction between strong magnetic and electronic fields. Typically, they have a standard frequency of 2.45 GHz, a wave length of 12.4 cm and a photon energy of 10^{-5} electron volts which cause dielectric oscillation of water molecules 2.45 billion times a second.[2,3] There are two types of microwave ovens: the domestic oven, used in almost every household, and the specially designed ones for the laboratory, such as the ones produced by Energy Beam Sciences, Inc. The specially designed ovens are superior to the domestic ovens in that it is controlled by a set temperature rather than power output. When the desired temperature is reached as measured by an internal thermoprobe, the machine will retain the temperature by a feed back mechanism to automatically switches the irradiation on and off for a preset period, ranging from 2 seconds to 59 minutes. The total time count can start from either the beginning of irradiation or the moment the heated object reaches the preset temperature. However, for most purposes, the domestic oven is adequate to achieve the goals of tissue fixation, antigen retrieval, or accelerated staining required in the lab. A number of points need to be kept in mind when using the domestic oven because of its inherent limitations. For example, this type of oven does not generate microwaves continuously. An "on" and "off" mechanism is built in by the manufacturers producing the ovens in order to achieve radiation uniformity and to control the extent of irradiation in food heating. In other words, the presumably continuous irradiation of an oven may, in fact, consists of cycles of "on and off" by the machine itself. A lower power setting actually entails an increase in the "off" period and a decrease in the "on" period. The only parameter changed for different power settings is the irradiation duration although this is not illustrated to the unaware users. This feature may sometimes affect the real irradiation duration of an experiment. Before operating a domestic microwave oven, it is advisable to check the relationship between the power settings (e.g. high, medium, low, etc.) and the durations of "on" and "off" irradiation. This can be performed by using a stop watch to measure the humming sound of the oven during operation. You will notice the on and off of the distribution fan. This fan synchronizes the "on" and "off" periods during microwave irradiation. It comes on and off following a preset interval for each power setting, the lower the power setting the longer the "off" period and the shorter the "on" period. By recording the on and off durations in two minutes for each power setting, the relationship among different settings can be calculated.

It should also be noted that different locations within the cavity of the oven are not irradiated with the same energy strength resulting in relatively hot and cold spots in the same oven. We tested two domestic microwave ovens (Sharp Household Microwave Oven, Model No. R-5A50 and Whirlpool Model MW 1200XW). Using the same bottle of water (20ml) irradiated at the same power setting for the same duration, different locations resulted in temperature differences of up to 30% within 2 minutes. This is the reason why a turntable is preferred for cooking as well as for morphological preparations. The microwave energy deposited in the target is largely in reverse proportion to the volume of the target. Increasing the target volume by putting a beaker of water in the microwave oven, for example, can effectively lower the irradiating level of the target sample and prevent overheating as well as provide a humid environment necessary for most antibody incubations. For a detailed discussion of the physical properties of microwaves and microwave ovens, readers are referred to "Microwave Cookbook of Pathology" by Boon and Kok.[1]

MICROWAVE FIXATION

The theoretical basis for microwave fixation is that raised temperatures can cause proteins to coagulate. Thus, the structure and steric relationship among tissue molecules can be preserved at a near life state allowing both morphologic evaluation and antigenic epitope recognition. Ideal fixation can be achieved when the optimal temperature is reached evenly throughout the tissue sample at the fastest rate. Microwave heating is well suited for this purpose. Different tissue components absorb microwave energy at different rates, thereby raising their temperature at varying speeds.[1] For this reason, different irradiation levels are required for different types of tissue structures and antigens to achieve optimal fixation.

The idea that tissue structure can be preserved by heat was first introduced about a century ago by Ehrlich and Lazarus, working with blood film.[4] It was found that raising the tissue temperature to 45-55°C caused cessation of all biological activities within the tissue sample and made protein coagulate. Not until 1960, was it briefly noted again by Baker.[5] In 1970, when microwave technology gained ground in domestic cooking, Mayers experimented with freshly dissected postmortem materials and reported that heating with a microwave oven resulted in acceptable fixation of tissues samples.[6] Peracchia and Mitler reported in 1972 that fixation in glutaraldehyde could be improved by beginning at room temperature and then warming to 45°C.[7] In 1974, Bernard reported a microwave fixation experiment in which he fixed whole anesthetized mice with a microwave oven and found that different tissue types required different temperatures for optimal fixation.[8] In 1976, Lillie and Fullmer reported that small tissue samples, 5-10 mm thick, may be satisfactorily fixed by boiling in isotonic saline for 2-3 minutes although this led to some tissue shrinkage.[9] It was then systematically experimented by many authors, particularly Login et al,[10-18] Leong et al[19-27] and Hopwood et al.[28] It was confirmed that microwave treatment of fresh tissue achieved a fixation effect similar to conventional chemical fixatives in preserving both morphology and antigenicity. In some cases, it was found that microwave fixation was better. In one study, Patterson and Bulard tested the protein content of cultured tissue samples fixed with either microwave heating or chemical (formaldehyde) fixation. They reported that microwave fixation preserved virtually all the proteins in the samples while chemical fixation led to about 40% protein loss.[29] It was also reported that microwave fixation often gave a better preservation of tissue antigens, particularly for monoclonal antibodies.[1] Many enzyme activities that catalyze rapidly metabolizing compounds such as catecholamines, acetylcholine, GABA, etc., can be effectively and instantly arrested by microwave irradiation.[29] For example, it was reported that acetylcholine esterase was better preserved by microwave irradiation alone, rather than by chemical fixation.[1] In our hands, microwave fixation with 0.01 M phosphate buffered saline (PBS), pH 7.2, alone gave better immunostaining for insulin and glucagon in the pancreas and for atrial natriuretic peptide (ANP) in the heart when detected using monoclonal antibodies. When polyclonal antibodies were used, however, the superiority in antigen preservation for the same peptide antigens was not evident.

Microwave fixation or stabilization of tissue samples can be divided into two categories. The first is microwave fixation or stabilization of tissue structure using water, saline, or buffer without fixative chemicals. When their temperature is raised by microwave irradiation, good fixation can be achieved. These solutions cannot be used alone to fix tissue samples at lower temperature. This type of microwave fixation achieves its results purely by homogeneously raising the temperature. It stabilizes or fixes the proteins within the tissue by coagulation in much the same way as boiling an egg. Raising the temperature also causes the enzymatic and other biological activities to stop. Hopwood et al. suggested that microwave irradiation caused protein denaturation and disulfide bond formation leading to protein aggregation.[28] A number of solutions including saline, cacodylate buffer, PBS, TBS, and water have been tested at different temperatures in a microwave oven.[19,29-31] Excellent results were reported using only normal saline.[19,29] It is generally agreed that a temperature between 45-55°C is optimal for most tissue structures and antigen fixations. However, different tissue samples may require different temperatures or solutions for optimal results and should be tested individually.

Microwave fixation can also be performed with conventional chemical fixatives. This process can accelerate the speed of fixation often yielding better fixing results. The above two approaches can also be combined by using microwave irradiation of a highly diluted fixative. In our laboratories, we have found that the combined method was superior to either of the above two used alone. This combination can also be performed by first using non-chemical microwave fixation followed by a brief chemical post fixation.[29,31] Fast fixation takes place because the rise in temperature can increase molecular motion, facilitating fixative diffusion. Microwave induced protein aggregation leads to cell membrane changes which also favor chemical penetration. It was speculated that microwaves help to create a framework of coagulated proteins along which the chemicals diffuse easily.[1]

Erythrolysis is one of the limitations of microwave-only fixation.[28] As early as 1865, Schultz noted that heating human red blood cells initially produced spherocytes and then fragmentation at 51-52°C.[32] This was confirmed by Hopwood et al who found that heating to 40°C produced no damage to red blood cells in the blood film but heating to 50°C produced fragmentation of many red blood cells.[28] Granulocytes also appeared damaged. At 60°C, most red blood cells were lysed and a proportion of the granulocytes destroyed. Heating to 70°C lysed all the red blood cells. At this temperature, lymphocytes, platelets and granulocytes were mostly destroyed. Recently, it was suggested that temperature increase might not be the sole reason for blood cell destruction. It was observed that erythrolysis occurred at 37°C.[1] When microwave irradiated at about 10 GHz, erythrolysis happened without a raise in temperature[33,34] indicating that other mechanisms may also play a role in this phenomenon.

Another disadvantage of using microwave fixation without chemical fixative is that the membrane structures are not fixed as well as their chemical counterparts. This is because lipid, which makes up most of the components in cellular membranes, absorbs approximately twice as much microwave energy as the surrounding tissue.[35] For a short pulse of irradiation, membrane temperature may be four times that of the cytoplasm. Therefore, microwave fixation is better suited for protein rich structures than lipid rich structures. Lipids can not be stabilized by microwave irradiation and are easily lost in lipid solvents during subsequent processing. As a result, lipid rich neuronal tissue such as brain, spinal cord and synaptic vesicles of the neurons cannot be satisfactorily fixed by microwave irradiation alone.[35,36]

For this reason, attempts at microwave preservation of ultrastructure have encountered some difficulties, as the lipid rich membrane structure is one of the most visible elements for conventional electron microscopic staining. For ultrastructural preparations, there is an abrupt change in morphology between 50° and 60°C.[28,36] At 50°C, minor changes take place in the organelle membranes. Mitochondria are slightly disorganized and the nuclear membrane is interrupted. At 60°C, the above changes are much more pronounced. At 70°C, these changes deteriorate and the ultrastructure is disrupted.[28] As the surrounding proteins heat up much slower, when the lipid components reach the damaging temperature, the cytoplasmic proteins have not been stabilized.

The above shortcomings of microwave fixation can be largely overcome by using a combination of microwave treatment with chemical fixatives simultaneously or sequentially. The fixatives are often more diluted than the normal concentrations when used alone. In 1983, Chew *et al.* demonstrated that microwave irradiation of mouse tissue in a solution containing 2.5% glutaraldehyde produced good ultrastructural preservation.[37] Login *et al.* reported fine ultrastructural preservation with microwave irradiation in a solution mixture of aldehyde containing 2% paraformaldehyde and 0.05-2.5% glutaraldehyde. The results were found to be comparable to standard chemical immersion fixation.[10,11,14,16,18] Microwave fixation of lipid rich hippocampal slices of rat brain in a mixture of aldehydes containing 2% paraformaldehyde and 6% glutaraldehyde achieved perfusion-quality preservation.[38] Morphometric analysis reveals no significant differences in the surface area to volume ratio of mitochondria from rat liver[10] or in the surface area of rat brain mitochondria[38] that were fixed in seconds by microwave-aldehyde methods compared with control specimens fixed for hours by immersion or perfusion in Karnofsky's aldehyde fixative. Primary fixation by microwave irradiation of tissue blocks of around 0.5 cm^3 of rat heart and liver, followed by immersion in 2% 0.2 mol symcollidine-buffered osmium tetroxide for a few seconds, showed equivalent fine structural preservation as compared with samples fixed by immersion in osmium for hours.[12] In all these cases, similar immersion fixation in aldehyde solution for the same or even much longer duration, without microwave treatment, produced inadequate ultrastructural preservation. Different additives may be used in the fixation solution such as 1% Malachite green, 0.1% tannic acid, 30 to 60 mM K-oxalate, pH 7.4, small amounts of $CaCl_2$ and $MgCl_2$ or sucrose to achieve ultrastructural preservation for specific cellular components. Login et al developed an ultrafast fixation method for EM preparations. They used a specially designed microwave irradiation device capable of irradiating at very high power. With this device, tissue samples can be fixed in fixative in milliseconds and microphotographs published rival the best morphology achievable by conventional fixation. Samples fixed by microwave-chemical fixation are stable in buffer at 4°C for weeks, prior to processing in alcohols. They appear well preserved for histological, ultrastructural and antigenic examinations.[31]

Can conventional heating to the same temperature achieve the same fixation effects as those obtained by microwave oven? To a large extent, the answer is "yes". Biological activities will cease and proteins coagulate at the equivalent temperature no matter how it is reached. Microwave radiation can achieve it instantaneously while conventional heating takes much longer. The latter gives the proteolytic enzymes and other self destructing enzymes a chance to play their roles in altering the tissue morphology and diminishing antigenicity. Microwave fixation is particularly advantageous for preserving certain diagnostic features such as mitoses which have been shown to complete their cycle after removal of tissue from the body,[39] this could conceivably occur during the slow process of chemical fixation.

The uniformity of microwave heating as a means for tissue fixation was tested by Login and Dvorak who used an Agar-Saline-Giemsa mixture as an artifactual tissue sample.[31] Such a tissue block can indicate temperature difference at different positions in the tissue by changing colors. It was found that microwave fixation was more uniform than chemical fixation. As a distinct advantage, Login and Dvorak found that microwave fixation can handle EM tissue samples up to 1 cm^3 in size, more than 100 times larger than the tissue size used by conventional chemical fixation for EM.[31] Leong and associates tested 23 antigens in normal saline, heated in the microwave to 63°C. All antigens had good preservation and good immunostaining results without using detergent.[21]

In our hands, the best results were obtained with a fixative 2-20 times more diluted at a power setting that brings the solution to about 55°C in about 2-4 min. This represents a manifold increase in speed from the conventional procedures. Comparing with the ultrafast microwave fixation with high power settings that fix tissue samples in seconds or milliseconds,[12-18,31,40] we prefer a slightly longer fixation period which does not require a specially designed oven and is relatively easier to control. The difference between a few seconds and a few minutes is insignificant when compared to hours or days required by conventional protocols.

The optimal procedure that we recommend for microwave fixation is listed below.

Microwave Fixation for Light Microscopic Morphology and Immunohistochemistry

1. Prepare 1% paraformaldehyde in 0.01 M PBS, Ph 7.2-7.4. Dissolve the white powder completely with a magnetic stirrer on a heater under a fume hood. Cool the fixative to room temperature and use within 30 min.
2. Immerse freshly dissected tissue samples of about 0.5-1 cm^3 in the fixative of about 100-200 ml. The volume ratio between the fixative and the sample size should be at least 20:1.
3. Place the container containing the fixative and the sample in a domestic microwave oven (maximum 700 W) and irradiate for 2 minutes at medium power setting. The temperature of the fixative should reach approximately 55°C.
4. Transfer the sample to PBS-containing 17% sucrose and wash for 2 hr. This step can also be shortened by irradiating in a microwave oven for about 2 min in the same solution. The sample is then ready to be processed for sectioning.

Microwave Fixation for Conventional and Immunogold Electron Microscope

1. Immerse freshly collected tissue samples of approximately 1 mm^3 in size into Karnovsky's fixative (4% formaldehyde, 5% glutaraldehyde and 0.05% calcium chloride in 0.08 M sodium cacodylate buffer pH 7.2-7.4) and 1% Malachite Green (w/v) in 0.08 cacodylate buffer pH 7.2-7.4.
2. Place the samples in the fixative of about 50 ml in the microwave oven. Irradiate at low power setting for 2 minutes and the temperature should reach approximately 45°C.
3. Wash the samples in 0.1 M sodium cacodyl buffer for 2 hr at room temperature. This step can be sped up by irradiating in a microwave oven for about 2 min at medium power setting. The preparation may be postfixed in 1% osmium tetroxide for 20 minutes if no immunostaining is to be performed. The samples are ready to be further processed.

NOTE: When using the Energy Beam Sciences H2500 microwave oven, the solution temperature and the duration can be precisely controlled and is better suited for EM fixation. Many versions of microwave fixation techniques have been introduced and each antigen and tissue type should be experimented individually to establish its optimal condition.

MICROWAVE ANTIGEN RETRIEVAL

Microwave antigen retrieval technique was first introduced by Shi and associates in 1991.[41] This was a very important contribution to pathology as it opened up the possibility of retrospective investigation of a vast collection of formalin fixed archived tissue samples for immunohistochemical diagnosis and research. The authors speculated that raising the temperature would break the crosslinking between proteins which are the chemical characteristics for formaldehyde fixed, particularly overfixed, tissue samples.[42-45] Microwave heating is ideal for achieving this goal. Once the crosslinking is disrupted, the antigenic epitopes become available for detection.

The widely used fixation methods for immunohistochemistry are often compromises between a reasonable but not ideal morphologic preservation and an acceptable but not optimal antigenicity protection. Formalin fixation, although known to diminish many antigenicities and be inadequate for immunohistochemistry, remains the most popular fixative for routine histopathologic specimens. Formalin has been proven to crosslink proteins and cause other molecules to link to antigenic epitopes,[45] preventing some of them from being detected by antibodies. Leong and Gilham demonstrated a progressive loss of antigenicities for 25 commonly examined antigens during formalin fixation for periods of 6 hours to 30 days.[46] Some important antigens for tumor diagnosis such as intermediate filaments, cytokeratin, and many lymphocyte membrane antigens were lost. This severely limits the ability of pathologists to analyze tissue elements immunohistochemically. False negative immunostaining can be partially reversed by protein digestion with enzymes which can break the crosslinks and unmask the antigenic epitopes. However, tissue morphology suffers from enzyme digestion.[47-49]

It was reported that heavy metal salts such as zinc or lead in combination with formalin can yield superior results in antigen preservation.[41,50,51] When formalin fixed tissue samples were immersed in zinc-formalin for the second time, immunoreactivity was improved.[52] It was hypothesized that heavy metal salts act as protein precipitants, forming insoluble complexes with polypeptides.[50] Protein-precipitating fixatives often produce antigenic preservation superior to cross-linking aldehyde fixatives.[51]

Shi and associates tested 52 monoclonal and polyclonal antibodies on formalin-fixed, paraffin-embedded tissues using microwave heating of slides immersed in metal solutions to 100°C.[41] Thirty-nine antibodies showed a significant increase in immunostaining, nine showed no change and four had reduced immunostaining. The most drastic improvements were observed with monoclonal antibodies to vimentin and keratin which otherwise were either negative or immunostained weakly. Microwave treatment unveiled antigenic epitopes that proteolytic digestion failed to recover. In addition, the incubation duration and primary antibody concentration were significantly reduced with the microwave antigen retrieval treatment. False positivity was not created in any case. Noticeably, this beneficial effect was not achievable in alcohol fixed tissue samples.[41] This confirms that the antigen retrieval effect of heating on formalin fixed samples is due to alteration of crosslinking reaction caused by formalin.

This discovery was further expanded by other groups testing more tissue types, fixation methods and antibodies.[29,53,54] It was found that heavy metal solutions may be desirable but not necessary for some antigens. This was a welcome development, as the vapor generated by heavy metal solution is hazardous to health. Water, buffer and saline have all been tested and found to benefit antigen detection to varying extent. Leong *et al* reported good antigen retrieval results with 10% sodium chloride solution and most antigens showed improved staining after boiling in a solution of 10 mM citrate buffer.[29] Momose *et al* reported that lead thiocyanate did not lead to superior immunostaining over protease digestion.[55] Gerdes and associates generated monoclonal antibodies to Ki67, a marker for proliferating cells, that required microwave irradiation to work on formalin fixed paraffin sections.[53,56,57] It provided a simple and reliable way to evaluate cellular, particularly tumorous, proliferative features in formalin fixed tissue samples. Munakata and Hendricks found that the strength of microwave irradiation required to retrieve the antigens was proportional to the length of formalin fixation.[54] Over-fixed tissue requires prolonged microwave treatment. For example, a two day formalin fixation needs at least 14 minutes of microwave heating to obtain good antigen detection. The improving effects, however, reach a plateau at about 49 minutes. It was also observed that prolonged heating in boiling solution did not affect the tissue morphology.[54] The authors cautioned that background staining might increase in some cases.[41,54]

We experimented on microwave antigen retrieval procedures in formalin-fixed, porcine heart tissue to study PCNA[58] and Ki67[56] antigens. It worked excellently. The antigens were specifically detected in the nuclei of dividing cells but were absent in other cell types. These antigens were otherwise undetectable without the microwave antigen retrieval step. It should be kept in mind, however, that although many reports of these methods testified to specificity of the recovered immunoreaction, the possibility of creating false positivity by high temperature treatment has not been convincingly ruled out. Strict positive and negative controls are always necessary to verify the specificity of positive staining obtained after microwave treatment.

Recommended Procedures for Antigen Retrieval

1. Dewax and rehydrate formalin fixed, paraffin embedded tissue sections and bring to water.
2. Immerse the slides in 10 mM sodium citrate buffer, pH 6.
3. Heat the buffer with the slides in a domestic microwave oven at high power setting for 20 mins. The buffer should reach the boiling point in about 2-4 mins. Top it up with a small amount of water as necessary to replenish the liquid evaporated during boiling.
4. Wash the slide briefly in 0.01 M PBS, pH 7.1-7.4. The slide is then ready for immunostaining.

NOTE: A solution of saturated lead thiocyanate may be used in place of sodium citrate buffer and may yield even better results for some antigens on overfixed tissue samples. Lead thiocyanate is hazardous and should be handled under a fume hood and disposed of properly.

MICROWAVE ACCELERATED IMMUNOHISTOCHEMISTRY

Recently, microwave technology has been employed in accelerating immunohisto-chemistry.[59-62] In a sense, this is just an alternative approach to raising incubation temperatures which is known to accelerate immunoreaction. The distinct advantage of using a microwave oven to raise the incubation temperature is speed and the possibility of having well controlled intermittent cycles of heating and cooling. Results obtained by our group and others indicate that microwave treatment can accelerate immunostaining procedures by at least 20 fold and still retain the quality of staining results. It should also be noted that some chemical reactions accelerated by microwave irradiation cannot be explained solely by temperature rise.[1] It is possible that the rapid molecular rotation may facilitate rapid and firm antibody-antigen bonding.[29] Immunohistochemistry requires a very delicate control of the temperature as too high a temperature will destroy the antibodies' activity. A drop of antibodies on the tissue section will evaporate very quickly yet its temperature is difficult to monitor accurately due to its very small volume. Thirty-five to 40°C appears to be a good range for rapid antibody-antigen reaction.

We tested a wide range of antibodies, concentrations, incubation durations and procedures. Each parameter was experimented individually with the other parameters remaining constant. We found that microwave immunocytochemistry was applicable at both the light and the electron microscopic levels and optimal procedures have been developed. In general, we found that a lower power output with an interrupted irradiation pattern gave the best results. The key element of our procedure is repeated cycles of "on" and "off" radiation to allow repeated rapid heating followed by cooling to facilitate affirmative antibody-antigen bonding. We have developed a procedure that allows irradiation of all three layers of antibodies/chromogen incubations with interrupted "on" and "off" cycles at low power output for the avidin-biotin-peroxidase complex (ABC) and the peroxidase-anti-peroxidase (PAP) methods. The whole procedure can be completed within 60 minutes with results comparable or superior to the overnight incubation procedures. The immunoreaction was specific and intense. The background was low, typically lower than the conventional procedure with much longer incubation. This was possibly due to the shortened incubation period that reduced the chance for antibody or chromogen to precipitate. We found that microwave immunocytochemistry can be used routinely in the laboratory and significantly improves efficiency without sacrificing the quality.

Microwave Procedure for Light Microscopic Immunostaining

1. Process the paraffin or cryostat sections to the step of primary antibody incubation.
2. Apply primary antibody on slide. The antibody dilution can be the same as the conventional procedure or more diluted.
3. Incubate in a domestic microwave oven at low power setting with 5 cycles of 1 minute "on" and 1 minute "off", followed by 3 minutes incubation at room temperature. A beaker of water (200ml) should be placed in the oven.
4. Wash in PBS 3 times, 2 min each. This step can be accelerated by heating at medium power setting for 1 min with 3 changes of PBS.
5. Apply the secondary antibody, optimally diluted, to the slides and irradiate in the same fashion as the primary antibody in a microwave oven.
6. Wash as before.
7. Apply third layer antibody or chromogen and treat the slide in the same fashion as the first and the second layers in a microwave oven.
8. Wash as before.
9. Develop the chromogen for 1-10 mins and monitor the reaction under a microscope. This step can be accelerated with microwave irradiation at low power setting. However, a longer developing period at room temperature is preferable for better control of the staining intensity.
9. Wash in water and counterstain.

NOTE: The whole procedure can be completed in 60 minutes. The results are comparable to those obtained with routine procedures which typically involve an overnight incubation and half to one hour each for the second and the third layer incubations. Theoretically for conventional immunostaining, an overnight incubation with maximally diluted antibodies gives the best contrast between the specific staining and the background. Over the years, many quick immunostaining procedures have been developed, each with its own compromises. Typically, the antibodies are applied at higher concentrations. These procedures should all be able to be further accelerated by microwave treatment without any compromises in quality or cost.

Microwave Procedure for Postembedding Indirect Immunogold Staining for Electron Microscope

1. Process the EM tissue sample and place the thin sections on EM grids.
2. Place the grids gently on top of a drop of primary antibody with the section facing the solution.
3. Place the above in a microwave oven and irradiate in the same fashion as for light microscopic immunohistochemistry. It is crucial to place a beaker of water in the oven to provide a humid atmosphere and to lower the energy absorption by the tissue section.
4. Wash manually as for the conventional procedure.
5. Apply the secondary gold-labeled antibody and irradiate the same way as for the primary antibody.
5. Wash as before and counterstain.
6. The gold particle may be further enlarged with silver enhancement solution (see Chapters 3, 14 of this book). The silver enhancement and the counterstaining steps can also be accelerated with microwave irradiation.

OTHER APPLICATIONS

Microwave treatment has been used in many other morphological procedures, including heating paraffin, dehydration, wax penetration, resin polymerization, section drying, washing and counterstaining.[1,29] Microwave irradiation technique has been applied successfully in many histochemical stainings.[1,29,63-65] It helps in both the diffusion of the dye and the rate of the chemical reactions. Accelerated histochemical procedures include hematoxylin and eosin stain, the combined alcian blue-PAS stain,[63] Grimelius silver stain,[64] elastica Van Gieson stain,[1] Fontana-Masson stain,[65] to name a few. Microwave irradiation is very advantageous for cryostat sections. It was found that 5-20 seconds microwave irradiation at high power setting not only resulted in firm attachment of the section, but also improved the general morphology of the frozen tissue. It was reported that immersing freshly cut cryostat sections in Kryofix[66] or Wolman's solution (95% ethanal and 5% glacial acetic acid) and irradiating for 15 seconds at high power setting in a domestic microwave oven significantly improved the morphology more than other methods.[29] We have found that a brief (15-60 seconds) microwave irradiation at high power output of cryostat sections significantly improved the histochemical staining for nitric oxide synthase, with increased specific reaction and lowered background staining67.

Microwave irradiation has also been applied to speed up *in situ* hybridization procedures.[68-71] Denaturation of double stranded DNA requires high temperature. Microwave heating is ideal for this step.[69] The temperature range of 37-52°C for hybridization reaction can be generated and maintained by a microwave oven. The subsequent detection steps for non-radioisotope-labelled probes resemble immunohistochemistry and, therefore, can be accelerated by microwave treatment. In fact, the entire procedure of *in situ* hybridization can be performed in a microwave oven.[71]

It should be noted that microwave irradiation has also been used in polymerase chain reaction (PCR) which requires rapid and precise increase and decrease of temperatures in many repetitive cycles. A microwave assisted thermocycler has been developed and was found to be superior to the conventional PCR machine.[1] It should not be long before microwave-assisted *in situ* PCR machines appear on the market that will help to popularize the extremely sensitive and powerful *in situ* PCR technique, promising to revolutionize many areas of morphological research and diagnosis.[71]

With increased laboratory automation, it is foreseeable that microwave technology will be incorporated into existing and future automatic machines and significantly improve the efficiency of tissue processing and staining. As the ideal temperature and duration for accelerating each step of processing and staining for various protocols is established, a specially designed microwave irradiating device can help to reach and maintain this temperature using a feedback temperature controlling mechanism. Those steps can be programmed into a computerized program panel to drastically shorten many time-consuming procedures with the ultimate objective of achieving even better morphological preparations. It holds the promise for second generation laboratory automation.

ACKNOWLEDGEMENT

The author wishes to acknowledge the excellent technical assistance of Michele Forte, Harry Hance, Mary Englund and Craig Bromley in experimenting the microwave procedures and Gayle Englund for her untiring effort in preparing this chapter and the book. The critical review of the content by Dr. David R. Kersten and Dr. Gerhart W. Hacker is also greatly appreciated.

REFERENCES

1. L.P. Kok and M.E. Boon, Microwave Cookbook for Microscopists. Art and Science of Visualization, Coulomb Press, Leyden (1992).
2. E.J. Valtonen, The effects of microwave radiation on the cellular elements in the peritoneal fluid and peripheral blood of the rat, *Acta Rheumatcl Scand* 12(4):291-299 (1966).
3. A.L. Moore, P.R. Rich, W.D. Bonner Jr, and W.J. Ingledew, A complex EPR signal in mung bean mitochondria and its possible relation to the alternate pathway, *Biochem Biophys Res Commun* 72(3):1099-1107 (1976).
4. P. Ehrlick and A. Lazarus, Die Anämie I. Abt Holder, Wien (1898).
5. J.R. Baker. Cytological Technique. The Principles Underlying Routine Methods, Methuen, London (1958).
6. C.P. Mayers, Histological fixation by microwave heating, *J Clin Pathol* 23:273-275 (1970).
7. C. Peracchia and B.S. Mittler, New glutaraldehyde fixaiton procedures, *J Ultrastruct Res* 39:57-64 (1972).
8. G.R. Bernard, Microwave irradiation as a generator of heat for histological fixation, *Stain Technol* 49:215-224 (1974).
9. R.D. Lillie and H.M. Fullmer, Histopathologic technic and practical histochemistry, 4th edn, McGraw-Hill, London (1976).
10. G.R. Login and A.M. Dvorak, Microwave energy fixation for electron microscopy, *Am J Pathol* 120:230-243 (1985).
11. G.R. Login and A.M. Dvorak, Microwave fixation provides excellent preservation of tissue, cells and antigens for light and electron microscopy, *Histochem J* 20:373-387 (1988).
12. G.R. Login, B.K. Dwyer, and A.M. Dvorak, Rapid primary microwave-osmium fixation. I. Preservation of structure for electron microscopy in seconds, *J Histochem Cytochem* 38:755-762 (1990a).
13. G.R. Login, B.K. Dwyer, and A.M. Dvorak, Immunocytochemical localization of histamine in

secretory granules of rat peritoneal mast cells with conventional or rapid microwave fixation and an ultrastructural post-embedding immunogold technique, *J Histochem Cytochem* **40**:1247-1256 (1992a).

14. G.R. Login, S.J. Gali, E. Morgan, N. Arizono, L.B. Schwartz, and A.M. Dvorak, Microwave fixation of rat mast cells. 1. Localization of granule chymase with an ultrastructural post-embedding immunogold technique, *Lab Invest* **57**:592 (1987).

15. G.R. Login, S. Kissell, B.K. Dwyer, and A.M. Dvorak, A novel microwave device designed to preserve cell structure in millisecond, In: Microwave Processing of Materials II, W.B. Synder, Jr., W.H. Sutton, M.F. Iskander, and D.L. Johnson, Eds., Materials Research Society, Pittsburgh, pp329-346 (1991).

16. G.R. Login, S.J. Schnitt, and A.M. Dvorak, Rapid microwave fixation of human tissues for light microscopic immunoperoxidase identification of diagnostically useful antigens, *Lab Invest* **57**:585-591 (1987b).

17. G.R. Login, W.B. Stavinoha, and A.M. Dvorak, Fast and ultrafast microwave energy fixation techniques demonstrated by immunofluorescence, light and electron microscopy, In: Microwave irradiation for histological and neurochemical investigations, C.L. Blank, S. Howard, and Y. Maruyama, Eds., Soft Science Publications, Tokyo 27-53 (1989).

18. G.R. Login, W.B. Stavinoha, and A.M. Dvorak, Ultrafast microwave energy fixation for electron microscopy, *J Histochem Cytochem* **34**:381-387 (1986).

19. A.S.Y. Leong, M.E. Daymon, and J. Milios, Microwave irradiation as a form of fixation for light and electron microscopy, *J Pathol* **146**:313 (1985).

20. A.S.Y. Leong and C.G. Duncis, A method of rapid fixation of large biopsy specimens using microwave irradiation, *Pathology* **18**:222 (1986).

21. A.S.Y. Leong, J. Milios, and C.G. Duncis, Antigen preservation in microwave-irradiated tissues: A comparison with formaldehyde fixation, *J Pathol* **156**:275-282 (1988).

22. A.S.Y. Leong, Applications of microwave irradiation in histopathology, *Path Ann* **2**:213 (1988).

23. A.S.Y. Leong and D.W. Gove, Applications of microwave irradiation in electron microscopy, In: Proc. XIIth Internat Congr Electr Micros, San Francisco Press, San Francisco, pp 140-141 (1990a).

24. A.S.Y. Leong and D.W. Gove, Microwave techniques for tissue fixation, processing and staining, *EMSA Bull* **20**:61 (1990b).

25. A.S.Y. Leong, Microwave fixation and rapid processing in a large throughput histopathology laboratory, *Pathology* **23**:271 (1991a).

26. A.S.Y. Leong, Microwave irradiation-applications in tissue fixation, processing and staining for light microscopy and electron microscopy. World Health Organization Bi-Regional Training Course on Electron Microscopy in Biomedical Research and Diagnosis of Human Diseases, University of Adelaide Press, Adelaide, pp 47-60 (1991b).

27. A.S.Y. Leong, the scope of microwave applications in a large throughput laboratory, *Scanning* **14(Suppl. II)**:65 (1992).

28. D. Hopwood, G. Coghill, J. Ramsay, G. Milne, and M. Kerr, Microwave fixation: its potential for routine techniques, histochemistry, immunocytochemistry and electron microscopy, *Histochemical J* **16** 1171-1191 (1984)

29. M.K. Jr. Patterson and R. Bulard, Microwave fixation of cells in tissue culture, *Stain Technol* **55**:71 (1980).

30. M.E. Daymon and A.S.Y. Leong, Microwave fixation for surgical and autopsy tissues, *Pathology* **16**:418 (1984).

31. G.R. Login and A.M. Dvorak, A review of rapid microwave fixation technology: Its expanding niche in morphologic studies, *Scanning* **15**:58-66 (1993).

32. M. Utze, Ein heizbarer objecttisch und seine vervendung bei untersuchungen der utes, *Archiv Mikrosk Anat* **1**:1-42 (1865).

33. S.F. Cleary, F. Garber, and L.M. Liu, Effects of X-band microwave exposure on rabbit erythrocytes, *Bioelectromagnetics* **3**:453-466 (1982).

34. S.F. Cleary, L.M. Liu, and F. Garber, Erythrocyte hemolysis by radiofrequency field, *Bioelectromagnetics* **6**:313-322 (1985).

35. M.M. Webber, F.S. Barnes, L.A. Seltzer, T.R. Bouldin, and K.N. Prasad, Short wave pulses cause ultrastructural membrane damage in neuroblastoma cells, *J Ultrastruct Res* **71**:321-330 (1980).

36. P. Kasa, K. Bansaghy, and K. Gulya, Ultrastructural changes and diffusion of acetylcholine in rat brain after microwave irradiation, *J Neurosci Meth* **5**:215-220 (1982).

37. E.C. Chew, D.J. Riches, T.K. Lam,and H.J. Hou Cahn, A fine structural study of microwave fixation of tissues, *Cell Biol Int Rep* **7**:135-139 (1983).

38. F.E. Jensen, K.M. Harris, Preservation of neuronal ultrastructure in hippocampal slices using rapid microwave-enhanced fixation, *J Neurosci Methods* **29**:215-230 (1989).

39. N. Graem and K. Helweg-Larsen, Mitotic activity and delay in fixation of human tissue, *Acta Path Microbiol Scand* **A87**:375-378 (1979).

40. A.S.Y. Leong, A review of microwave techniques for diagnostic pathology, *MSA Bulletin* **23(4)**:253-263 (1993).

41. S-R. Shi, M.E. Key, and K.L. Kalra, Antigen retrieval in formalin-fixed, paraffin-embedded tissues: An enhancement method for immunohistochemcial staining based on microwave oven heating of tissue sections, *J Histochem Cytochem* **39(6)**:741-748 (1991).

42. D.P. Kelly, M.K. Dewar, R.B. Johns, W.L. Shao, and J.F. Yates, Cross-linking of amino acids by formaldehyde. Preparation and 13C NMR spectra of model compounds. *In*: Protein cross-linking. Symposium on protein crosslinking. San Francisco, M. Friedman, ed. Plenum Press, New York, pp 641 (1976).

43. J.W. Harlan, S.H. Feairheller, Chemistry of the crosslinking of collagen during tanning, *In*: Protein crosslinking. Symposium on protein crosslinking, M. Friedman, ed., Plenum Press, New York, pp 425 (1976).

44. C.H. Fox, F.B. Johnson, J. Whiting, P.P. Roller, Formaldehyde fixation, *J Histochem Cytochem* **33**:845 (19850.

45. P.B.Jr. Bell, I. Rundquist, I. Svensson, V.P. Collins, Formaldehyde sensitivity of a GFAP epitope, removed by extraction of the cytoskeleton with high salt, *J Histochem Cytochem* **35**:1375 (1987).

46. A.S.Y. Leong and P.N. Gilham, The effects of progressive formaldehyde fixation on the preservation of tissue antigens, *Pathology* **21**:266-271 (1989).

47. S.N. Huang, H. Minassian, J.D. Moore, Application of immunofluorescent staining on paraffin sections improved by trypsin digestion, *Lab Invest* **35**:383 (1976).

48. H. Battifora, M. Kopinski, The influence of protease digestion and duration of fixation on the immunostaining of keratins, *J Histochem Cytochem* **34**:1095 (1986).

49. *Histochemical Journal*. Two special microwave issues, **20**, 311-40, 1988 and **22** 311-393, 1990.

50. M.D. Jones, P.M. Banks, and B.L. Caron, Transition metal salts as adjuncts to formalin for tissue fixation, *Lab Invest* **44**:32A (1981).

51. F.W. Herman, E. Chlipape, G. Bochenski, L. Sabin, and E. Elfont, Zinc formalin fixative for automated tissue processing, *J Histotechnol* **11**:85 (1988).

52. S.L. Abbondanzo, D.C. Allred, S. Lampkin, and P.M. Banks, Enhancement of immunoreactivity in paraffin embedded tissues by refixation in zinc sulfate-formalin, Proc Annu Meeting US and Canadian Acad Pathol, Boston: March 4-9, 1990.

53. G. Cattoretti, M.H.G. Becker, G. Key, M.Duchrow, C. Schlüter, J Galle, and J. Gerdes, Monoclonal antibodies against recombinant parts of the Ki 67 antigen (MIB 1 and MIB 3) detect proliferating cells in microwave-processed formalin-fixed paraffin sections, *J Pathol* **168**:357 (1992).

54. S. Munakata and J.B. Hendricks, Effect of fixation time and microwave oven heating time on retrieval of the Ki-67 antigen from paraffin-embedded tissue, *J Histochem Cytochem* **41(8)**:1241-1246 (1993).

55. H. Momose, P. Mehta, and H. Battifora, Antigen retrieval by microwave irradiation in lead thiocyanate. Comparison with protease digestion retrieval, *Appl Immunohistochem* **1**:77 (1993).

56. J. Gerdes, U. Schwab, H. Lemke, H. Stein, Produciton of a mouse monoclonal antibody reactive with a human nuclear antigen associated with cell proliferation, *Int J Cancer* **31**:13 (1983).

57. J. Gerdes, M.H.G. Becker, G. Key, G. Cattoretti, Immunohistological detection of tumour growth fraction (Ki-67 antigen) in formalin-fixed and routinely processed tissue, *J Pathol* **168**:85 (1992).

58. O.W. Kamel, D.P. LeBrun, R.E. Davis, G.J. Betty, and R.A. Warnke, Growth fraction estimation of malignant lymphomas in formalin-fixed paraffin-embedded tissue using anti-PCNA/cyclin 19A2. Correlation with Ki-67 labelling, *Am J Pathol* **138**:1471 (1991).

59. A.S.Y. Leong and J. Milios, Rapid immunoperoxidase staining of lymphocyte antigens using microwave irradiation, *J Pathol* **148**:183 (1986).

60. A.S.Y. Leong and J. Milios, Accelerated immunohistochemical staining by microwave, *J Pathol* **161**:327 (1990).

61. A. Hjerpe, M.E. Boon, and L.P. Kok, Microwave stimulation of an immunological reaction (CEA/anti-CEA) and its use in immunohistochemistry, *Histochem J* **20**:388 (1988).

62. B.L. Giammara, J. Pierce, A. Rustioni, M. Borowitz, D. Chandler, P. Yates, and J. Hanker, Microwave acceleration of the metalization and intensification of cytochemical and immunocytochemical reaction products with silver methenamine, *Proc 43rd Ann Meet Electr Microsc Soc Amer* pp704 (1985).

63. K. Matthews and J.K. Kelly, A microwave oven method for the combined alcian blue-periodic acid-Schiff stain, *J Histotechnol* **12**:195 (1989).

64. T.C. Staples and W.E. Grizzle, Effect of temperature on argyrophil impregnation: Development of a high temperature rapid argyrophil procedure, *Stain Technol* **62**:41-49 (1987).

65. A.S.Y. Leong and P. Gilham, A new, rapid microwave-stimulated method of staining melanocytic lesions, *Stain Technol* **64**:81 (1989).

66. L.P. Kok and M.E. Boon, Microwave fundamentals for ultrastructural research, *Scanning* **14(Suppl II)**:55-56 (1992).

67. M. Forte and J. Gu, Microwave irradiation of cryostat sections accelerates and improves nitric oxide synthase staining,(Abstract), Cell Vision - Journal of Analytical Morpholgy, (in press) (1994).

68. C. Evans and K.J. Towner, A note on the use of microwaves in an improved non-radioactive DNA hybridization procedure for the detection of bacteria in foodstuffs. *Let in Appl Microbiology* **10**:233-235 (1990).

69. P. J. Coates, P.A. Hall, M.G. Butler, and A.J. D'Ardenne, Rapid technique of DNA-DNA *in situ* hybridization on formalin fixed tissue sections using microwave irradiation, *J Clin Path* **40**:865-869 (1987).

70. A.S. Bourinbaiar, V. Zacharopoulos, D.M. Phillips, Microwave irradiation-accelerated *in situ* hybridization technique for HIV detection, *J Virol Meth* **35**:49-58 (1991).

71. A. S. Bourinbaiar, Microwave irradiation-stimulated *in situ* hybridization procedure with biotinylated DNA probe, *Eur J Morphol* **29**:213-218 (1992).

72. Kreateach, The AmpliWave™, Product information sheet, Amsterdam (1991).

CHAPTER 7

BASIC PRINCIPLES OF IMMUNOHISTOPATHOLOGY

Gerhard W. Hacker,[1,2] **Anton-Helmut Graf,**[1,3]
Otto Dietze[1]

[1]Institute of Pathological Anatomy, Immunohistochemistry and
Biochemistry Unit, General Hospital, Salzburg, Austria
[2]University of Salzburg, Institute of Zoology, Salzburg, Austria
[3]Department of Gynecology and Obstetrics, General Hospital
Salzburg, Austria

SUMMARY

Analytical methods, especially immunocytochemical techniques, have revolutionized
diagnostic histopathology during the last years. Such techniques are remarkably sensitive,
specific, and can often be applied even on routinely processed, formalin-fixed and
paraffin-embedded material. Polyclonal and monoclonal antibodies to various
immunocytochemical markers can be used to classify tumors (tumor classification
markers), or to ensure better therapy (e.g. steroid receptors in breast cancer). Possibly,
they may also give an outlook on the progression (e.g. proliferation markers) and
metastatic potential of a given tumor (some oncogenes and receptors). *In situ*
hybridization techniques may be applied to specifically differentiate between "benign" and
"oncogenic" (cancer-related) types of viruses e.g. human papilloma virus (HPV), or
simply to prove a virus infection. *In situ* applications of molecular biological methods
such as the polymerase chain reaction (PCR *in situ* hybridization) allow the detection of
single copies of nucleic acids (e.g. viruses) in tissue sections and cytological preparations.
In this way, they add further potential to the fascinating possibilities of modern
histopathological examination.

[1]Address for correspondence: Univ.-Doz. Dr. Gerhard W. Hacker, Salzburg General Hospital, Institute of
Pathological Anatomy, Immunohistochemistry and Biochemistry Unit, Muellner Hauptstr. 48, A-5020
Salzburg, Austria. Tel:++43-662-4482-4730; Fax: ++43-662-4482-88

The present text introduces the principles of today's use of immunocytochemistry and related techniques in histopathology. Reference is also given to some applications in breast cancer, one of the fields where immunocytochemistry has most promise to give better orientation concerning therapy and prognosis.

INTRODUCTION

Immunocytochemistry and related techniques open new insights into tumor pathology and may provide new concepts in cancer research. Since Coons used a fluorescent label coupled to immunoglobulins to detect tissue-bound antigens for the first time in 1941,[1,2] a new group of techniques with the potential to revolutionize our knowledge in biology and medicine was born. Only in recent years, have modifications of Coons' immunofluorescence techniques been widely applied in diagnostic histopathology. The discovery of monoclonal antibodies by Köhler and Milstein[3] has lead to a series of new concepts in pathology. Today, we have specific probes available which allow better and more accurate tumor classification in tissue sections and cytological material from various kinds of neoplasms. This also yields increasing information for prognosis and therapy of tumors. In addition, these techniques, including *in situ* hybridization and *in situ* polymerase chain reaction (PCR *in situ* hybridization) methods, can be applied to the detection of viruses and bacteria and, therefore, increase our knowledge in microbiology.

In the present text, a basic introduction of the use of immunocytochemistry in diagnostic histopathology is given. We are well aware that in such a short text, no highly comprehensive information is possible. Therefore, it is our intention to describe aspects of experience from our own laboratories and to clarify which tumor marker antibodies were found to perform well in routine formalin-fixed paraffin sections.

TISSUE PROCESSING

For the accurate localization of tissue antigens, a good preservation of tissue morphology and antigens, and the use of sensitive immunocytochemical techniques in combination with specific antisera are essential conditions.[4] Fixation has a great influence on the intensity of both histochemical and immunocytochemical stainings. It is necessary to treat as fresh as possible tissue samples with chemical fixatives such as formalin. This allows the preservation of tissue architecture and prevents tissue constituents from disintegrating during the histochemical staining procedures. In routine histopathology, the standard formalin fixation process can sometimes cross-link proteins in such a strong way that antigenic sites are obscured and are not accessible to the antibody. Fixation can also induce non-uniform alterations in the tertiary structure of tissue antigens. Therefore, a compromise between an acceptable morphological preservation and antigen availability must be found.

Effects of various primary fixatives based on cross-linking and/or precipitation on the sensitivity of immunocytochemical staining have been extensively discussed.[4,6] The usual immersion-type fixation yields heterogeneous results and is a rather complex procedure that depends on the intrinsic properties of the fixative, pH, osmolarity, temperature, length of treatment, and type of tissue.[5] Sometimes, special ways of processing tissue such as liquid or vapor fixation in Stefanini's/Zamboni's fluid, diethyl-pyrocarbonate (DEPC) or para-benzoquinone are required, especially to detect fixation-labile regulatory peptides.[7-11] However, routine fixation in neutral buffered formaldehyde (NBF) solutions and paraffin embedding are often quite satisfactory with the sensitive techniques and antibodies available today. In some cases, protease predigestion may reveal "overfixed" antigenic sites.[12-13] Other methods, such as refixation/reprocessing of tissue, may also help but in our experience, they appear to be somewhat complicated and time-consuming.[15] A newly described "antigen retrieval" method using microwave boiling of the deparaffinized sections in citrate buffer, saturated lead thiocyanate, or in "target unmasking fluid" (TUF, Kreatech, Amsterdam, NL) in our hands proved to be of immense benefit for the detection of substances in overfixed tissue sections.[14] It must be kept in mind that the effects of the same fixative on the retention of immunoreactivity on the same antigen are not always consistent in different anatomic sites, even when used for the same length of time.[5]

Conventional NBF is an easily available and comparatively cheap universal fixative. It provides consistent and adequate preservation of tissue morphology but is not innocuous for immunocytochemical purposes because of its reaction with protein. Ideally, fixation should take place in 4% NBF and not be performed for longer than 12 to 24 hours. Still, even optimal fixation in NBF differentially compromises the immunoreactivity of various antigens such as lymphocyte membrane antigens, cytokeratins, vimentin, fibronectin, peptides, and immunoglobulins. Preservation of such antigens is often better in Stefanini's/Zamboni's, Bouin's or B-5 fixative.

For many purposes, antibodies are now available that can also be successfully applied on formalin-fixed, paraffin-embedded tissue, allowing the widespread use of diagnostic immunocytochemistry.[5] However, most of these antibodies will only perform well if tissue processing has been mild and moderate, possibly at a lower temperature (4°C). For many purposes, optimal fixation in NBF and paraffin-embedding is an acceptable compromise in the view that at least some degree of immunoreactivity is retained. For the examination of fixation-susceptible peptides and proteins, other fixation and tissue processing techniques are sometimes necessary including acetone-postfixed cryostat sections of freshly frozen tissue specimens (e.g. various kinds of lymphoma markers), or the use of specialized immersion or instillation/perfusion fixation before freezing for cryostat (for many regulatory peptides).[4,5] In a given tissue, e.g. an uncharacterized tumor in routine pathology, it is advisable to apply different fixation and tissue processing protocols in parallel in order to gain optimum results. In our lab, we always try to have NBF-fixed paraffin sections available for routine histopathology. In addition, we use Stefanini's/ Zamboni's-fixed paraffin- and cryostat sections, as well as primarily unfixed but acetone or Stefanini's/Zamboni's postfixed cryostat sections for specialized immunocytochemical and molecular biological purposes.

Microwave fixation is a way of drastically shortening fixation time while retaining immunoreactivity for many purposes.[5,6] Several authors claim that when microwave irradiation is properly used, standardization of microscopic results may be achieved. It has been reported that with some antigens (such as cytokeratins, factor VIII-related antigen) which normally require proteolytic digestion to restore immunoreactivity, this pretreatment does not appear to be necessary with microwave-fixed tissues. However, the full value of microwave fixation in immunocytochemistry is not yet fully determined.[5]

The currently applied dehydration and clearing schedules appear to affect immunostaining properties less than the preceding fixation treatments, although the alcohols also clearly play their role as fixatives completing the fixation process. The effects of water replacement by organic solvents before embedding are scarcely known, but some denaturation occurs which may alter the immunoreactivity of some tissue constituents.[5] There are also reports that prolonged soaking of the tissue in water, alcohol or chloroform has little influence on immunocytochemical performance.[5,9]

The influence of paraffin or parablast on immunoreactivity has provided a controversial discussion. Compared to cryosections, immunoreactivity appears to be lowered and the unwanted background staining is sometimes higher. There is general agreement about the importance of the complete deparaffinization of slides to ensure uniform staining. It has even been recommended that slides be soaked for 30 minutes in fresh xylene at 60°C to ensure that all the paraffin is removed.[5]

IMMUNOCYTOCHEMICAL TECHNIQUES

All immunocytochemical techniques available today are entirely based on Coons' ingenious invention of immunofluorescence techniques, described in 1941.[1,2] Direct or indirect immunofluorescence techniques are still widely used because they are very easy to perform, reliable, and specific. In direct methods, the antibody raised against the antigen in question is directly labeled (one-layer or single-step methods), whereas in indirect methods, the primary antibody molecule is recognized by labeled or unlabeled antibodies directed against the primary antibody species (two-step or bridging procedures).

One of the main fields where immunofluorescence is of great help, is in the detection of chemical substances within nerve fibers or neuroendocrine structures in Stefanini's/ Zamboni's-prefixed cryostat sections, especially when a combination of different fluorescent labels as double or even triple immunofluorescence is used. This allows an elegant double-localization of different antigens within one single cell simply by switching the fluorescence filters in the microscope (see also chapter by Dr. Hauser-Kronberger in this volume). Immunofluorescent methods are also frequently used in immunocytochemical dermatohistopathology, e.g. for the demonstration of different types of cytokeratins or collagens, or for the detection of bacterials. Another large field where immunofluorescent methods are heavily used is that of flow cytometry.

Immunofluorescent methods also have disadvantages, e.g. tissue morphology cannot be assessed properly, and the antibodies have to be used in comparatively higher concentrations. Also, most fluorescent labels used including fluorescein isothiocyanate (FITC) or rhodamine isothiocyanate (RITC), will fade during prolonged exposure to UV light when no special anti-fading treatment has been carried out, and, therefore, yield non-permanent stains. Immunofluorescence requires a special fluorescence-equipped microscope and this makes the technique more expensive. Some of these problems were solved by Nakane et al.[16] with the introduction of immunoperoxidase methods. Nakane was the first to use an enzyme (i.e., horseradish peroxidase) coupled to antibodies for immunocytochemistry. The peroxidase-antiperoxidase (PAP) method, published by Sternberger only few years later,[17,18] is also based on Coons' and Nakane's work. The PAP method is an unlabeled antibody-bridging technique, consecutively applying three layers of antibodies to the section. In contrast to Nakane's indirect immunoperoxidase method, where peroxidase is covalently bound to the secondary antibody (thereby possibly affecting the activity of the antibody), an unlabeled bridging antibody binding to both the first and the third layer is used in the PAP method. This was made possible because the primary and tertiary antibodies are raised in the same species. The secondary antibody is directed against immunoglobulin of the primary (and tertiary) antibody species. A specialty of the PAP method is the third layer, which is apparently built up by a complex of peroxidase and antibodies to peroxidase. The enzyme peroxidase is held together by antigen-antibody-binding. Immunoperoxidase or PAP methods require the use of a revealing substrate to make the peroxidase label visible in the microscope. In this additional reaction, diaminobenzidine-tetrahydrochloride (DAB) in the presence of hydrogen peroxide is most widely used,[19] yielding a brownish and permanent staining, visible in a normal bright field light microscope. DAB and comparable alternatives giving other colors[4] are hazardous to health and require careful handling. Due to their possible carcinogenic potential, the development reaction should only be carried our in a fume cupboard, using gloves to protect the skin. Everything contaminated with DAB, as well as the DAB solution itself, should be neutralized after use with household disinfectant chlorine which breaks the DAB molecule.

With immunoperoxidase methods, it became possible to greatly improve assessment of morphology by using hematoxylin nuclear counterstaining. Thus, it is no longer required to use a special microscope. This is also true for various other immunoenzyme methods which evolved from immunoperoxidase, i.e. those applying alkaline phosphatase, beta galactosidase, or glucose oxidase as the enzyme label.[4] With bridging methods such as PAP- or APAAP- (*alkaline phosphatase-anti alkaline phosphatase*),[20] a high sensitivity and detection efficiency can be achieved. In a peroxidase-DAB-system, the signal obtained may be further enhanced by application of heavy metal salts such as cobalt chloride, cobalt acetate, or imidazole.[4]

The fluorescent and enzyme labels briefly discussed can also be used in streptavidin-biotin based methods,[21,22] which apparently further increase the sensitivity of immunocytochemistry. Avidin-biotin-peroxidase-complex (ABC) methods appear to be ideal techniques for both cryostat and paraffin sections. They can also be successfully used for pre-embedding immunoelectronmicroscopy (see chapter by Kummer in this volume).

However, the highest sensitivity and detection efficiency available today is fulfilled by the great amplifying potential of *immunogold-silver staining* (IGSS) methods (see chapters by Hacker and by Krenács in this volume).[23-27] In the indirect IGSS technique, the specific primary antibody binding site is revealed by a secondary antibody adsorbed to colloidal gold and directed against immunoglobulin of the species in which the primary antibody had been raised. The initially small gold particles (0.8 to 5 nm in diameter) are enlarged by shells of metallic silver in a subsequent silver amplification reaction.[27] Provided that high quality antibodies are used, IGSS methods give a greatly contrasting grayish to black color of the immunostained structures and, therefore, allow conventional hematoxylin and eosin/nuclear fast red counterstaining, making it the method of choice for the histopathologist. Immunostaining can be observed even with low power objectives in the microscope; this saves time as preparations can be more rapidly screened. The sensitivity is very high. As with IGSS methods, positive reactions are often obtained where other methods have failed.[24,25] The detection efficiency is high as many antibodies can be diluted far higher than in other methods; we have used some polyclonal antibodies in dilutions up to 1:50.000 with this method.[23-25] In order to obtain the highest sensitivity and detection efficiency, it is crucial to use an efficient silver enhancement procedure, such as silver acetate autometallography.[27] This method allows amplification in daylight under visual microscopic control. It is cheaper and more specific than most commercial developers. For use of IGSS methods in diagnostic immunohistopathology, it has to be kept in mind that some (few) primary antibodies are not recognized well by immunogold reagents currently available (which is true for some antibodies directed against chromogranin A), and that some intranuclear antigens (such as PCNA, MIB-1, or p53) are better detected with ABC-peroxidase-based xylene rather than using IGSS probably because of penetration or charge problems.

ANTIBODIES

Immunocytochemical techniques completely rely on the availability of antibodies that will react in a specific way with the tissue antigen. It is often difficult to obtain such antibodies. Nevertheless, there is an increasing number of companies selling these antibodies. Usually, not all antibodies obtained commercially will fulfill the quality criteria for highly specific and sensitive immunocytochemistry even if the companies promises that all of their products are of the highest quality. Therefore, it is advisable not to rely solely on commercial sources alone, but also to ask one's colleagues which antibodies satisfy their demands best. Raising antibodies for diagnostic immuno-histopathological purposes in our own laboratory does not seem necessary anymore, apart from those circumstances where special scientific targets are to be reached.

Basically, two different sorts of antibodies are available: *polyclonal* and *monoclonal* antibodies. Polyclonal antibodies are raised in host animals by conventional methods: immunogen is injected into host animals such as rabbit, guinea pig, donkey, sheep, goat and others. Subsequently, booster injections are given. After about two weeks or more, the blood is taken and centrifuged. The diluted serum is then used in immunocyto-chemical and other specificity tests in order to find out which antiserum is specific and useful for immunocytochemistry. It will always contain a mixture of different antibodies. Some of them recognize the antigen in question (via different epitopes on the antigen). Others are "unspecific" in terms of immunocytochemistry as they recognize various substances and particles in which the immune system of the host animal has produced antibodies previously by its natural response to foreign antigens. Therefore, unpurified polyclonal antibodies are not very "clean" antibodies. In a good animal bleed, the amount of "specific" antibodies produced will only be about 20%, whereas all the rest consist of unwanted antibodies. One method for "cleaning" polyclonal antibodies is by affinity purification. Still, each one of the remaining specific antibodies will have a different binding affinity and avidity. It is also almost impossible to produce a consistent reagent on a large scale when polyclonal antiserum raising techniques are used.[28,29]

On the other side, the introduction of a process to produce monoclonal antibodies by Köhler and Milstein[3] has made it possible to obtain completely "clean" antibodies, all of the same sort, recognizing the same antigenic determinant. This is because a monoclonal antibody is always the product of a single clone of immortalized B-lymphocytes, and, therefore, is uniform in molecular structure, specificity, affinity, and avidity. Such antibodies can be produced consistently in large amounts. Monoclonal antibodies have revolutionized many areas of experimental, clinical, and industrial work. The process of producing monoclonal antibodies is complex and time-consuming.[29] However, once a particular clone has been found whose antibodies specifically recognize the antigen in question and can also give good results in immunocytochemical tests, it is comparatively cheap to produce large amounts of monoclonal antibody. Therefore, it appears difficult to understand why so many pharmaceutical companies charge such high prices for monoclonal antibodies.

Polyclonal and monoclonal antibodies, even when commercially purchased, always need to be tested for performance. For this purpose, specificity tests are of crucial importance. Positive and negative controls have been recommended.[4] The first test is to use sections of tissue known to contain the antigen in question within well defined morphological structures. Various dilutions of antiserum must be tested: for every antiserum, the correct working dilution for a particular staining method must be found. The dilution required will vary with the temperature and duration of incubation. For example, acceptable polyclonal antibodies will require (in overnight incubation at 4°C) the testing of doubling dilutions from 1/50 to 1/3,200 using the indirect immunofluorescence method, and dilutions from 1/100 to 1/32,000 for the PAP method. The "optimal" working dilution may vary from tissue type to tissue type. Sometimes individual variations can be found in sections from the same time and type, even when the specimens have been processed in the same way. The optimum dilution is that which gives the best possible staining intensity, together with lowest levels of unwanted background staining, in a particular tissue.

In histopathological sections, especially in tumor tissues, the best working dilution is often lower than in normal test tissues. This is because pathological tissues may contain lower quantities or molecularly altered antigen which will require comparatively higher concentrations of antibody. Therefore, a "suboptimal" antibody dilution is recommended in such cases, e.g. double of the concentration from the optimal working dilution.

The principle of finding the optimal antibody working dilution, briefly described here for polyclonal antibodies, is essentially the same for monoclonal antibodies. However, with monoclonal antibodies, one sometimes must start testing dilutions from the concentrated version. It also has to be kept in mind that some antibody suppliers sell prediluted antibodies, often suggesting that these are already optimally diluted. Based on the principles of immunocytochemistry and the fact that every method and every tissue (from same or different location, as well as different processing) requires different "optimal" dilutions of antibody, the histochemist should always order concentrated antibodies and perform the dilution and specificity tests in the immunostaining systems applied. Antibodies performing well in other techniques like radio- or enzyme-immunoassays often cannot be used for immunocytochemistry since the conditions in such *in vitro* tests are quite different. The golden rule for immunocytochemistry, therefore, is to assess the usefulness of antibodies by immunocytochemistry.[4]

Testing different dilutions is already a "positive" control, because a positive preparation known to contain the antigen in question is used. For the histopathologist, ideally, a specimen that stains weakly should be used; concentrations containing higher amounts of antigen will always stain with a "suboptimal" dilution. Negative controls include omission of primary antibody or its replacement by antibody diluent alone or nonimmune serum (at the same dilution) on positive control tissue. These should not yield any staining.

The best test of the specificity of a given antibody is the absorption control. Optimally diluted antiserum should be reacted with antigen concentrations of 0.001, 0.01, 0.1, 1, and 10 nMol per ml optimally diluted antiserum, at 4°C, 24 hours prior to immunostaining. Ideally, at 1 and 10 nMol/ml, all antibody molecules should have built up antigen/antibody complexes with the antigen molecules added in excess. The mixtures are then reacted with the tissue sections instead of with the primary antibody alone. The rest of the staining procedure is carried out as usual. The result should be a complete abolishment of the immunostaining with a concentration of 1 nMol or at least with 10 nMol antigen per ml antibody solution. Antigen concentrations at or below 0.1 nMol usually show some degree of positive immunostaining.

It cannot be overemphasized that specificity controls should be applied for all antibodies used in immunocytochemistry, even with antibodies obtained from commercial sources. In immunohistopathology, it is also crucial to use positive control tissue sections in every staining run, to check whether the staining procedure and the antibody used have performed well. As positive controls, sections from a specially designed multiblock-system can be recommended (see Chapter 10).

IMMUNOCYTOCHEMICAL TUMOR MARKERS

In immunocytochemistry, the term *tumor marker* is confined to those chemical substances secreted or stored by the tumor or its stroma which might help to characterize a given tumor more accurately than with conventional histochemical staining techniques. These substances are detected with antibodies using immunocytochemical procedures in tissue sections or cytological specimens. A large selection of different antibodies to tumor markers is available on the market. Comparatively few are of practical use, i.e. giving intense specific staining even in problematical cases. So far, no perfect tumor marker has been found. The so-called mesenchymal markers like vimentin are also present in epithelial or glial structures. Some "melanoma markers" and some "lymphoma markers" are also present within tissue of epithelial differentiation. Therefore, selections of several antisera should always be used in order to further define differentiation by exclusion of other tissue types. To properly assess immunocytochemical histopathology, it is necessary to have years of experience and up-to-date training in the field of modern analytical techniques, obviously requiring review of the most recent literature in the related fields.

Epithelial Markers

Antibodies to "epithelial tumor markers" should facilitate the identification and characterization of poorly differentiated and metastatic epithelial tumors in tissue sections even though the morphological appearance of such neoplasms may be uncharacteristic.[30] The epithelial cell system is highly complex and expressed by the variety of cell forms of differing histogenetic origin, varying widely in structure and function. Normal epithelial cells have several features in common. Some of the various broad spectrum tumor markers which can be used to identify epithelial differentiation of a tumor are listed in *Table 1*. Mainly, they include cytokeratins and some membrane or secretion-associated substances such as epithelial membrane antigen, human milk fat globulin, and carcinoembryonic antigen.

Cytokeratins

Cytokeratins, a family of cytoskeletal proteins, have been found to be expressed in almost all types of epithelial cells. They are the epithelial type of the intermediate filament proteins. Intermediate sized filaments (8-10 nm diameter) are generally present in all human cells. They can be divided into at least six classes on the basis of their protein composition and include cytokeratins (present in epithelial cells), vimentin (mainly present in most mesenchymal cells), desmin (contained in mesenchymal myogenic cells), neurofilament proteins (present in neurons and their processes), glial fibrillary acidic protein (mainly present in central glia) and peripherin (present in some neuroendocrine structures).[31] Using antibodies to these intermediate filament protein subclasses, it is often possible to determine the cellular origin of human tumors. However, it has been learned during the past few years that the biology of intermediate filaments is far more complex than originally supposed. Intermediate filament protein profiles of cells can change during the hyperproliferative-versus-hypoproliferative states, as well as in response to drug exposure. Also, cells can occasionally co-express several types of intermediate filament proteins.[32]

The family of cytokeratins is a most diverse subgroup of intermediate filament proteins, presently comprising more than 20 polypeptides. They can be divided into type I (acidic) and type II (basic) subfamilies on the basis of constitutive amino acids.[33] The cytokeratin proteins are differentially expressed in specific combinations characteristic of particular epithelia. Very often, cytokeratin spectra tend to be preserved in carcinomas.[31,34,35] Cytokeratin filaments are also present in most neuroendocrine cells and neuroendocrine tumors,[36] and in mesothelial cells.[30] Sometimes cytokeratins may also be demonstrated in myogenic tumors such as leiomyosarcomas.[37] Numerous studies have shown that most nonepithelial cells, with a few exceptions (e.g. some muscle and lymphoma cells), lack cytokeratins using immunocytochemical detection assays. Therefore, in immunohisto-pathology, cytokeratin (as recognized by broad spectrum antibodies) is often regarded as being general, relatively specific, and perhaps the best marker of epithelial differentiation.[30,33]

Several studies of the distribution of cytokeratins in normal cells have been published.[32] Cytokeratins of low molecular weight first appear in the early stages of embryogenesis. Later, high molecular weight cytokeratin proteins appear during the development of stratified epithelia. This progressive expression is recapitulated during the maturation of cells of the adult epidermis, where the basal cells express low molecular weight cytokeratins and the maturing keratinocytes express high molecular weight cytokeratins. Nonstratified epithelia, such as gastrointestinal epithelia, contain low molecular weight cytokeratins. More complex epithelia, such as respiratory epithelium, exhibit a more complex pattern of cytokeratin expression. The highest molecular weight subgroups of cytokeratins are typically found only in the upper levels of stratified squamous epithelia. In the liver, the hepatocytes usually give only minimal labeling with cytokeratin antibodies after paraffin processing. The same has been observed in the salivary glands (acinar tissue), where usually only duct cells are labeled.[32]

The expression of cytokeratins is a constant feature of all carcinomas, irrespective of their degree of differentiation. Therefore, cytokeratins are also a useful marker for undifferentiated and anaplastic carcinomas, even for metastasizing carcinoma cells in lymph nodes and bone marrow.[30,31] In such tumors, cytokeratin immunocytochemistry may allow the determination of the origin of the lesion, the distinction of carcinomas from non-epithelial malignant tumors, and, if freshly frozen material is available, the subclassification of epithelial lesions relative to the expression of cytokeratin subgroup profiles. Broad spectrum antibodies to cytokeratins are very useful to distinguish epithelial from lymphoid differentiation (carcinoma vs lymphoma, in combination with antibodies to leucocyte common antigen = LCA = CD-45), epithelial from mesenchymal differentiation (carcinoma vs sarcoma), or carcinoma from melanoma (by application of antibodies to S-100 proteins). Sarcomas are generally not labeled by anti-cytokeratin antibodies. The negative staining, therefore, may give negative evidence for an epithelial origin of any such lesion.[38] However, there are two exceptions to this rule, the synovial sarcoma and the epithelioid sarcoma, both of which may be labeled for cytokeratins. Anti-cytokeratin antibodies may in some cases also help to distinguish mesothelioma from adenocarcinoma. For possible mesotheliomas, the usefulness of the combination of cytokeratin-evaluation with using antibodies to carcinoembryonic antigen (CEA), epithelial membrane antigen (EMA), and Leu-M1-Antibodies (Becton-Dickinson, Mountain View, CA, USA) has been discussed. Leu-M1 has been reported to be lacking in mesotheliomas but present in most adenocarcinomas.[39,40] Mesotheliomas are mostly negative for CEA, while adenocarcinomas are positive for this marker.[41]

In typical epithelial tumors, cytokeratin filaments are usually the only intermediate filament class present. In malignant mesotheliomas and neuroendocrine tumors, vimentin immunoreactivity may be additionally present. This feature is also seen in granulosa cells of the ovary, granulosa cell tumors, and in most renal cell carcinomas. Testing for the presence or absence of cytokeratin-vimentin co-expression is of value for differential diagnosis, as it appears to be restricted to certain carcinoma types.[30] Certain epithelial neuroendocrine tumors also may express neurofilament proteins in addition to cytokeratins, e.g. some Merkel cell tumors of the skin, and some medullary thyroid carcinomas. Co-expression of cytokeratins with glial filaments is very rare and may appear in choroid plexus carcinomas.[30]

A profile of the different proteins of the cytokeratin family in a tumor may give evidence of the origin of the keratins and, in some instances, the origin of epithelial tumors.[33] However, it must be emphasized that most antibodies available today to detect cytokeratin subclasses do not perform well on paraffin sections. Therefore, the reliable assessment of the expression of cytokeratin subgroups is best made in freshly frozen and acetone-postfixed cryostat sections. Some antibodies to cytokeratin subclasses may also work in formalin-fixed paraffin sections. However, false-negative results can be observed, depending on the fixation and sensitivity of the immunostaining method used. Sometimes protease predigestion (trypsin, pepsin, or pronase) is necessary. The antibodies we also have found to perform well in paraffin-embedded tissue are listed in *Table 1*. Of particular value, especially in sections from human skin, are the two monoclonal antibodies *AE1* and *AE3*.[42] These two antibodies can distinguish between epidermal and epithelial origin from skin glands e.g. sweat glands or sebaceous glands. *AE1* mostly binds from low to medium molecular weight cytokeratins and can be used as a general screening antibody, while *AE3* recognizes higher molecular weight (basic) cytokeratins. The two antibodies *Z-622* (rabbit polyclonal antibody, Dako, Carpinteria, CA, USA) and *KL-1* (Dianova, Hamburg, FRG) are broad spectrum anti-cytokeratin antibodies and give, in our hands, positive immunostaining even on overfixed paraffin sections. Such antibodies have also proved very useful in detecting the epithelial part of thymus (Hassal's corpuscles) or of thymomas.

Epithelial Membrane Antigens

Carcinoembryonic antigen (CEA) was the first identified oncofetal antigen.[43] It is an extensively glycosylated, high molecular weight glycoprotein and is found mainly in colonic carcinomas and fetal gut. Subsequently, it was found that the serum levels of immunoreactive CEA may also be increased in individuals without cancer.[44,45] Immunoreactive CEA can be localized in many normal, reactive, and neoplastic tissues other than adenocarcinoma of the colon. However, for monitoring patients with colonic adenocarcinoma, CEA levels in serum accurately indicate residual tumor, but poorly mark local recurrence.[44] Several CEA-like peptides have been characterized, and some of them were called nonspecific crossreacting antigens (NCA).[45] The demonstrated heterogeneity of CEA and CEA-like substances makes it particularly difficult to give general statements on the usefulness of antibodies against these substances. There is a great potential diversity of immunogen preparations used for raising antibodies against CEA. Therefore, many reported localizations of CEA may not be valid because the nature of the molecule localized with poorly characterized polyclonal antibodies has often not been known. The development of monoclonal antibodies against the "true" CEA will probably lead to a more precise characterization of the pattern of cell expression of the various CEA molecules.[45]

Usually, CEA seems to be localized at the membrane of epithelial cells in a variety of tissues. Cells that normally lack CEA include mesothelial cells and normal squamous cells of cervix, esophagus, and skin. Malignant transformation is associated with variable redistribution of CEA, in basolateral membranes and to the cytoplasm of carcinoma cells.[45]

Epithelial Membrane Antigen (EMA) is derived from delipidized tumor milk globule membranes and is a carbohydrate-rich, protein-poor, high molecular weight substance on the surface of epithelial cells. Originally, it was isolated as the most abundant protein fraction from over 35 proteins in human milk fat globules (HMFG). Since then, various EMA-like substances between 180 kD and 420 kD have been extracted.[45] Antibodies raised against different human milk fat globules have been also called *HMFG-1*, *HMFG-2*, or *"carcinoma-associated antigen"* (clone MAM-6).

EMA immunoreactivity is usually well preserved in formalin-fixed paraffin-embedded tissues (Fig.1). If staining problems occur, antigen retrieval using microwave radiation proved to be very helpful.[14] The antibodies available react with carbohydrate groups of the EMA proteins which vary remarkably in their degree and type of glycosylation. The labeling properties of antibodies raised against EMA proteins are most often comparable, however, sometimes differential labeling of tumor cells may be observed. In normal epithelial tissue, e.g. the gastrointestinal epithelium, as well as in some carcinomas, the positive immunostaining is located mainly on the apical or luminal surface of cells. The cytoplasm is also sometimes stained (Fig.1). The intensity of immunostaining varies with the degree and type of differentiation and fixation. Labeling is most intense in the human breast. Cells other than glandular epithelia may contain EMA. Antibodies to EMA may stain plasma cells, some histiocytes, and some lymphomas, and, therefore, should only be applied in combination with the use of cytokeratin antibodies.[46]

Leu-M1, monoclonal antibodies which are mainly used to localize neutrophils, monocytes, histiocytes, some T-lymphocytes, and Reed-Sternberg cells, recognize an epitope also present in epithelial cells, but not in mesothelium. Therefore, such antibodies are of help in the differential diagnosis of mesothelioma and adenocarcinoma.[45]

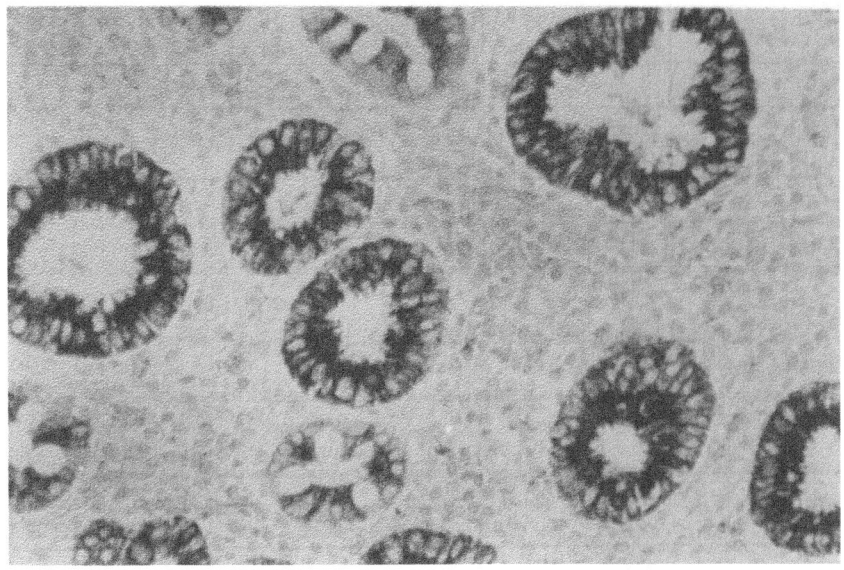

Figure 1. Gastric gland epithelia demonstrated with a monoclonal mouse antibody to epithelial membrane antigen and the IGSS method. Formalin-fixed 4 μm thick paraffin section, counterstained with nuclear fast red. Original magnification: x360.

In addition to the markers mentioned, a variety of more or less organ-specific antigens can be used as markers for the detection of the site of origin. These also include prostatic markers [*prostate-specific antigen (PSA)* and *prostatic acidic phosphatase (PAP)*], markers to differentiate between different tumors of the thyroid [thyroxin, thyroglobulin, calcitonin, calcitonin gene-related peptide (CGRP)], or parathyroid markers (parathyroid hormone (PTH) and PTH-related substances).

Mesenchymal Markers

Intermediate Filament Markers for General Mesenchymal or Myogenic Differentiation. The majority of mesenchymal tumors can be classified by their typical morphological patterns in routine stains. However, in many cases of mesenchymal tumors, immunocytochemistry may provide additional information on the histogenesis or differentiation. Of special value here are mainly the two types of intermediate filament proteins desmin and vimentin, plus various other markers. Some valuable antibodies to mesenchymal markers are summarized in Table 2. For the detection of desmin and vimentin, antigen retrieval methods are of great benefit.[14]

Table 1. Antibodies to tumor markers for epithelial differentiation working on formalin-fixed paraffin sections

Antibody	Donor	Antigen Spectrum	Commercial Source
Z-622	rabbit	broad spectrum keratins	Dako, Carpinteria, CA, USA
KL-1	mouse	several cytokeratins	Dianova, Hamburg, FRG
CAM 5.2	mouse	cytokeratins 8 + 18	Becton-Dickinson, San Jose, CA, USA
AE-1	mouse	low MW cytokeratins	BioGenex, San Ramon, CA, USA
AE-3	mouse	basic type II cytokeratins	BioGenex
E29	mouse	epithelial membrane antigen	Dako
MAM-6	mouse	human milkfat globulins	Sanbio, Uden, NL
A-115	rabbit	carcinoembryonic antigen	Dako

Table 2. Antibodies to tumor markers for mesenchymal differentiation for use with formalin-fixed paraffin sections.

Antibody	Donor	Clone	Antigen or Tissue Spectrum	Commercial Source
M-725	mouse	V9	vimentin	Dako, Carpinteria,CA,USA
7230	mouse	BMA120	endothelial cells	Behring, Marburg, FRG
A-082	rabbit	-	factor-VIII-related antigen	Dako
A-099	rabbit	-	lysozyme, histiocytes	Dako
M-747	mouse	MAC387	macrophages	Dako
M-814	mouse	KP1	CD-68, macrophages	Dako
MONF3001	mouse	n.k.	desmin, muscle	Sanbio, Uden, NL
F-796	mouse	mixed	LCA, Lymphocytes	Dako
PU058UP	rabbit	-	S-100, melanomas, etc.	BioGenex, San Ramon, CA, USA
Z-311	rabbit	-	S-100, melanomas, etc.	Dako
MON7003	mouse	NK1-C3	melanomas	Sambio
MA930	mouse	HMB45	melanomas	Enzo, NY, NY, USA

Desmin is the intermediate filament protein typical of skeletal, visceral and some vascular muscle cells.[47] It was discovered independently by two groups and was initially called "skeleton".[48,49] Desmin has been found to be the only intermediate filament type present in mature muscle cells. The situation is different in vascular smooth muscle, which can express vimentin but not desmin.[47]

A positive reaction with antibodies to desmin is characteristic for rhabdomyosarcoma (*Fig. 2*), rhabdomyoma, leiomyosarcoma and leiomyoma. In routine formalin and paraffin sections, it might be useful to apply several different antibodies (e.g. polyclonal and monoclonal) with different epitopes in order to obtain a positive reaction. In alveolar rhabdomyosarcoma, a co-expression of desmin and vimentin is often found, especially in more undifferentiated tumors. On the other hand, well differentiated rhabdomyosarcomas are characterized exclusively by desmin.

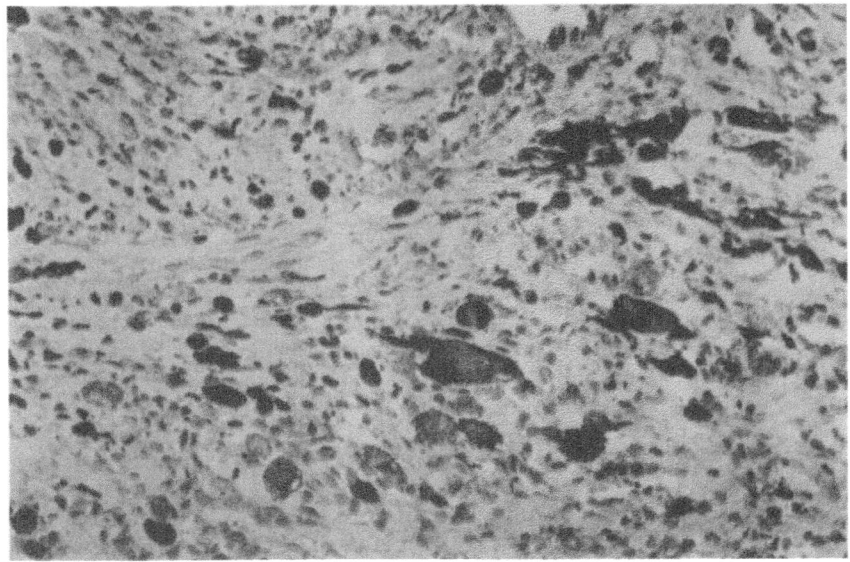

Figure 2. Desmin-immunoreactive structures in a case of rhabdomyosarcoma, detected with monoclonal mouse antibodies and the IGSS method. Formalin-fixed 4 μm thick paraffin section, lightly counterstained with hematoxylin and eosin. Original magnification: x360.

Vimentin filaments are the only intermediate filament type characteristically present in nonmuscular mesenchymal cells. First references to the existence of this protein were given independently by two groups,[50,51] and the protein detected in fibroblasts was later called vimentin.[52] In normal tissue, almost all non-myogenic mesenchymal cells express vimentin. Vimentin-positive cell types include: fibrocytes and fibroblasts, osteocytes and osteoblasts, chondrocytes and chondroblasts.[47] Even cells derived from the neuroectoderm such as Schwann cells or some neuroendocrine cells express vimentin. Also, at least a proportion of lymphatic cells also express vimentin. Vimentin is present in granulosa cells of the ovary, Sertoli cells of the testis, and Langerhans cells and melanocytes of the skin.[47]

Tumors expressing vimentin immunoreactivity include the great majority of non-muscular soft tissue tumors (plus granular cell tumor and glomus tumor), bone tumors, some cases of lymphomas and leukemias, malignant melanomas, schwannomas (neurinomas) and some cases of neuroendocrine tumors. Synovial sarcomas, epithelioid sarcomas and mesotheliomas co-express vimentin and cytokeratins, as also seen in some bone tumors. Most Hodgkin and non-Hodgkin lymphomas are vimentin-positive, and in most of these cases, a heterogeneous staining distribution is seen. Differential diagnosis of tumors expressing vimentin requires the use of additional markers, including endothelial markers, histiocytic markers, and lymphocytic markers.[47]

Mesenchymal Tumors Expressing Cytokeratins. It was surprising to find that mesenchymal tumors exist which express cytokeratins. Earlier, cytokeratins were thought to be restricted to carcinomas,[47] but cytokeratin immunoreactivity was soon detected in synovial sarcomas,[53] in epithelioid sarcomas,[54] in adamantinoma of tibia,[55] and in chordomas.[56] In most of these tumors, a co-expression of cytokeratins and vimentin can be found. This co-existence could be an expression of a transformation of mesenchymal cells to epithelial cells, which is also seen in nephroblastomas.[57]

Myogenic Markers

There are numerous reports on the use of antibodies to actins, myosins and myoglobin in the diagnosis of myosarcomas.[58,59] Unfortunately, most of the antibodies available to actin and myosin also react with isoforms present in the microfilament system of all cells in the organism and, therefore, are of limited value. Regarding the situation with myoglobin antibodies, only a small proportion of rhabdomyosarcomas could be positively stained.[60] Thus, desmin currently appears to be the best available marker for the diagnosis of myosarcoma. However, neither desmin antibodies nor antibodies to other discussed myogenic markers seem to allow the differentiation between rhabdomyosarcomas and leiomyosarcomas because both types express desmin. Specifically stained fibers or cellular buds in a tumor must not always be regarded as intrinsic parts of the tumor; they might well be degenerating or regenerating non-neoplastic muscle fibers. In case of unidirectional arrangement, such fibers or cells should be regarded as remnants of normal muscle.[60]

Endothelial Markers

Endothelial cells and their tumors can be easily visualized using antibodies to vimentin. In addition, antibodies to specific endothelial markers should be used. These include *factor VIII-related antigen* (Van Willebrand factor),[60,61] as well as *BMA-120* and the *endothelins* (Table 2). Factor VIII-related antigen is regarded as a reliable marker for highly differentiated angiogenic tumors. Another marker often recommended for endothelial tumors is the lectin *UEA-1* (Ulex europaeus agglutinin one) which appears to be more sensitive than antibodies to factor VIII-related antigen. However, a positive UEA-1 reaction is also present in other cell types, such as various epithelial cells.[60]

It is often difficult to demonstrate factor VIII-related antigen in routine formalin-paraffin sections and antibodies to this marker also often stain only a subpopulation of normal endothelial cells. We found the antibody *BMA-120* (Behring, Marburg, FRG) to be a more reliable routine marker for endothelial differentiation. This antibody also often gives positive reactions in overfixed tissues,[62] and can be successfully applied to obtain information on the overall vascularization of various tumors and tissues (*Fig.3*).

Endothelins are a newly discovered group of peptides present in endothelial cells. To our knowledge, these substances can presently only be specifically detected in cryostat sections.[63]

Histiocytic Markers

The literature does not provide uniform findings with regard to the reaction range of "histiocytic" markers. Several markers have been discussed including *lysozyme, alpha-1-antitrypsin, MAC-387,* and *CD-68.* Lysozyme (muramidase) among mesenchymal cells, is generally restricted to the monocyte/macrophage/epithelioid cell system. It has been reported as a marker for neoplastic histiocytes of malignant histiocytoma.[64] Later reports have shown that only a very small proportion of fibrohistiocytic lesions express lysozyme[65] which appears to be the most reliable conventional marker for "true" histiocytic lymphomas. It has a low sensitivity for histiocytic sarcoma, but in the case of positive immunoreactivity, it may aid discrimination between malignant fibrous histiocytoma and other tumors of comparable morphology. Tumor cells of histiocytosis-X do not contain lysozyme[60] but can be identified using antibodies to S-100 proteins and to CD-43.

The protease inhibitors *alpha-1-Antitrypsin* (AAT) and *alpha-1-antichymotrypsin* (ACT) have also been discussed as markers for histiocytic origin of tumor cells. However, both substances turned out to be of minor value in the differential diagnosis of soft tissue tumors, as various sharply contrasting findings have been reported in the literature. Otto *et al.*[60] have observed AAT in various kinds of soft tissue tumors including rhabdomyosarcoma, liposarcoma, hemangiopericytoma, endothelioma and even schwannoma. Therefore, they doubt whether AAT is likely to be a helpful tool for a specific differential diagnostic problem. A similar situation can be found with antibodies to ACT.

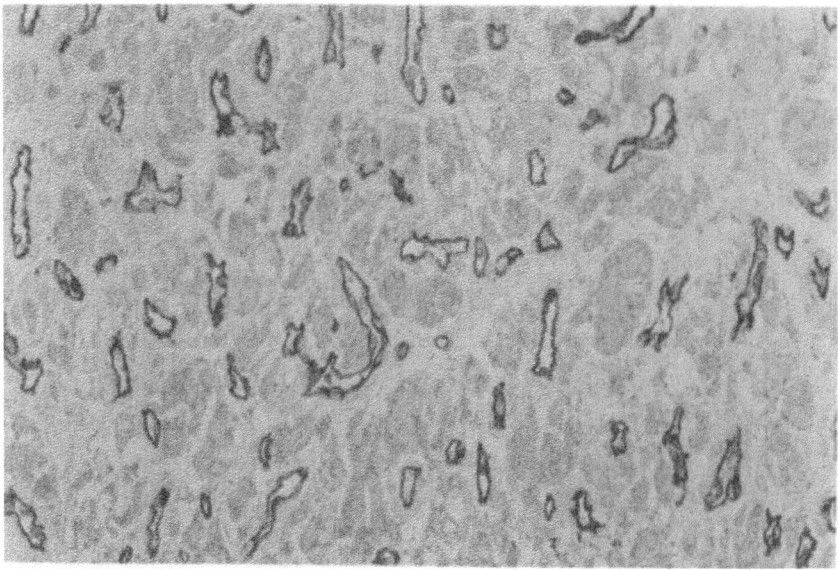

Figure 3. BMA-120 immunoreactivity in endothelial cells shows the dense vascularization in a case of a glomus tumor, immunostained with mouse monoclonal antibodies and the IGSS method. Formalin-fixed 4 μm thick paraffin section, counterstained with nuclear fast red. Original magnification: x360.

More potential markers for a subpopulation of macrophages seem to be *CD-68* (Dako, Carpinteria, CA, USA), *MAC-387* (Dako), and *HAM-56* (Enzo, New York, NY, USA). *CD-68* (clone KP1) is a monoclonal mouse antibody to a macrophage-specific 110 kD glycoprotein, presumably associated with lysosomal granules.[66] It stains macrophages in a wide variety of human tissues and is also positive in blood monocytes. CD-68 antibodies show a variable pattern of staining that can be correlated with the histopathological process and presumably with the state of macrophage activation. Using CD-68 antibodies in a study on the central nervous system, ramified microglia in normal white matter were well stained, and reactive astrocytes were lightly positive. It was noticed that macrophages showed less staining with increasing distance from the blood vessel in infarcted areas.[67] The mouse monoclonal antibody *MAC-387* recognizes a cytoplasmic antigen that is reportedly expressed in monocytes and some subtypes of tissue histiocytes from normal and neoplastic tissues.[68] *HAM-56* is a mouse monoclonal antibody to a macrophage-specific antigen.[67,69] This antibody seems to be very useful for intensely and specifically delineating macrophages of various diseases in routinely processed tissue sections.[67]

Melanoma Markers

The diagnosis of metastases of malignant melanoma is a difficult differential diagnostic problem, especially when only small amounts or no melanin are present. Melanomas contain a variety of immunocytochemical markers, including S-100 proteins, HMV-45, NKI-C1, vimentin and sometimes also neuronspecific enolase.[70,71]

The most well-known melanoma-marker *S-100*, a group of Ca-binding proteins, was its first form discovered in brain astrocytes in 1965.[72] Several S-100 isoforms can be distinguished (S-100 a_o, S-100a and S-100b) that are characterized by about the same molecular weight (21 kD) and similar pH optimum (about 4.3). They differ in their subunit composition.[73] Antibodies specific to subunits of S-100 proteins that can differentiate between different cell types (mainly glial and neuronal structures) in cryostat sections have been discussed at recent conferences.

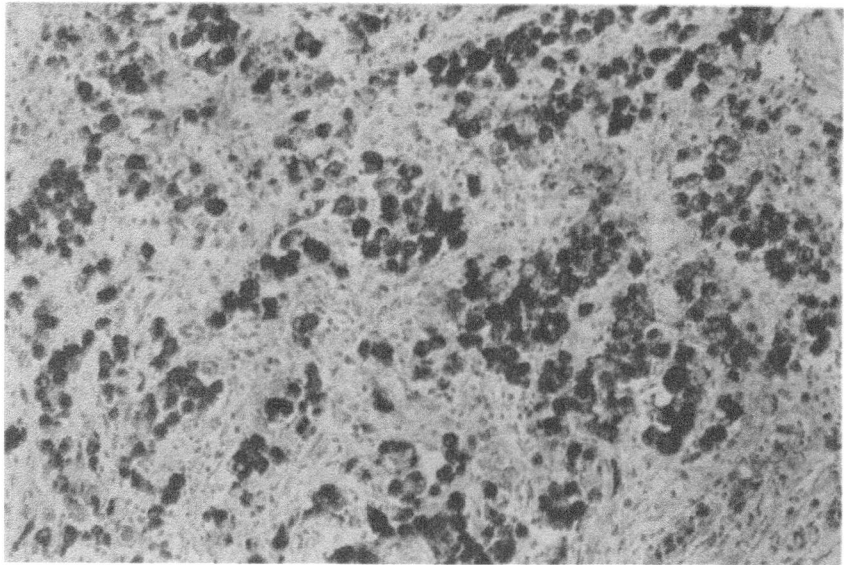

Figure 4. Tumor cells of a malignant melanoblastoma immunostained with rabbit polyclonal antibodies to S-100 proteins and the IGSS method. Formalin-fixed 4 μm thick paraffin section, lightly counterstained with hematoxylin and eosin. Original magnification: x225.

Most antibodies to S-100 working on paraffin material available today derive from crude preparations of S-100 protein subunit mixtures and, therefore, cannot distinguish these cell types. Reactions with such antibodies were originally reported to appear in brain astrocytes, oligodendrocytes, and ependymal cells. S-100 immunoreactivity was then observed in Schwann cells and supportive cells of the peripheral and autonomic nervous system.[74] In addition, S-100-positivity was found in other cell types,[60] including chondrocytes, fat cells and some of their tumors, interdigitating reticulum cells, Langerhans cells of the epidermis, melanocytes and their tumors (*Fig. 4*),[70] and in satellite cells of adrenal medulla and pheochromocytomas.[75] S-100 can also be detected in cells of histiocytosis X.[76] According to various reports, S-100 can be regarded as a reliable marker in the differential diagnosis between malignant melanoma and carcinoma.[60,70] However, as the number of S-100-positive tumor types is still increasing, a highly active search for more specific melanoma markers has started.

Amongst the most promising new markers for melanoma are the monoclonal antibodies *NKI-C3* (Sanbio, Uden, NL) and *HMB-45* (BioGenex, San Ramon, CA, USA). *NKI-C3* recognizes a glycoprotein located mainly at the inner side of membranes of cytoplasmic vesicles in melanoma cells. We found that the antibody gives strong positive staining in most cases, even of overfixed melanomas. However, it must be used in combination with other markers, as it is also demonstrable in some other tumors and tissues, such as some carcinomas and neuroendocrine tumors, mast cells, histiocytes and cells with secretory functions such as salivary glands, bronchial glands, sweat glands and pancreas.

The antibody *HMB-45* recognizes cytoplasmic melanoma antigens. It reacts with tumor cells and most, but not all, melanomas and with junctional nevus cells, but not with normal melanocytes or intradermal nevus cells. It appears to be unreactive with other neoplasms, including other sarcomas, carcinomas, and glial tumors.

Lymphoid Markers

Enormous progress has been made during the last years regarding our ability to diagnose and classify non-Hodgkin lymphomas and Hodgkin's disease. Immunocyto-chemical analysis in this respect has proven invaluable, as antigen expression and cellular morphology can be evaluated simultaneously *in situ*. Today, a variety of antibodies recognizing formalin-resistant epitopes is available, allowing the effective immunophenotyping for many cases of malignant lymphoma. Recently, very exciting reviews on the laboratory diagnosis of malignant lymphoma were given in the literature.[77,78] In the following paragraphs, only some antibodies which proved to work on formalin-fixed and paraffin-embedded material are briefly reviewed. When using these antibodies, we recommend the use of the immunogold-silver staining (IGSS) method in order to obtain a highly sensitive and specific labeling together with H&E counterstaining for an improved assessment of morphology.[23-27] For many antibodies to lymphoma markers, antigen retrieval methods globally improve immunostaining performance.

Probably the single most useful application is the use of panleucocyte antibodies, the so-called leucocyte common antigen (LCA, CD-45),[79] permitting us to distinguish lymphomas from other anaplastic malignancies. Immunophenotyping with B and T lymphocyte specific antibodies can establish lineage, aiding a better characterization of lymphomas. Though a more complete immunophenotypic analysis of lymphoid neoplasms can be performed on freshly frozen tissue, paraffin section immunocytochemistry has several advantages. The greatest advantage is the preservation of both tissue architecture and cytomorphology, allowing a more precise identification of the neoplastic cells.[78] Using double immunocytochemical staining techniques (the best combination seems to be the IGSS method plus immunoperoxidase or APAAP methods),[23-27] it is possible to visualize the B- and T-cell subsets simultaneously in one paraffin section.

Some commercially available antibodies which have proved useful for lymphoma diagnosis are listed in Table 3. Reliable determination of B cell clonality using antibodies to heavy and light chain immunoglobulins is best performed in frozen sections. However, some authors have reported the successful demonstration of monotypic immunoglobulin in 80% of B-cell non-Hodgkin lymphomas.[80] LCA, a cocktail of two monoclonal antibodies, will solve most diagnostic dilemmas where the differential diagnosis includes lymphoma versus a nonlymphoid neoplasm.[78,79] LCA can also be demonstrated in granulocytic sarcomas and, therefore, should be used in a panel that includes antibodies to T- and B-cell subsets.[80] Some, especially large cell lymphomas, cannot be labeled with LCA antibodies.

Table 3. Antibodies to lymphoma markers for use in formalin-fixed paraffin sections.

Antibody	Donor	Clone	Antigen or Tissue Spectrum	Commercial Source
F-796	mouse	mixed	CD-45, lymphocytes	Dako, Carpinteria, CA, USA
M-742	mouse	UCHL1	CD-45RO,T- cells	Dako
812 255	mouse	MT1	CD-43, T-Cells	Biotest, Denville, NJ, USA
812 260	mouse	MT2	T-cells, B-cells	Biotest
A-452	rabbit	-	CD-3, T-cells	Dako
M-755	mouse	L-26	CD-20, B-cells	Dako
M-754	mouse	4KB5	CD-45R, B-cells	Dako
812 260	mouse	MB2	predominantly B-cells	Biotest
812 225	mouse	LN1	Follicle center B-cells	Biotest
812 230	mouse	LN2	follicle center and mantle zone B-cells	Biotest
M-751	mouse	Ber-H2	Ki-1, CD-30, RS-cells	Dako
CKIGS	rabbit	-	IgG	DPC, Los Angeles, CA, USA
CKIMS	rabbit	-	IgM	DPC
CKIAS	rabbit	-	IgA	DPC
CKLAS	rabbit	-	lambda light chains	DPC
CKKAS	rabbit	-	kappa light chains	DPC

DPC = Diagnostic Products Corporation

The first B-cell associated antibodies working on paraffin sections were published in 1984[81] and called "LN-1" and "LN-2". They react with a high percentage of B-cell lymphomas, but also with other hematopoietic and non-hematopoietic cells. A few years later, first reports were written on the usefulness of "L26", later classified as CD-20,[78,82] an antibody that turned out to be really B-cell specific and highly useful. Other B-cell associated antibodies for paraffin sections are MB1 and MB2, but these appear to be much less specific than L26.

"*UCHL1*" was the first antibody reported to specifically label a subset of T-lymphocytes in paraffin sections,[83] reacting with a restricted epitope of LCA (CD-45RB) and also present in granulocytes. Other antibodies gaining acceptance as T-cell markers in routine sections are "*L-60*" (Leu-22) and"*MT1*" reacting with CD-43, an antigen preferentially, but not exclusively, expressed on T-lymphocytes: some B-cells, myeloid cells, histiocytes, and granulocytic sarcoma also express CD-43 immunoreactivity. Therefore, co-expression of CD-43 and CD-20 may be used as an indicator for a B-cell proliferation.[78] Another antibody, "*MT2*", is expressed in interfollicular T-cells and mantle-zone B-cells in normal and reactive lymph nodes, but not in germinal center cells.[84] On the other hand, in most follicular lymphomas, the neoplastic germinal centers express MT2-positivity, a feature that may be used to distinguish follicular lymphoma from reactive follicular hyperplasia.[85]

Positive labeling with polyclonal antibodies to CD-3 provides strong evidence of T-cell histogenesis.[86] Immunocytochemical studies of non-Hodgkin lymphomas should include more than one T-cell antibody in the diagnostic panel. The use of a combination of UCHL-1- and CD-3 antibodies could identify more than 98% of T-cell non-Hodgkin lymphomas.[78] Anaplastic large cell Ki-1 positive lymphoma is a newly recognized clinicopathologic entity which, by definition, can be positively labeled with antibodies to the Hodgkin-s disease-related antigen "Ki-1" (CD30). However, this antigen is also expressed in other neoplasms, such as Hodgkin's disease and lymphomatoid papulosis.[78] Ki-1 antigens, can now be immunostained in paraffin sections using the "Ber-H2" antibody.[78] For best results with this monoclonal antibody, it is important that enzyme pretreatment and/or microwave antigen retrieval are carried out. In our experience, this is better achieved with pepsin digestion than with pronase or - the least effective - trypsin. Interestingly, the Ber-H2 antibody also labels Purkinje cells of the cerebellum.

In Hodgkin's disease, Reed-Sternberg (RS) cells and variants often comprise a very small percentage of cells. Several antigens present in RS cells can now be detected in paraffin sections. These mainly include CD-15 (T9 and Leu-M1, Fig.5) and CD-30 (Ki-1, Ber-H2).[78] In some cases, also CD-45 (LCA), CD-20 (L-26), CD75 (LN1) and/or CD74 (LN2) can be detected in RS- or Hodgkin's cells. Using panels of such antibodies, different categories of Hodgkin's disease may be distinguished (personal communication, C. Sandström, Uppsala, Sweden).[78]

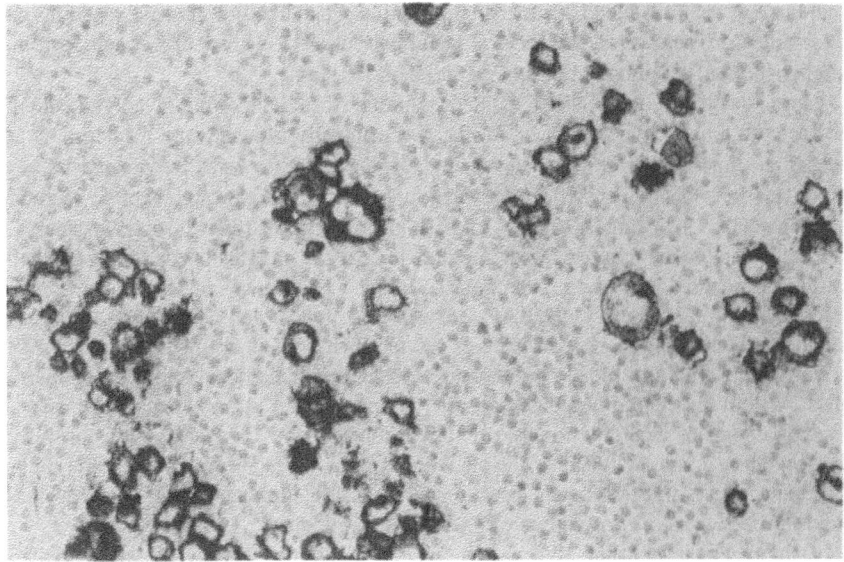

Figure 5. Leu-M1-immunoreactive Reed-Sternberg and Hodgkin cells in a case of Hodgkin-s disease, detected with mouse monoclonal antibodies and the IGSS method. Formalin-fixed 4 μm thick paraffin section, lightly counterstained with hematoxylin and eosin. Original magnification: x360.

Markers for the Nervous, Glial and Neuroendocrine System

Histopathology of the nervous-, glial- and diffuse neuroendocrine system (DNES) has been greatly advanced by the introduction of immunocytochemical methods. The assessment of antigenic marker proteins and peptides has generally led to a higher level of diagnostic accuracy. Although the spectrum of available antibodies with proven diagnostic usefulness in neuropathology is still limited, help in various cases can be achieved. This is true for the distinction of gliomas and embryonal central nervous system tumors from metastatic lesions of epithelial and mesenchymal origin, as well as from malignant lymphomas. In the case tumors of the DNES (carcinoids and other neuroendocrine tumors), immunocytochemistry may be the only way to optimally characterize these neoplasms.

The most well-known markers for the central and peripheral nervous system include neurofilament proteins (NFP), protein gene-product (PGP) 9.5, neuronspecific enolase (NSE) and neuronal cell adhesion molecule (NCAM).[74] Also sometimes, synapsin, GAP-43, synaptophysin or microtubule-associated protein I can be demonstrated. Central glial markers include glial fibrillary acidic protein (GFAP) and S-100 protein. In peripheral glia (Schwann cells), S-100, and vinertria can be detected.[74,87]

Neurofilament proteins (NFP) are the intermediate filament protein type of the neuronal cytoskeleton. They are composed of a triplet of polypeptides, display species-specific variation of the molecular weight (approximately 68, 150 and 200 kD) and can be regarded as specific markers for neurons and nerve fibers in the central and peripheral nervous system.[74,88] The three subunits are differentially expressed during brain development and have a differential distribution within the neuronal cell. However, all three subunits are found in the axon and in the cell body. NFP have a hypervariable amino-terminal head and carboxy-terminal tail on both sides of a 40 kD alpha-helix. The tail ends contain a domain with a high content of acidic residues. This "b" domain represents the binding site for the Bodian silver stain, which specifically stains neurofilament proteins.[87,89] Polyclonal and monoclonal antibodies recognizing one, two, or all three of the NFP subunits are available. With such antibodies, positive reactions can be obtained in peripheral and central neuronal tumors, e.g. in pheochromocytoma, ganglioneuroma, ganglioneuroblastoma, and neuroblastoma.[75,87]

Neuron-specific enolase (NSE), an acidic soluble isoenzyme of the glycolytic enzyme enolase with a molecular weight of 78 kD, was originally isolated from rat brain.[90,91] As a set of 5 dimeric isoenzymes can be distinguished. The γγ-form is reported to be a unique marker for the specific visualization of both neurons and peptide-secreting endocrine cells.[74,92] Within the CNS, γ-enolase is almost exclusively localized in neurons, whereas α-enolase expression is found in glial, meningeal, endothelial, and Schwann cells and has been called a non-neuronal enolase (NNE).[92] High levels of NSE are present in the adrenal and the pituitary gland, and also in a variety of non-neuronal and non-neuroendocrine cell populations.[93] In spite of the fact that NSE is not neuron-specific, it is currently employed as a broad-range marker for neoplastic neuronal and neuroendocrine cells.[87] Within the central nervous system, it is mainly used as an indicator of neuronal differentiation in neuroblastomas, retinoblastomas, and a variety of primitive neuroectodermal tumors including the medulloblastoma. Special care in the interpretation of staining results must be taken, because positive NSE immunostaining has also been found in reactive astrocytes and in various types of gliomas.[87]

Synaptophysin is an acidic glycoprotein with a subunit molecular weight of 38 kD first isolated from rat and bovine brain.[94,95] It has been identified as an integral calcium--binding membrane protein of presynaptic vesicles in neurons and similar vesicles in neuroendocrine cells.[95,96] The original monoclonal antibody "SY 38" recognizes an epitope of synaptophysin, also preserved in paraffin material.[95,97] However, commercially available SY 38 antibodies are often prediluted, a fact which sometimes makes the detection in paraffin sections difficult. With sensitive methods and antigen retrieval, it is possible to detect synaptophysin in neurons, the adrenal medulla, and in certain cells of the DNES.[97] Differentiated and undifferentiated neuronal tumors can be immunostained, i.e. neuroblastoma, ganglioneuroblastoma, glial-hamartomas and primitive neuroectodermal tumors, including medulloblastoma. Synaptophasin is further expressed in neural and epithelial types of neuroendocrine tumors.[98]

Protein gene-product 9.5 (PGP 9.5) is another relatively new marker for nervous structures, neuronal and neuroendocrine tumors.[99-101] This soluble cytoplasmic protein which is different from NSE, has been isolated from the brain and has a molecular weight of 27 kD. It can be used as a reliable marker for the delineation of the peripheral nervous system and is present in neuronal and some neuroendocrine structures and tumors. Within the DNES, it is a good marker for C-cells and medullary carcinoma of the thyroid, and for Sertoli and Leydig cells and their tumors (Hacker, unpublished). It can also be reliably demonstrated in pheochromocytomas and extra-adrenal paragangliomas.[75,98]

Glial fibrillary acidic protein is the major polypeptide of glial filaments and has a molecular weight of 45 kD. It is an intermediate filament protein originally found in astrocytes of the central nervous system.[102] Subsequently, it was shown that a GFAP immunoreactive substance is also present in a subpopulation of peripheral glia.[74] GFAP has been extensively investigated and has turned out to be the most useful marker for normal and diseased astroglial cells. Co-expression of GFAP and vimentin is not uncommon. In addition, GFAP is co-expressed with S-100 proteins in central glia. In several neoplasms, however, immunoreactivities to these marker proteins dissociate, which may provide a valuable diagnostic sign.[87,103] In our experience it has turned out valuable to always use several antibodies (polyclonal and mixed monoclonal) in parallel when immunocyto-chemically typing brain tumors. With a higher degree of undifferentiation, GFAP or some epitopes for antibodies to GFAP may be lost.

The calcium-binding S-100 proteins were originally thought to be nervous-system specific, but have later on been detected in various other cell types and their neoplastic counterparts (see above). Within the normal central nervous system, it is present in astrocytes and oligodendrocytes, ependymal cells and for some neurons. In peripheral glia, Schwann cells are labelled by S-100 antibodies.[74,87] Within CNS tumors, S-100 is expressed in astrocytomas, ependymomas, glioblastomas, glial hamartomas, and the adamantinous areas of craniopharyngiomas.[87]

Within the spectrum of soft tissue tumors, S-100 immunoreactivity has been detected in benign tumors of the nerve sheath, including neurinoma or schwannoma, neurofibroma, nerve sheath myxoma and traumatic neuroma. S-100 is only expressed in a small proportion of malignant schwannomas and neurogenic sarcomas. In neurogenic sarcomas associated with von Recklinghausen's disease and malignant nerve sheath tumors originating from a nerve trunk, the great majority of tumors express S-100. However, in neurogenic sarcomas which were exclusively diagnosed on histological grounds, the number of tumor cells containing S-100 was markedly reduced.[60,104]

To achieve optimal typing of neuroendocrine tumors, it is crucial to apply a panel of various antibodies to regulatory peptides and amines in addition to the classical histochemical silver staining methods (see also the chapter by Grimelius in this book). Many peptides can now be routinely immunostained in optimally processed paraffin sections [vasoactive intestinal polypeptide (VIP), neuropeptide tyrosine (NPY), calcitonin gene-related peptide (CGRP), calcitonin, somatostatin, pancreatic polypeptide, helodermin, helospectin, insulin and others] using sensitive detection systems such as the IGSS method.[23-27]

With peptide antibodies, it should be easy to differentiate between different entities of neuroendocrine tumors. Antibodies to human chorionic gonadotrophins (HCG) and to proliferation markers (see below) may help to more exactly define the malignant potential of neuroendocrine tumors. See also the chapter by Wilander in this book.[71,105,106]

Proliferation-Associated Markers

During the last years, several antibodies have been discovered that can be reliably applied to detect proliferating nuclei and therefore permit an estimation of the proliferation rate of a given tumor. The most well known monoclonal antibody is "*Ki-67*", reacting with all nuclei of cells in the active parts of the cell cycle, i.e. G_1, S, G_2, and M phase, but not in the G_0 phase.[107] With the help of Ki-67, a rapid determination of the growth fraction of a given human cell subset has become possible. Tumors of high malignancy mostly show a higher number of Ki-67-labelled tumor cell nuclei, whereas low malignant or even benign tumors show few or hardly any labelled nuclei. A comparison of the mean values of Ki-67-positive tumor cells with the histological grade of malignancies showed a highly significant correlation between the two variables.[108] In the beginning, Ki-67 detection was restricted to cryostat sections. However, new techniques and a new antibody now make it possible to detect Ki-67 antigen even in paraffin-embedded material. This can be achieved by antigen retrieval in a microwave oven and by the use of the new antibody "*MIB-1*" produced by Dianova (Hamburg, FRG)(*Fig. 6*).[14] In our experience, the new antibody also gives a more reliable detection of proliferating cells in cryostat sections than the original Ki-67 antibody.

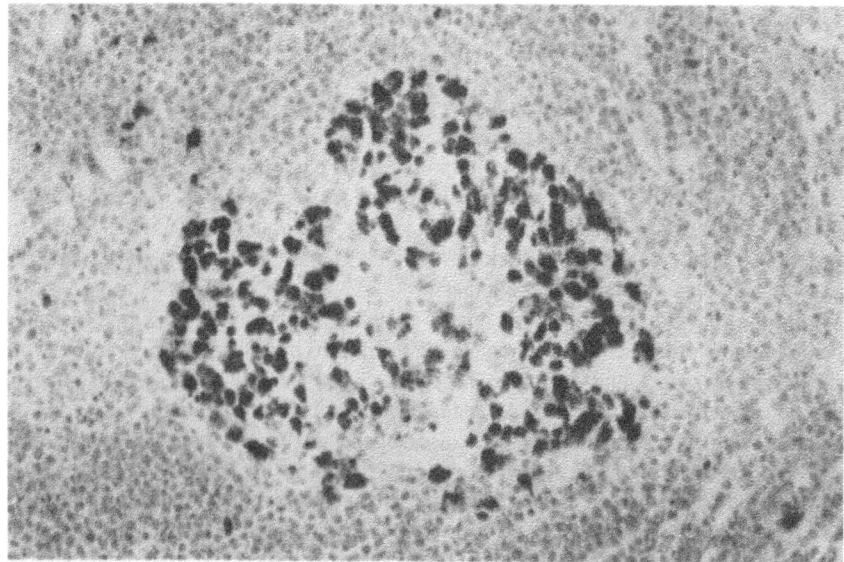

Figure 6. The high proliferation rate in a lymph node follicle is detected with MIB-1 monoclonal mouse antibodies to Ki-67 and the ABC-peroxidase method using microwave antigen retrieval in a formalin--fixed 4 μm thick paraffin section, lightly counterstained with hematoxylin. Original magnification: x360.

Another marker which may be useful for the estimation of the tumor growth fraction is the "*proliferation cell nuclear antigen*"(PCNA),[109,110] PCNA offers an advantage over Ki-67 in that it can be easily detected, even in overfixed paraffin sections, without the need for a special treatment (*Fig. 7*). Microwave antigen retrieval even further improved the staining performance. In some tumor types, PCNA positivity correlates with mitotic activity and tumor grade.[109] In non-Hodgkin lymphomas, a linear correlation between Ki-67 and PCNA immunostaining was demonstrated. However, in some other forms of neoplasia including breast and gastric cancer, the simple relation between PCNA expression and cell proliferation is lost. We found that in breast cancer, usually more nuclei are labelled with PCNA antibodies that with Ki-67 or MIB-7- antibodies (up to

102

three times more nuclei) (Hacker, unpublished). PCNA should be used in parallel to Ki-67-immunostaining, therefore, to give an idea of the proliferation rate in cases where Ki-67 cannot be detected due to overfixation. However, which phases of the nucleic cycle are labelled by PCNA-antibodies and if one/some of these phases may be over represented with this staining remains to be established in reference cell culture preparations .

KiS1 is a new antibody which showed promising results for the assessment of the proliferation rate.[111] However, attempts at proving the usefulness of this antibody to aid prognostic information are still going on. The antibody has also been used on formalin-fixed paraffin-embedded material and gave a significant correlation with the disease-free interval, overall survival, and post-relapse survival in breast cancer.[111]

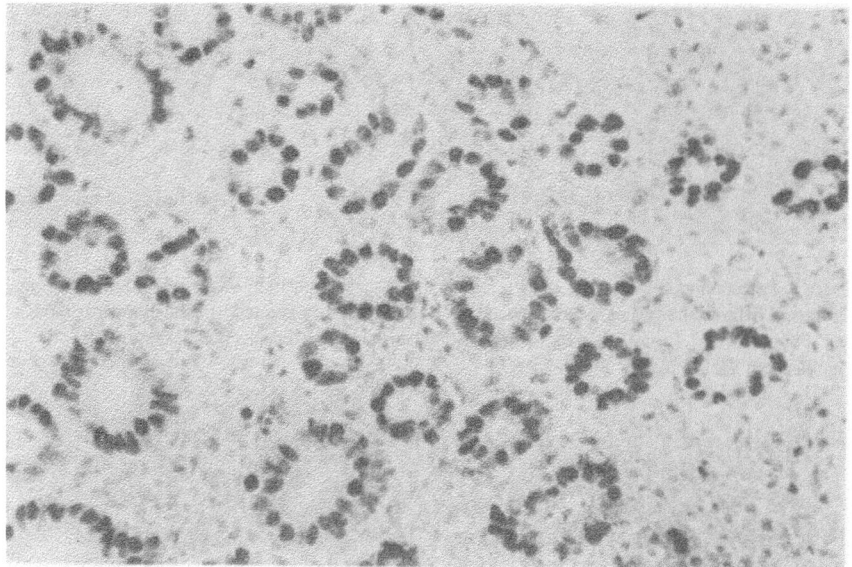

Figure 7. Presumably proliferating nuclei in gastric glands demonstrated with monoclonal antibodies to proliferating cell nuclear antigen (PCNA) and the IGSS method using ultrasmall gold particles (0.8 nm in diameter). Formalin-fixed 4 μm thick paraffin section, lightly counterstained with hematoxylin and eosin. Original magnification: x360.

Viral Tumor Markers

For the detection of viral antigens, several antibodies are available. In our hands, most of these antibodies appear to be unsensitive and sometimes also unspecific. We found it much more reliable to use *in situ* hybridization or even *in situ* polymerase chain reaction (PCR *in situ* hybridization) methods (see also the chapters by Terenghi and by Zehbe in this book). Currently, it is widely believed that viruses are etiological candidates for various human malignancies. Neoplasms under discussion include T-cell leukemias, Burkitt-lymphomas, nasopharyngeal carcinomas and papillomas, squamous cell carcinomas at different mucocutaneous sites, and liver carcinomas.[112] Viruses considered to influence, if not cause tumor genesis, include Epstein-Barr virus [EBV, associated with various types of lymphomas, Fig.8)], herpes simplex virus (HSV, associated with some genital cancers), hepatitis B virus (HBV, associated with hepatocellular carcinoma) and human papillomavirus [HPV, mainly associated with condylomas (Fig.9)] and papillomas, as well as with cervical cancer). All of these DNA virus types and subtypes can be easily detected using simple protocols of DNA *in situ* hybridization.

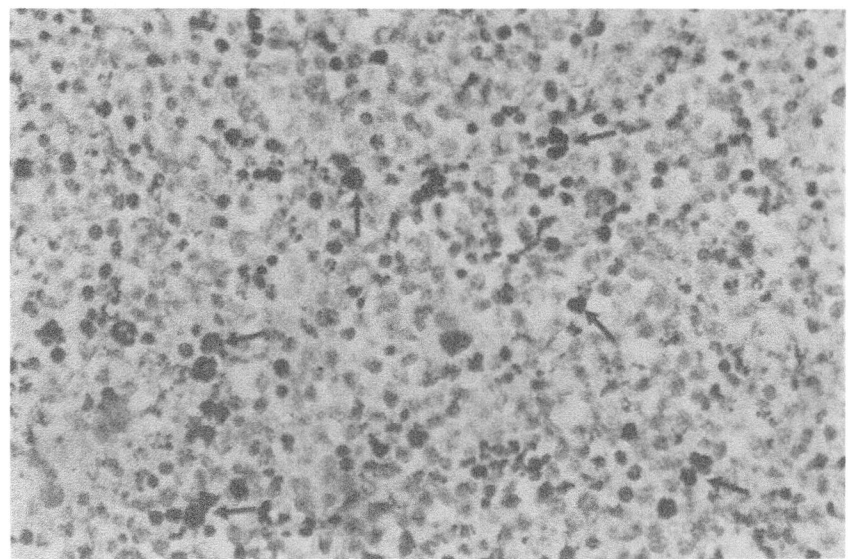

Figure 8. Detection of Epstein-Barr-virus (EBV) in a cell suspension from a case of Burkitt lymphoma with in situ hybridization using a biotinylated cDNA probe and a subsequent streptavidin-peroxidase layer. Formalin-fixed 4 μm thick paraffin section, counterstained with hematoxylin. Original magnification: x450.

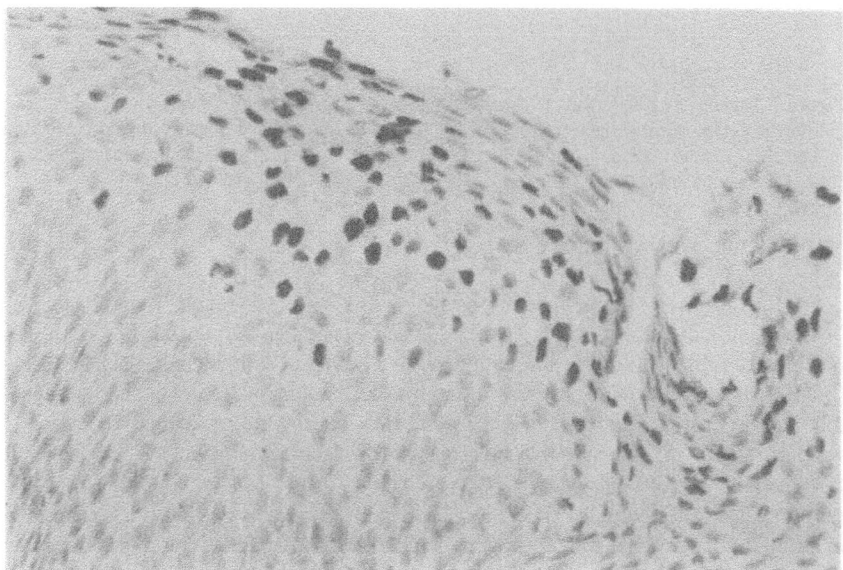

Figure 9. Detection of human papillomavirus types 6 and/or 11 in a case of genitoanal condylomata acuminata with *in situ* hybridization using a biotinylated cDNA probe and a subsequent direct IGSS method with 1 nm gold-labeled anti-biotin antibodies. Formalin-fixed 4 μm thick paraffin section, counterstained with hematoxylin. Original magnification: x360.

Such methods may even be applied in routinely formalin-fixed paraffin-embedded tissues and cytological preparations as long as enough virus DNA copies per cell are available. (The average detection limit with sensitive detection systems is currently about 10-50 DNA virus molecule copies; Zehbe, personal communication). In cases where only a very few virus copies are available, *in situ* PCR methods should be applied.[113]

APPLICATION TO BREAST CANCER

One of the most promising applications of immunocytochemistry is in its use in breast cancer. The high number of investigations in this field is also caused by the fact that this group of neoplasms constitutes a major cause of morbidity and mortality among women in most of the world.

Several prognostic factors have been used to identify those patients at greatest risk for recurrence of their disease. These include size, grade, histologic type, stage of the tumor, presence or absence of metastases, DNA-ploidy, the expression of oncogene products, proliferation markers, estrogen receptors, and progesterone receptors. A most useful and fascinating review on the techniques and markers used has recently been given by Elias.[5]

Classification Markers

In most cases, classification into several types of breast cancers is easily possible using histological criteria. In doubtful cases, antibodies to the above mentioned epithelial and other tumor markers can help to verify epithelial differentiation. In most instances, it is reasonable only to immunostain for cytokeratins using broad spectrum antibodies. It should be kept in mind that in about 30% of breast cancer tumor cells, the mesenchymal intermediate filament protein vimentin can also be expressed.[114] Often, NSE can be detected immunocytochemically, not always reflecting neuroendocrine differentiation. This can be clarified using the Grimelius argyrophilic silver stain, as well as with antibodies to chromogranin A in immunocytochemistry. Using these techniques, up to 5% of human breast cancers can be identified as possibly real neuroendocrine tumors. In such cases, we have also identified regulatory peptides and other hormone immunoreactivities including metenkephalin, calcitonin gene-related peptide (CGRP), vasoactive intestinal polypeptide (VIP), prepro-VIP,[111-122] galanin, bombesin-flanking peptide, neuropeptide tyrosine (NPY) and human chorionic gonadotropin (HCG) by using the IGSS method (Hacker and Graf, unpublished).

Tumor Progression Markers in Breast Cancer

Tumor progression markers in breast cancer include the well-known proliferation markers (Ki-67, MIB-1, PCNA) as well as substances discussed as invasion and metastasis markers [cathepsin D, collagenase IV, laminin-receptor, epidermal growth factor (EGF) receptor]. Concerning the proliferation markers Ki-67 and PCNA, we distinguish between negative to low grade positive (0-10% of tumor cell nuclei are labeled), medium grade positive (10-20% of the tumor cell nuclei are labeled) and high grade positive cases (more than 20% of the tumor cell nuclei are labeled). In a series of about 400 cases of breast cancers (acetone-postfixed cryosections), we found high grade positivity for Ki-67 in about 32 % of the cases, a medium grade number of positive nuclei in 11%, and a low number of Ki-67-positive nuclei in about 44% of our cases. About 13% of our cases could not be immunostained. However, for this examination the "old" original Ki-67 antibody (Dako) was used. It remains to be elucidated if the situation is the same with the new MIB-1 antibody to Ki-67 antigen (Dianova). In the first experiments, we found a highly superior and stronger labeling with the new MIB-1 antibody. In our opinion, therefore, it can be predicted that the number of negative cases could be drastically reduced with this antibody. With respect to PCNA, in most cases, we found a higher number of PCNA-labeled nuclei than with Ki-67 or MIB-1-antibodies. However, a few cases stained for PCNA and were MIB-1-negative. On the other hand, there were also cases that were positively stained with MIB-1 and negative with PCNA antibodies.

The usefulness of immunostaining for other markers mentioned above (cathepsin D, EGF-receptor etc.) is still a matter of controversy.[115,116] In about 150 cases studied in our laboratory so far, very low correlations, if any, of EGF-R and cathepsin D with grade, receptor state and proliferation rate were found.

Markers for Therapy of Breast Cancer

The clinical usefulness of steroid hormone receptor determination in breast cancer is well documented in literature. The estrogen receptor (ER) state is an important therapeutical indicator. The prognostic value of the ER is, amongst other prognostic factors, settled at the third place, behind the axillary lymph nodal state and the histological tumor grading according to Bloom and Richardson. The value of the progesterone receptor (PgR) status as a prognostic parameter is controversial. It does seem to have a great value concerning the treatment with endocrine therapy. The American National Institute of Health (NIH, Bethesda, MD) recommends determining the ER- and PgR state in all patients with primary breast cancer. ER-positive, and especially ER- plus PgR-positive patients, should be treated with anti-estrogen therapy (e.g. tamoxiphen), which blocks the estrogen binding site at the tumor cell ER and, therefore, is likely to block tumor cell proliferation.

For many years, ER and PgR detection in breast cancer had been carried out using biochemical methods (e.g. dextran coated charcoal assay, DCC). The direct *in situ* determination of the ER- and PgR status offers a number of advantages over the DCC method. Using immunocytochemistry, the receptor proteins are detected within the tumor cell nuclei. This detection method, therefore, is independent from the relation between tumor-, connective- and fat tissue. It is also independent from the presence of tumor necroses. Positive reactions within tumor cells can be objectively distinguished from reactions in normal ductules or lobules or hyperplastic epithelial cells. It is also possible to use the immunocytochemical method within very small tissue samples, e.g. fine needle punctations. In our hands, the best method to determine the ER- and PgR state is by using the ER-ICA (estrogen receptor immunocytochemical assay) and the PgR-ICA (progesterone receptor immunocytochemical assay) kits in cryostat sections from the tumor (Abbott, Chicago, IL, USA). We have carefully tested other alternative commercially available antibodies and methods. However, none of them were proven as reliable or give the reproducible results that the Abbott kits do. Also, the ER-ICA kit is the only one acknowledged and recommended by the American Food and Drug Association as the approved diagnostic test.

We use these Abbott kits in a slightly modified version: unfixed tumor tissue is freshly frozen on a cryostat. Frozen sections are air dried for one hour at room temperature and then postfixed in Stefanini's-Zamboni's fixative[7] for 10 minutes followed by washing the phosphate-buffered saline (PBS) pH 7.2. Then, the ER-ICA- or PgR-ICA-protocol is used with an overnight primary antibody incubation (diluted 1:2 to 1:10) at 4°C. Development of the peroxidase-labeled sections is carried out using freshly dissolved diaminobenzi-dine-tetrahydrochloride in PBS (without sodium azide!) in the presence of hydrogen peroxide. It is controlled in the microscope and interrupted at its optimal level. In ER- and PgR-immunostainings, the positive and negative control sections should always be stained in parallel. Evaluation of the microscopical receptor state is then carried out by estimation of the percentage of positive nuclei. A 20% + nuclei has been accepted as the threshold, independent from the staining intensity. Above this value, tamoxifen or other anti-estrogen and/or anti-progesterone treatment is recommended. Deduction of ER and PgR in paraffin sections is also possible but special techniques are required.[5,117,118] These methods are currently under evaluation and, at present, no final protocol for every day use has been settled. It should also be mentioned that in most studies, not all cases positive for ER or PgR in cryostat sections are also positive in paraffin sections (false-negative cases). In our lab, we successfully use a combined trypsin- and proteinase-K predigestion of deparaffinized sections, overnight incubation of the primary ER- antibodies from the Abbott kit, and a subsequent streptavidin-biotin-peroxidase method with nickelacetate-- enhanced DAB development (*Fig. 10*). PgR can be detected in paraffin sections by application of microwave antigen retrieval.[14]

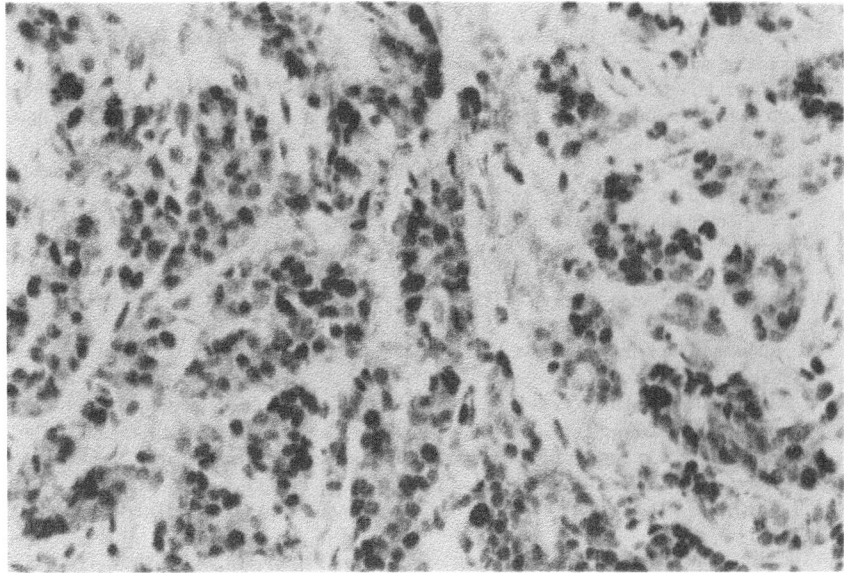

Figure 10. Detection of estrogen receptors in nuclei of breast cancer using Abbott monoclonal antibodies from the ER-ICA kit and subsequent ABC-peroxidase immunocytochemistry. The formalin-- fixed 4 μm thick paraffin section has been predigested with trypsin and proteinase K. Lightly counterstained with hematoxylin Original magnification x360.

Genetic Markers for Breast Cancer

The possible usefulness of various genetic markers in breast cancer has been discussed in the literature.[116,119-121] Genetic alterations in tumors may appear as changes in total DNA content which can be measured by interactive image analysis of DNA ploidy, (see chapter by Mack in this book). Changes in individual chromosomes, single genes, or gene expression may be investigated using molecular biological methods such as *in situ* hybridization. In such cases, the nature of produced protein(s) or peptide(s) may also be altered. Genetic markers investigated in breast cancer include some oncogens and oncoproteins, and some tumor suppressor genes. For the routine laboratory, the investigation of oncogenes or oncoproteins is most relevant presently. The most well-known oncoprotein in breast cancer is "*c-erbB-2*" (also called "Her-2","neu" or "p-185"), a protein showing molecular similarities to the ERF-receptor protein. Overex- pression of c-erbB-2 has been correlated with poor prognosis. However, the detection method has not yet been standardized, and time will show if this is a really useful marker. Own studies in more than 200 routine cases of breast cancer using the IGSS method showed positivity in about 45% and only a light correlation to the Ki-67 proliferation rate.

CONCLUSIONS

In the last few decades, immunocytochemistry has become an accepted adjuvant technique in diagnostic surgical pathology. It has proved to be especially valuable in cases where the limits of morphological diagnosis have been reached. However, classical histological techniques will still be the actual basis for histopathological diagnosis. Immunocytochemistry and related techniques can be used to support classical histochemistry, but they will probably never fully substitute for them.

REFERENCES

1. A.H. Coons, H.J. Creech, and R.N. Jones, Immunological properties of an antibody containing a fluorescent group. *Proc Soc Exp Biol Med* **47**:200 (1941).
2. A.H. Coons, E.H. Leduc, and J.M. Connolly, Studies on antibody production. I. A method for the histochemical demonstrations of specific antibody and its application to a study of the hyperimmune rabbit, *J Exp Med* **102**:49 (1955).
3. G. Köhler and C. Milstein, Continuous cultures of fused cells producing antibodies of predefined specificity, *Nature* **256**:495 (1975).
4. S. Van Nooden, Tissue preparation and immunostaining techniques for light microscopy. *in*: "Immunocytochemistry - Modern Methods and Applications", J.M. Polak, S. Van Noorden, eds., Wright, Bristol, UK 26 (1986).
5. J.M. Elias. "Immunohistopathology - A Practical Approach to Diagnosis," ASCP Press, Chicago, USA (1990).
6. M.E. Boon and L.P. Kok. "Microwave Cookbook of Pathology," Coulomb Press, Leiden, Netherlands (1988).
7. M. Stefanini, C. De Marfino, and L. Zamboni, Fixation of ejaculated spermatotzoa for electron microscopy, *Nature* **216**:173 (1967).
8. L. Zamboni and C. De Martino, Buffered picric acid-formaldehyde: a new rapid fixative for electron microscopy, *J Cell Biol* **35**:148A (1967).
9. W.D. Kuhlmann. "Imunoenzyme Techiques in Cytochemistry." Verlag Chemie, Deerfield Beach, FL, (1984).
10. A.G.E. Pearse and J.M. Polak, Bifunctional reagents as vapour and liquid phase fixatives for immunohistochemistry, *Histochem J* **7**:179 (1975).
11. A.E. Bishop, J.M. Polak, S.R. Bloom, and A.G.E. Pearse, A new universal technique for immunocytochemical localization of peptidergic innervation, *J Endocrinol* **77**:25 (1976).
12. S. Huang, H. Minassian, and J.D. More, Application of immunofluorescent staining in paraffin sections improved by trypsin digestion, *Lab Invest* **35**:383 (1976).
13. J.W.C. Finley and P. Petrusz. The use of proteolytic enzymes for improved localization of tissue antigens with immunocytochemistry, *in*: Techniques in Immunocytochemistry. Vol. 1, G.R. Bullock and P. Petrusz, eds., Acacemic Press, New York, 239 (1983).
14. S.R. Shi, M.E. Key, K.L. Kalra, Antigen retrieval in formalin-fixed, paraffin-embedded tissues: an enhancement method for immunohistochemical staining bases on microwave oven heating of tissue sections, *J Histochem Cytochem* **39**:741 (1991).
15. G. Goss, R.E. Petras, A. Perkins, and M. Miller, Effects of refixation and reprocessing on the quality of slides prepared from paraffin embedded tissues, *J Histotechnol* **15**:43 (1992).
16. P.K. Nakane and G.B.Jr. Pierce, Enzyme-labeled antibodies: preparation and application of the localization of antigen, *J Histochem Cytochem* **14**:929 (1966).
17. L.A. Sternberger and J.J. Cuculis, The unlabeled antibody-enzyme method of immunohistochemistry. Preparation and properties of soluble antigen-antibody comples (horseradish peroxidase-antihorseradish peroxidase) and its use in identification of spirochetes, *J Histochem Cytochem* **18**:315 (1970).
18. L.A. Sternberger. Immunocytochemistry, 2nd ed. Churchill Livingstone, Edinburgh (1976).
19. R.C. Graham and M.J. Karnovsky, The early stages of absorption of injected horseradisch peroxidase in the proximal tubules of mouse kidney, *J Histochem Cytochem* **14**:291 (1966).

20. J.L. Cordell, B. Falini, W.N. Erber, A.K. Ghosh, Z. Abdulaziz, S. Macdonald, K.A.F. Pulford, H. Stein, and D.Y. Mason, Immunoenzymatic labeling of monoclonal antibodies using immune complexes of alkaline phosphatase and monoclonal anti-alkaline phosphatase (APAAP complexes), *J Histochem Cytochem* **32**:219 (1984).

21. J.L. Guesdon, T. Ternynck, and S. Avrameas, The use of avidin-biotin interaction in immunoenzymatic techniques, *J Histochem Cytochem* **27**:1131 (1979).

22. G. Coggi, P. Dell'Orto, and G. Viale, Avidin-biotin methods, *in*: "Immunocytochemistry - modern methods and applications," J.M. Polak, S. Van Noorden, eds., Wright, Bristol, UK, 54 (1986).

23. C.S. Holgate, P. Jackson, P.N. Cowen, and C. Bird, Immunogold-silver staining: new method of immunostaining with enhanced sensitivity, *J Histochem Cytochem* **31**:938 (1983).

24. D.R. Springall, G.W. Hacker, L. Grimelius, and J.M. Polak, The potential of the immunogold-silver staining method for paraffin sections, *Histochemistry* **81**:603 (1984).

25. G.W. Hacker, D.R. Springall, S. Van Noorden, A.E. Bishop, L. Grimelius, and J.M. Polak, The immunogold-silver staining method. A powerful tool in histopathology, *Virch Arch A* **406**:449 (1985).

26. G. W. Hacker, Silver-enhanced colloidal gold for light microscopy, *in*: "Colloidal gold - principles, methods, and applications." Vol. 1, M.A. Hayat, ed., Academic Press Inc., San Diego, USA, 297 (1989).

27. G.W. Hacker, L. Grimelius, G. Danscher, G. Bernatzky, W. Muss, H. Adam, and J. Thurner, Silver acetate autometallography: an alternative enhancement technique for immunogold-silver staining (IGSS) and silver amplification of gold, silver, mercury and zinc in tissues, *J Histotechnol (USA)* **11**:213 (1988).

28. J. De Mey and M. Moeremans. Raising and testing polyclonal antibodies for immunocytochemistry, *in*: "Immunocytochemistry - modern methods and applications", J.M. Polak, S. Van Noorden, eds., Wright, Bristol, UK, 3 (1986).

29. M.A. Ritter. Raising and testing monoclonal antibodies for immunocytochemistry, *in*: "Immunocytochemistry - modern methods and applications", J.M. Polak, S. Van Noorden, eds., Wright, Bristol, UK (1986).

30. R. Moll. Epithelial tumor markers: cytokeratins and tissue polypeptide antigen (TPA), *in*: Morphological tumor markers - general aspects and diagnostic relevance, G. Seifert, ed., Springer-Verlag, Berlin and Heidelberg, FRG, 71 (1987).

31. M. Osborn and K. Weber, Biology of disease. Tumor diagnosis by intermediate filament typing: a novel tool for surgical pathology, *Lab Invest* **48**:372 (1983).

32. K.W. Barwick, Intermediate filaments and keratin, *in*: Atlas of diagnostic immunohistopathology, L.D. True, ed., J.B. Lippincott, Philadelphia, USA (1990).

33. S.H. Shapiro, Keratin as a factor in histopathologic diagnosis, *J Histotechnol* **14**:51 (1991).

34. C. Cooper, A. Schermer, and T.T. Sun, Classification of human epithelia and their neoplasms using monoclonal antibodies to keratins: strategies, applications and limitations, *Lab Invest* **52**:243 (1985).

35. J. Bartek, B. Vojtesek, J. Bartkova, Z. Kerekes, A. Rejthar, and J. Kovarik, A series of 14 new monoclonal antibodies to keratins: characterization and value in diagnostic histopathology, *J Pathol* **164**:215 (1991).

36. H. Hoefler and H. Denk, Immunocytochemical demonstration of cytokeratin in gastrointestinal carcinoids and their probable precursor cells, *Virch Arch (Path Anat)* **403**:235 (1984).

37. J.W. Said. Immunohistochemical localization of keratin proteins in tumor diagnosis, *Hum Pathol* **14**:1017 (1983).

38. H. Denk, R. Krepler, U. Artlieb *et al.*, Proteins of intermediate filaments: an immunohistochemical and biochemical approach to the classification of soft tissue tumors, *Am J Path* **110**:193 (1983).

39. G.S. Pinkus and J.S. Said, Leu-M1 immunoreactivity in non-hematopoietic neoplasms and myoproliferative disorders. An immunoperoxidase study of paraffin sections, *Am J Clin Pathol* **85**:278 (1986).

40. K. Shelbani, H. Battifora, and J.S. Burke, Antigenic phenotype of malignant mesotheliomas, *Lab Invest* **54**:57A (1986).

41. C.N. Otis, D. Carter, S. Cole *et al.*, Immunohistochemical evaluation of pleural mesothelioma and pulmonary adenocarcinoma: a bi-institutional study of 47 cases, *Am J Surg Pathol* **11**:445 (1987).

42. D. Cooper, A. Schermer, and T.T. Sun, Classification of human epithelia and their neoplasms using monoclonal antibodies to keratins: strategies, applications and limitations, *Lab Invest* **52**:243 (1985).

43. P. Gold and S.O. Freedman, Demonstration of tumor-specific antigens in human colonic carcinomata by immunological tolerance and absorption techniques, *J Exp Med* **121**:439 (1965).

44. R.H. Fletcher, Carcinoembryonic antigen, *Ann Int Med* **104**:66 (1986).

45. L.D. True, Epithelial membrane antigens, *in*: Atlas of diagnostic immunohistopathology, L.D. True, ed., J.B. Lippincott, Philadelphia, USA (1990).

46. G.S. Pinkus, L.L. Etheridge, and E.M. O'Connor, Are keratin proteins a better tumor marker than epithelial membrane antigen? *Am J Clin Pathol* **85**:269 (1986).

47. M. Altmannsberger and M. Osborn. Mesenchymal tumor markers: intermediate filaments, *in*: Morphological tumor markers, G. Seifert, ed., Springer-Verlag, Berlin, 155 (1987).

48. E. Lazarides and B.D. Hubbard, Immunological characterization of the subunit of the 100 A filaments from muscle cells, *Proc Natl Acad Sci USA* **73**:4344 (1976).

49. J.V. Small and A. Sobieszek, Studies on the function and composition of the 10 nm (100 A) filaments of vertebrate smooth muscle, *J Cell Sci* **23**:243 (1977).

50. S. Brown, W. Levinson, and J.A. Spudich, Cytoskeletal elements of chick embryo fibroblasts reveiled by detergent extraction, *J Supramol Struct* **5**:110 (1976).

51. M. Osborn and K. Weber, The detergent-resistant cytoskeleton of tissue culture cells includes the nucleus and the microfilament bundles, *Exp Cell Res* **106**:339 (1977).

52. W.W. Franke, E. Schmid, M. Osborn, and K. Weber, Different intermediate-size filaments distinguished by immunofluorescence microscopy, *Proc Natl Acad Sci USA* **75**:5034 (1978).

53. M. Miettinen, V.P. Lehto, and I. Virtanen, Keratin positivity in the epithelial-like cells of classical biphasic synovial sarcoma, *Virchows Arch (Cell Pathol)* **40**:157 (1982).

54. D.R. Chase, S.W. Weiss, F.M. Enzinger, and I.M. Langloss, Keratin in epitheloid sarcoma, *Am J Surg Pathol* **8**:435 (1984).

55. J. Rosai and G.S. Pinkus, Imunohistochemical demonstration of epithelial differentiation in adamantinoma of the tibia, *Am J Pathol* **6**:427 (1982).

56. M. Miettinen, V.P. Lehto, D. Dahl, and I. Virtanen, Differential diagnosis of chordoma, chondroid, and ependymal tumors as aided by anti-intermediate filament antibodies, *Am J Pathol* **112**:160 (1983).

57. M. Altmannsberger, M. Osborn, H.J. Schäfer, A. Schauer, and K. Weber, Distinction of nephroblastomas from other childhood tumors using antibodies to intermediate filaments, *Virchows Arch (Cell Pathol)* **45**:113 (1984).

58. K. Mukai, J. Rosai, and B.E. Hallaway, Localization of myoglobin in normal and neoplastic human skeletal muscle cells using an immunoperoxidase method, *Am J Surg Pathol* **3**:373 (1979).

59. A.S.H. De Jong, W. Raamsdonk, M. Van Vark, P.A. Voute, and C.E. Albus-Lutter, Myosin and myoglobin as tumor markers in the diagnosis of rhabdomyosarcoma, *Am J Surg Pathol* **8**:521 (1984).

60. H.F. Otto, R. Berndt, K. Schwechheimer, and P. Möller, Mesenchymal tumor markers: special proteins ands enzymes. *in*: Morphological tumor markers, G. Seifert, ed., Springer-Verlag, Berlin, 179 (1987).

61. E.A. Jaffe, Endothelial cells and the biology of factor VIII, *N Engl J Med* **296**:377 (1977).

62. J.E. Thurner Jr., A.H. Graf, A. Walter, and G.W. Hacker, Small bleeding hemangiosarcoma of the jejunum: case report and immunocytochemical findings, *J Histotechol* **13**:141 (1990).

63. M. Yoshizumi, H. Kurihara, T. Morita *et al.*, Interleukin 1 increases the production of endothelin-1 in cultured endothelial cells, *Biochem Biophys Res Commun* **166**:324 (1990).

64. P. Meister, D. Huhn, and W. Nathrath: Malignant histiocytosis, Immunohistological characterization on paraffin-embedded tissue, *Virchows Arch (Pathol Anat)* **385**:233 (1980).

65. S. Nakanishi, S. Shinomiya, T. Sano, and K. Hizawa, Immunohistochemical observation of intracytoplasmic lysozyme in proliferative and neoplastic fibrohistiocytic lesions, *Acta Pathol Jpn* **32**:949 (1982).

66. K.A.F. Pulford, R.M. Rigney, K.J. Mickin, *et al.*, KPI: a new monoclonal antibody that detects monocyte/macrophage associated antigens in routinely processed tissue sections, *J Clin Pathol* **42**:414 (1989).

67. C.M. Hulette, B.T. Downey, and P.C. Burger, Macrophage markers in diagnostic neuropathology, *Am J Surg Pathol* **16**:493 (1992).

68. D.J. Flavell, D.B. Jones, D.H. Wright, Identification of tissue histiocytes on paraffin sections by a new monoclonal antibody, *J Histochem Cytochem* **35**:1217 (1987).

69. A.M. Gown, T. Tsukoda, and R. Ross, Human arteriosclerosis II: Immunocytochemical analysis of the cellular composition of human arteriosclerotoic lesions, *Am J Pathol* **125**:191 (1986).

70. D.R. Springall, J. Gu, D. Cocchia, F. Michetti, A. Levene, M.M. Levene, P.J. Marangos, S.R. Bloom, and J.M. Polak, The value of S-100 immunostaining as a diagnostic tool in human malignant melanomas, *Virchows Arch (Pathol Anat)* **400**:331 (1983).

71. Ph.U. Heitz, Neuroendocrine tumor markers, *in*: Morphological tumor markers, G. Seifert, ed., Springer-Verlag, Berlin, 279 (1987).

72. B.W. Moore, A soluble protein characteristic of the nervous system, *Biochem Biophys Res Comm* **19**:739 (1965).

73. T. Isobe, N. Ishioka, T. Masuda, Y. Takahashi, S. Ganno, and T. Okuyama, A rapid separation of S-100 subunits by high performance liquid chromatography: the subunit composition of S-100 proteins. *Biochem Int* **6**:419 (1983).

74. G.W. Hacker, J.M. Polak, D.R. Springall, J. Ballesta, A. Cadieux, J. Gu, J.Q. Trojanowski, D. Dahl, and

P.J. Marangos, Antibodies to neurofilament protein and other brain proteins reveal the innervation of peripheral organs, *Histochemistry* **82**:581 (1985).

75. G.W. Hacker, A.E. Bishop, G. Terenghi, I.M. Varndell, J. Aghahowa, K. Pollard, J. Thurner, J.M. Polak, Multiple peptide production and presence of general neuroendocrine markers detected in 12 cases of human phaeochromocytoma and in mammalian adrenal glands, *Virchows Arch (Pathol Anat)* **412**:399 (1988).

76. S. Watanabe, T. Nakajima, Y. Shimosato, Y. Sato, and T. Ise, A case report of histiocytic medullary reticulosis defined as a neoplasm of T-zone histiocytes, *Jpn J Clin Oncol* **11**:411 (1981).

77. A. S.Y. Leong (ed.) *et al.*, The laborarory diagnosis of malignant lymphoma. *15th anniversary special issue of J Histotechnol* **15**:170 (1992).

78. R.W. Cartun, W.N. Rezuke, and W.T. Pastuszak, Immunohistochemistry of malignant lymphoma, *J Histotechnol* **15**:199 (1992).

79. R.A. Warnke, K.C. Gatter, B. Fallini *et al.*, The diagnosis of human lymphoma using monoclonal anti-leucocyte antibodies, *N Engl J Med* **309**:1275 (1983).

80. A.J. Norton and P.G. Isaacson, Detailed phenotypic analysis of B-cell lymphoma using a panel of antibodies reactive in routinely fixed wax-embedded tissue, *Am J Pathol* **128**:225 (1987).

81. A.L. Epstein, R.J. Marder, J.N. Winter *et al.*, Two new monoclonal antibodies reactive in B5 formalin-fixed, paraffin-embedded tissues with follicular center and mantle zone human B lymphocytes and derived tumors, *J Immunol* **133**:1028 (1984).

82. R.W. Cartun, F.B. Coles, and W.T. Pastuszak, Utilization of monoclonal antibody L26 in the identification and conformation of B-cell lymphomas, *Am J Pathol* **129**:415 (1987).

83. A.J. Norton and P.G. Isaacson, An immunocytochemical study of T-cell lymphomas using monoclonal and polyclonal antibodies effective in routinely fixed wax-embedded tissues, *Histopathology* **10**:1243 (1986).

84. S. Poppema, H. Hollema, L. Visser, and H. Vos, Monoclonal antibodies (MT1, MT2, MB1, MB2, MB3) reactive with leucocyte subsets in paraffin-embedded tissue sections, *Am J Pathol* **127**:418 (1987).

85. G. Browne, B. Tobin, D.N. Carney, and P.A. Dervan, Aberrant MT2 positivity distinguishes follicular lymphoma from reactive follicular hyperplasia in B5- and formalin-fixed sections, *Am J Clin Pathol* **96**:90 (1991).

86. D.Y. Mason, J. Cordell, M. Brown, *et al.*, Detection of T cells in paraffin wax embedded tissue using antibodies against a peptide sequence from the CD3 antigen, *J Clin Pathol* **42**:1194 (1989).

87. P. Kleihues, M. Kiessling, and R.C. Janzer, Morphological markers in neuro-oncology. *In*: Morphological tumor markers, G. Seifert, ed., Springer-Verlag, Berlin, 307 (1987).

88. P.N. Hoffman and R.J. Lasek, The slow axonal transport. Identification of major structural polypeptides of the axon and their generality among mammalian neurons, *J Cell Biol* **66**:351 (1975).

89. P. Gambetti, L. Autilio-Gambetti, and S.C. Papasozomenos, Bodian's silver method stains neurofilament proteins, *Science* **213**:1521 (1981).

90. B.W. Moore, A soluble protein characteristic of the nervous system, *Biochem Biophys Res Commun* **19**:739 (1965).

91. P.J. Marangos, C. Zomely-Neurath, and C. York, Immunological studies of a nerve specific protein, *Arch Biochem Biophys* **170**:289 (1975).

92. D. Schmechel, P.J. Marangos, and M. Brightman, Neuron specific enolase is a molecular marker for the peripheral and central neuroendocrine cells, *Nature* **276**:834 (1978).

93. H. Haimoto, Y. Takahashi, T. Koshikawa, H. Nagura, and K. Kato, Immunohistochemical localization of gamma enolase in normal human tissues other than neurons and neuroendocrine tissues, *Lab Invest* **52**:257 (1985).

94. R. Jahn, W. Schiebler, C. Quimet, and P. Greengard, A 38,000-dalton membrane protein (p38) present in synaptic vesicles, *Proc Natl Acad Sci USA* **82**:4137 (1985).

95. B. Wiedenmann and W.W. Franke, Identification and localization of synaptophysin, an integral membrane glycoprotein of MW 38,000 characteristic of presynaptic vesicles, *Cell* **41**:1017 (1985).

96. H. Rehm, B. Wiedenmann, and H. Beth, Molecular characterization of cynaptophasin, a major calcium-binding protein of the synaptic vesicle membrane, *EMBO J* **5**:535 (1986).

97. B. Wiedenmann, W.W. Franke, C. Kuhn, R. Moll, and V.E. Gould, Synaptophysin: a marker protein for neuroendocrine cells and neoplasms, *Proc Natl Acad Sci USA* **83**:3500 (1986).

98. R. Jovanovic, G.W. Hacker, U.G. Falkmer, S. Falkmer, L. Mendel, A.H. Graf., A. Höög, V. Kanjuh, C. Silfverswärd, and L. Grimelius, Paragangliomas: neuroendocrine features and cytometric DNA distribution patterns, *Virchows Arch (Pathol Anat)* **419**:455 (1991).

99. J.F. Doran, P.J. Jackson, P.A.M. Kynoch, and R.J. Thompson, Isolation of PGP 9.5, a new human neuron specific protein detected by high resolution two dimensional electrophoresis, *J Neurochem* **40**:1542 (1983).

100. R.J. Thompson, J.F. Doran, P. Jackson, A.P. Dhillon, and J. Rode, PGP 9.5 - a new marker for vertebrate neurones and neuroendocrine cells, *Brain Res* **278**:224 (1983).

101. J. Rode, A.P. Dhillon, J.F. Doran, P. Jackson, and R.J. Thompson, PGP 9.5, a new marker for human neuroendocrine tumours, *Histopathol* **9**:147 (1985).

102. J.E. Goldman, H.H. Schaumburg, and W.T. Norton, Isolation and characterization of glial filaments from human brain, *J Cell Biol* **18**:426 (1978).

103. T. Kimura, H. Budka, and S. Soler-Federspiel, An immunocytochemical comparison of the glia-associated proteins glial fibrillary acidic protein (GFAP) and S-100 protein in human brain tumors, *Clin Neuropathol* **5**:21 (1986).

104. Y. Daimaru, H. Hashimoto, M. Enjoji, Malignant peripheral nerve sheath tumors (malignant schwannomas). An immunohistochemical study of 29 cases, *Am J Surg Pathol* **9**:434 (1985).

105. J.M. Polak and S.R. Bloom (eds.), Endocrine tumors: the pathobiology of regulatory peptide-producing tumours. Churchill Livingstone, Edinburgh, (1985).

106. L.D. True, Neuroendocrine antigens, *in*: Atlas of diagnostic immunohistopathology, L.D. True, ed., J.B. Lippincott, Philadelphia, USA, chapter 11 (1990).

107. J. Gerdes, U. Schwab, H. Lemke, and H. Stein, Production of a mouse monoclonal antibody reactive with a human nuclear antigen associated with cell proliferation, *Int J Cancer* **31**:13 (1983).

108. J. Gerdes, S. Pileri, H. Bartels, and H. Stein, Proliferation marker Ki-67: correlation with histological diagnosis, histological tumor grading and prognosis, *in*: G. Seifert and K. Hbner (eds.), Pathology of cell receptors and tumor markers. Application of immunocytochemistry and hybridization in tumor diagnosis. G. Fischer Verlag, Stuttgart, New York, 145 (1987).

109. B.A. Robbins, D. de la Vega, K. Ogata, E.M. Tan, and R.M. Nakamura, Immunohistochemical detection of proliferating cell nuclear antigen in solid human malignancies, *Arch Pathol Lab Med* **111**:841 (1987).

110. P.A. Hall, D.A. Levison, A.L. Woods, C.C.W. Yu, D.B. Kellock, J.A. Watkins, D.M. Barnes, C.E. Giletti, R. Camplejohn, R. Dover, N.H. Wasoom, and D.P. Lane, Proliferating cell nuclear antigen (PCNA) immunolocalization in paraffin sections: an index of cell proliferation with evidence of deregulated expression in some neoplasms, *J Pathol* **162**:285 (1990).

111. S.A. Sampson, H. Kreipe, C.E. Gillett, P. Smith, M.A. Chaudary, A. Khan, K. Wicks, R. Parwaresch, and D.M. Barnes, KiS1 - a novel monoclonal antibody which recognizes proliferating cells: evaluation of its relationship to prognosis in mammary carcinoma, *J Pathol* **168**:179 (1992).

112. Th. Lning and K. Milde, Viral tumor markers, *in*: Morphological tumor markers, G. Seifert, ed., Springer-Verlag, Berlin, 307 (1987).

113. I. Zehbe, G.W. Hacker, E. Rylander, J. Sällström, E. Wilander, Detection of single HPV copies in SiHa cells by *in situ* polymerase chain reaction (*in situ* PCR) combined with immunoperoxidase and immunogold-silver staining (IGSS) techniques. *Anticancer Res* **12**:2165 (1992).

114. B. Alexiev, I. Valkov, and A. Popov, Immunohistochemical evidence of vimentin in cases of fibrocytic breast disease and mammary carcinoma, *Zentralbl Pathol* **138**:284 (1992).

115. W. Domagala, G. Striker, A. Szadowska, A. Dukowicz, K. Weber, M. Osborn, Cathepsin D in invasive ductal NOS breast carcinoma as defined by immunocytochemistry. No correlation with survival at 5 years, *Am J Pathol* **141**:1003 (1992).

116. Y. Umekika, N. Enokizono, Y. Sagara, K. Kuriwaki, T. Takasaki, A. Yoshida, and H. Yoshida,: Immunohistochemical studies on oncogene products (EGF-R, c-erbB-2) and growth factors (EGF, TGF-a) in human breast cancer: their relationship to oestrogen receptor status, histological grade, mitotic index and nodal status, *Virchows Arch (Pathol Anat)* **420**:345 (1992).

117. L. Cheng, S.W. Binder, Y.S. Fu, and K.J. Lewin, Methods in laboratory investigation. Demonstration of estrogen receptors by monoclonal antibody in formalin-fixed breast tumors, *Lab Invest* **58**:346 (1988).

118. J.M. Elias, A. Heimann, T. Cain, F. Gallery, and C. Gomes, Estrogen receptor localization in paraffin sections by enzyme digestion, repeated applications of primary antibody, and imidazole, *J Histotechnol* **13**:29 (1990).

119. S.R. Wolman, R.J. Pauley, A.N. Mohamed, P.J. Dawson, D.W. Visscher, and F.H. Sarkar, Genetic markers as prognostic indicators in breast cancer, *Cancer* **70**:1765 (1992).

120. P. Shrestha, K. Yamada, T. Wada, S. Maeda, M. Watatani, M. Yasutomi, H. Takagi, and M. Mori, Proliferating cell nuclear antigen in breast lesions: correlation of c-erbB-2 oncoprotein and EGF receptor and its clinicopathological significance in breast cancer, *Virchows Arch (Pathol Anat)* **421**:193 (1992).

121. G. Gasparini, W.J. Gullick, P. Bevilacqua, J.R.C. Sainsbury, S. Meli, P. Boracchi, A. Testolin, G. La Malfa, and F. Pozza, Human breast cancer: prognostic significance of the c-erbB-2 oncoprotein compared with epidermal growth factor receptor, DNA ploidy, and conventional pathologic features, *J Clin Oncol* **10**:686 (1992).

CHAPTER 8

CLASSIFICATION OF ENDOCRINE TUMORS BY HISTOCHEMICAL TECHNIQUES

Erik Wilander, M.D., Ph.D.[1]

Department of Pathology
University Hospital
Uppsala, Sweden

SUMMARY

Human endocrine tumors can be grouped into two main types, neuroendocrine and endocrine (non-neuroendocrine) neoplasms. The neuroendocrine tumors display neurosecretory properties. Their hormonal or regulatory products, mainly peptide hormones and biogenic amines, are stored in membrane-bound granules in the cytoplasm. Other types of endocrine tumors lack visible secretory granules. Their hormones are of steroid nature (adrenal cortex) or thyroxin (follicular epithelium of the thyroid gland). Neuroendocrine tumors can be separated from the other endocrine tumors by applying various light microscopical stains for visualization of secretory granules. To achieve this, chromogranin immunocytochemistry and the argyrophil reaction of Grimelius are most frequently employed. Specific methods for general identification of the other endocrine tumors are not yet available. Neuroendocrine tumors can be further subgrouped into those of neuroectodermal and endodermal nature. Endodermal tumors, like their tissues of origin, usually express intermediate filaments of cytokeratin type, while the neuroectodermal tumors and their corresponding normal endocrine cells lack such chromocytoma properties. The adrenal medulla derived pheochromocytomas, neuroblastomas, and the gangliomas are neuroectodermal tumors. Pituitary and parathyroid gland tumors, medullary thyroid carcinomas, pancreatic islet cell tumors, and most gastrointestinal carcinoids express endodermal characteristics and often produce regulatory

[1]Address for correspondence: Dr Erik Wilander, Department of Pathology, University Hospital, S-751 85 Uppsala, Sweden, Tel: 46-18-66-3000; Fax: 46-18-55-2739

peptides and/or biogenic amines. However, some carcinoids appear to be of neuroectodermal origin and display a tumor marker profile similar to that observed in gangliomas, for example.

INTRODUCTION

Endocrine tumors can be classified according to various principles. From the clinical point of view, concentrations of hormonal substances in serum and the clinical picture or syndrome are of major importance. In oncology, knowledge of the biological behavior of tumors is essential, for example, when choosing the mode of tumor suppressive therapy. In histopathology, the light-microscopic classification of tumors is based chiefly on tumor histogenesis, i.e., identification of the tumor phenotype disclosing from which normal tissue the tumor has developed by neoplastic transformation. By comparing various clinical and morphological procedures for tumor classification, a considerable (though not absolute) degree of coherence can be achieved. A classification of human endocrine tumors based chiefly on histogenetic criteria and staining methods useful for identification of individual tumors is presented. During recent decades, increased knowledge has erased to some extent, the distinction between endocrine and non-endocrine tumors. This problem is also considered in the following text.

Regular Light Microscopy

A prerequisite for accurate evaluation is to possess basic knowledge of endocrine tumor morphology since available diagnostic methods cannot be applied to all tumor samples. In most cases, endocrine tumors can be easily identified. As a rule, they are highly differentiated and display typical and regular growth patterns in contrast to carcinomas. However, when reading large series of microscopic slides, it is evident that sometimes tumors occur despite a tentative diagnosis of "endocrine tumor", that later prove to be non-endocrine and vice versa. Furthermore, poorly differentiated endocrine tumors also occur which can be difficult to distinguish from non-endocrine carcinomas. The most typical examples of this phenomenon are the small-cell lung carcinoma and the neuroblastoma, but similar poorly differentiated neuroendocrine tumors may also occur at other locations.

Endocrine and Neuroendocrine Tumors

All human endocrine tumors can be grouped into two main categories: endocrine (non-peptide hormone producing) and neuroendocrine (regulatory peptide-producing) tumors. The first type mainly comprises tumors arising in the adrenal cortex and from the follicular epithelial cells of the thyroid gland. The hormones of the adrenal cortex, mainly cortisone and aldosterone, are of a steroid nature and are synthesized by chemical transformation of cholesterol. Thyroxin, produced by the follicular epithelial cells of the thyroid gland, is formed by a combination of two iodinated tyrosine molecules. Tumors arising from endocrine tissues, such as adrenal cortical adenoma and carcinoma, and those deriving from the thyroid gland follicular epithelium, carcinomas of papillary, follicular, and undifferentiated type, can be diagnosed in routinely processed histological sections in most cases, without the addition of any special histochemical stains.

By contrast, more sophisticated diagnostic procedures are usually employed in the evaluation of peptide-amine-containing neuroendocrine tumors. All these tumors demonstrate several common features, among which the presence of intracytoplasmic secretory granules is the most constant and can be easily analyzed. Although secretory granules storing peptide hormones and/or biogenic amines can only be directly observed at the ultrastructural level, staining methods can be applied to visualize them, even by light microscopy.

Neuroendocrine tumors arise from diffusely dispersed neuroendocrine cells or organs, i.e. the pituitary gland, parathyroid gland, parafollicular C-cells of the thyroid gland, pancreatic islets, suprarenal medulla and the diffuse neuroendocrine cell system of the gastrointestinal tract and bronchial tree. From all these sites, neuroendocrine tumors can develop with common as well as particular diagnostic characteristics which can be used for their identification. The most reliable method for the visualization of neuroendocrine differentiation in tumors is the demonstration of an intracytoplasmatic hormone granular component.

Detection of neuroendocrine granules is most appropriately accomplished with the argyrophil reaction of Grimelius[1,2] or by chromogranin immunocytochemistry.[3-11] The Grimelius silver staining method exploits the property of hormonal granules possessing an avidity for silver ions that can be converted to metallic silver (argyrophil reaction) by adding a reducing agent. Small silver particles thus accumulate to the secretory granules as a result visible in the light microscope as brownish to black granular silver deposits, distinctly outstanding against yellowish background. Grimelius silver staining can be used for reliable and sensitive identification of neuroendocrine cells already at the light-microscopic level. One advantage of the Grimelius stain is that it is almost non-discriminating, yet quite specific. This means that most neuroendocrine cells and tumors at different topographic locations are argyrophil, even though they contain various hormonal products because non-neuroendocrine tissues are non-argyrophil. Now, the Grimelius stain is a widely-used routine diagnostic aid in many laboratories.

Neuroendocrine tumors can also be identified immunocytochemically by applying antibodies to chromogranins. The latter are a family of acidic proteins present in the hormonal secretory granules. In higher animals, they appear to be the most abundantly secreted proteins. The functional role of the chromogranins is obscure but they are widely distributed in neuroendocrine cells, irrespective of their individual hormone production. So far, three different species in the chromogranin family have been identified: chromogranin A, B, and C, of which A is most extensively used as a marker for the demonstration of neuroendocrine cells.

It has been shown that comparable results can be obtained when employing the argyrophil reaction of Grimelius and chromogranin immunocytochemistry. At electron microscopical level, the silver grains resulting from Grimelius silver staining and the immunoreaction after applying chromogranin antibodies are both concentrated over the electronlucent space in the A (glucagon) cells of human pancreatic islets. Furthermore, purified chromogranin displays an argyrophil reaction *in vitro*. This is why the two staining methods give similar - though not absolutely identical - results.[12-13] Disparities can be attributed to differences at the reaction site in the chromogranin molecule between the two methods and to the fact that they are not equally affected by fixation and tissue preparation. Grimelius staining and chromogranin immunoreactivity can be used as alternative methods for visualization of neuroendocrine cells and tumors, but since they do not give identical results it is recommended to use them in combination. In the chapter by Grimelius *et al.* in this book, a modification of the Grimelius stain utilizing microwave irradiation is described. This can be completed within a few minutes.

Cytokeratin Expression

Neuroendocrine-differentiated cells can be grouped into two main categories: those of epithelial or endodermal origin, and those of neuroectodermal origin. Both general types can be identified by their not completely identical intermediate filament content.

A fibrillar intracytoplasmatic matrix is present in all eukaryotic cells. Within this matrix, one can distinguish filaments measuring 8-11 nm in diameter, also called the intermediate-sized filaments from other types of cytoskeleton. At least five different types of intermediate filament protein clones have been identified. Among these, neurofilament proteins are confined mainly to nerve tissue and neuroectodermally derived neuroendocrine tumors, such as tumors originating from the adrenal medulla and the paraganglia.

Another type of intermediate filament, cytokeratin, is present in epithelial cells derived from both ecto- and endoderm. Cytokeratins are characterized by a remarkable biochemical diversity and can be subgrouped into at least 20 different polypeptides. Since they are not randomly distributed, but display selective topographic combinations, they are commonly used as markers to identify the epithelial differentiation of tumors at different sites.

Neuroendocrine tumors of ecto- and endodermal origin usually express cytokeratin filaments. Such tumors are the pituitary gland adenomas, parathyroid adenomas, medullary carcinoma of the thyroid gland, pancreatic islet cell tumors, and lung and gastrointestinal neuroendocrine tumors including classical carcinoids. It is emphasized that the expression of cytokeratin filaments is useful for distinguishing between epithelially and neuroectodermally derived neuroendocrine tumors - but not for distinguishing neuroendocrine tumors from the more common carcinomas at various topographic sites, since these are also of epithelial origin and mostly express cytokeratin filaments.[14,15]

It is noteworthy that some neuroendocrine tumors of the lung lack cytokeratin immunoreactivity but contain a population of elongated satellite-like cells with evidence of S-100 protein immunoreactivity, a phenomenon often observed in pheochromocytomas. For that reason, it has been questioned whether some of these lung tumors might be of non-epithelial nature, but rather belong to the paraganglioma tumors. Further, the typical appendiceal carcinoids are also cytokeratin-negative and contain intermingled S-100 protein-reactive cells. It has been speculated, therefore, that they may be derived from the peripheral nervous system and arise from subepithelial cells occurring beneath the epithelial layer of the appendiceal mucosa. This could explain why appendiceal carcinoids do not have the same biological behavior as the classical carcinoids of the small intestine.[15-21]

Specific Neuroendocrine Tumor Markers

After defining a tumor as endocrine or non-endocrine, and as epithelial or neuroectodermal, it is of interest to identify the specific properties, i.e. the peptide hormone production of the individual tumor. Several silver stains and antibodies to regulatory peptides are preferably applied for that purpose.[13]

Silver Stains

Some silver stains are useful for the characterization of neuroendocrinal tumors. The classical mid-gut carcinoids almost invariably contain serotonin and this biogenic amine can easily and reliably be identified in tumors by application of Masson's argentaffin reaction, provided that the tissue has been fixed in formalin. Since tumors at other locations are non-argentaffin or contain only a few scattered argentaffin cells, the method can be used to distinguish classical carcinoids from neuroendocrine tumors of other origin. Serotonin can also be visualized in carcinoids by using serotonin antibodies, for some only vaguely understood reason, the serotonin immunoreactivity is sometimes weak or even negative, despite a clinically evident serotonin production from the tumor.

Another useful silver stain is the argyrophil reaction of Sevier-Munger. This method is used preferably to identify the enterochromaffin-like cell tumors (ECLomas) that arise in the acid-secreting part of the gastric mucosa. Since the development of ECLomas is frequently preceded by endocrine cell proliferation and hyperplasia in the mucosal crypts, even premalignant changes can be revealed with the stain.[22-25] The importance of the Sevier-Munger method is also emphasized by the fact that the ECL-cells do not contain many of the known peptide hormones and can therefore not yet be visualized with antibodies to regulatory peptides.

Peptide Hormone Immunocytochemistry

For adequate characterization, it is important to identify the regulatory peptide or biogenic amine content of an individual tumor. In general, differentiated neuroendocrine tumors produce the same secretory products as the cell of origin. Accordingly, calcitonin is characteristic of medullary thyroid carcinoma and insulin, glucagon, somatostatin and/or pancreatic polypeptide (PP) are often present in pancreatic islet cell tumors. For primary diagnosis, only a limited number of antisera need to be applied which are selected with respect to the confirmed or presumed primary site of the tumor. However, it must be remembered that ectopic hormone production or production of derivatives from peptide precursors other than the well known peptides may also occur. This is especially evident in pancreatic islet cell tumors which secrete gastrin or vasoactive intestinal polypeptide (VIP) relatively frequently.

Amyloid Related to Neuroendocrine Tumors

Several forms of localized amyloidosis occur in association with neuroendocrine cells and tumors. Stromal amyloid is a characteristic component of the C-cell derived medullary carcinoma of the thyroid gland. This amyloid consists of a calcitonin- or procalcitonin-related peptide.[26] The localized cardiac atrial amyloidosis in humans is formed of B-pleted sheets of atrial natriuretic peptide (ANP), a natural hormonal constituent of the atrial myocytes.[27] However, neuroendocrine tumors of the heart with amyloid content are not found possibly because the atrial myocytes do not possess proliferative properties. Amyloid is also present in about half of all pancreatic islet cell tumors of B-cell type, often producing insulin.[28]

The isolation, purification and amino-acid sequence analysis of the amyloid in a pancreatic islet cell tumor, achieved in 1986, revealed the existence of a native peptide which was designated "islet amyloid polypeptide" (IAPP). Other authors called this amylin. This peptide consists of 37 aminoacid and has about 50% homology with "calcitonin gene-related peptide" (CGRP). IAPP appears to be a major component of islet amyloid, both in tumors and in the islets, seen primarily in individuals with type II diabetes mellitus.[29-31] At the ultrastructural level, amyloid has been observed intracellularly and in close proximity to the hormonal granules, which is why it is conceivable that amyloid develops by some abnormal granular protein process.[32] IAPP is co-stored with insulin in normal pancreatic B-cells and co-secreted on glucose stimulation. The functional role of IAPP is enigmatic.

Amyloid displays a yellow-green birefringence after staining with alkaline Congo red and examination in polarized light. Its occurrence in neuroendocrine tumors is of some significance. About 75% of all medullary thyroid carcinomas and about 50% of the

B-cell-derived islet cell tumors contain amyloid. Amyloid may also be present in other types of neuroendocrine tumors but it is extremely rare. For instance, a few lung and gastric carcinoids and some pituitary adenomas with amyloid stroma have been reported in the literature.

General Considerations

The present description of methods for the identification and further characterization of endocrine tumors has been presented as a four-stage procedure for didactic purposes. Of course, the entire evaluation can be carried as a single sequence. The method is especially useful when metastatic lesions are being scrutinized and the primary tumor still is unidentified. Since many endocrine tumors are highly differentiated, they often display a low degree of heterogeneity. This is of value when small biopsies are being studied and the staining properties of the tumor fragment can be considered as representative of the whole tumor.

Endocrine activity, in terms of hormone synthesis, storage and release, is usually more pronounced in highly differentiated than in the poorly differentiated tumors. This is demonstrated by weaker reactions of undifferentiated neuroendocrine tumors with various silver stains and immunocytochemistry for the presence of peptides. It is also important to be aware of the not infrequent occurrence of endocrine-differentiated tumor cells in typical adenocarcinomas. Thus, the diagnosis of endocrine tumors involves both strictly morphological and more advanced histochemical considerations.

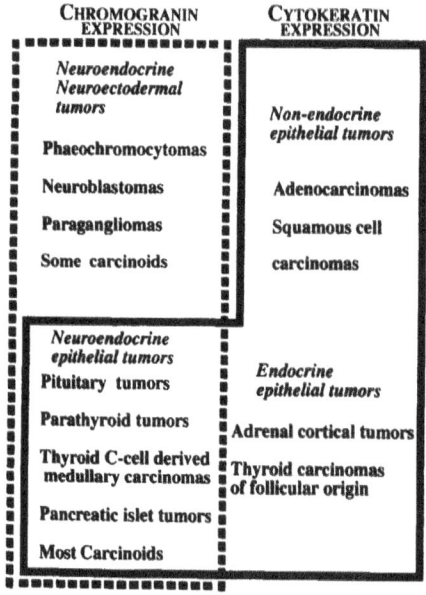

Figure 1. Illustration of the organization of endocrine cell derived tumors, which can be applied for tumor classification.

When evaluating histochemical stainings of endocrine tumor materials, qualitative and quantitative interpretations are necessary. For instance, in classical small intestinal carcinoids, most of the tumor cells are argyrophil and chromogranin-immunoreactive, while only a smaller proportion of tumor cells display an argentaffin reaction. On the other hand, carcinoid-like intestinal carcinomas in routinely processed histological sections, have been misinterpreted and could only be separated from the classical carcinoids after quantitative assessment of positive staining for endocrine differentiated tumor cells.[33]

ACKNOWLEDGEMENT

This work is supported by grants from the Swedish Medical Research Council

REFERENCES

1. L. Grimelius, A silver nitrate stain for α_2 cells of the human pancreatic islets, *Acta Soc Med Upsal* **73**:243 (1968).
2. L. Grimelius and E. Wilander, Silver stains in the study of endocrine cells of the gut and pancreas, *Invest Cell Pathol* **3**:3 (1980).
3. A.D. Smith and H. Winkler, Purification and properties of an acidic protein from chromaffin granules of bovine adrenal medulla, *Biochem J* **103**:483 (1967).
4. H. Blaschko, R.S. Compline, F.H. Schneider, M. Silver, and A.D. Smith, Secretion of a chromaffin granule protein, chromogranin, from the adrenal gland after splanchnic stimulation, *Nature* **215**:58 (1983).
5. R. Fischer-Colbrie, H. Lassmann, C. Hagn, and H. Winkler. Immunological studies on the distribution of chromogranin A and B in endocrine and nervous tissues, *Neuroscience* **16**:547 (1985).
6. D.T. O'Connor, F.P. Frigon, and R.F. Sokoloff, Human chromogranin A: Purification and characterization from catecholamine storage vesicles of pheochromocytoma, *Hypertension* **6**:2 (1983).
7. L.J. Deftos, S.S. Murray, D.W. Burton, R.J. Parmer, and D.T. O'Connor, A cloned chromogranin A (CgA) cDNA detects a 2.3 KbmRNA in diverse neuroendocrine tissues, *BBRC* **137**:418 (1986).
8. A. Icangelo, H-U. Affolter, L.E. Eiden, E. Herbert, and M. Grimes, Bovine chromogranin A sequence and distribution of its messenger RNA in endocrine tissues, *Nature* **323**:82 (1986).
9. S.S. Murray, L.L. Daven, D.W. Burton, D.T. O'Connor, P.L. Mellon, and L.J. Deftos, The gene for human chromogranin A (CgA) is located on chromosome 14. *Biochem Biophys Res Commun* **142**:141 (1987).
10. D.T. O'Connor and L.J. Deftos, Secretion of chromogranin A by peptide producing endocrine neoplasms, *N Engl J Med* **314**:1145 (1986).
11. B.S. Wilson and R.V. Lloyd, Detection of chromogranin in neuroendocrine cells with a monoclonal antibody, *Am J Pathol* **115**:458 (1984).
12. G. Rindi, R. Buffa, F. Sessa, O. Tortora, and E. Solcia, Chromogranin A, B and C immunoreactivities of mammalian endocrine cells. Distribution, distinction from costored hormones/prohormones and relationship with the argyrophil component of secretory granules, *Histochemistry* **85**:19 (1986).
13. E. Wilander, Diagnostic pathology of gastrointestinal and pancreatic neuroendocrine tumors, *Acta Oncol* **28**:363 (1989).
14. H. Höfler, H. Denk, E. Lackinger, G. Helleis, J.M. Polak, and Ph.U. Heitz, Immunocytochemical demonstration of intermediate filament cytoskeleton proteins in human endocrine tissues and neuro-endocrine tumors, *Virchows Arch (Pathol Anat)* **409**:609 (1986).
15. E. Wilander and L. Scheibenphlug, Cytokeratin expression in small intestinal and appendiceal carcinoids. A basis for classification, *Acta Oncol* (in press).
16. J. Rode, A.P. Dillon, L. Papadaki, and D. Griffiths, Neurosecretory cells of the lamina propria of the appendix and their possible relationship to carcinoids, *Histopathol* **6**:69 (1982).
17. L. Auböck and H. Höfler, Extraepithelial intraneural endocrine cells as starting points for gastrointestinal carcinoids, *Virchows Arch (Pathol Anat)* **401**:17 (1983).
18. J. Rode, A.P. Dhillon, and L. Papadaki, Serotonin immunoreactive cells in the lamina propria plexus of the appendix, *Hum Pathol* **14**:464 (1983).
19. E. Wilander, M. Lundqvist, and T. Movin, S-100 protein in carcinoid tumors of the appendix, *Acta Neuropathol* **66**:306 (1985).
20. M. Lundqvist and E. Wilander, Subepithelial neuroendocrine cells and carcinoid tumors of the human small intestine and appendix. A comparative immunohistochemical study with regard to serotonin, neuron-specific enolase and S-100 protein reactivity, *J Pathol* **148**:141 (1986).

21. M. Lundqvist and E. Wilander, A study on the histopathogenesis of carcinoid tumors of the small intestine and appendix, *Cancer* **60**:201 (1987).

22. W.C. Black and H.E. Haffner, Diffuse hyperplasia of gastric argyrophil cells and multiple carcinoid tumors, *Cancer* **21**:1080 (1968).

23. J.R. Hodges, P. Isaacson, and R. Wright, Diffuse enterochromaffin-like (ECL) cell hyperplasia and multiple gastric carcinoids: a complication of pernicious anaemia, *Gut* **22**:237 (1981).

24. K. Borch, H. Renvall, and G. Liedberg, Gastric endocrine cell hyperplasia and carcinoid tumors in pernicious anemia, *Gastroenterology* **88**:638 (1985).

25. K. Borch, H. Renvall, E. Kullman, and E. Wilander, Gastric carcinoid associated with the syndrome of hypergastrinemic atrophic gastritis, *Am J Surg Pathol* **11**:435 (1987).

26. K. Sletten, P. Westermark, and J.B. Natvig, Characterization of amyloid fibril proteins from medullary carcinoma of the thyroid, *J Exp Med* **143**:993 (1976).

27. B. Johansson, C. Wernstedt, and P. Westermark, Atrial natriuretic peptide deposited as atrial amyloid fibrils, *Biochem Biophys Res Commun* **148**:1087 (1987).

28. P. Westermark, L. Grimelius, J.M. Polak, L.I. Larsson, S. Van Noorden, E. Wilander, A.G.E. Pearse, Amyloid in polypeptide hormone producing tumors, *Lab Invest* **37**:212 (1977).

29. P. Westermark, C. Wernstedt, E. Wilander, and K. Sletten, A novel peptide in the calcitonin gene-related peptide family as an amyloid fibril protein in the endocrine pancreas, *Biochem Biophys Res Commun* **140**:827 (1986).

30. P. Westermark, C. Wernstedt, E. Wilander, D.W. Hayden, T.D. O'Brien, K.H. Johnson, Amyloid fibrils in human insulinoma and islets of Langerhans of the diabetic cat are derived from a neuropeptide-like protein also present hormonal islet cells, *Proc Natl Acad Sci USA* **84**:3881 (1987).

31. P. Westermark, E. Wilander E, G.T. Westermark, and K. H. Johnson, Islet amyloid polypeptide-like immunoreactivity in the B cells of Type 2 (non-insulin-dependent) diabetic and non-diabetic individuals, *Diabetologia* **30**:887 (1987).

32. A. Lukinius, E. Wilander, G.T. Westermark, U. Engström, and P. Westermark, Co-localization of islet amyloid polypeptide and insulin in the B cell secretory granules of the human pancreatic islets, *Diabetologia* **32**:240 (1989).

33. E. Wilander, L. Scheibenphlug, B. Eriksson, and K. Öberg, Diagnostic criteria of classical carcinoids, *Acta Oncol* **30**:469 (1991).

CHAPTER 9

MODERN METHODS IN CYTOLOGY

**Siew-Khin Tang[1] MD,FRCPA,FIAC and
Anthony S-Y Leong[2] MBBS,MD,FRCPA,FRCPath,FCAP**

[1]Principal Specialist, Cytopathologist
Department of Anatomical Pathology
Alfred Hospital, Melbourne, Victoria, Australia
Visiting Senior Pathologist, Institute of Medical & Veterinary Science
Adelaide, South Australia, Australia
[2]Clinical Professor and Director, University of Adelaide, and
Division of Tissue Pathology, Institute of Medical & Veterinary Science
Adelaide, South Australia, Australia

SUMMARY

The analyses of exfoliated cells in peritoneal, pleural, synovial, vaginal and cerebro-spinal fluids, and in urine and sputum have formed the mainstream of cytopathologic examination. The ability to obtain cells at endoscopy, lavage, and directly through scrapings of cutaneous and mucosal surfaces and operative specimens has increased the diagnostic potentials of cytopathology. This specialty has enjoyed a resurgence of popularity largely through the introduction of fine needle aspiration techniques which provide the ability to retrieve cells from almost any anatomical site in the body.

In recent years, the introduction of several new technologies has provided important and powerful adjuncts to morphologic examination, and most of these procedures can be adapted to cytologic preparations. In this chapter, the use of immunocytochemistry for the identification of poorly differentiated neoplasms, the delineation of newer biological markers which aid in tumor prognostication, and the analysis of molecular parameters which assist in our understanding of the genesis of neoplasms are reviewed. As in histologic sections, the application of immunohistochemical techniques to cytologic preparations has the distinct advantage of allowing concurrent

---

[1] Address for Correspondence: Siew-Khin Tang, M.D., Department of Anatomical Pathology, Alfred Hospital, Prahran, Victoria, 3181, Australia, Phone No: 61 3 276 3157, Fax: 61 3 521 3113

Modern Methods in Analytical Morphology, Edited by
J. Gu and G.W. Hacker, Plenum Press, New York, 1994

morphologic correlation as opposed to cytosolic or biochemical assays performed on homogenized tissues containing neoplastic cells, non-neoplastic cells and supporting tissue. The use of flow cytometry, image analysis and cytophotometry in tumor analysis, particularly for the definition of tumor growth fractions and ploidy, as well as for the rapid sorting of cells identified by antigen-specific fluorescent markers is also discussed. The adaptation of molecular biological techniques such as *in situ* hybridization and, more recently, *in situ* polymerase chain reaction to tissue sections and cytologic preparations provide additional powerful tools for the identification of gene mutations, viral DNA and messenger RNA, further extending the potentials of cytologic examination. Applications of these modern analytical techniques to cytologic specimens and methodological details of each of these procedures are presented.

INTRODUCTION

Cytologic examination has been an important modality in the early diagnosis of cancer and for the confirmation of suspected pathological processes, especially those of an infective nature.[1-4] The specialty of cytopathology has enjoyed a resurgence of popularity with the introduction of the fine needle aspiration technique which allows retrieval of tissue samples from almost every anatomical site in the body.[5,6] In addition, cells obtained at endoscopy, lavage and from direct scrapings of skin, mucosa and operative material have supplemented the examination of exfoliated cells in sputum, effusions, urine, cerebro-spinal and joint fluids. Not only has cytologic examination provided a means of rapid diagnosis, but it has also increased diagnostic accuracy by providing tissues for various adjunctive analyses including immunocyto-chemistry for the identification of cell and tumor markers,[7-13] ultrastructural examination for the identification of fine structural characteristics and features of differentiation,[14,14a,15-18] for morphometric parameters, image analysis, flow cytometry for rapid cell sorting and DNA analysis, *in situ* hybridization and polymerase chain reaction.[19-21] The latter two procedures are particularly useful when only small quantities of viral protein are present, or when point mutation and gene rearrangement is suspected as with the malignant lymphomas.[22-31]

Besides the diagnosis and classification of tumors, prognostication is an important aspect of cytologic and histologic examination. The prediction of tumor behavior has long been imprecise and estimated with uncertainty. Until recently, this was based largely on morphologic features and dependent on therapeutic response. With the increasing repertoire of cancer treatment regimes and protocols, there is additional impetus to identify subsets of patients with similar histologic types of tumors who may respond to different therapies. In recent years, much new information has been gleaned from molecular techniques. Many tumor prognostic parameters can be assessed in cytologic preparations employing immunocyto-chemical procedures, *in situ* hybridization and polymerase chain reaction. These techniques allow the study of cell kinetics which assists in prediction of tumor behavior, the assessment of hormone receptors which indicate tumor responsiveness to hormonal manipulations, and the assessment of oncogenes and their products which not only provide prognostic information but also widen our understanding of tumor genesis.[32-34] The combination of cell image analysis and cytophotometry coupled with immunostaining, allows simultaneous quantitative and qualitative measurements which have the potential of providing more refined prognostic information.[19,35-37]

The sorting of disaggregated tumor cells by flow cytometry followed by *in situ* hybridization of target DNA sequences allows the detection of viral protein and tumor markers in different cell compartments or groups.[20,21,38-40] While surgical biopsy material was the main source of tissue for such studies, it has become increasingly apparent that fine needle aspirates, tissue scrapings as well as effusions, lavage, urine and cerebro-spinal fluid provide sufficient material for such analyses without the risks of general anesthesia, operative morbidity and the need for hospitalization.[41] Many modern biological analytical methods have been adapted for application to cytologic material.[42,43] In the past, tissues submitted for tumor cell protein and receptor assays, studies of oncogenes and their proteins, and other biological assessments were based on minced and homogenized material which contained both tumor and non-tumorous tissue. The tests designed to detect the substances of interest were not controlled by morphologic criteria and thus did not selectively identify neoplastic cells in the homogenization and extraction process. In contrast, *in situ* analyses in cytologic and histologic preparations have the major advantage of allowing morphologic identification of the cells of specific interest. The increasing sensitivity of such procedures makes analysis in cytologic preparations a viable proposition.

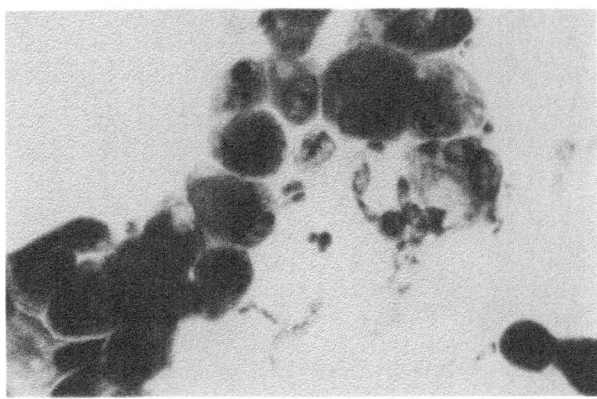

Figure 1. Malignant cells from an adenocarcinoma (Papanicolaou stain X 10).

CELLULAR MORPHOLOGY

Cellular morphology is the sole criteria for determining the nature of cells in cytologic preparations and for identifying the presence of neoplastic change. This contrasts with histologic diagnosis which is based not only on cytomorphologic features but also on cellular associations or architectural patterns.[44] Cytologic examination of disaggregated cells is thus handicapped by the absence of the tissue architecture. Where operative biopsy is contraindicated, often the only tissue available for examination would be that obtained by fine needle aspiration, particularly of deep structures in intrathoracic, intra-abdominal or intracranial locations. As the sample obtained is limited, it has to be carefully allocated for various studies so that its examination is optimized, particularly in the investigation of neoplastic tissue.[45,46]

ELECTRON MICROSCOPY

A benign or atypical cell closely resembles its normal counterpart. A transformed or mutated cell, especially when neoplastic, becomes progressively different in cytologic appearance with increasing degrees of malignancy. Ultimately, its histogenetic origin becomes unidentifiable by light microscopy. The determination of cell lineage or differentiation can be aided by electron microscopy and immunohistochemistry, both of which have been established as important tools to extend the morphological examination of cytologic preparations.[47-49]

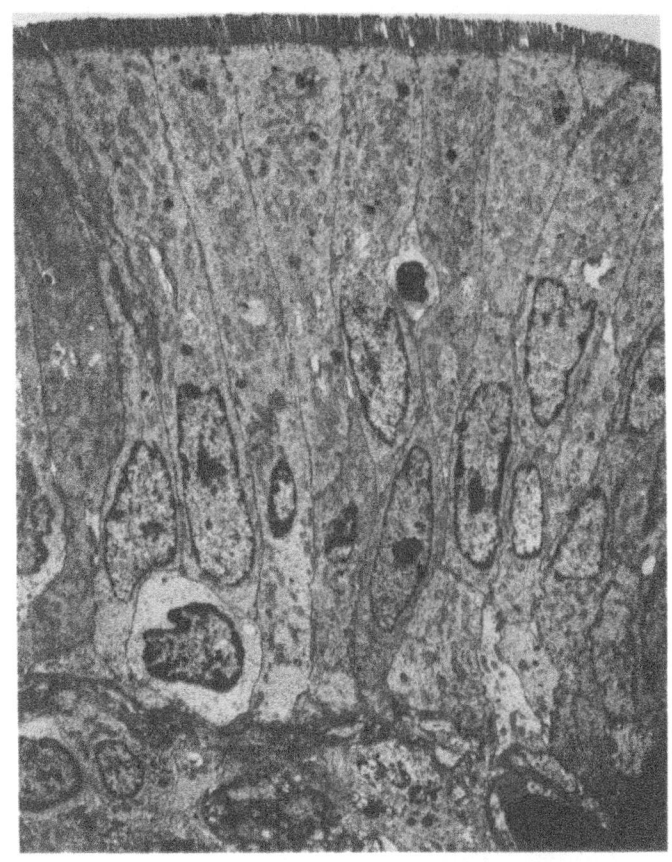

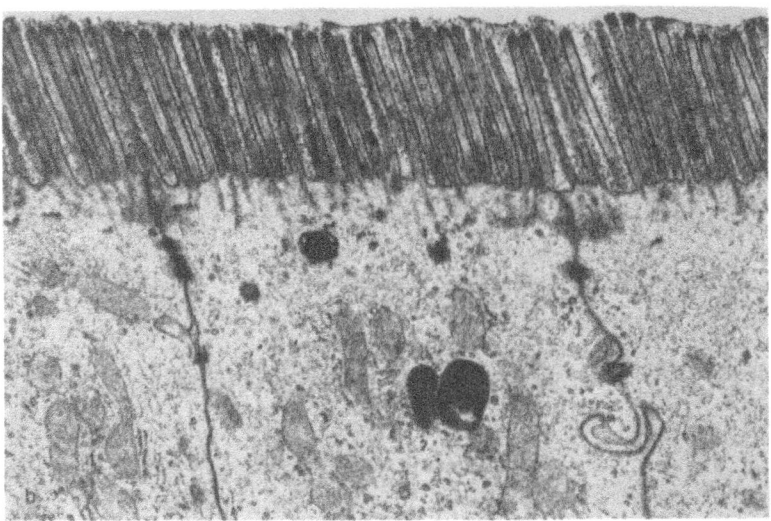

Figure 2a. Intestinal mucosa showing brush border, tight-junction complex and (TEM X 5,600).
Figure 2b. Microvilli of brush border showing glycocalyceal filaments (TEM X 23,000).

Epithelial cells are identified ultrastructurally by their well-defined cytoplasmic boundaries, attachment apparatus, cytoskeletal components and secretory products.[43,44]

Mesothelioma cells possess long complex microvilli which are identifiable by electron microscopy. These microvillous protrusions correspond to the long surface processes which can be accentuated by immunostaining for epithelial membrane antigen (EMA), allowing differentiation from glandular epithelial cells with luminal microvilli at light microscopic level. The complex, long and slender microvillous processes are aberrant and circumferentially located in the mesothelioma cells.

Non-epithelial cells include cells of mesenchymal and hemopoietic origin. Such cells do not have cell junctions or attachment apparatus. Their cytoplasmic boundaries may display pseudopodia which interdigitate with neighboring cells, or they may have filopodia as observed in lymphoid cells or in hairy cell leukaemia. Filaments and specialized structures such as membrane-bound crystals of alveolar soft-part sarcoma, Birbeck granules of Langerhans' cells, Weibel-Palade bodies of endothelial cells and dense core membrane-bound granules of neuroendocrine cells permit identification of cell lineage and differentiation at the ultrastructural level.

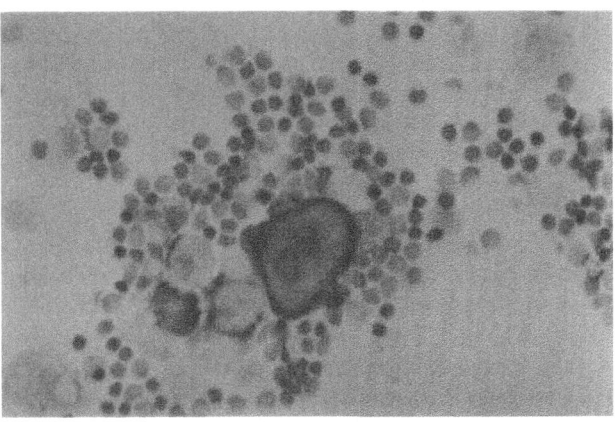

Figure 3. Mesothelial cell with membrane staining of epithelial membrane antigen (EMA) (indirect immunoperoxidase stain X 400).

Microwave-stimulated fixation of aspirated material for transmission electron microscopy and microwave-stimulated staining have considerably accelerated turn-around times to provide a more precise and rapid diagnosis particularly in the examination of neoplasms.[50]

SPECIAL MICROSCOPIC TECHNIQUES

Scanning laser confocal microscopy bridges the gap between light and electron microscopy. Point probing by a combined laser source and a beam scanning system allows high resolution of up to 0.2 μm. This allows "optically sectioned" images at different focal planes to be generated and by composite imaging, three-dimensional pictures can be produced. Combined with immunocytochemistry employing fluorescent dyes, details of subcellular organelles can be examined with increased resolution and precision. More recently, the very powerful atomic force microscope has been developed permitting the scanning of molecules on cell membranes as well as the imaging of DNA. Its ability to operate in a vacuum or submerged in liquid allows imaging to be performed under physiological conditions and in living cells.[51]

IMMUNOCYTOCHEMISTRY

Sensitive immunoenzyme techniques together with the increasing range of sensitive and specific monoclonal antibodies have rapidly established immunocytochemistry as a powerful diagnostic adjunct to morphologic examination. Its role in tumor diagnosis is well accepted and it has made important contributions in other areas of human diseases.

Tissue-Associated Antigens

Cells have a cytoskeletal network that provides integrity and matrix for cytoplasmic organelles. The cytoskeleton is made up of structural proteins which vary with cell function. It is made up of filaments of 6 to 25 μm comprising 6 μm actin filaments, 8-10 μm intermediate filaments, 15 μm myosin filaments and 26 μm microtubules. Although subtypes of intermediate filaments are indistinguishable at ultrastructural level, they show distinct biochemical and antigenic characteristics and are divided into at least five groups, reflecting their tissue of origin (Table 1). Neurofilament proteins are found in neurons, glial fibrillary acidic proteins in astrocytes, vimentin in non-myogenic mesenchymal cells, desmin in muscle, and cytokeratins in epithelial cells. The expression of these filamentous proteins is restricted and conserved in both benign and malignant neoplastic proliferation so that intermediate filament analysis has been a useful means of classifying tumors, particularly those which are undifferentiated, lack identifying features, or appear morphologically similar.

The cytokeratins are a multigene-coded group of at least 20 different cytokeratin proteins which are present in virtually every true epithelial cell. As a general rule, high-molecular weight cytokeratins are expressed in stratified epithelial cells, whereas low-molecular weight cytokeratins are found in simple epithelial cells. While cytokeratins are powerful tools in the identification of epithelial neoplasms, it is now recognized that some non-epithelial tumors may also express this intermediate filament subtype. Such tumors include chondroblastoma, leiomyosarcoma, synovial sarcoma, epithelioid sarcoma, chordoma and epithelioid angiosarcoma. Antibodies to desmosomes and tight junction complexes react with specific antigens at these sites and are readily revealed by immunostaining of frozen sections employing an antibodies to desmoplakin. The demonstration of desmin identifies mesenchymal tumors manifesting muscle differentiation such as rhabdomyosarcomas and leiomyosarcomas. Three types of neurofilament proteins (68, 150 and 200 kd) have been found in differentiated ganglion cells of central and peripheral nervous system neoplasms, in tumors of mixed cell origin such as gangliogliomas, ganglioneuro-blastomas, neuroblastomas, medulloblastomas, differentiated retinoblastomas and the neural elements of teratomas. In addition, neurofilament proteins have been found in some tumors of the peripheral neuroendocrine system including those of the adrenal medulla, extra-adrenal paragangliomas and Merkel cell carcinomas. Glial fibrillary acidic protein (GFAP) expressed by astrocytes may also be found in non-astrocytic central nervous system tumors such as ependymomas, subependymomas, oligodendro-gliomas, choroid plexus papillomas, as well as the stromal cells of hemangioblastomas. GFAP may also be found in non-myelinating Schwann cells, cells of the myenteric plexus, folliculostellate cells of the pituitary, epithelial cells of the eye lens, Kupffer cells, cartilage cells of the epiglottis and in tumors such as schwannomas, neurofibromas, paragangliomas, salivary gland tumors and neoplasms of Müllerian origin.

Table 1. Intermediate Filament Proteins

Cytokeratins	Epithelial	40,000-68,000 daltons	Keratinizing and non-keratinizing squamous cells
Vimentin	Mesenchymal	57,000	Fibroblast, chondrocytes, endothelial cells, mesothelium
Desmin	Muscle	53,000	Skeletal and smooth muscle
GFAP	Glial	55,000	Glial tissue
Neurofilament protein	Neutral	68,150,200kD	Neurons, peripheral nerve

Antibodies to tissue-associated antigens are particularly useful in tumor diagnosis as they often allow the specific identification of anaplastic tumors, especially when they occur as metastases from unknown primary sites (Tables 2 and 3). Unfortunately, very few antigens are truly tissue specific and, at best, are only tissue-associated. This is important to remember as the demonstration of such antigens may not specifically indicate the source of the tumor and the exceptions, although less common, should also be considered. For example, while prostatic acid phosphatase is most commonly

Table 2. Morphologic and Immunocytochemical Features of Mesothelioma and Adenocarcinoma

	Mesothelioma	Adenocarcinoma
Morphology	Thick cytoplasm Peripheral blebs Anisonucleosis	Delicate vacuolated cytoplasm Indistinct or irregular borders Pleomorphic, anisocytosis
Stains	PAS +, Intra- and extracellular Diastase digestible Alcian blue + Hyaluronidase digestible	PAS + intracellular in secretory vacuole Diastase resistant
Immunocytochemistry	CEA negative EMA membrane staining Cam 5.2 (LMW) + Keratin (HMW) +	CEA + EMA cytoplasmic and membrane staining Cam 5.2 (LMW +)
Ultrastructure	Slender microvilli all around cell No glycocalyceal bodies Desmosomes + No tight junctions Tonofilaments surround nucleus Glycogen granules No secretory vacuoles	Short, stubby microvilli on luminal aspect Glycocalyceal filamentous bodies, core to villi Terminal bars near luminal aspect Tight junctions Irregularly distributed intermediate filaments Variable glycogen granules Numerous secretory granules

LMW = low molecular weight keratin
HMW = high molecular weight keratin
CEA = carcinoembryonic antigen
EMA = epithelial membrane antigen

expressed by tumors of prostatic origin, the antigen may also be found in islet cell tumors, albeit less commonly; transitional cell carcinoma of the bladder, renal cell carcinoma and carcinoid tumors. Other tissue-associated antigens include prostatic specific antigen which is expressed by prostatic cells and their tumors, factor VIII-associated protein expressed by endothelial cells of blood vessels and their tumors such as angiosarcoma and Kaposi's sarcoma, HMB45, a melanoma-specific antigen which is expressed by epidermal naevus cells and malignant melanomas and rarely in other tumors, muscle-specific antigen expressed by cells and tumors showing myogenic differentiation including myoepithelial cells, chromogranin by neuro-endocrine cells, and placental alkaline phosphatase by germ cells and their tumors. With neoplastic conversion, some tumors develop abnormal protein synthesis and revert to the production of oncofetal antigens such as carcinoembryonic antigen, alpha fetoprotein, human chorionic gonadotrophin and human placental lactogen, particularly in cases of tumors of germ cell origin at both gonadal and extragonadal sites.[49]

The leucocyte common antigen is, to date, one of the most tissue-specific markers and labels hemopoietic and lymphoid cells. Neoplastic transformation of B-lymphocytes often displays light chain immunoglobulin restriction and allows a method of identifying monoclonal proliferations of B-lymphocytes. The immunotyping of lymphomas has been pursued with great interest and our current understanding of the maturation sequence and differentiation of normal lymphocytes has largely resulted from the study of neoplastic lymphoid cells. However, many of the surface markers of lymphoid cells are sensitive to fixation so that fresh cell suspensions or frozen tissue are generally required for immunocytochemistry.

Some hormones are tissue-specific and their demonstration in metastatic deposits are invaluable pointers to the primary source of the tumor. For example, thyroglobulin is expressed only by cells of the thyroid and their tumors. Other hormones, however, such as calcitonin may be secreted by a variety of tumors besides the parafollicular cells of the thyroid gland. Calcitonin may also be produced by endocrine tumors such as Merkel cell carcinoma and oat-cell carcinoma of the lung.

Table 3. Differential Diagnosis of Small Round Cell Tumors

	LMW-K	LCA	NSE	NFP	Vimentin	DES
Small Cell carcinoma	+	-	+	-	-	-
Lymphoma	-	+	-	-	+	-
Neuroblastoma	-	-	+	+	-	-
Rhabdomyosarcoma	-	-	-	-	+	+

LMW-K	= Low molecular weight cytokeratin
LCA	= leucocyte common antigen
NSE	= neurone specific antigen
NFP	= neurofilament protein
DES	= desmin

Antibodies to viral antigens such as hepatitis B, cytomegalovirus, herpes virus and human papilloma virus have been useful markers particularly in the immune compromised patient who may be subjected to a myriad of infectious agents. Immunostaining provides a rapid method for the detection of small quantities of antigens associated with micro-organisms and monoclonal antibodies have been raised to various bacteria including *B abortus, E coli, H influenzae, M tuberculosis, M leprae, N gonorrhoea, N meningitidis, P aeruginosa, T palladium* and the Staphylococcus species.

Table 4. Differential Diagnosis of Germ Cell Tumors

	CK	CEA	Vimentin	LCA	PLAP
Germ cell tumor	+	±	-	-	+
Lymphoma	-	-	+	+	-
Carcinoma	+	+	-	-	-

CK = broad spectrum cytokeratin
CEA = carcinoembryonic antigen
LCA = leucocyte common antigen
PLAP = placental alkaline phosphatase

Cell Proliferation Markers

Several factors regulating cell kinetics have recently been identified and they have contributed to a better understanding of tumor behavior as well as allowing an important method of tumor prognostication.[52-54] One of these is the nucleolar organizer region protein. A simple silver stain can be performed on cytologic specimens to display nucleolar organizer regions which represent loops of DNA in interface nuclei. These are found in five of the acrocentric chromosomes and are thought to transcribe DNA. They have been shown to correlate with the number of aneuploid cells measured by DNA cytometry.[55-58] Nucleolar organizer region proteins have an affinity for silver and are readily stained by an argentaffin silver nitrate procedure. Their size and number reflect cell proliferative activity and appear to correlate with nuclear grade as in the non-Hodgkin's lymphomas, although the relationship between interphase nucleolar organizer region counts, S-phase fraction and DNA index of tumor cells is controversial.[59]

Table 5. Enzyme Reaction and Chromogen Staining

Enzyme	Chromogen	Reaction Product	Counterstain	Mounting Medium
Horseradish Peroxidase (indirect peroxidase PAP, ABC)	3.3 Diaminobenzidine (DAB)	Brown (insoluble)	Hematoxylin	Depex
	3-amino-ethylcarbazole	Red (soluble in alcohol)	Hematoxylin	Glycerogel
Alkaline Phosphatase (APAAP)	5-bromo-4-chloro-3 indoloylphosphate (BCIP)/nitro blue	Purple (partly soluble)	0.5% fast green FCF in DH_2O or 0.3% pyronine in 0.05M acetate buffer pH 4.8	Depex
	Fast Red	Red (soluble in alcohols)	Hematoxylin	Glycerogel
	New Fuchsin	Red (insoluble)	Hematoxylin	Depex

129

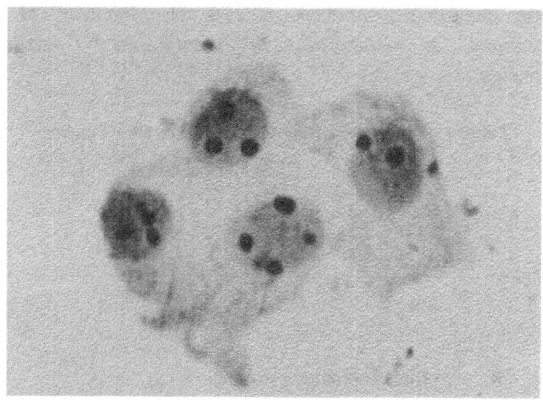

Figure 4. Proliferating cell nuclear antigen in breast carcinoma (PAP X 400).

The tumor growth fraction is one of the more important factors which has been demonstrated to influence prognosis. While mitotic activity has long been employed as one indicator of tumor proliferation, its analysis is fraught with inconsistencies because of the subjective nature of identifying mitotic cells. Furthermore, delays in tissue fixation significantly influence the mitotic count. Measurements of DNA content (DNA ploidy) and proliferation indices by flow cytometry and image cytophotometry allow the identification of diploid and aneuploid tumors, the latter showing a poorer prognostic outcome. The influence of tumor growth fraction on the behavior of neoplasms has been widely studied,[60-64] and aneuploidy and abnormal DNA content have also been demonstrated to indicate shorter survival or shorter disease-free intervals in patients with breast cancer.[65] The recent production of antibodies to cycling cells (Ki-67) and to proliferating cell nuclear antigen (PCNA) allows the analysis of these parameters in frozen and even formalin-fixed tissue. Assessment of these antigens can also be performed on cytologic smears.[66]

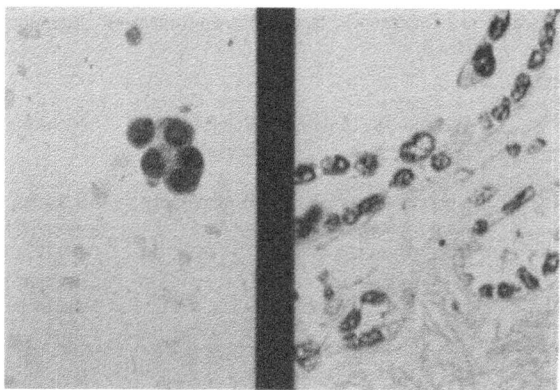

Figure 5. Nucleolar organizer regions (NOR) in breast carcinoma (X 400).

Oncogenes and Oncogene Proteins

The discovery of oncogenes has led to considerable interest in their biological role and value as markers of neoplastic behavior. When oncogenes become amplified and undergo mutation or translocation, their inappropriate activity is reflected by increased levels of expression of mRNA and oncoprotein. The alteration of DNA level can presently be demonstrated only by the techniques of molecular biology, and *in situ* hybridization can be used to demonstrate the mRNA, but the signal is often weak. The method is not reliable for routine application. Oncoproteins often have a short half-life, which often makes their detection by immunohistochemistry difficult. However, the mutant oncoprotein can be detected in several instances.

The proteins of *c-erbB-2 (HER2/neu)*, *c-myc* and *ras* have been the most widely studied. Expression of *c-erbB-2* in invasive breast carcinoma has been shown to be associated with a relatively poor survival, as in the case of ovarian carcinomas.[67-71] The oncoprotein can be demonstrated in fine needle aspiration samples of breast carcinoma.[72,73]

Several tumor suppressor genes have been identified, the three more important ones in human tumors being the retinoblastoma susceptibility gene, deleted in colonic cancer (DCC) gene and p53. The p53 protein binds to specific DNA sequences and appears to be a transcription factor that blocks the progression of cells through the G_1 phase of the cell cycle. Mutation in one allele of this tumor-suppressor gene renders a normal cell susceptible to transformation, and subsequent loss of the other allele presumably removes all growth restraints in the tumor cells. Mutant p53 genes have also been shown to co-operate with *ras* to transform cells *in vitro*. The mutant form of p53 protein accumulates in the nuclei of cells of breast carcinoma and has been shown to be associated with poor histologic grade, reduced overall survival, reduced metastasis-free status and inversely with oestrogen receptor status.[74-75]

Steroid Hormone Receptors

The interplay between hormone levels and target organs and the stimulation by trophic mediators are finely tuned and not fully understood. Cells of the ovary, endometrium and breast are responsive to sex steroid hormone levels.[76-80] Other organs and tissues responsive to hormone manipulation have been reported and require further studies.[81] The biochemical method of employing dextran-coated charcoal for the estimation of hormone receptors in breast cancer has been in practice for many years. While this method provides quantitative measurements and interlaboratory observations can be standardized to a certain degree, only larger institutions are able to afford the hardware required because of the prohibitive costs in purchase and maintenance of the machine. Recently, it has been shown that immunocytochemical methods of detection of hormone receptors are reliable and can be readily performed at low cost, with rapid turn around times.[82,83] With the wide practice of fine needle aspiration cytology, breast carcinoma can be readily assessed, particularly in inoperable, recurrent or metastatic disease where surgery may be contraindicated.[84] Faced with a choice of treatment modalities which include non-surgical intervention with chemotherapy and hormone manipulation, or chemotherapy or radiotherapy, selection of the optimal regime may be guided by the hormone receptor status of the tumor.

Steroid hormones regulate cell activity and proliferation by interaction with specific proteins through binding sites or receptors. Estrogen receptors in breast and gynecologic tissues, and androgen receptors in prostate and germinal epithelium of the testis are located in the epithelial cell nuclei. Binding with the appropriate hormone converts the receptors to an active form and the estrogen-receptor-complex binds to chromatin to regulate gene expression and synthesis of mRNA specific for the cell, initiating the synthesis of many specific proteins. In addition, the estrogen-receptor-complex initiates transcription of progesterone-receptor mRNA and the binding of progesterone to activated progesterone-receptor similarly induces synthesis of specific proteins including growth factors. Thus, the expression of estrogen, progesterone receptor-related proteins, and other steroid hormone receptors in breast carcinoma are useful indicators of tumor behavior and influence the selection of treatment modalities. Breast cancers which are estrogen and progesterone receptor positive respond better to hormonal manipulation than those negative for these receptors, particularly in the post-menopausal patient. Cell kinetic studies show S-phase proliferation to be highest in estrogen receptor and progesterone receptor negative breast cancers.[85-89]

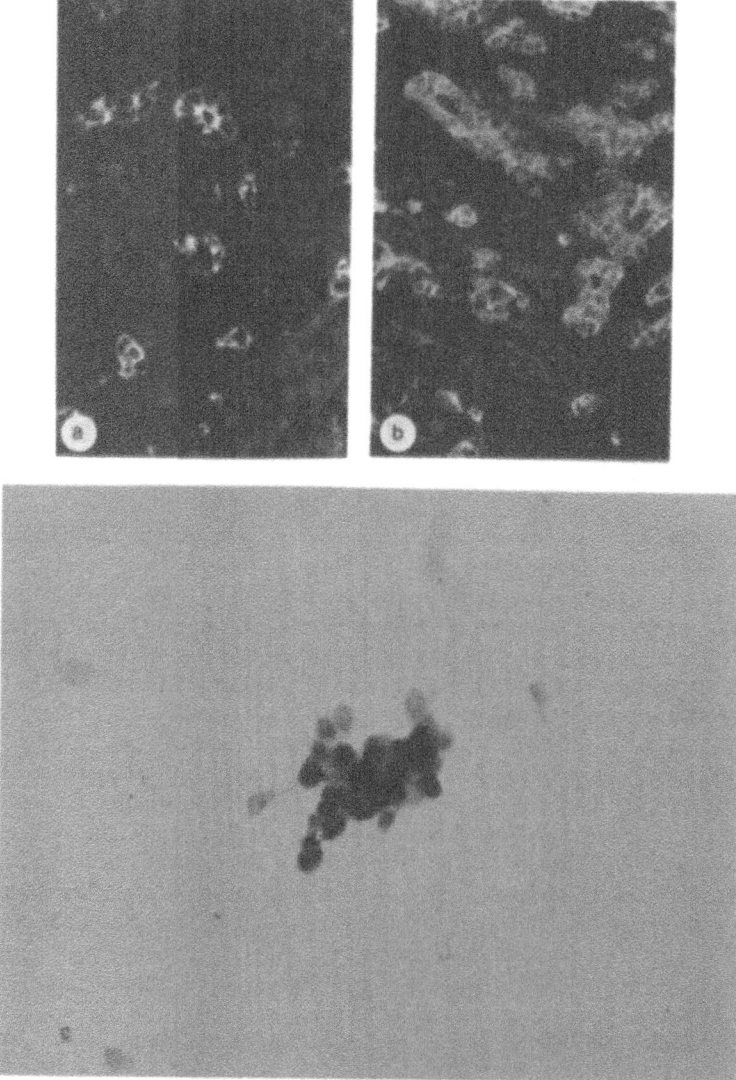

Figure 6a. Oestrogen receptors in breast carcinoma (FITC X 400).
Figure 6b. Progesterone receptors in breast carcinoma (TRITC X 400).
Figure 6c. Oestrogen receptors in nuclei of malignant cell from breast carcinoma, ER-ICA (Abbott) (PAP X 400).

Immunocytochemical assay of hormone receptors is a less expensive and more rapid procedure compared to radioimmunoassay methods, including the dextran coated charcoal method for receptor quantitation. The results obtained by immunocytochemical stains correlate well with those of cytosolic assays and, therefore, have been increasingly used in many institutions.[90] The ability to selectively identify cells which are cytologically malignant in immunostains for hormone receptors provides a much more accurate method over cytosolic assays which measure both homogenized neoplastic and non-neoplastic tissues in the sample.

The lack of quantitative measurement and standardization of immunocytochemical stains poses a disadvantage which may be overcome by the use of image analysis and cytophotometric techniques as they retain the ability for morphologic selection of the malignant cells for measurement.[91,92]

Other Biological Markers

Epidermal growth factor receptor is an estrogen-dependent protein found in epithelial cells of the breast and reproductive organs. An inverse relationship between oestrogen receptor status and the epidermal growth receptor has been demonstrated in breast cancer with elevated levels of the latter receptor in patients who demonstrate decreased survival or shorter disease-free intervals.[97] The levels of Cathepsin D, an estrogen-induced lysosomal protease, correlates with aneuploidy in breast carcinoma but is not related to tumor size or the oestrogen and progestogen receptor status.[98] Other receptors and protein products such as metastasis-associated laminin receptor and collagenase IV have been shown to be related to stromal invasion and angiogenesis in carcinoma. P-glycoprotein is a transmembrane protein which functions as an efflux drug transporter and recent studies have shown that patients who demonstrate good response to chemotherapy have significantly lower levels of p-glycoprotein in pre-treatment tumor cells when compared to those displaying residual disease. A multidrug resistant gene may be responsible for poor response to chemotherapy or hormone manipulation[90-101] and this protein, as well as those discussed above, are detectable by immunostaining in cytologic preparations. We are, therefore, moving into an era where many biological markers are clearly assessable by immunostaining, providing a potential to identify subsets of patients who may respond to different modalities of cancer treatment.

While there is a potential to assess many biological and molecular markers with immunocytochemical techniques, it cannot be overemphasized that the technique is largely fixation-dependent. The inability to label the tumor cell for a specific antigen implies an absence of the marker or its destruction due to suboptimal preservation. While some antigens may be rescued by prior enzyme digestion, an antigen which is destroyed by fixation is unlikely to be retrievable. Detailed discussion of suitable fixatives is beyond the scope of this chapter. Briefly, acetone is suitable for the preservation of lymphocyte membrane antigens, alcohol for intermediate filament proteins and, in some situations such as the staining of oestrogen receptor proteins, a sequential fixation in various reagents may be necessary. The choice of fixative is thus dependent on the nature of the antigen to be detected. In some situations, it may be necessary to choose between fixatives which preserve antigens and those which preserve cytomorphology as the two are often inversely related.

Antibody specificity must be taken into consideration in planning diagnostic antibody panels. The latter are determined largely by the pathologic entities considered in differential diagnosis. In all instances, appropriate positive and negative controls must be included.

FLOW CYTOMETRY

The speed with which millions of cells can be analyzed makes flow cytometry a most attractive diagnostic tool. Cells in suspension (obtained by fine needle aspiration, scraping or washings, and body fluids), are allowed to pass across a beam of light, in a single file in a stream or column of fluid. Light scatter is captured by photometric tubes, the stimulus being translated by a computer to digital signals which are interpreted as a frequency histogram.

The DNA content of a cell, its surface contour and the number of cells with specific markers identified by fluorescent labelled antibodies can be determined. Cell groups can also be sorted according to phases in the cell cycle, and further analyses or tests may be performed on cells in these compartments. Such cell populations may be identified as being diploid or aneuploid based on DNA content. Cells from benign tumors are mainly diploid, while malignant cells are more often aneuploid. Fluorescent activated cell sorting (FACS) can be achieved with an accuracy of 98%, attaining what was once an impossible task of automated separation of cells with specific markers identified by fluorescent tagged antibodies.[80]

IMAGE ANALYSIS AND CYTOPHOTOMETRY

Cell morphology and DNA content can be measured by computerized microscopy on cytological smears.[94,95] The equipment comprises a computer-enhanced, high resolution digital video camera mounted on a microscope. By employing one of many software programs that have been designed by experts in the field, images of cells can be studied in detail. At least some of these programs ensure objective uniformity in measurement of observations in the fields of histopathology and cytopathology, allowing a more disciplined approach to analyses which for many years have been based on subjective observations. Thus, some form of quantitative immunocytochemistry is now possible. Besides, image analysis allows examination of small numbers of cells, a distinct advantage over flow cytometry which requires a minimum number of 10^{x6} cells/ml. Additional requirements in the form of optimal staining, essential controls and standards, and precision microscopic equipment should not detract from the use of morphometric technique of cell analysis.

MOLECULAR BIOLOGY

The importance and contributions of molecular biology to modern medicine are well reflected in the burgeoning number of papers describing the technology and by the increasing number of DNA probes and reagents available commercially. The identification and characterization of normal as well as abnormal genes has led to elucidation of their function and dysfunction. This has contributed to our understanding of the pathogenesis of many diseases, allowing more confident selection of treatment and more accurate determination of prognosis. *In situ* hybridization is rapidly establishing its importance as a useful and powerful adjunct to diagnostic tests, especially in the identification of viruses and the detection of gene rearrangements.[91-93]

Immunocytochemistry employs a primary antibody for the identification of a specific cellular or nuclear protein. The antigen-antibody complex is visualized by a detection system such as the avidin-biotin complex, peroxidase antiperoxidase or alkaline phosphatase anti-alkaline phosphatase systems. In situ hybridization requires a DNA or RNA probe to identify a specific nucleic acid sequence with either dyes or radioactive tags as the signal molecule. Its increased accuracy is inherent in the fact that it identifies genetic material before it is expressed as a protein. The specific DNA or RNA probe is allowed to hybridize to the target of interest. It is visualized by a reporter molecule which employs streptavidin-biotin enzyme labels or digoxigenin. For detection of specific DNA sequences, the double stranded DNA must be denatured to a single strand and this is achieved by heating. The labelled probe is allowed to hybridize with the single strand target DNA, and the hybridized sequences are then identified by a detection system.

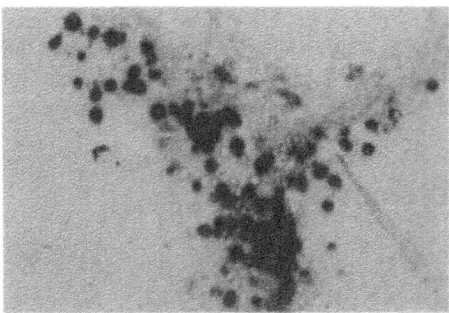

Figure 7. Cervical condylomata with HPV 6 infection (ISH X 400).

Many viruses can be detected using this system. Gene expression can also be studied by detecting the messenger RNA responsible for transcribing the gene expression. To allow penetration of labelled probes into the nucleus, digestion of the nuclear membrane is required. Fortunately, nucleic acids are relatively resistant to proteolytic digestion and although heat denatures strands of DNA, the actual DNA sequences themselves are resistant to heat.

Many viral nucleic acid targets including DNA viral particles, unpackaged DNA and mRNA can also be detected. Oligonucleotide probes tailored to hybridize to any one of these targets can be artificially synthesized.

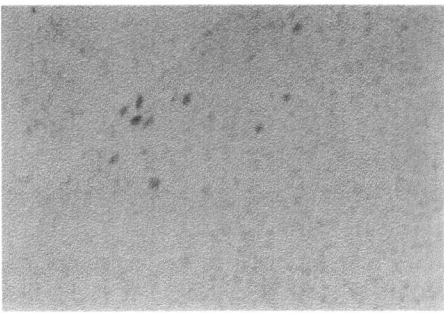

Figure 8. Carcinoma of cervix with HPV 18 infection. DNA copy numbers too low for detection by *in situ* hybridization. Positive staining obtained by *in situ* polymerase chain reaction (IS-PCR X 400).

A test needs to be specific and yet sensitive enough such that small copy numbers may be detected. It has been shown that ISH can produce signals when there are more than 20 copy numbers of DNA per cell. Any further decrease in number will fail to emit a signal.[96] To overcome this, the polymerase chain reaction (PCR) can be employed. The process involves binding a pair of artificially synthesized oligonucleotide primers to opposite strands of the denatured double helix flanking the segment of target DNA. The orientation of these primers are such that by heat denaturation, annealing of primers and extension of the primers by DNA polymerase, DNA synthesis proceeds across the target sequence. The incorporation of the primer, itself acting as a template for its partner, means that with each cycling of denaturation - annealing-extension/amplification, a defined fragment of DNA is doubled with exponential amplification of 10^7 to 10^{11} increase over 40 cycles. It is now possible to conduct this reaction *in situ*, i.e. within the cell nucleus in a tissue section, hence the term *in situ* PCR (IS-PCR). The hybridization and detection steps are similar to that described for ISH.

This technique of DNA amplification is especially useful for detecting low copy numbers of HPV-DNA and latent HIV infection, and for studying of gene rearrangement and point mutation in lymphoma and leukemia.

PREPARATORY TECHNIQUES

General Considerations

All smears are prepared on clean dry glass slides. Fine needle aspirate smears, brush smears obtained at endoscopy, and imprint smears of surgical specimens are all suitable for immunocytochemistry.

Imprint smears are made on clean glass slides by firmly imprinting a freshly cut surface. If there is excessive blood, the tissue is gently imprinted once on soft clean absorbent or blotting paper prior to imprinting on glass slides. Two or three imprints are made along the length of the glass slide.

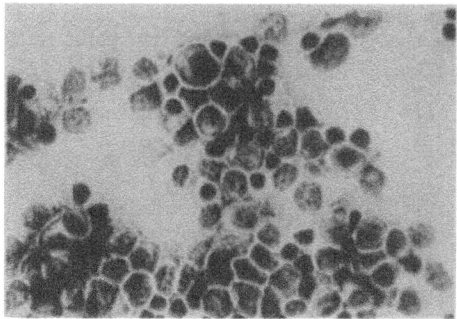

Figure 9. Imprint smears of surgical specimen on glass slide.

Fine needle aspirates (FNA) are made at the operational procedure following the recommendations cited in text books of cytology. As thin a film of cells (monolayer) as possible is made without crushing the cells with excessive pressure.

Cells in suspension from effusions, cerebrospinal fluid or other body fluids and from needle washes of fine needle aspiration are centrifuged and harvested. Cells from scrapes of fresh tissue are often rich in cells and suitable for cell harvesting. The suspension may contain blood, cell "debris" or disrupted cells. To remove excessive blood cells and cell debris, a Ficoll-Hypaque gradient is employed.

Smears on plain glass slides are usually adequate. However, for increased cell adhesion, glass slides coated with poly-L-lysine may be used. Freshly prepared smears are placed in fixative, depending on the analytical tests to be done, or storage medium if not required for immediate use.

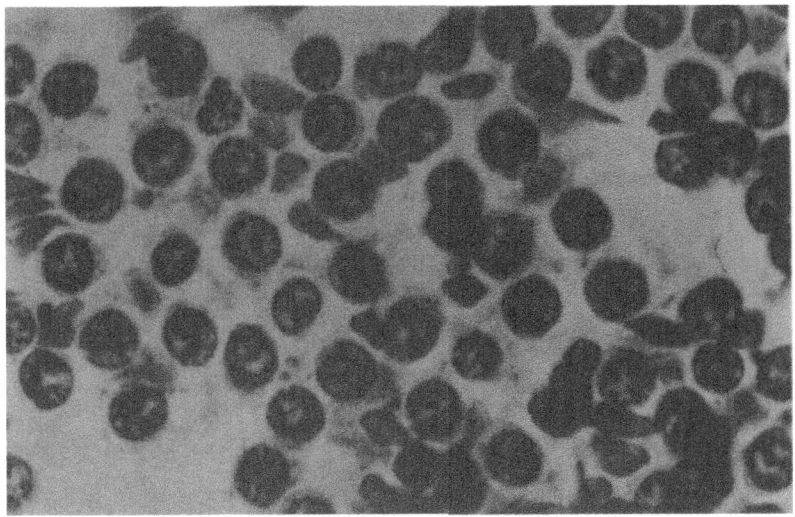

Figure 10. Fine needle aspirate smears.

Smears for Immunocytochemistry (ICC)

1. Air dried smears are stored in sealed plastic bags or containers at 4°C for use within 24 to 48 h.
2. For immunofluorescence staining, smears are placed in frost-free refrigerator at 4°C for 1 h.
3. For *in situ* hybridization, smears are prepared on aminopropyl ethylsilane (APES) coated glass slides.
4. For steroid hormone receptor assay (ER and PgR), fresh smears are stored in sucrose-glycerol medium at -20°C until required.

Smears for image analysis, DNA content estimation and for identification of nucleolar organizer regions (AgNOR) are fixed in 95% ethyl alcohol.

Smears for immunocytochemistry or DNA ploidy measurement may also be fixed in 95% ethyl alcohol for 5 min, or in 3.7% formaldehyde-phosphate buffered saline for 30 sec followed by 5 min wash in PBS pH 7.4 prior to storage in glycerol-sucrose solution at -20°C.

Cytospin Smears

All cells in suspension (i.e., CSF, effusions, tissue scrapes in PBS or TBS/transport medium, needle washes, lavage fluid) are washed at least once in PBS or TBS to reduce background staining from undesirable proteinaceous substances in solution.

1. Place cell suspension in conical centrifuge tube and centrifuge for 10 min at 2,400 rpm.
2. Discard supernatant and resuspend cells in PBS or TBS until the solution just begins to appear a little hazy.

3. Prepare cytospin buckets with mounted clean glass slides and absorbent filter paper.

4. Into cytospin bucket, place 4 drops of 2% bovine serum albumin (BSA), such that it fills the horizontal arm of the bucket.

5. Add 2 to 4 drops of cell suspension and centrifuge for 5 min (at mark 4 on dial of Shandon Cytocentrifuge for lymphoid cells, i.e. 400 rpm, at mark 6, i.e. 600 rpm for epithelial and other cells).

6. At completion of spin, remove slide; be careful not to smear circular area on glass slide containing cells.

7. For hormone assay, place immediately in glycerol-sucrose storage medium (recipe according to Abbott ER-ICA kit) and store at -20°C.
   For immunocytochemistry, allow to air-dry, then pack in sealed plastic bags and store at 4°C. For immunofluorescence, store air-dried at -20°C in sealed plastic bags or in glycerol-sucrose storage medium.

Ficoll-Hypaque Gradient Separation (for heavily blood stained fluids and suspensions containing cell-debris)

1. Gently layer cell suspension on the surface of Ficoll-Hypaque medium using a disposable pipette. Centrifuge at 2,000 rpm for 20 min.

2. With clean pipette, aspirate cells at interphase and place in conical centrifuge tube. Add sufficient PBS or TBS to resuspend and wash cells (3-5 ml).

3. Centrifuge at 24,000 rpm for 5 min, discard supernatant and resuspend cells in sufficient PBS or TBS to obtain a hazy cell suspension (0.5 ml to 1 ml).

4. Prepare cytospin smears as indicated in Cytospin Smears.

Cell Blocks for Ultrastructural Examination

Cytological material received as FNA/needle wash or cells suspended in fluids can be used for examination by transmission electron microscopy. Fragments of tissue obtained at fine needle aspirates can be picked out for fixation with glutaraldehyde. Cell blocks from cell suspensions and body fluids can be embedded in agar and processed as for tissue fragments.

Cell Blocks for EM (use plastic centrifuge tube)

1. Centrifuge cell suspension at 2400 rpm for 5 min.
2. Remove supernatant and resuspend in PBS . Centrifuge at 2400 rpm for 5 min and discard supernatant.
3. Resuspend cells in 0.5 ml of 8% bovine serum albumin (BSA).
4. Centrifuge at 2400 rpm for 5 min.
5. Add 4 drops of 25% glutaraldehyde and allow to gel for 10 min.
6. Cut tip off centrifuge tube and remove cell block in BSA gel.
7. Place cell block in 3% glutaraldehyde for 30 min.
8. Change 2 X 0.1 M cacodylate buffer, 5 min each change.
9. Place in 2% osmium tetroxide (OsO_4) for 60 min.
10. Change 2 X 0.1 M cacodylate buffer, 5 min each change.
11. Dehydrate in 50% ethanol for 5 min.
12. Place in 2% uranyl acetate in 75% ethanol for 5 min.
13. Dehydrate in 95% ethanol 2 X 10 min each.
14. Dehydrate in 100% ethanol 2 X 10 min each.
15. Place in propylene oxide (PO) 2 X 15 min each.
16. Place in 1.1 PO/epoxy resin (EPON) 30 min.
17. Place in 1:3 PO/epoxy resin (EPON) 30 min.

18. Place in 2 changes epoxy resin 30 min each.
19. Place in rubber molds.
20. Polymerize resin at appropriate temperature as suggested by manufacturer.
21. Cut sections.

Alternatively, the cells in suspension are fixed in glutaraldehyde prior to cell-block preparation.

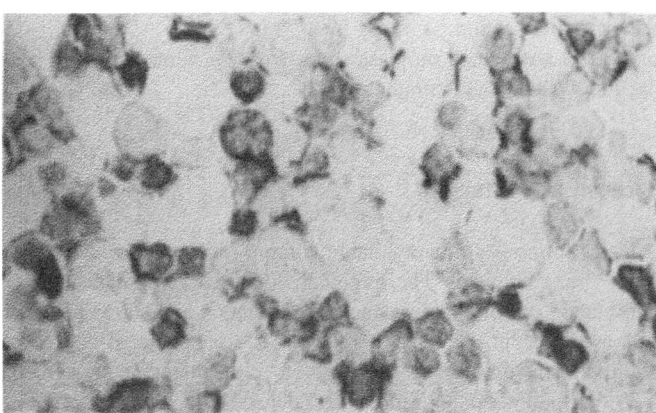

Figure 11. Fine needle aspirate smears.

Pre-fixation

1. Plastic microcentrifuge tubes are used. Centrifuge cell suspension at 1000 rpm for 10 min.
2. Discard supernatant and resuspend cells in 3% glutaraldehyde for 30 min.
3. Centrifuge at 2,400 rpm for 5 min, discard supernatant and resuspend in 1% osmium tetroxide (OsO_4) for 60 min.
4. Rinse with distilled water.
5. Dehydrate with increasing concentration of acetone - 30%, 75%, 90%, 100% - for 5 minutes each.
6. Discard acetone and add mixture of 2:1 EPON-acetone for 5 minutes.
7. EPON cure at 80°C for 12 to 18 h.
8. Cut tube, remove block for cutting.

Rapid Fixation (2 h)

1. Centrifuge cell suspension at 2400 rpm for 5 min.
2. Discard supernatant and resuspend cells in 10 ml of 1:1 of 1% glutaraldehyde: 4% formaldehyde (McDowell's fixative) in cacodylate buffer.
3. Place specimen in centre of 600-Watt domestic microwave oven and irradiate on "high" setting (50°C) for 25 s.
4. Centrifuge at 2400 rpm for 5 min, discard supernatant and post-fix in 2% osmium tetroxide (OsO_4) for 10 min.
5. Centrifuge at 2400 rpm for 5 min, discard supernatant and stain *en bloc* in 2% aqueous uranyl acetate for 10 min.
6. Centrifuge at 2400 rpm for 5 min, discard supernatant and fix in increasing concentrations of ethanol - 70%, 90%, 100% for 20 s each.

7. Centrifuge at 2,400 rpm for 5 min, discard supernatant and add epoxy propane for 20 s.

8. Add 1:1 epoxy propane - epoxy resin mixture for 2 min.

9. Centrifuge at 2,400 rpm for 2 min, discard supernatant. Add 100% epoxy resin for 10 min.

10. Centrifuge at 2400 rpm for 5 min, discard supernatant and add fresh 100% epoxy resin. Transfer to rubber molds and allow to polymerize at 95°C for 60 min.

11. Cool blocks and section.

12. Thin sections are stained on glass slides with 2% aqueous uranyl acetate for 1 min, followed by lead citrate.

Immunocytochemical Techniques

There are many antibodies now available commercially with which various staining techniques may be employed. The protocol for these techniques are listed with additional information on microwave accelerated reaction and stain enhancement.

Reactions are carried out in moist chamber. It is recommended that each laboratory determine the optimal dilution for reagents to suit their method of fixation and processing. Working dilutions of antibodies depends on the source of supply and dilutions indicated are a guide.

Avidin-Biotin Peroxidase Complex Technique (ABC technique)

1. Fix smears in cold acetone or Stefanini fixative, 10 min.

2. Block endogenous peroxidase with 0.3% H_2O_2 in methanol, 20 min.

3. Wash in distilled H_2O, 3 X 5 min each.

4. Permeabilize in 0.1% Triton in PBS, 20 min.

5. Wash in PBS pH 7.4, 2 X 5 min each.

6. Apply 10% normal goat serum (NGS), 5 min each.

7. Drain off excess NGS. Incubate with diluted primary antibody, at least for 30 min at room temperature or overnight at 4°C.

8. Wash in PBS, 3 X 5 min each.

9. Apply biotinylated secondary antibody diluted 1:100 with PBS for 30 min. For mouse monoclonal primary antibody use sheep anti-mouse biotinylated antibody.
   For rabbit polyclonal primary antibody use donkey anti-rabbit biotinylated antibody.

10. Wash in PBS, 3 X 5 min.

11. Apply streptavidin-biotinylated horseradish peroxidase complex at the appropriate dilution in PBS for 30 min.

12. Wash in PBS, 3 X 5 min each.

13. Add diaminobenzidine tetrachloride (DAB) (7.5 mg in 10 ml PBS) containing 3 drops 30 $^w/_v$ H_2O_2 for 2 to 10 min, checking on staining reaction.

14. Wash in PBS, 2 X 5 min, then distilled H_2O, 2 X 5 min each.

15. Counterstain with Mayer's hematoxylin and wash in running tap water.

16. Dehydrate, clear and mount. Positive staining has a brown color.

Indirect Immunoperoxidase Technique

1. Block with 10% normal rabbit serum (NRS) for 10 min.
2. Drain off excess NRS and add primary antibody (mouse monoclonal) diluted appropriately with 5% NRS; allow to incubate at room temperature for 30 min.
3. Wash in PBS pH 7.4, 2 X 5 min each.
4. Block endogenous peroxidase with 0.3% H_2O_2 in PBS for 10 min.
5. Wash with PBS, 2 X 5 min each.
6. Incubate with peroxidase-conjugated rabbit anti-mouse antibodies for 20 min.
7. Wash, PBS, 2 X 5 min each.
8. Apply freshly prepared DAB substrate solution containing 5 mg diaminobenzidine tetrachloride (DAB Sigma Cat No D5637) in 10 ml PBS to which 3 drops of 30 $^w/_v$ H_2O_2 added. Allow staining with chromogen for 2 to 10 min, monitoring colour reaction microscopically/macroscopically. Rinse in cold, running tap water.
9. Counterstain in Mayer's hematoxylin and wash in running tap water.
10. Dehydrate in alcohol, clear in xylene and mount in Depex. Positive staining has a brown color.

For primary antibody raised in rabbit (polyclonal), use normal swine serum (NSS) instead of normal rabbit serum (NRS) and peroxidase-conjugated swine anti-rabbit antibodies instead of peroxidase-conjugated rabbit anti-mouse antibodies.

Peroxidase Anti-Peroxidase (PAP) Technique

1. Apply 10% normal serum of the species providing the second layer antibody for 10 min.
2. Drain off excess normal serum and incubate with appropriately diluted primary antibody using 10% normal serum in PBS at room temperature for 30 min.
3. Wash in PBS, 2 X 3 min.
4. Block endogenous peroxidase with 0.3% H_2O_2 in methanol, 20 min.
5. Wash in PBS, 2 x 3 min.
6. Apply appropriately diluted bridging antibody, raised against the primary antibody at room temperature for 30 min.
7. Wash in PBS, 2 x 3 min.
8. Apply diluted PAP complex raised in the same species as the primary antibody.
9. Wash in PBS, 2 x 3 min.
10. Apply DAB substrate containing 5 mg DAB in 10 ml PBS to which 3 drops of 30 $^w/_v$ H_2O_2 have been freshly added. Check staining after 2 min to 10 min.
11. Rinse in running tap water.
12. Counterstain with Mayer's hematoxylin and wash in running tap water.
13. Dehydrate in alcohol, clear in xylene and mount in Depex.

Double Immunofluorescence Staining

1. Block smears with normal goat serum (NGS) in PBS pH 7.4 containing 0.1% bovine serum albumin (BSA) for 10 min.
2. Apply mixture of appropriately diluted primary antibodies raised in different species (e.g. mouse monoclonal and rabbit polyclonal) at room temperature for 4 hours or at 4°C overnight.
3. Rinse in PBS, 3 X 5 min.

4. Apply simultaneously biotinylated anti-mouse antibody and FITC-conjugated anti-rabbit antibody. Place in dark for 60 min.
5. Wash in PBS, 3 x 5 min.
6. Apply streptavidin-Texas red conjugated complex and incubate at room temperature for 60 min in dark.
7. Wash in PBS, 3 x 5 min.
8. Mount in 1:1 PBS:glycerol or glycergel (Dakopatts) and examine under UV light microscope.

Texas red gives an orange red fluorescence. FITC antibody gives a green fluorescence.

Alkaline Phosphatase Anti-alkaline Phosphatase Technique (APAAP Technique)

1. Prefix smears prior to staining in 1:1 methanol-acetone or 19:19:2 of acetone-methanol-formaldehyde (40%) fixative for 90 s.
2. Allow smears to dry, then apply 1:5 of BSA:TBS, 5 min.
3. Drain off excess and apply appropriately diluted primary antibody for 30 min at room temperature.
4. Wash in TBS, 3 X 3 min each.
5. Apply diluted bridging antibody directed against immunoglobulin of the species providing the primary antibody at room temperature for 30 mins.
6. Wash in TBS, 3 x 3 min each.
7. Apply alkaline phosphatase anti-alkaline phosphatase immune complex from the same species as the primary antibodies at room temperature for 30 min.
8. Wash in TBS, 2 x 3 min each.
9. Apply new fuchsin substrate solution.

Reagent Solution A:	Reagent B:
Naphthol AS-MX phosphate	Fast-Red TR salt
2 mg (Sigma Cat. No. N4875),	10 mg (Sigma Cat. No. F1500)
Dimethylformamide 0.2 ml,	
0.1 M TBS pH 8.2, 9.8 ml,	
1 M levamisole, 10 µl	

Add solution A to B, mix well and filter directly onto smears. Incubate for 20 min at room temperature.

10. Rinse with distilled water, 5 min.
11. Counterstain with Mayer's hematoxylin.
12. Wash in distilled water and mount in aqueous medium.

Double Staining with PAP and APAAP Techniques

All steps are carried out at room temperature.
1. Block endogenous peroxidase with 0.5% H_2O_2 in methanol, 20 min.
2. Apply 10% normal goat serum (NGS) for 10 min.
3. Drain off excess NGS and incubate with appropriately diluted primary mouse monoclonal antibody for 30 min.
4. Wash in PBS pH 7.4, 2 X 3 min each.
5. Apply biotinylated goat anti-mouse secondary antibody diluted 1:200 with 10% NGS in PBS and incubate for 30 min.

6. Wash in PBS, 2 X 3 min each.
7. Apply peroxidase conjugated streptavidin (Piesce, USA) diluted 1:1500 with 10% NGS in PBS and incubate for 30 min.
8. Wash in PBS, 2 X 3 min each.
9. Apply DAB substrate: 5 mg DAB in 10 ml PBS to which 3 drops of 30 $^w/_v$ H_2O_2 have been freshly added. Control reaction microscopically/macroscopically, 2 to 10 min or longer.
10. Rinse in distilled water, 3 X 5 min, then wash in PBS, 5 min.
11. Block with 10% normal swine serum (NSS) for 10 min.
12. Drain off excess NSS and incubate in appropriately diluted primary rabbit polyclonal antibody for 30 min.
13. Wash in PBS, 2 X 3 min each.
14. Apply diluted secondary swine anti-rabbit immunoglobulin antibodies, 30 min.
15. Wash in PBS, 2 X 3 min each.
16. Apply rabbit alkaline phosphatase anti-alkaline phosphatase immune complex diluted 1/50 with 10% NSS for 30 min.
17. Wash in PBS, 2 X 3 min.
18. Add new fuchsin substrate solution as in previous protocol. Incubate for 10-20 min at room temperature, controlling the reaction.
19. Wash in distilled H_2O.
20. Counterstain with Mayer's hematoxylin and wash in running tap water.
21. Wash in distilled H_2O and mount in aqueous medium.

Results: Monoclonal antibody-peroxidase complex: brown.
             Polyclonal antibody alkaline phosphatase complex: red.

Imidazole Enhancement of DAB Reaction Product

Add 100 µl imidazole to DAB solution (5 mg imidazole in 10 ml Tris HCl pH 7.6).

Results: The antigen-antibody reaction is seen as intense brown or grayish color.
Note: Imidazole may cause background staining if incubation is prolonged or washings inadequate.

Microwave Accelerated Immunostaining

Household Phillips Whirlpool microwave oven at "low" setting (90 Watts) can accelerate steps for primary antibody and secondary antibody incubations.

For antibody incubations, place smears in moist chamber on rotating carousel.
   (a)   45 s at 90 Watts ("low" setting)
   (b)   1 min pause with smears/slides on freezer pack/ice-block (placed on bench top).
Repeat cycle (a) and (b) seven times (Also see Chapter 6 by Jiang Gu in this book).

Heat Accelerated Immunostaining

Steps for primary antibody and secondary antibody can be shortened by half by placing moist chamber with smears in 60°C incubator.

Immunogold Silver Staining (IGSS) Technique

Use double distilled deionized H_2O (d_3H_2O) and carefully cleaned glassware only. All steps are carried out at room temperature. See also Chapter 3 by Hacker *et al* in this book.

1. Rehydrate air-dried smears in distilled water.
2. Place in Lugol's iodine (1% iodine in 2% potassium iodide).
3. Wash in running tap water for 1 min, then distilled H_2O for 1 min.
4. Decolorize in 2.5% sodium thiosulfate till colorless (approximately 2 min).
5. Wash in running tap water, 1 min, then distilled H_2O, 1 min.
6. Wash in TBS, pH 7.6 containing 0.1% Triton X-100 and 2.5% NaCl, 2 X 5 min, then place in moist chamber.
7. Apply 10% normal goat serum (NGS) in TBS, 5 min.
8. Drain excess NGS and incubate with primary antibody appropriately diluted in PBS or TBS pH 7.2-7.6 containing 0.1% bovine serum albumin (BSA) for 60 min.
9. Wash in TBS pH 7.6 containing 0.1% fish gelatin (TBS-gelatin), 3 x 5 min.
10. Apply 10% NGS 1:10 in TBS-gelatin for 5 min.
11. Incubate with secondary 1-5 nm gold labelled antibody diluted 1/50 in TBS-gelatin containing 0.1% BSA, 60 min.
12. Wash in TBS-gelatin, 2 X 5 min.
13. Change to PBS, pH 7.2, 5 min.
14. Immerse in 2% glutaraldehyde in PBS for 10 s.
15. Wash thoroughly in distilled H_2O (10 times or more).
16. Wash in double distilled deionized H_2O, 2 X 5 min.
17. Silver acetate autometallography:
    Solution A.   200 mg hydroquinone in 40 ml citrate buffer pH 3.8.
    Solution B.   80 mg silver acetate in 40 ml double distilled H_2O.
    Just before use, mix and incubate smears in Tris solution for 3 to 10 min, check the staining intensity under microscope every minute.
18. When adequately stained, place smear in dish containing 5% photographic fixer.
19. Wash in running tap water 5 min, then distilled water 2 X 5 min.
20. Counterstain with Mayer's hematoxylin and eosin, or nuclear fast red.
21. Dehydrate, clear and mount.

Steps 8 and 11 can be accelerated by microwave with 7 cycles of (a) 45 s at 90 Watt ("low" setting), (b) 1 min pause on ice-block/freezer pack.

The advantage of IGSS is the ability to use very high dilutions of primary antibody. Staining is specific and especially demonstrative in neural and in renal tissue also at ultrastructural level. See also chapter 3 by Hacker *et al* and Chapter 14 by Krenács *et al* in this book.

Estrogen & Progesterone Receptor Immunocytochemical Assay (ER-ICA, PgR-ICA)

Smears prepared and stored at -20°C in sucrose-glycerol storage medium as

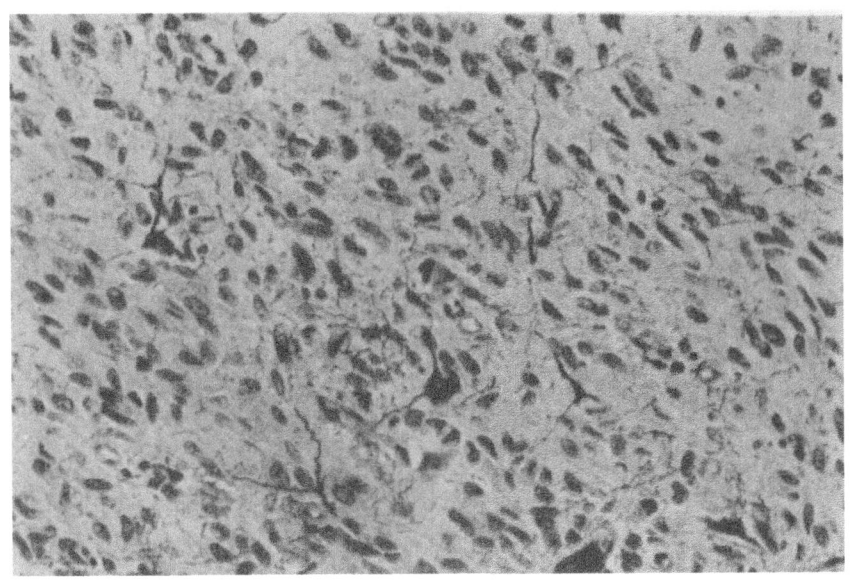

Figure 12. Hypothalamus showing axonal processes (IGSS neurofilament, X 632).

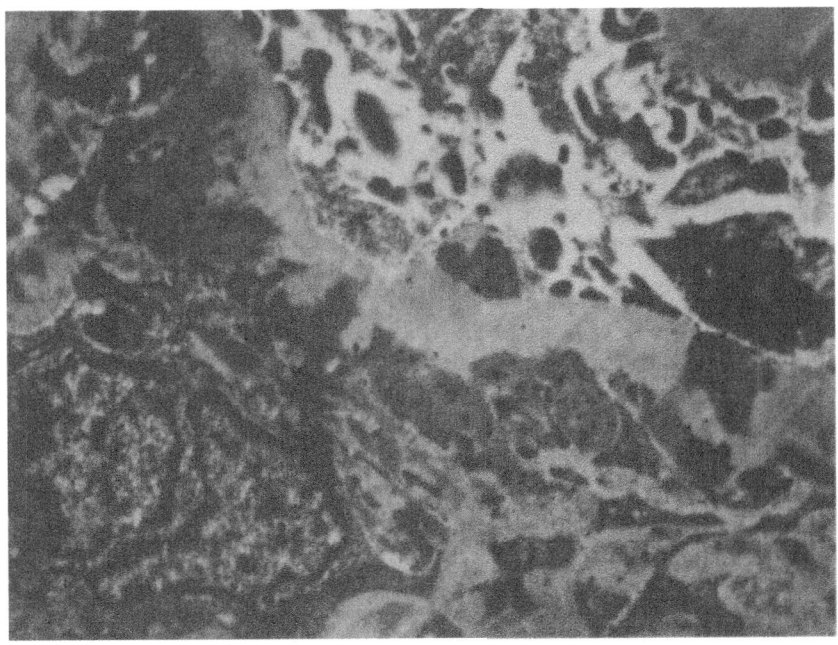

Figure 13. Renal tissue showing mesangial deposits with silver enhanced C1q staining (IGSS X TEM 25,000).

indicated under "Cytospin Smears". The smears are removed from storage medium and rinsed for 2-3 min in cold PBS (4°C) pH 7.2. The fresh smears that are not placed in storage medium may be submitted to first step of the following fixation procedure.

Fixation

1.	Place slides in 3.7% formaldehyde - PBS	10 min
2.	PBS rinse	5 min
3.	Absolute methanol (cooled -10°C to -20°C)	3 min
4.	Acetone (cooled -10°C to -20°C)	1 to 3 min
5.	PBS rinse	5 min

Staining Procedure (according to Abbott protocol, in moist chamber)

1.	PBS wash	2 x 5 min
2.	Blocking reagent	15 min
3.	Primary antibody/control antibody	30 min
4.	PBS wash	2 x 5 min
5.	Link antibody	30 min
6.	PBS wash	5 min
7.	PAP complex	30 min
8.	PBS wash	2 x 5 min
9.	Chromogen-substrate solution	5 min
10.	Rinse gently in running distilled or tap water	5 min

Counterstaining

1.	Lillie-Mayer's hematoxylin	3 s
2.	Rinse in running water till clear	3 min
3.	Acid alcohol	1 s
4.	Rinse in running water	1 min
5.	Scott's solution	few dips
6.	Rinse in running water	3 min
7.	Light green SF (0.02% in 0.01% acetic acid)	10 s
8.	Dehydrate, clear and mount in Depex	

Receptor positive cells show brown staining of nuclei. Benign and malignant cells are distinguishable by green cytoplasmic staining, size and intercellular relationship.

Two hundred cells are examined for each receptor staining. Receptor status of malignant cells may be scored by intensity of staining of positive tumor cells and grouped by percentage of cells with mild/moderate/marked (or graded 1 to 4, intensity of chromogen staining. Only nuclei with moderate or marked chromogen staining examined under light microscopy are regarded as "receptor positive".

Only when the score is >30% is the tumor regarded as oestrogen receptor or progesterone receptor positive. (This corresponds with the accepted limit of 10 fm units being "receptor positive" status as analyzed by dextran coated charcoal radioimmunoassay method.)

Semi-quantitation H-Score
% Nuclei stained X intensity =   4 x 10%   = 40
              (1 to 4+)
                               3 x 20%   = 60
                               2 x 30%   = 60
                               <u>1 X 40%</u>   <u>= 40</u>
                               H-Score   = 200

Cytochemical Detection of Estrogen and Progesterone Receptors

Using a fluorescent cytological technique, lower affinity cytoplasmic oestrogen binding proteins can be identified (in contrast to high affinity nuclear oestrogen receptor). Smears prepared are stored at -20°C or in glycerol-sucrose storage medium at -20°C. Smears in storage medium are washed briefly in PBS. Smears not in storage medium are allowed to thaw at 4°C for 10 min. Staining is performed with fluorescent labelled steroid hormones (Zeus Scientific Inc., Fluoro-Cep)

1.   Place smears in moist chamber and apply 2% BSA 10 min
2.   Drain BSA and add FITC-oestrogen receptor/TRITC - progesterone receptor preparation 2 h
3.   Wash in PBS 2 x 10 min
4.   Mount in glycerol and examine under UV light

Presence of estrogen receptor protein is identified by green fluorescence.
Presence of progesterone receptor protein shows bright red fluorescence.

In Situ Hybridization (ISH)

(*In Situ* Hybridization Detection System for Biotinylated Probes. Dakopatts, Cat No: K0602)

Smears on APES or Poly-L-lysin coated glass slides.

1.   Digest smears in pre-warmed 4% pepsin HCl for 10 min at 37°C.
2.   Apply 10 µl of test DNA probe to one smear, positive control to second smear negative control to third smear. Gently float coverslip over solution on smear. Denature by placing on heating block at 85°C for 5 min.
3.   Transfer to moist chamber and incubate at 37°C for 60 min.
4.   Float coverslip off by immersion in beaker of TBS, then immerse smears in stringent wash solution pre-warmed to 48°C for 30 min.
5.   Wash in TBS, 1 min.
6.   Apply streptavidin-AP and incubate at room temperature for 20 min.
7.   Wash in TBS, 2 X 5 min.
8.   Apply substrate *BCIP/NBT and incubate in dark for 60 min.
9.   Rinse in distilled H_2O, 5 min.
10.   Coverslip using aqueous mounting medium.

Results: Nuclei stain dark purple when positive for DNA under test.
*   5-bromo-4-chloro-3indoloylphosphate (BCIP), Nitroblue tetrazolium (NBT).

In Situ Polymerase Chain Reaction (IS-PCR)

Use double distilled deionized d_3H_2O (autoclaved) and autoclaved glass pipettes. Gloves must be worn when handling nucleotides. Reagents supplied by Perkin Elmer (NJ, USA) may be applied. Thermocycling must be performed preferably in a clean, dust-free room, away from bench top where solutions are prepared. Use APES slides for fine needle aspirate or imprint smears and cytospin smears. See also Chapter 18, by Zehbe *et al.* in this book.

Fixation

Smears must be fixed in 10% buffered formal saline for 24 h. Prior to procedure immerse smears in absolute ETOH, 2 X 5 min.

Hot Start Set thermocycler at 82°C.

1. Digest smears in 0.2% pepsin 0.2 N HCl preheated at 37°C for 12 min.
2. Wash in d_3H_2O, 2 X 5 min.
3. Block endogenous peroxidase with 0.5% H_2O_2.
4. Wash in d_3H_2O, 2 X 5 min.
5. Immerse in 100% Analar ETOH, 3 to 5 min.
    Dry well in air.
6. Prepare "Nucleotide Mix" cooled on ice.
    50 µl d_3H_2O, 12.5 µl dATP, 12.5 µl dCTP, 12.5 µl dGTP, 12.5 µl, dTTP
7. Prepare "Master Mix" cooled on ice. 5.0 µl 10 X PCR buffer, 1.0 µl primer 1, 1.0 µl primer 2, 1.0 µl 11-d-UTP (digoxigenin), 6.4 µl nucleotide mix, 9.6 µl $MgCl_2$. Total 24.0 µl. Add 26.0 µl d_3H_2O = 50 µl.
8. Place slides on thermocycler at 82°C and warm "Master Mix" to 95°C (5 min).
9. Hot start: Add 1.7 µl Taq polymerase to "Master Mix" at 95°C and rapidly apply "Master Mix-Taq" to smear. Cover with coverslip and line with mineral oil.
10. Shut lid of thermocycler and program to (a) 82°C for 10 min, then (b) 94°C for 3 min, then (c) 58°C for 2 min, and (d) 94°C for 1 min. Repeat cycle (c) and (d) 25 times.
11. Remove slides and float off coverslip in xylene.
12. Immerse in 100% ethanol with 2 changes.
13. Block with BSA (or 2% fat free milk power), 30 min.
14. Wash in TBS, 2 X 5 min.
15. Apply peroxidase labelled anti-sheep digoxigenin antibodies, diluted 1:100 in azide-free TBS for 30 min.
16. Wash thoroughly in TBS, 3 X 30 min each.
17. Apply DAB substrate, 5 mg in 10 ml TBS.
18. Wash in running tap water, 5 to 10 min.
19. Counterstain with Mayer's hematoxylin, dehydrate, clear and mount.
20. To further increase sensitivity and specificity, digoxigenin incorporation may be omitted, and the amplified sequence can be detected by *in situ* hybridization.

Cell Suspension for Image Analysis

Cells in suspension from body fluids, lavages, fine needle aspirates and tissue scrapings are centrifuged at 2,000 rpm for 5 min. The cells are washed in PBS twice and finally placed in special solution for DNA staining. When immunocytochemistry is required or when cells are to be sorted, fluorescent labelled antibody is added to the prepared cell suspension and allowed to incubate at 37°C for 30 min. The immuno-stained cells are washed once in PBS and finally suspended in special solution for analysis.

BUFFERS AND DILUENTS

PBS pH 7.2-7.4   1,000 ml distilled deionized H_2O
8.7 g sodium chloride (NaCl)
0.272 g KH_2PO_4
1.136 g Na_2HPO_4

TBS pH 7.2   stock: 1,000 ml double distilled deionized H_2O
14.8 g $Na_2PO_4 \cdot H_2O$
70 g NaCl
4.8 g KH_2PO_4
50 g Tris
Working solution 1:10, dilute 100 ml stock solution in 1,000 ml double distilled deionized H_2O

Citrate Buffer pH 3.8   500 ml double distilled deionized H_2O
23.5 g trisodium citrate dihydrate
25.5 g citric acid
Adjust pH with citric acid solution

"Diluent" for Antibodies   100 ml PBS pH 7.2
100 mg BSA (bovine serum albumin)
10 mg sodium azide
100 μL Triton X-100 may be added for some antibodies

Lugol's Iodine   1% Iodine
2% Potassium iodide
Both dissolved in distilled water using a magnetic stirrer for 24 hours. Shield working solution from light.

2.5% sodium thiosulfate   5 g $NaHSO_4$ in 200 ml d_3H_2O

CONCLUSION

The practice of fine needle aspiration cytology has added immensely to the current practice of cell harvesting from body fluids, scraping of skin and mucosal surfaces, brushings of internal linings and other fluids which contain exfoliated cells. The ability to obtain fresh tissue or cells with minimal discomfort to the patient, devoid of major complications and in a very cost-effective manner has popularized the use of cytologic material for almost every conceivable testing technique for diagnosis. With only minor adaptations in processing, both simple histological and biochemical studies as well as highly sophisticated molecular analyses can be performed on tissue or cells in cytologic preparations.

Fine needle aspiration has often been likened to a microbiopsy and although not so many years ago it was suggested that "fine needle aspiration cytology cannot replace a frozen section tissue assessment". This opinion has proven to be untenable. The diversity of cytologic material for use in immunocytochemistry, ultrastructural examination, cell imaging and molecular studies has added new dimensions to morphologic observations.

Cytologic examination has not only allowed the diagnosis of infectious agents without the patient having to be subjected to a surgical biopsy procedure, it can now also be employed to delineate cell origin of tumors (Table 1).

Immunocytochemistry has enabled identification of tumor markers, the most important indicating cell lineage assisting in tumor classification. Of greater importance are the cell proliferation markers which are associated with tumor aggression and progression, and other prognostic markers which indicate response to chemotherapy or hormone manipulation. With the intelligence choice of a diagnostic panel of antibodies for immunocytochemical reactions, tumors of similar morphological features may be separated and the optimal therapy be instituted (Tables 2 and 3).

When dedifferentiation occurs in a malignant neoplasm, the identification of filaments, specialized organelle structures, specific receptors or cell products allows the correct cell lineage to be established and prognosis to be determined (Table 4). A variety of detection systems and chromogens are available for immunocytochemical staining (Table 5).

The detection of hormone and secretory products or proteins by immunocyto-chemistry can now be augmented by the detection of pre-hormones and DNA sequences at a molecular level utilizing *in situ* hybridization (ISH) techniques. The recent development of *in situ* polymerase chain reaction (IS-PCR) enables even more sensitive detection of small amounts of DNA.

Thus cytopathology has now come of age and many newer analytical techniques can be adapted for application to cytologic material.

APPENDIX I

Miscellaneous Fixatives and Antibodies, Sources and Working Dilutions
APES (aminopropyl ethylsilane) Coated Glass Slides

1. Stack clean glass slides on side in slide carrier.
2. Immerse in glass dish containing:
        80 ml acetone
        1.6 ml APES
    (Solutions reusable if stored in dark bottle at 4°C.)
3. Wash in acetone, 2 X 5 min.
4. Wash in glass distilled H$_2$O, 2 X 5 min.
5. Dry in oven, away from dust, at 60°C. Store dust-free.

Poly-L-Lysin Slides

Place 1 drop of 0.1% solution on clean slide and spread as in making a blood film/smear. Allow to dry and store dust free.

Zamboni-Stephanini Fixative (for neuropeptides, hormone receptor assay)

1. 20 g paraformaldehyde.
2. 150 ml picric acid (saturated).
    Mix and stand at 60°C for 2 h.
3. Add drop by drop of concentrated sodium hydroxide (NaOH) till solution is clear; filter.
4. Add to 850 ml PBS pH 7.3. Store at 4°C up to 1 year.

Acetone Fixative

Smears for lymphoid surface markers and Ki-67 antigen detection are prefixed in cold acetone for 5 s prior to staining procedure. If staining procedure is delayed, smears can be kept in sealed plastic bags at -20°C.

Periodate-Lysine-Paraformaldehyde (PLP) Fixative

50 ml 3% paraformaldehyde (dissolve powder in heated distilled H_2O, adding drops of 5N NaOH to clear solution).
$\quad$ 100 ml distilled H_2O
$\quad$ 0.9 g lysine
$\quad$ 0.146 g sodium periodate
$\quad$ 1.46 g disodium hydrogen orthophosphate
Solution to be used within 24 hours.

ACKNOWLEDGEMENTS

The authors gratefully acknowledge the excellent secretarial assistance of Mrs M. Elemer, and thank Drs. G.W. Hacker, T. Mukherjee and P. Smith for their help with Figures 12 and 13. We also thank Dr. I. Zehbe for showing her *in situ* PCR protocol before publication, and Dr. J. Gu for his protocol of microwave technology.

REFERENCES

1. L.A. Brown and S.B. Coghill, Fine needle aspirate for the diagnosis of granulomatous diseases, *Cytopathol* **3**:9 (1992).
2. R. Arora, R. Rewari, and S. Bertharias, Fine needle aspiration cytology of orbital and adnexal masses, *Acta Cytol* **36**:483 (1992).
3. N. Shabb and R. Katz, Exfoliative and fine needle aspiration cytology of human immunodeficiency virus infection: A systems review, *in* "Cytopathology Annual", W. Schmidt, ed., Williams and Wilkins, Baltimore (1992).
4. K. Wehle, M. Blanke, G. Koenig, and P. Pfitzer, The cytological diagnosis of *Pneumocystis carinii* by fluorescent microscopy of Papanicolaou stained bronchoalveolar lavage specimens, *Cytopathol* **2**:113 (1991).
5. M. Akhtar, M.A. Ali, and M. Bakry, Fine needle aspiration biopsy of pediatric neoplasms, *Diagn Cytopathol* **8**:258 (1992).
6. M. Nadjib and D. Lubis, The technical procedure and value of FNA biopsy of nasopharynx, *Pathology* **25**:35 (1993).
7. M. Nadjib and P. Gonjei, Immunocytochemistry in diagnostic cytology: A 12 year perspective, *Am J Clin Pathol* **94**:470 (1992).
8. K. Nance and J. Silverman, The utility of ancillary techniques in effusion cytology, *Diagn Cytopathol* **8**:185 (1992).
9. J.P. Wazir, E. Martin-Bates, G. Woodward, and D. Coleman, Evaluation of immunocytochemical staining as a method of improving diagnostic accuracy in a routine cytopathology laboratory, *Cytopathol* **2**:75 (1991).
10. S. Perez, M. Perez-Guillermo, A.B. Bernal, and C. Lopez, Malignant rhabdoid tumor of soft tissue: A cytopathologic and immunocytochemical study, *Diagn Cytopathol* **8**:369 (1992).
11. G. Gherardi and C. Marveggio, Immunocytochemistry in head and neck aspirates: Diagnostic application on direct smears in 16 problematic cases, *Acta Cytol* **36**:687 (1992).
12. J.A. Stoop, J.G.M. Hendriks, and D. Berends, Identification of malignant cells in serous effusions using a panel of monoclonal antibodies - Ber EP4, MCA 6-12 and EMA, *Cytopathol* **3**:297 (1992).
13. C. Gottlieb, S. Kini, and M. Feingold, Cytomorphologic analysis of small cell neoplasm of lung from specimens obtained via bronchoscopy, *Anal Quant Cytol Histol* **14**:41 (1992).

14. P. Strausbauch, J. Neill, T. Benning, and J. Silverman, Application of electron microscopy in fine needle aspiration cytology, *in:* "Cytopathology Annual", W. Schmidt, ed., Williams and Wilkins, Baltimore (1992).

14a. P. Strausbauch, Application of electron microscopy in fine needle aspiration biopsy in a diagnostic electron microscopy laboratory, *Arch Pathol Lab Med* 113:1354 (1989).

15. M. Akhtar, M. Bakry, A. Al-jeaid, and J. McClintock, Electron microscopy of fine needle aspiration biopsy specimens: A brief review, *Diagn Cytopathol* 8:278 (1992).

16. C.W.M. Bedrossian, Electron microscopy: The neglected tool of cytopathology, *Diagn Cytopathol* 8:179 (1992).

17. R. Arisic, New technique for cell-block preparation after FNA of breast lumps, *Diagn Cytopathol* 8:424 (1992).

18. J. Neill and J. Silverman, EM of fine needle aspiration biopsies of mediastinum, *Diagn Cytopathol* 8:272 (1992).

19. M. Bibbo, P.H. Bartels, H.E. Dytch, and G.L. Weid, Cell image analysis, *in:* "Comprehensive Cytopathology", M. Bibbo, ed. W.B. Saunders, Philadelphia (1991).

20. M. Veilh. "Guide to Clinical Aspiration Biopsy: Flow Cytometry", Igaku-Shoin, New York (1991).

21. A. Rijken, A. Dekker, S. Taylor, P. Hoffman, M. Blank, and J.R. Krause, Diagnostic value of DNA analysis in effusions by flow cytometry and image analysis, *Am J Clin Pathol* 95:6 (1991).

22. M. Stoler, C. Rhodes, A. Whitbeck, S.M. Wolinsky, L.T. Chow, and T.R. Broker. Human papilloma virus type 16 and 18 gene expression in cervical neoplasia, *Human Pathol* 23:117 (1992).

23. Y. Toh, H. Kuwano, and S. Tanaka, Detection of human papilloma virus DNA in oesophageal carcinoma in Japan by polymerase chain reaction, *Cancer* 70:2234 (1992).

24. J. Fraga-Fernandez and B. Vicandi-Plaza, Diagnosis of hairy leucoplakia by exfoliative cytologic methods, *Am J Clin Pathol* 97:262 (1992).

25. G. Higgins, D. Uzelin, G. Phillips, L.L. Villa, and C.J. Burrell, Differing prevalence of human papilloma virus RNA in penile dysplasias and carcinomas may reflect differing etiologies, *Am J Clin Pathol* 97:272 (1992).

26. N. Iwa, M. Sasaki, C. Yutani, and K. Wakasa, Detection of cytomegalovirus DNA in pulmonary specimens; confirmed by *in situ* hybridization, *Diagn Cytopathol* 8:357 (1992).

27. T. Greiner, Polymerase chain reaction: Uses and potential applications in cytology, *Diagn Cytopathol* 8:61 (1992).

28. H. Clark, D. Jones, and D. Wright, Cytogenetic and molecular studies of t(14;18) and t(14;19) in nodal and extranodal lymphoma, *J Pathol* 166:129 (1992).

29. F. Cotter, The role of the bcl-2 gene in lymphoma, *Brit J Haem* 75:449 (1990).

30. M. Volkenandt, L. Cerroni, L. Rieger, H.P. Soyer, O. Koch, J. Wienecke, J. Atzodien, J.R. Bertino and H. Kerl, Analysis of 14;18 translocation in cutaneous lymphoma using PCR, *J Cut Pathol* 19:353 (1992).

31. J.P. Baak, D.M. Chin, PJ van Diest, R. Ortiz, P. Matze-Cok and S.S. Bacus, Comparative long-term prognostic value of quantitative HER-2/neu protein expression, DNA ploidy and morphometric and clinical features in paraffin-embedded invasive breast cancer, *Lab Invest* 64:215 (1991).

32. C.A. Finlay, P.W. Hinds and A.J. Levine. The p53 proteo-oncogene can act as suppressor of transformation, *Cell* 57:1083 (1989).

33. M. Ramael, G. Lemmens, C. Zerdekens, C. Buysse, I. Deblier, W. Jacobs, and E. van Marck, Immunoreactivity for p53 protein in malignant mesothelioma and non-neoplastic mesothelium, *J Pathol* 168:371 (1992).

34. M. Roncalli, C. Doglionic, D. Springall, M. Papotti, A. Pagani, J.M. Polak, N.B.N. Ibrahim, G. Coggi, and G. Viale, Abnormal p53 expression in lung neuroendocrine tumors, *Diagn Mol Pathol* 1:129 (1992).

35. C. Charpin, L. Andraic, M.L. Habib, H. Vacheret, L. Xerri, B. Devictor, M.N. Lavaut, and M. Toga, Immunodetection in fine needle aspirate and multiparametric image analysis (SAMBA), *Cancer* 63:863, 1989.

36. H. Neal and P. Hurst, The estimation of mean nuclear volume in diagnosis of breast carcinoma, *Diagn Cytopathol* 8:293 (1992).

37. R. Detweiker, D.J. Zahniser, G.L. Gardic, and M. Hutchinson, Contextual analysis compliments single cell analysis of breast cancers in fine needle aspirates, *Anal Quant Cyto Histol* 10:10 (1988).

38. V. Keshyap, N. Kaushik, S. Bhambhanis, D.K. Das, and U.K. Luthra, Supportive role of image analysis and DNA ploidy pattern in diagnosis of thyroid tumors, *Diagn Cytopathol* 8:228 (1992).

39. P. Zeppa, H.G. Van der Poel, M.E. Boon, A.N. Kurniawan, and L.P. Kok, A model for quantitative follow-up studies of cervical lesions, *Diagn Cytopathol* 8:8 (1992).

40. J. Barba, Y. Li, and J. Gil, Cell contour extraction on multithreshold images, *Pathol Res Pract* **188**:449 (1992).

41. L.A. Brown and S.B. Caghill, Cost-effectiveness of FNA clinic, *Cytopathol* **3**:275 (1992).

42. D.Y. Mason. Immunocytochemical analysis of human tissues, *in:* "Oxford Textbook of Pathology", J. McGee, P. Isaacson and N.A. Wright, eds., Oxford Medical Publications, Oxford (1992).

43. J. Neill and J. Silverman, EM of fine needle aspiration biopsies of mediastinum, *Diagn Cytopathol* **8**:272 (1992).

44. S. Hajdu, The value and limitations of aspiration cytology in diagnosis of primary tumors: A symposium, *Acta Cytol* **33**:741 (1989).

45 M. Nadji and P. Gaijei, Immunocytochemistry in diagnostic cytology: a 12 year perspective, *Am J Clin Pathol* **94**:470 (1992).

46. W. Domogala, Diagnosis of major tumor categories in FNA is more accurate when light microscopy is combined with intermediate filament typing: A study of 403 cases, *Cancer* **63**:504 (1989).

47. M. Akhtar, M.A. Ali, M. Bakry, M. Hug and K. Sackey, Fine needle aspiration biopsy diagnosis of rhabdomyosarcoma: A cytologic, histologic and ultrastructural study, *Diagn Cytopathol* **8**:465 (1992).

48. A.S-Y. Leong, M. Stevens, and T. Mukherjee, Malignant mesothelioma: Cytologic diagnosis with histologic, immunohistochemical and ultrastructural correlation, *Sem Diagn Cytopathol* **9**:141 (1992).

49. G. Dell'Antonio, G.L. Taccagni, M.R. Terreni, B.E. Leone, and A. Cantaboni, Electron microscopy of fine needle aspiration biopsy from extragonadal germ cell tumors, *Diagn Cytopathol* **8**:283 (1992).

50. D.W. Gove, L.K. Waterhouse, and A.S-Y. Leong, Rapid microwave-stimulated fixation of fine needle aspiration biopsies for transmission electron microscopy, *Diagn Cytopathol* **6**:68 (1990).

51. J. Hoh and P. Hansen, Atomic force microscopy for high resolution cell imaging in cell biology, *Trends Cell Biol* **2**:208 (1992).

52. F.A. Carey, G. Fabboni, and D. Lamb, Expression of proliferating cell nuclear antigen in lung cancer: a systematic study and correlation with DNA ploidy, *Histopathol* **20**:499 (1992).

53. M. Akhtar, M. Ali, A Haider, J. Antonius, B. Hainau, and F. Al Dayel, Fine needle aspiration biopsy of Ki-1 positive anaplastic large-cell lymphoma, *Diagn Cytopathol* **8**:242 (1992).

54. M. Turbiana and A. Courdi, Cell proliferation kinetics in human solid tumors: relation to probability of metastatic dissemination and long term survival, *Radio Oncol* **15**:1 (1989).

55. M. Colecchia and O. Leopardi, Evaluation of AgNOR count in distinguishing benign from malignant mesothelial cells in pleural fluid, *Pathol Res Pract* **188**:541 (1992).

56. W.A. Raymond and A.S-Y. Leong, Nucleolar organizer regions relate to growth fractions in human breast carcinoma, *Hum Pathol* **20**:741 (1989).

57. C. Bayindir, O. Dogan, F. Unal, and S. Dervisoglu, Nucleolar organiser regions in pituitary adenomas, *Cytopathol* **3**:223 (1992).

58. C. Lesty, C. Chleg, G. Contisso, and C. Jacquillat, Nucleoli and AgNOR proteins in primary breast carcinoma: Spatial patterns of interaction between clinical and histochemical criteria, *Anal Quant Cytol Histol* **14**:175 (1992).

59. K. Misra and N. Kumas, Nucleolar organizer regions in FNAC smears of breast lesions, *Diagn Cytopathol* **8**:346 (1992).

60. P.A. Hall, D.A. Levison, A.L. Woods, C.C. Yu, D.B. Kellock, J.A. Watkins, D.M. Barnes, C.E. Gillett, R. Camplejohn, and R. Dover, Proliferating cell nuclear antigen (PCNA) immunolocalization in paraffin sections: An index of cell proliferation with evidence of deregulated expression in some neoplasms, *J Pathol* **162**:285 (1990).

61. C.M. Quinn and N. Wright, The clinical assessment of proliferation and growth in human tumors: Evaluation of methods and applications as prognostic variables, *J Pathol* **160**:93 (1990).

62. K. Sheibani, J. Eskeban, M. Bailey, H. Battifora and L.M. Weiss, Immunopathologic and molecular studies as an aid to diagnosis of malignant mesothelioma, *Hum Pathol* **23**:107 (1992).

63. I.D. Buley, E.H. Morrison, and L. Kaklamanis, Measuring proliferation in routine FNA. Immunocytochemical detection of bromodeoxyuridine incorporation and Ki-67 expression in breast aspirates, *Cytopathol* **3**:149 (1992).

64. A.S-Y. Leong, J. Milios, and S.K. Tang, Is immunostaining of PCNA in paraffin sections a valid index of cell proliferation? *Appl Immunohistochem* (In press, 1993).

65. A.D. Thor, Prognostic factors in breast cancer. Integrating the cytology laboratory, *Diagn Cytopathol* **8**:319 (1992).

66. N. Kawakita, Immunocytochemical identification of proliferating hepatocytes using monoclonal antibody to proliferating cell nuclear antigen (PCNA/cyclin). Comparison with

immunocytochemical staining for DNA polymerase-alpha, *Am J Clin Pathol* **97(Suppl. 1)**:514 (1992).

67. M. Barbareshi, E. Leonardi, F. Mauri, G. Serio, and P.D. Palma, p53 and c-erbB-2 protein expression in breast cancer, *Am J Clin Pathol* **98**:408 (1992).

68. W.L. McGuire, A.K. Tandon, D.C. Allred, G.C. Chamness, and G.M. Clark, How to use prognostic factors in axillary node-negative breast cancer patients, *J Natl Cancer Inst* **82**:1006 (1990).

69. H. Tsuda, S. Hirohashi, Y. Shimosato, T. Hirota, S. Tsugane, S. Watanabe, M. Terada, and H. Yamamoto, Correlation between histological grade of malignancy and copy number of c-erbB-2 gene in breast carcinoma, *Cancer* **65**:1794 (1990).

70. D. Visscher, F. Sarkar and D. Crissman, Correlation of DNA ploidy with c-erbB-2 expression in preinvasive and invasive breast tumors, *Anal Quant Cytol Histol* **13**:418 (1992).

71. McCann, P.A. Dervan, M. O'Reagen, M.B. Codd, W.J. Gullick, B.M. Tobin, and D.N. Carney, Prognostic significance of c-erbB-2 and estrogen status in human breast cancer, *Cancer Res* **51**:3296 (1991).

72. C. Wright, B. Angus, S. Nicholson, J.R. Sainsbury, J. Cairns, W.J. Gullick, P. Kelly, A.L. Harris, and C.H. Horne, Expression of c-erbB-2 oncoprotein: A prognostic indicator in human breast cancer, *Cancer Res* **49**:2087 (1989).

73. S.K. Tang, K. Scott and C. Ozga, Detection of oestrogen and progesterone receptors in breast cancer - a preliminary study to compare fluorescent cytochemical method and biochemical radioligand assay, Presented at the Annual Scientific Meeting of the Royal College of Pathologists, Melbourne, Australia, Oct 1991.

74. A.D. Thor, Accumulation of tumor suppressor gene p53 is an independent prognostic marker in breast cancer, *Breast Cancer Res Treat* **19**:157 (1991).

75. N.J. Levine, J. Momad, and C.A. Finlay, The p53 tumor suppressor gene, *Nature* **351**:453 (1991).

76. S. Masood, Oestrogen and progesterone receptors in cytology: A comprehensive review, *Diagn Cytopathol* **8**:475 (1992).

77. S. Masood, Immunocytochemical localisation of estrogen and progesterone receptors in imprint preparations of breast carcinoma, *Cancer* **70**:2109 (1992).

78. S. Masood, The use of monoclonal antibody for assessment of estrogen and progesterone receptors in malignant effusions, *Diagn Cytopathol* **8**:161 (1992).

79. S. Masood, Application of oestrogen receptor immunocytochemical assay to aspirates from mammographically guided fine needle aspiration biopsy of non-palpable lesions, *South Med J* **84**:857 (1991).

80. S. Masood, Fluorescent cytochemical detection of oestrogen and progesterone receptors in breast fine needle aspirates, *Am J Clin Pathol* **95**:35 (1991).

81. J. Brolin, T. Lowhagen, and L. Skoog, Immunocytochemical detection of androgen receptors in fine needle aspirates from benign and malignant prostate, *Cytopathol* **3**:351 (1992).

82. R. Katz, S. Patel, N. Sneige, HA Fritsche, Jr, GN, Hortobagyi, FC Ames, T. Brooks, and NG Ordonez, Comparison of immunocytochemical and biochemical assays for estrogen receptor in fine needle aspirates and histologic sections from breast carcinomas, *Breast Cancer Res Treat* **15**:191 (1990).

83. F. De Negri, D. Compani, R. Sarnelli, L. Martini, A. Gigliotti, R. Bonacci, R. Fabbri, F. Squartini, A. Pinchera, and C. Giani, Comparison of monoclonal immunocytochemical and immunoenzymatic methods of steroid hormone evaluation in breast cancer, *Am J Clin Pathol* **96**:53 (1991).

84. M. Silverstein, J.R. Waisman, P. Gamagami, E.D. Gierson, W.J. Colburn, R.J. Rosser, P.S. Gordon, B.S. Lewinsky, and A. Fingerhut, Intraduct carcinoma of breast (208 cases): Clinical factors influencing treatment choice, *Cancer* **66**:102 (1990).

85. S. Lockett, K. Jacobson, and B. Herman, Quantitative precision of an automated fluorescent based image cytometer, *Anal Quant Cyto Histol* **14**:187 (1992).

86. K. Preston and R. Sideris, New techniques for 3-dimensional data analysis in histopathology, *Anal Quant Cyto Histol* **14**:398 (1992).

87. P. Vielh, S. Chevillard, V. Mosseri, B. Donatini, and H. Magdelenat, Ki-67 index and S-phase fraction in human breast cancer: Comparison and correlations with prognostic factors, *Am J Clin Pathol* **94**:681 (1990).

88. R.J. Sklarew, S.C. Bodmer, and L.P. Perkschuk, Quantitative imaging of immunocytochemical (PAP) estrogen staining patterns in breast cancer sections, *Cytometry* **11**:359 (1990).

89. M. Dowsett, Rationale for endocrine treatment of breast cancer, *in:* "Endocrine Aspects of Breast Cancer", M. Dowsett, ed, Parthenon Publishing Group, Parkridge, New Jersey (1991).

90. S.K. Tang and C Ozga, Immunocytochemical assay of oestrogen and progesterone receptors in breast cancer. Presented at the Annual Scientific Meeting of the Royal College of Pathologists, Melbourne, Australia, Oct 1991.

91. Y. Chardonet, B. Thivolet, and Guerin-Reverchou, Human papilloma virus detection in cervical cells by *in-situ* hybridization with biotinylated probes, *Cytopathol* **3**:341 (1992).

92. T. Kobayashi, Comparison of immunocytochemistry and *in-situ* hybridization in cytodiagnosis of genital herpetic infection, *Diagn Cytopathol* **8**:53 (1992).

93. N. Iwa, N. Sasaki, C. Yutani, and K. Wakasa, Detection of CMV DNA in pulmonary specimen, confirmed by *in-situ* hybridization, *Diagn Cytopathol* **8**:357 (1992).

94. L.W. Dalton, Computer-based image analysis of prostate cancer, *Hum Pathol* **23**:280 (1992).

95. I. Salmon, R. Kiss, and B. Fran, Comparison of morphonuclear features in normal thyroid tissue by digital cell image analysis, *Anal Quant Cyto Histol* **14**:47 (1992).

96. T. Greiner, Polymerase chain reaction: Uses and potential applications in cytology, *Diagn Cytopath* **8**:61 (1992).

97. L. Pertschuk, J. Feldman, D. Kim, K. Nayeri, K.B. Eisenberg, A.C. Carter, W.T. Thelmo, Z.T. Thelmo, Z.T. Rhong, P. Benn, and A. Grossman, Steroid hormone receptor immunohistochemistry and amplification of c-myc proto-oncogene: Relationship to disease free survival in breast cancer, *Cancer* **71**:162 (1993).

98. S. Tandon, Cathepsin D and prognosis in breast cancer, *New Engl J Med* **322**:297 (1990).

99. J. Ro, A. Sahin, J.Y. Ro, H. Fritsche, G. Hortobagyi, and M. Blick, Immunohistochemical analysis of p-glycin expression correlated with chemotherapy resistance in locally advanced breast cancer, *Hum Pathol* **21**:787 (1990).

100. M. Dietel, What is new in cytostatic drug resistance and pathology?, *Pathol Res Pract* **187**:892 (1991).

101. S.M. Grunberg, Treatment of non-resectable meningiomas with anti progesterone agent Mifepristone, *J Neurosurg* **74**:861 (1991).

CHAPTER 10

THE USE OF THE MULTIBLOCK IN DIAGNOSTIC IMMUNOHISTOCHEMISTRY

Wolfgang Kraaz[1] and Lena Scheibenpflug

Department of Pathology
University Hospital
Uppsala, Sweden

SUMMARY

In immunohistochemistry, the availability of a steadily increasing number of antibodies allows for a more comprehensive characterization of tissue components. Often, very large numbers of slides are easily generated when panels of antibodies are applied to a series of cases, as in clinical research.

To save time and space and prevent generation of large numbers of slides, we have developed the multiblock technology. The multiblock is a conventional size paraffin block, composed of up to 24 different tissue samples. These samples are punched from preexisting paraffin blocks and reembedded together. The multiblock is then routinely sectioned and immunostained.

The multiblock method facilitates the immunostaining and quicker evaluation of samples. As a result of the ability to decrease the number of immunostaining rounds, certain variations are eliminated. Multiblocks can also be used for routine laboratory tests on available antibodies and quality control.

INTRODUCTION

Immunohistochemistry has expanded the pathologists' ability to characterize tissues. We are able to identify many different cell components with antibodies and immunohistochemistry. With the number of antibodies being developed consistently on the rise, the

---

[1] Dr. Wolfgang Kraaz, Department of Pathology, University Hospital, S-751 85 Uppsala, Sweden, Telephone: (+46) 18-66 38 48; Fax: (+46) 18-55 27 39

possibilities appear to be endless. Oftentimes, an individual case, a tumor for example, may require several immunostainings. It may be necessary to use different epithelial, mesenchymal, hematopoietic, neuroectodermal, and endocrine markers. In addition, it may be necessary to test the same or very similar antibodies from different suppliers. It is not hard to imagine that it would be very easy to generate more than ten immunostained slides per tissue or case. In clinical research, large numbers of cases are studied, and each new antibody creates a new set of slides, easily generating hundreds of slides.

Using the multiblock method as described below, would make it possible to mount the selected relevant tissue areas from different cases together on a single slide, thereby greatly reducing the number of slides to be handled.

PROCEDURE

To prepare a multiblock, slides of relevant cases representative of the paraffin block face are collected and examined. The areas of tissue to be studied are identified from the slides and correlated to the paraffin block face (*Fig. 1*). The areas are then punched out of the block using a standard skin punch instrument with a knife diameter of 4 mm, that has been bored through and equipped with a springy 3 mm wooden stick that helps the samples to be ejected (*Fig. 2*).

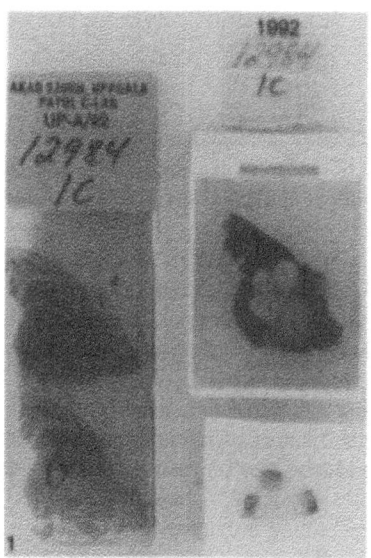
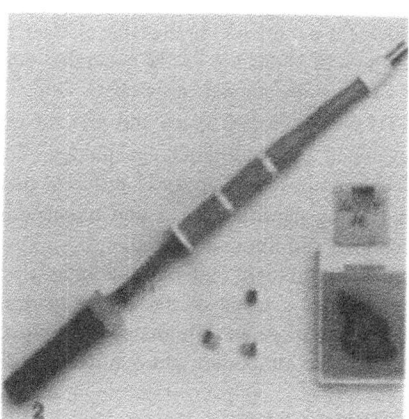

Figure 1. Slide with delineated areas, matching paraffin block and punched biopsies.
Figure 2. Punch instrument with punch biopsies and punched paraffin block.

Before the samples are placed together in a cast, they are assigned specific positions in the cast (*Fig. 3,4*). The cast consists of a metal plate with 3x1 mm spikes, at a distance of 5 mm apart. In the example exhibited, we have a plate with 24 spikes. The punch biopsies are mounted on the spikes to keep them in their respective assigned position. The block is cast as a routine block and the metal cast along with the spike plate can be removed after the paraffin has solidified. From the resulting multiblock, 5 μm thick sections can be cut and stained as for any other routine block (*Fig. 5*).

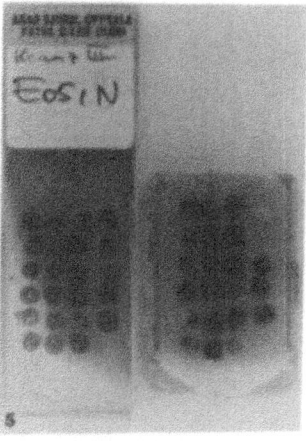

14645/91 A8 Skiv	12222/91 D8 Skiv	23401/91 G8 Skiv	12959/91 H8 Skiv	23946/90 H7 Skiv	4340/91 G7 Skiv
3124/92 A1 AdCa	15649/91 B1 AdCa	17201/91 D1 AdCa	13131/91 F1 AdCa	4117/91 D2 AdCa	13665/91 E3 AdCa
10060/92 C5 AdCa	11083|92 A5 Ca mono	16531/91 A4 Coid	7160/91 C4 Coid	3718/91 D4 Coid	8595/92 E4 Coid
1204/89 A3 CaSa	14028/92 B3 Meso	13824/92 C3 Meso	8084/92 A6 Parag	12239/92 F2 AdCa	O

Figure 3. Cast equipped with spike-plate, and mounted punch biopsies.
Figure 4. Graphic protocol showing the predetermined position of each single case in the cast. The "0"-position facilitates the orientation when mounting the section and reading the slides.
Figure 5. Multiblock containing 23 punched biopsies from different tumors (note the "0"-position), and slide with corresponding section.

Figure 6. Archival storing box for plastic tubes, containing punched biopsies from different cases.

To avoid loss of tissues or even eventually whole tissue samples, it is advisable to avoid repeated trimming. Cutting as many sections as possible in one sitting and store them until use is also a good idea. It is also possible to store punch biopsies from many tissues in labelled plastic tubes, allowing further rapid processing and designing of new multiblocks (*Fig. 6,7*).

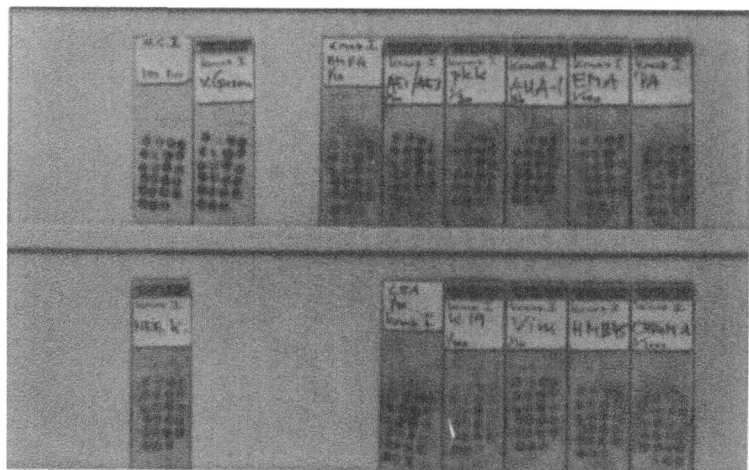

Figure 7. A typical example of an investigation series of 23 different tumors: eleven immunostainings, one negative control slide, two routine stainings. With the conventional method this investigation would have required cutting and staining of 23 x 14 = 322 slides.

DISCUSSION

The multiblock method facilitates rapid and standardized immunohistochemical study of many tissues with different antibodies simultaneously. This will save a lot of valuable technician time and supplies. The number of rounds will be reduced greatly thereby eliminating a cause for variation. With a group of cases to be studied on one slide instead of just one case, the slide does not need to be changed for every case, and the frame of reference can be more easily maintained.

The size of a multiblock biopsy is about 12.5 mm^2, which exceeds many other endoscopic and diagnostic biopsies. The size of the area to be evaluated is not a disadvantage when compared with other methods. Battifora[1] previously described a somewhat similar method in which even smaller pieces of tissue are used (about 1 mm^2). This method was used primarily to screen the supernatant of hybridomas for antibody production. It is not reliable for patient related clinical studies, as the case cannot be safely traced. The multiblock method guarantees the proper identification of a single case, which is a definite advantage over the other methods. One of the most important applications of using the multiblock method is for comparative studies. Larger series of tumors can be characterized using an antibody panel. The multiblock can also be used to set a panel of tissues with a variety of antigen concentrations, ranging from negative to full positivities which could be used as a control for a particular antibody or quality control for the laboratory. Other purposes for the multiblock would be to compare laboratory working steps such as fixatives, fixation time, and education.

Acknowledgement

Skilful technical assistance was given by Maj-Lis Book.

REFERENCES

1. H. Battifora, The Multitumor (Sausage) Tissue Block: Novel Method for Immunohistochemical Antibody Testing, *Lab Invest* **55(2)**:244 (1986).
2. W. Kraaz, B. Risberg, and A. Hussein, Multiblock: an aid in diagnostic immunohistochemistry (Letter to the Editor), *J Clin Pathol* **41**:1337 (1988).

CHAPTER 11

IMMUNOHISTOCHEMICAL, BIOCHEMICAL AND
PHYSIOLOGICAL CHARACTERIZATION OF CALCIUM-
BINDING PROTEINS

A. Hermann[1] and H.H. Kerschbaum

University of Salzburg, Department of Animal-Physiology, Institute of
Zoology, Hellbrunnerstr. 35, A-5020 Salzburg, Austria

SUMMARY

Intracellular Ca^{2+} signaling in many cases is mediated by calcium-binding proteins
(CaBP). To investigate immunological and biochemical characteristics of CaBPs in
relationship to their functions, a combination of different techniques can be applied.
Immunocytochemistry (PAP, IGSS) was used to investigate the localization and cellular
distribution of CaBPs, and biochemical techniques (gel-electrophoresis 1D/2D-PAGE,
calcium blot, Western blot) were employed to further identify these proteins according
to their molecular weight, isoelectric point or calcium binding ability. Electrophysiology
(intracellular recording, voltage clamp, microiontophoresis) was used to study the
electrical discharge behavior and membrane currents. Immunostained cells and
intracellular free calcium changes were monitored by using imaging techniques (confocal
laser microscopy). The results show that CaBPs are distributed in a cell specific manner,
with some CaBPs being exclusively contained in a subpopulation of neurons which
exhibit a certain electrical behavior (silent or spontaneously active). In neurons, CaBPs
appear able to modulate calcium channel activity and in sensory hair cells, there is
indication that the expression of CaBPs is dependent on their activation by Ca^{2+}. Future

[1] Address for Correspondense: University of Salzburg, Department of Animal-Physiology, Institute of
Zoology, Hellbrunnerstr. 35, A-5020 Salzburg, Austria. Tel:0662-8044; Fax:0662-8044-5698

research, particularly on the function of CaBP, could gain substantially from an integrative approach using the different technologies outlined in this contribution.

INTRODUCTION

In many cases, the combination of different techniques for scientific research has led to new insights into basic biological and biomedical mechanisms. Immunocytochemistry, for example, constitutes a powerful tool for the identification and localization of proteins. It allows the investigation of pathways and local circuits in the brain as well as functional aspects such as protein processing, neuronal transport or exocytosis. Biochemical techniques allow a further detailed characterization of proteins, i.e. the identification of their molecular weight, their isoelectric point or their interaction with ions or other proteins. Electrophysiology, on the other hand is useful for the study of electrical properties (membrane resting potential, electrical discharge activity, calcium and other membrane conductances) of single cells or of cell assemblies and imaging techniques can be used to follow the distribution and kinetics of intracellular calcium signals. In the course of studying calcium-binding proteins (CaBPs), we became interested in the combination of these techniques in order to obtain further information about their distribution, localization, and function.

Calcium in Cellular Physiology

Intracellular calcium signaling constitutes a major transduction pathway in a great variety of functions in many different types of excitable and non-excitable cells. A key to the understanding of the importance of calcium in cell physiology is its extremely low intracellular concentration. The free intracellular calcium ion concentration in almost all cells is 10^{-7}-10^{-8}M compared to its extracellular concentration of 10^{-2}-10^{-3}M. This results in a concentration gradient of ~100.000 fold across the plasma membrane - which is much higher than for any other cellular ion, such as Na^+, K^+ or Cl^-. Due to its considerable driving force, the opening of calcium channels (voltage- or receptor gated) causes a massive influx of calcium ions and an elevation of the intracellular free calcium concentration. Calcium ions may also be liberated from intracellular stores, such as the endoplasmatic reticulum or mitochondria after appropriate second messenger activation (i.e. inositol-triphosphate). Hence the rise and fall of intracellular calcium appears as a rather complex, space and time dependent process which can proceed in waves or spirals through the cell[1].

To function as a fast, re-useable pathway, the intracellular calcium signal ought to be transient and, therefore, calcium has to be tightly regulated. This is accomplished by *calcium sequestration systems*, including the sarco- or endoplasmatic reticulum, mitochondria, membranes, nuclei, calcium-binding proteins and/or *calcium extrusion systems*, such as calcium-transporters and/or a sodium/calcium exchanger. A summary of contributing factors to *calcium homeostasis* is shown in *Fig. 1.*[2-5]

Impairment of the calcium homeostasis can damage or even destroy cells which may provide a basis of various degenerative disorders of the nervous system. Prolonged intracellular calcium increase may be responsible for the activation of various enzymes such as *phospholipases* which disintegrate membranes and produce toxic metabolites; *proteases*, which are involved in the disassembly of cytoskeletal elements; enzymes, receptors and ion channels or *endonucleases* which fragment DNA.

Calcium-Binding Proteins

CaBPs are important regulators of cellular activities. Important functions of these proteins are the *buffering of calcium ions* and the *operation of target proteins*. CaBPs have also been used as intracellular *markers* to explore the morphology and the distribution of neurons in the nervous system. Most studies have been conducted concerning the biochemical characterization and immunological localization of these proteins whereas, excluding a few like calmodulin or troponin C, little is known about their functions.

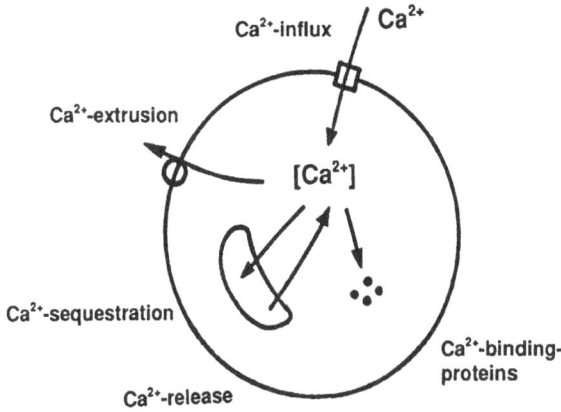

Figure 1. Cellular Calcium Homeostasis

1) Calcium-Influx
- voltage-activated calcium channels
- ligand-activated ion channels
- stretch-activated ion channels

2) Calcium-Release
- inositol-trisphospate (IP$_3$), caffeine
- calcium-induced calcium release

3) Calcium-Extrusion
- calcium-transporter (Ca-ATPase)
(high affinity/low transport capacity)
- sodium/calcium-exchanger
(low affinity/high transport capacity)

4) Calcium-Sequestration
- sarco- or endoplasmatic reticulum
- mitochondria
- **calcium-binding proteins**
- membranes, vesicles, nuclei etc.

Two major families of CaBPs are the *EF-hand* CaBPs and the *Annexins*. A characteristic of the EF-hand CaBPs is their helix-loop-helix configuration[6] with the loop as the calcium binding site, consisting of 12-14 amino-acids and 6-8 oxygen groups. There are 2-8 EF-hands in different types of proteins which may have different calcium affinity or may have lost their calcium binding ability. Calcium binding usually a mixed calcium magnesium affinity, induces conformational changes of the molecule which causes the exposition of hydrophobic sites and allows interaction with other proteins.[7,8] Well known examples of EF-hand CaBPs are *troponin C*, the first well characterized CaBP that is involved in muscle contraction; *calmodulin*, an ubiquitous cellular regulator protein; *parvalbumin* which occurs in fast contracting muscle fibers and in nerve cells functioning as a buffer or a transport system for calcium ions; *sarcoplasmic-calcium-binding protein* probably serving a similar function to parvalbumin in invertebrates; *calbindin-D28K*, also considered to act as calcium buffer in nerve cells; *calsequestrin*, a calcium buffer located in the sarcoplasmic or endoplasmic reticulum; *S-100*, originally found in glial cells but also occurring in many other cell types including nerve and muscle, reported to regulate protein phosphorylation, ATPase, cytoskeleton interactions, neurite extension etc. in addition to being released from cells and having extracellular effects or *calcineurin*, a calcium and calmodulin-dependent phosphatase. The number of different types of EF-hand CaBPs is presently at ~200.[9-12]

The other large family of CaBPs, the annexins, do not contain the EF-hand motive in their structure and bind calcium ions (up to 8) only in the presence of phospholipids.[11,13,14] Some of the annexins are localized at or in the plasma membrane and they are found in high concentration in the white matter of the brain. Although their functions have not been well determined, they appear to play a role in exocytosis, in crosslinking membrane receptors, or in membrane transport. Annexin V has been reported to exhibit calcium channel properties.

CaBPs also appear to be involved in pathological conditions such as Alzheimer's or Parkinson's disease, epilepsy or stroke.[7,8] They have been reported to reduce "calcium toxicity", i.e. neurons which contain CaBPs as calcium buffer have a better chance to survive excessive calcium loading, for example, after excessive stimulation during an epileptic seizure.[15] Up regulation of calbindin D-28K is indeed found after induction of "non-toxic" experimental seizures.[16] Calbindin D-28K (and its mRNA) is reduced, however, in brain regions affected in Alzheimer's or Parkinson's disease. The S-100ß gene is located on chromosome 21 which is tripled in Down's syndrome. The over expression of this CaBP may, therefore, be involved in the degenerative brain disorders. S-100, which interferes with the assembly and disassembly of microtubules, may be involved in Alzheimer's disease and ß-amyloid potentiates the increase of intracellular calcium suggesting an effect on CaBPs, the Na/Ca exchanger or the Ca-pump.[11,17]

In an attempt to study various aspects of CaBPs using different experimental approaches, we took advantage of relatively simple preparations. For example, the nervous system of molluscs has large neurons which are relatively stable in their position in the ganglia and can be identified in different animals. Recordings from these cells can be readily accomplished and electrical characteristics of some of the cells are well defined. Some cells exhibit a slow oscillatory activity with regular periods of action potential discharge and silent periods (bursting pacemakers), others discharge action potentials in a continuous regular mode (beating pacemakers) or they are silent in the absence of external synaptic or hormonal input.[18,19] In models concerning the mechanism of endogenous pacemaker activity, oscillations of the intracellular, free calcium ion concentration and calcium-dependent ion conductances play an essential role.[20-23] Therefore, we hypothesized that cells with different electrical discharge properties are differently equipped with CaBPs (or other calcium sequestering and extrusion systems). In order to investigate immunocytochemical and biochemical characteristics of some selected CaBPs [sarcoplasmic calcium-binding protein (SCP), S-100, parvalbumin, calmodulin, calbindin] in relationship to electrical activity or in response to electrical stimulation, we used nerve and muscle cells of molluscs (*Aplysia californica, Helix pomatia*), the sensory cells (inner ear, lateral line organ) of the clawed frog, *Xenopus laevis* and cultured cells.

Immunocytochemistry

Fixation. The selection of the proper fixative depends on the nature of the antigen and upon the technique of how the antiserum was produced. In the production of antisera, small molecules such as neurotransmitters (5-HT, GABA, etc.) are often coupled to a carrier protein by glutaraldehyde, and, therefore, better staining results may be obtained when the tissue had been fixed with an aldehyde-containing fixative. Large proteins are generally good immunogens and do not need to be conjugated to a carrier protein (see preparation of antiserum). Tissues used in our study were fixed in Bouin's or

Stefanini's/Zamboni's solution (1-2 hours) or freeze-dried followed by formaldehyde vapor fixation. Bouin's-fixed specimens were washed in PBS, dehydrated in ethanol (70%, 80%, 90%, 96%, 100%), cleared with xylene and embedded in paraffin. Freeze-dried specimens were vacuum-embedded in paraplast using benzene and benzene-paraplast (1:1) as intermedium. Seven μm thick sections were mounted on chrome-alum gelatine or poly-L-lysine (PLL, Sigma P1274) coated glass slides. For quick-freezing, the tissue specimens were placed on small containers of aluminum foil and inserted into melting isopentane prechilled with liquid nitrogen.

Chemicals, Proteins and Antibodies

Chemicals of analytical grade were obtained from Fluka, Merck or Sigma. Parvalbumin, calbindin D-28K, SCP I, SCP II and S-100 were isolated as indicated earlier,[24] bought or obtained as gifts (C. Heizmann/Zürich; J. Cox/Geneva; R. Donato/Perugia). The following antibodies were used: a) affinity-purified rabbit antibodies against HPLC-purified rat muscle parvalbumin (PV), b) rabbit polyclonal antibodies against rat muscle PV (for preparation of antibodies see below), c) rabbit antiserum against carp II PV and pure antigen from carp muscle d) rabbit antiserum against recombinant rat calbindin D-28K, e) affinity-purified rabbit antibodies against amphioxus SCP I and SCP II, and f) rabbit polyclonal antibodies against S-100 probes.

For affinity purification of antisera, the immunoglobulins were first isolated using a 50% ammonium sulfate precipitation step with subsequent dialysis. The dialyzed antibody solution was then loaded onto a CNBr-activated Sepharose 4B (Pharmacia) previously coupled with the respective HPLC-purified antigen. Specific antibodies were eluted at low pH (0.1 M glycin-HCl, 75 mM NaCl, pH 2.7) in a broad peak which was pooled and then dialyzed (25 mM boric acid, 15 mM NaCl, pH 8.4). During the described purification procedure, the affinity purified antibodies became diluted by a factor of 3. Elution from the affinity column at low pH caused a slight decrease of activity.

Preparation of Parvalbumin Antiserum

Antisera against frog PV (Sigma) were raised in rabbits. Rabbits were injected subcutaneously with 200 μg PV (dissolved in 1 ml distilled water) together with 1 ml Freund's complete adjuvant. 100 μg PV in 1 ml Freund's incomplete adjuvant was injected intra-muscularly on the seventh day. Subsequently, 100 μg PV in 1 ml incomplete adjuvant was injected subcutaneously on day 21, intramuscularly on day 22, and intravenously on day 23. Blood samples were taken four days after the last boost. Blood was allowed to clot and the blood clot removed. The serum was centrifuged and stored in aliquots at -80°C. Specificity tests were run using Ouchterlony and Western blots; no cross reaction of the serum was observed with other calcium-binding proteins such as calmodulin, S-100, and calbindin D-28K.

Immunocytochemical Staining

For localization of the antigen, the immunogold-sliver staining technique (IGSS-technique)[25-27] and the peroxidase anti-peroxidase technique (PAP-technique)[28] were used. For application of the IGSS-method, the sections were dewaxed in a descending series of ethanols starting with xylene, rinsed in tap water for 5 minutes, transferred to Lugol's iodine for 5 minutes, again rinsed in running tap water (2 minutes), and immersed for 5 minutes in IGS-buffer I (2.5% NaCl, 0.5% Tween 80, dissolved in 0.05 M TrisHCl buffer; pH 7.2). The sections were then incubated for 15 minutes with normal goat serum (1:25, Dakopatts, Denmark), followed by specific primary antisera applied for 24 hours at 4°C. Antisera against PV (diluted 1:800), SCP (1:800) against rat calbindin D-28K (1:500-1000) or affinity purified antibodies directed against rat muscle PV (1:200), amphioxus SCP I (1:200), amphioxus SCP II (1:200), and S-100 proteins (1:500) were diluted in phosphate-buffered saline (PBS; pH 7.2) containing 0.1% sodium azide and 0.1% bovine serum albumin. Sections were incubated overnight with primary antibodies. The sections were then washed for 5 minutes in IGSS buffer I and for 2 x 5 minutes in IGSS-buffer II (0.05 M Tris HCl buffer, pH 8.2-8.4) followed by incubation with normal goat serum (1:25) for 5 minutes and a 1:1 mixture of goat anti-rabbit immunoglobulin antibodies adsorbed to 5 nm and to 1 nm gold particles (Amersham, UK), each diluted 1:50 in IGS-buffer II containing 0.8% bovine serum albumin for one hour.

After three washes in IGSS-buffer II for 4 minutes each, sections were postfixed in 2% glutaraldehyde (diluted in PBS, pH 7.2) for 2 minutes, and washed in 6 portions of distilled water, 2 minutes each. For silver enhancement, the method of silver acetate autometallography as described by Hacker[25] was used. The immunostained sections were counterstained with Mayer's hemalum and eosin, dehydrated in ethanol, cleared in xylene and mounted with DPX (BDH Chemicals, England).

The PAP-method was used according to Sternberger.[28] Sections were rehydrated, rinsed in PBS for 3 x 5 minutes, and incubated in normal goat serum (1:30, Dakopatts). Subsequently, the sections were incubated with diluted primary antiserum for 24 hours at 4°C. The sections were then washed for 3 x 5 minutes in PBS, incubated with affinity purified antibodies against rabbit IgG (BioMakor) for 30 minutes or with goat anti-rabbit immunoglobulins (1:100) (Dakopatts) as bridging antibody, rinsed again for 3 x 5 minutes in PBS, and incubated with peroxidase-antiperoxidase complex from rabbit (1:100) (Dakopatts) for 30 minutes. For visualization of peroxidase, H_2O_2 (0.03%) as a substrate and 3,3'-diaminobenzidine-HCl (0.5mg/ml) as a chromogen were used. Finally, sections were dehydrated and cleared in xylene and mounted.

Glutaraldehyde fixed specimens (2.5% for ~1 hour) for semithin sectioning were incubated with 1% OsO_4, dehydrated in acetone and embedded in spur. Sections were cut, mounted and etched with a saturated solution of NaOH in 100% ethanol for 15 minutes and then washed in ethanol.

Controls. In control sections the primary antiserum was replaced by normal serum or adsorbed with the homologous antigen (1 μM) under liquid phase conditions for 24 hours at 4°C. Rat muscle and brain tissue of known antigen content were used as positive controls.

Biochemistry

Tissue Extracts. For immunoblotting and transblot/[45]Ca-overlays, frozen tissue was thawed, and, if necessary, cut into small pieces and ground in liquid nitrogen. The frozen powder was suspended in 2 volumes of extraction buffer (4 mM EDTA, pH 7.0, containing the protease inhibitors: 1 μM pepstatin A, 0.1 mM phenylmethylsulfonyl fluoride, 150 μM L-1-(tosylamido)-2 phenylethyl chloromethyl ketone and 30U/ml leupeptin). The suspension was sonicated for 30 sec and stored at 4°C for 1 hr. After centrifugation (15,000 g), the pellet was re-extracted, sonicated and the suspension centrifuged. Supernatants were pooled and heated for 30 min at 85°C. Many proteins denature during heat treatment, whereas most CaBPs remain stable. After centrifugation, the protein concentration in the supernatant (heat-extract) was measured using a micro-biurett assay. Heat-treated extracts were lyophilized and then adjusted with distilled water to a protein concentration of approximately 10 mg/ml. Extracts were stored at -20°C.

The water insoluble membrane- and heat-pellets were resuspended in 50 mM Tris-HCl, pH 6.8, 4% SDS, 12% glycerol, 2% mercaptoethanol and 0.01% Serva Blue and heated for 2-5 min in boiling water. Insoluble material was removed by centrifugation, and the supernatants were stored at 4°C.

Single Cell Isolation for Gel-Electrophoresis

Aplysia neural ganglia were isolated, the connective tissue was removed by microdissection and frozen in isopentane, freeze dried under vacuum (less than 10^{-3} torr) and stored at -80°C. Single neuronal cell bodies, identified by their location and size (diameter 200-600 µm), were dissected from the freeze-dried ganglia in a humidity (about 30%) and temperature (+20°C) controlled room. The cells were stored in capillaries under vacuum at -80°C.[24]

Electrophoresis

1D-PAGE of heat-treated tissue extracts (100-150 µg) were separated according to Schaegger and Jagow[29] with slight variations: in place of two staged separating gels of different concentration, 15% acrylamide-gels were prepared (*Fig. 2a*). This procedure was chosen because of a better separation of low molecular weight proteins compared to the Laemmli gel system used in 2D-PAGE.[24] 2D-PAGE (*Fig. 2b*) was performed according to O'Farrell.[30] For isoelectric focusing, a mixture of 80% ampholines, pH 3.5-10 and 20% ampholines, pH 2.5-4 (LKB, Sweden) was used to give a pH gradient that was linear between pH 4 and 6.5. Before electrophoresis, heat-treated tissue extracts were first dialyses against extraction buffer in order to remove excess amounts of salt. Proteins were either visualized by fluorography or by Coomassie staining, using 50 µg or 250 µg, respectively. For molecular calibration of SDS-gels, either a pre-stained protein mixture ranging from Mr 17,000 to Mr 130,000 (Bio Rad, USA) or a mixture of ^{14}C-methylated proteins (Amersham, England) ranging from Mr 14,000 to 200,000 as well as ^{14}C-labeled rat PV, Mr 12,000, were used.

Labeling of Proteins *In Vitro*

For fluorography, proteins were ^{14}C-labeled by reductive methylation according to Jentoft and Dearborn,[31] using ^{14}C-formaldehyde with a specific activity of 52 mCi/mmol (New England Nuclear, USA). This labeling procedure permits detection of proteins in the nanogram range and was chosen mainly because virtually all proteins become tagged, and label does not cause observable changes in the net charge of proteins.[32,33] The tissue was labeled for 3 h at room temperature in a solution containing 32 µl extraction buffer, 8 µl HEPES buffer (0.6M), pH 7.5, 5 µl NaCNBH$_3$ (200 mM) and 6µl ^{14}C-formaldehyde (2.6 mCi/mmol). After labeling, 17 µl of 0.25M Tris-HCl, pH 6.8, 9.2% SDS, 40% glycerol, 20% mercaptoethanol and bromphenol blue was added and samples were incubated for 2-3 min in boiling water.

Transblot/^{45}Ca Overlay Assay

In order to screen for calcium binding proteins, a ^{45}Ca^{2+}-blot assay was performed according to Maruyama.[34] Proteins were first subjected to gel electrophoresis, electrophoretically transferred to nitrocellulose membrane and finally incubated with ^{45}Ca^{2+} (specific activity 10-40 mCi/mg, NEN, USA). Binding of ^{45}Ca^{2+} to proteins was visualized by autoradiography. This technique allows for the detection of high affinity CaBPs in tissue extracts in amounts lower than 100 ng per sample.

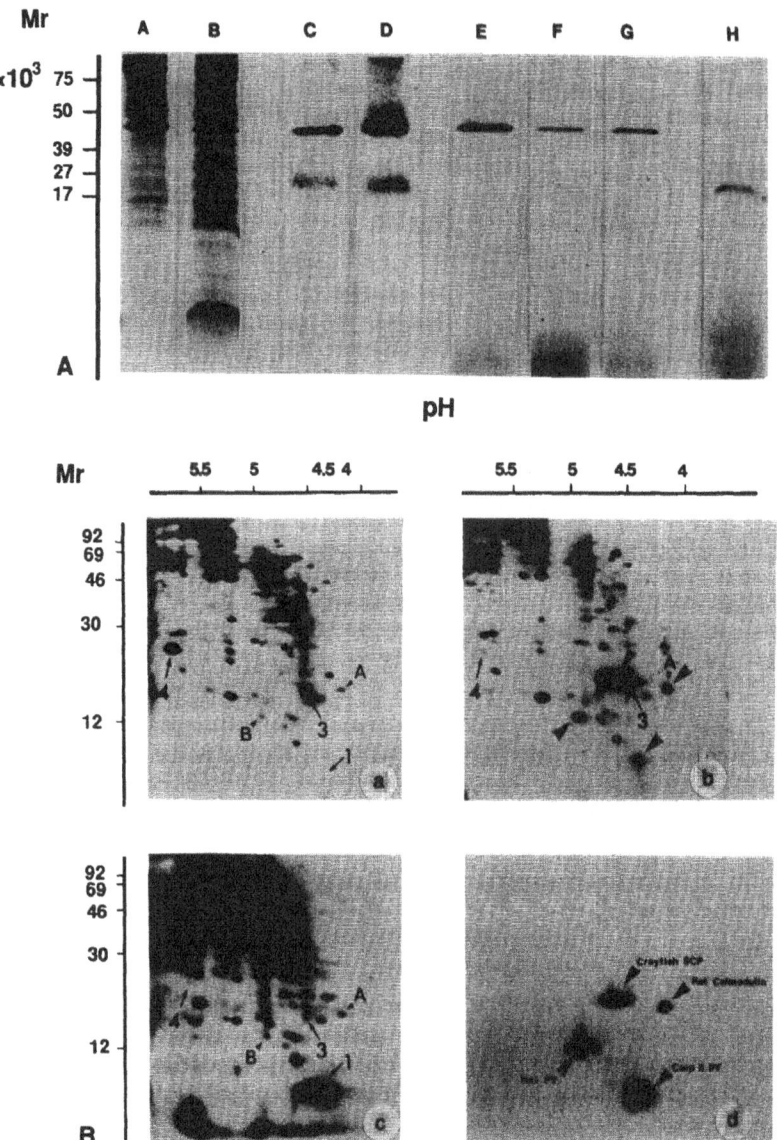

Figure 2. (A) 1D-PAGE Western blots and transblot/^{45}Ca^{2+}-overlays of *Aplysia* neural ganglia extracts. (A, B) Coomassie stained gels of separated proteins from circum-esophageal and abdominal ganglia. (C, D) Ca^{2+}-blots. (E-H) Western blots incubated with primary antibodies directed against (E) carp II parvalbumin, (F) rat parvalbumin, (G) rat calbindin D-28K and (H) *Amphioxus* SCP followed by a second antibody incubation with peroxidase-coupled goat anti-rabbit IgG.[24] (B) 2D-PAGE of single neurons and purified Ca^{2+}-binding proteins. 2D-PAGE: (a) of silent R2 cell, (b) of silent R2 cell with comigrating pure proteins (^{14}C-labeled rat parvalbumin (rat PV), carp II parvalbumin (carp II PV), rat calmodulin and crayfish sarcoplasmic Ca^{2+}-binding protein (SCP), (c) of bursting R15 cell and (d) of pure Ca^{2+}-binding proteins. Large arrowheads indicate the position of pure Ca^{2+}-binding proteins. Arrows 3 and 4 show characteristic proteins of silent cells, arrow 1 indicates specific proteins of bursting cell R15. Small arrowhead show comigrating *Aplysia* proteins to (A) rat calmodulin and (B) rat parvalbumin.[24]

Immunoblotting (Western Blot)

Western blot assays were used to further identify CaBPs found in Ca^{2+} blots. Extracts of nerve or muscle tissues were separated by 1D- or 2D-gel electrophoresis as described above, using 100-150 µg or 250-300 µg protein per sample, respectively. After electrophoresis, proteins were transferred onto nitrocellulose (Schleicher & Schüll, pore size 0.45 µm, or Bio Rad, pore size 0.2 µm) at a constant voltage of 20 V for 16 h.[35] After blotting, the nitrocellulose was first stained with Ponceau-Red, photographed and then destained. The membranes were incubated at 37°C for 3 h in a solution containing 3% bovine serum albumin and 1% fetal calf serum in 200 mM NaCl, 1 mM $CaCl_2$, 50 mM Tris-HCl, pH 7.4. Primary antisera and peroxidase coupled with goat anti-rabbit IgG were diluted 1:1,000 and 1:1,500, respectively, in the blocking solution described above. The affinity purified antibodies were used at a concentration of 1 µg/ml. Final staining was carried out with chloronaphthol as a substrate.

Comigration Experiments

Comigration experiments are carried out in order to compare characteristic protein spots with purified CaBPs. Single isolated neurons were treated for 2D-PAGE as described above and purified, ^{14}C-methylated Ca-binding proteins were added before isoelectric focusing (*Fig. 2b*).

Electrophysiology

Preparation and Solutions. Neuronal ganglia were dissected, pinned to the bottom of a recording chamber and the connective tissue removed, to expose the neuronal somata. Physiological bath solutions were: *Helix* solution, containing (in mM): NaCl 80; KCl 4; $CaCl_2$ 10; $MgCl_2$ 5; Tris-HCl 5; sucrose 10; pH 7.5 or *Aplysia* solution (artificial sea water) containing: NaCl 468; KCl 10; $CaCl_2$ 10; $MgCl_2$ 45 and Tris 25 at pH 7.7. To facilitate the removal of the connective tissue, protease (type XIV, Sigma) treatment was used for about 10-30 min in some cases. The ganglia were rinsed thoroughly afterwards with normal ASW solution to remove the protease. There was no difference in electrical activity or protein pattern with or without enzyme pretreatment. For electrophysiological experiments, the large, identified *Helix* neurons in the central nervous system[36] or *Aplysia* cells in the abdominal ganglion which exhibit various modes of electrical activity, were used (*Fig. 3a*).[18,19] These cells can be readily identified by their location within the ganglion, their size, and their color.

Intracellular Recording and Voltage Clamp Set Up. For intracellular recording of membrane potential, action potentials and membrane currents electrophysiological techniques were employed.[37-42] An Axoclamp-II (Axon Instr., USA) is the central recording unit. In this system, de- or hyperpolarizing pulses can be delivered either via a single electrode using a bridge circuit or with a second intracellular electrode. For measuring membrane currents, the two-electrode or single-electrode voltage clamp mode can be used and the current monitored either in the experimental bath (*Fig. 3b*) or in the feedback loop of the clamp system. Glass-microelectrodes used for recording (Clark, Reading, UK, GC150F-15, 1.5 mm diameter with filament) were filled with 3M KCl and had resistances of ~10 MΩ for voltage recording electrodes and 2-5 MΩ for current injection electrodes, the membrane potential was measured differentially between an intracellular microelectrode and an extracellular agar (3%)/bath solution electrode. The signals were digitized, stored on disk and further analyzed and processed off-line using p-clamp software (Axon Instr., USA). All experiments including those with cultured cells were performed at a room temperature of 19-21°C.

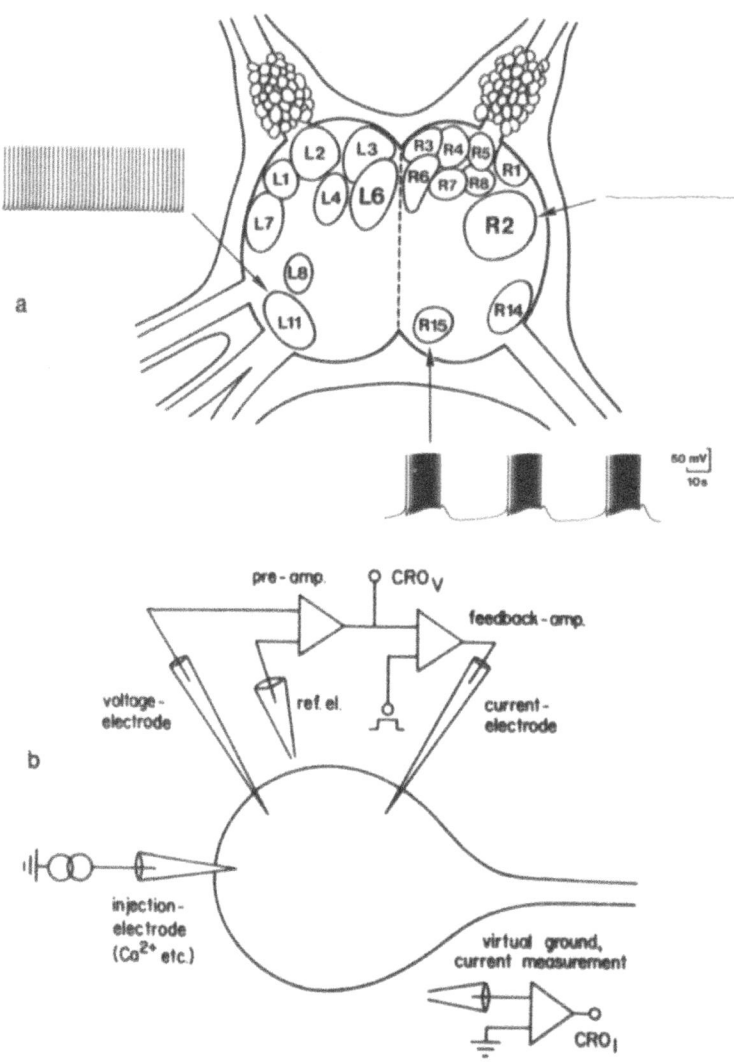

Figure 3. (a) Schematic of the <u>Aplysia</u> abdominal ganglion (dorsal view). Localization of some identified neurons together with examples of their electrical discharge activity.[24]
(b) Schematic of a two microelectrode voltage clamp set up and microiontophoretic injection.

Separation of Ion Currents. To separate membrane calcium currents from sodium and potassium currents, cells were bathed in a solution containing (in mM): $CaCl_2$ 40; $MgCl_2$ 5; KCl 4; tetraethylammoniumchloride (TEA) 45; 4-aminopyridine (4-AP) 5; glucose 5; Tris-HCl 5 at pH 7.5. The increased calcium concentration results in larger currents and

TEA and 4-AP are blockers of voltage dependent potassium channels. Barium ions may be used as charge carriers through calcium channels in place of calcium ions with the advantage of a higher permeability and an additional blocking effect on potassium channels. To inactivate the transient potassium outward current, the membrane holding potential was kept at -40 mV. Calcium currents were on-line corrected for leakage currents by subtracting membrane currents obtained by hyperpolarizing voltage pulses from membrane currents obtained by depolarizing voltage pulses.[21,38]

Isolation of single cells for electrophysiology. In order to obtain information from isolated single cells, neuronal ganglia were dissociated by incubation in pronase (10mg/ml) for 60-90 min with agitation. Then, ganglia were rinsed thoroughly in several volumes of normal solution. Ganglia were slit open and somata cut loose from the neuropil or cells were collected after triturating ganglia using pipettes of the approximate size of the cell batch. Single cells were collected with a pipette and stored in physiological solution supplemented with glucose (20 mM) at 4°C for 90 min or longer. Cells were then transferred onto coverslips coated with polylysine and left for at least 30 min to attach. For recordings, the solution was replaced by the appropriate experimental solution. Cultured cells (MDCK, Mardin-Darby Canine Kidney, an epithelium cell line or GH3 cells, a rat pituitary tumor cell line) were grown on appropriate sized, poly-L-lysine coated coverslips and used without pretreatment.

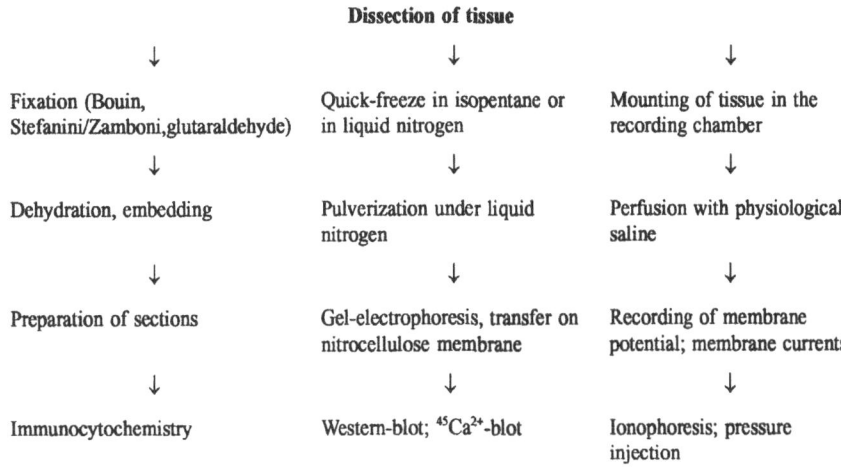

Table 1. Working diagram

Dissection of tissue		
↓	↓	↓
Fixation (Bouin, Stefanini/Zamboni,glutaraldehyde)	Quick-freeze in isopentane or in liquid nitrogen	Mounting of tissue in the recording chamber
↓	↓	↓
Dehydration, embedding	Pulverization under liquid nitrogen	Perfusion with physiological saline
↓	↓	↓
Preparation of sections	Gel-electrophoresis, transfer on nitrocellulose membrane	Recording of membrane potential; membrane currents
↓	↓	↓
Immunocytochemistry	Western-blot; $^{45}Ca^{2+}$-blot	Ionophoresis; pressure injection

Microiontophoresis

Charged compounds can be injected ionophoretically into voltage-clamped cells, using a constant current source (WPI, USA). This set up has the advantage that the membrane voltage is kept constant during injections and large amounts of current can be applied (several hundred nA). If the injection electrodes are filled with negatively charge BAPTA ions (50 mM), for example, negative current, or for the injection of Ca^{2+} positive current has to be applied to eject the ions. To assure a correct intracellular localization of the electrode, a small holding current (~5 nA), with an opposite sign to the ejection current should be applied. This current injected into the bath can be measured by the bath electrode and disappears immediately after insertion of the electrode into the cell.

The approximate amount of substance (M) injected can be estimated by using the equation: $M = I \, t \, n/z \, F \, V$, where I is the injection current, t is the duration of the injection, n is the transport or transference number of the ion, z is the charge of the ion, F the Faraday constant and V the volume into which the ion is injected. Transport numbers (between 0 and 1) can be obtained from the literature[43] or it can be estimated from the above equation after M, the amount of ejected substance, has been determined[37,43-45]. We have estimated a transport number for Ca^{2+} from microelectrodes of ~0.3.[43]

Pressure injection. Various substances, charged or non-charged, can be injected by applying positive pressure (Pneumatic PicoPump, WPI, USA) at the back of an electrode filled with the appropriate compound.[37,45] The approximate amount of fluid ejected can be estimated from a spherical fluid droplet injected into a drop of oil on a coverslip. The volume of the drop can be estimated by measuring the radius of the sphere after $V = 4/3 \pi r^3$. The amount of injection volume can also be estimated by observing the movement of an air bubble or the meniscus of a water/oil interface in the shank of the electrode. If cylinder shape of the electrode is assumed the ejected volume can be calculated from $V = \pi r^2 l$, where l is the length of the cylinder or if a truncated cone is assumed, $V = \pi l/3 (R^2+Rr+r^2)$, where R is the large radius and r the small radius and l the length of the cone.

Confocal Laser Microscopy

Confocal laser microscopy has been established as an important tool in light microscopy. In confocal laser microscopy, both the illumination and the detector part of the microscope are optimized. Confocal laser microscopy offers the advantages of a) an improved lateral resolution up to the theoretical optimum (~200 nm lateral resolution in confocal microscopy compared to 300 nm in conventional light microscopy; b) a ~3-fold increased depth (z)-resolution which c) allows for optical sectioning of the specimen. Only picture elements "in focus" are detected, out of focus regions of the preparation are not visible. Since the picture elements are rather small, the complete picture is composed after scanning a light beam across the area of interest.[46-50] However, even much smaller structures than light microscopy allows for can be visualized due to diffraction, although in this case no further resolution is obtained. In our system, we have been able to visualize, for example, 5 nm gold particles and 30 nm fluorescent beads.

The confocal system can be used in different epi-illumination modes: a) reflection, where light is reflected from the specimen, i.e. from gold particles (*Fig. 4a*) or from the PAP-complex b) fluorescence, where light is emitted from the preparation, i.e. from a calcium-dye complex (*Fig. 4b*) or c) transmission, which is not confocal but allows to image the preparation (*Fig. 4b*). For imaging, a Bio-Rad MRC 600 system equipped with an argon laser and a Leica inverted microscope (Fluovert) was used. Confocality is increased the higher the numerical aperture of the objective lens. Our objectives were Olympus SplanApo 60x, N.A. 1.4, DplanApoUV 40x, N.A. 0.85 and SplanApo 20x, N.A. 0.7. Confocal pinhole aperture was usually set to allow for optical sections of ~1 μm. For reflection imaging of IGSS- or PAP-processed specimens, as outlined under "Immunocytochemistry", were employed.

Measurement of the free, intracellular calcium ion concentration is of great interest in the context of studying CaBPs since the amount and the duration of the calcium signal is of paramount importance for the activation of the various follower reactions.[3-5,23,51] For calcium measurements, fluorescence imaging techniques were used. The optical filter was a BHS (exciter filter: 488nm/DF10, dichroic mirror: 510 nm/LP and emission filter: OG 515 nm/LP) in photomultiplier channel one. Photo-multiplier channel two was used for non-confocal transmission imaging or for fluorescence imaging of a different aspect, i.e. Fura-Red. For Fura-Red imaging, a K2 filter was used (exciter filter: 522nm/DF35, dichroic mirror: DC 560 nm/LP and emission filter: EF 600 nm/LP). Images were collected either by using a Kalman filter mode (which averages all frames after filtering is initiated) or by using the fast photon counting mode with 20-50 frames being integrated. Scanned images (768 x 512 pixels) were transferred to a framestore unit, further processed with appropriate computer software (SOM or CMOS, Bio-Rad), and photographed from a high resolution video monitor (VM1710 flat screen, black and white) or a Polaroid Freezeframe Recorder (color).

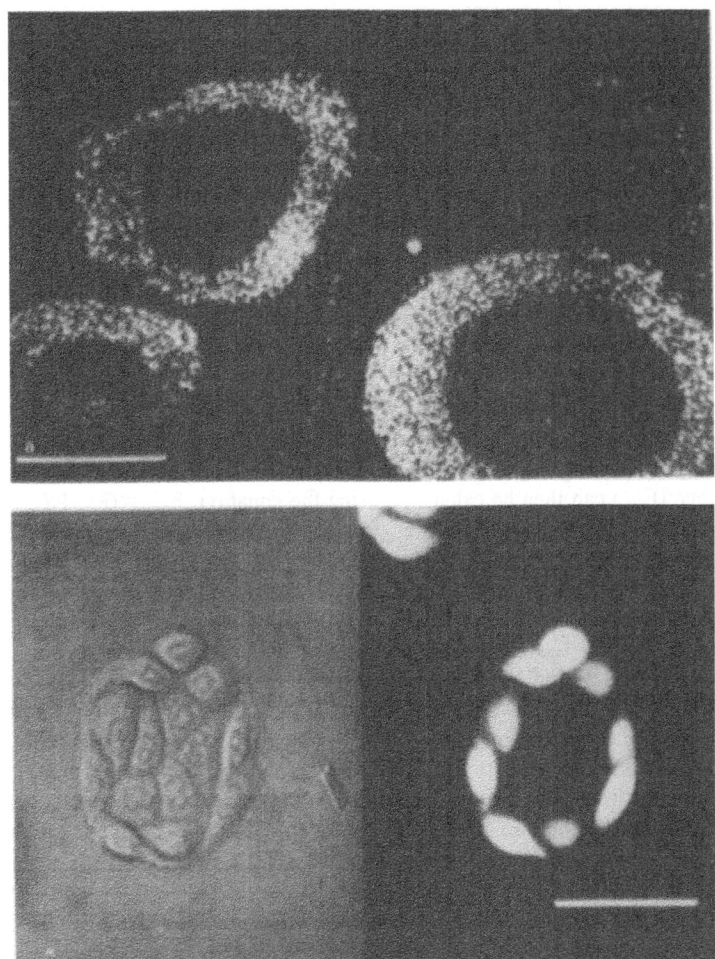

Figure 4. Confocal laser microscopy. (a) Immuno-gold-silver staining: Sarcoplasmic calcium-binding protein (SCP I) immunoreactivity of neurons in the abdominal ganglion of *Aplysia*. Confocal reflection images were obtained after labeling with a first antibody against SCP I and a secondary swine anti-rabbit antibody conjugated to 5nm colloidal gold particles. The cytoplasm but not the nucleoplasm is stained. Kalman filter (average of 20 frames). Bar: 250 µm. (b) right: confocal fluorescence image obtained by fast photon counting (average of 16 frames) taken from MDCK cells loaded with the calcium indicator Fluo-3. Cells at the periphery of the islet show strong fluorescence indicating a higher level of free calcium compared to cells in the center, left: non-confocal transmission image of the same cell islet. Bar: 50 µm.

For fluorescence measurements, cells were loaded with the membrane permeable calcium-indicator dye Fluo-3-AM (F-1242) or Fura-Red-AM (F3021) (Molecular Probes, USA) dissolved in dry DMSO (Sigma, D-8779). Since the calcium affinities of the dyes differ, a loading concentration of 10 μM for Fluo-3 and 20 μM for Fura-Red was used. To facilitate the uptake and dispersion of the indicator dyes in cells, the compound Pluronic F-127 (5 μM, BASF Wyandotte Corp., USA) may be added to the incubation solution. Before application of the dye, the incubator solution in the petri dish containing the cells was replaced by the appropriate physiological solution (500-1000 μl). Loading time for invertebrate cells was 60-120 min at room temperature and for cultured cell 45-60 min at 37°C. The Fluo-3 excitation maximum is at: 506 nm, emission at: 526 nm with a 40-fold increase of fluorescence intensity after binding of calcium (80-fold for new dyes). Its calcium affinity (K_D) is 400 nM which allows for calcium measurements in the range up to ~10μM. The AM-dyes after permeation through the cellular membrane are split by intracellular esterases into a dye component (Fluo-3 or Fura-Red) and an acetoxymethylester component. The dye is no longer membrane-permeable and therefore, is trapped inside the cell. After loading the cells, the external dye is removed by washout with normal physiological solution. The fluorescence of Fura-Red is <u>decreased</u> after binding of calcium. Its excitation maximum is at 488 nm, emission at 660 nm and its KD is 133 nM.

For *in vivo* calibration of the calcium signal, the method of Kao[52] was used. For calculations, the maximum and minimum fluorescence has to be determined. To obtain the maximum fluorescence, cells can be treated with the calcium-ionophore, ionomycin (2.5-10 μM) and a minimum fluorescence (down to 20% of the maximum fluorescence) can be obtained after quenching the calcium signal by using Mn^{2+} ($MnCl_2$: 2-10 mM). To further determine the autofluorescence of the preparation (F_{aut}), the cells can be permealized by adding digitonin (10-50 μM) to the bath solution. The maximum fluorescence (F_{max}) can then be calculated using the equation: $F_{max} = (F_{Mn}-Fa_{aut})/0.2+F_{aut}$, the minimum fluorescence (F_{min}) after: $F_{min} = (F_{max}-F_{aut})/40+F_{aut}$, and the intracellular free calcium concentration after: $[Ca^{2+}] = K_D (F-F_{min})/(F_{max}-F)$, where F is the actual fluorescence of the preparation, F_{Mn} the fluorescence after magnan quenching and the K_D for Fluo-3 is 400nM.

In vitro calibration techniques were employed by using different defined calcium solutions between 0.01 and 10 μM and different amounts of calcium indicator dye (Fluo-3 or Fura-Red, cell impermeable) from 7.5 to 20 μM. The results show that at 15 μM Fluo-3, a calcium concentration as low as 50 nM and at a dye concentration of 7.5 μM a calcium concentration down to 100 nM could be detected with our system. Fura-Red is even more sensitive. At a dye concentration of 20 μM a calcium concentration change from 10 to 20 nM can be measured.[53]

Figure 4b shows fluorescence measurements of MDCK cells loaded with Fluo-3. Fluorescence is high in some cells, particularly at the periphery of the cell islet.[33] The fluorescence could be caused either by a different amount of dye loading or a difference in the free intracellular calcium concentration. Since all cells in the islet show a similar fluorescence after treatment with the calcium ionophore ionomycin, the dye appears to be taken up by all cells which supports the notion that the high fluorescence indicates an increased steady state free calcium concentration in these cells.

Identification and Localization of Calcium-Binding Proteins

In this section, we will briefly outline some of the results we have obtained with the techniques described above. In the *Aplysia* nervous system, two proteins, one at Mr ~40,000 and another with a Mr of ~20,000, with high calcium affinity were found (Ca^{2+} blots) (*Fig. 2a*). The CaBP at Mr 40,000 cross-reacts with antibodies against parvalbumin and calbindin D-28K (1D-Western blots, *Fig. 2a*). Using immunocytochemistry, we found that cytoplasmic material contained in some neurons cross-reacts either with antibodies against parvalbumin (*Fig. 5b*) or calbindin D-28K or with both antibodies.[54]

Glial cells or cells in the connective tissue around the neuronal ganglia were not stained. A similar protein that cross-reacts with both PV and/or calbindin D-28K antibodies has been described in the *Drosophila and Helix* nervous system.[55,79] The 40 kD protein, therefore, appears to constitute a new type of calcium-binding protein in invertebrates. The protein at Mr 20,000 shares several characteristics with *sarcoplasmic calcium-binding protein* (SCP) from Amphioxus, such as isoelectric point, molecular weight, $^{45}Ca^{2+}$-binding ability and immunocrossreactivity with antibodies against SCP (1D-PAGE, *Fig. 2a*). In 2D-PAGE, it was shown that this protein consists of various isoforms (Mr 13,000-20,000).[24] SCPs originally described in muscle tissue have since been found in a variety of invertebrate tissue where they are abundant in muscle and exhibit strong polymorphism and high evolutionary drift between phyla.[56] Since the general properties of SCPs are different from those of PVs, they were classified into separate subfamilies of EF-hand calcium-binding proteins. PVs and SCPs have not been found to coexist in the same animal but both proteins occur with highest concentrations in fast contracting muscle,[57,58] which gave rise to the hypothesis that PV and SCP are involved in muscle relaxation.[59]

The present affinity purified antibodies directed against SCP isoforms I and II distinguished between muscle and nervous tissue in our immunocytochemical studies. It should be noted, however, that the antisera raised against SCP I and SCP II neither distinguish between SCP I and SCP II antigens under partially denatured conditions (Western-blot)[54] nor under more physiological conditions (Ouchterlony). This is not surprising when taking into account that antisera represent a pool of immunoglobulins that recognize various structural aspects at the surface of the antigen. Since SCP I and SCP II are isomers, they will share to some extent similar surface structures, and, therefore, will be recognized by both antisera. On the other hand, the difference in immunostaining of *Aplysia* tissue probably reflects the identification of unique structures of these proteins. It is possible that the immunocrossreactivity of the two affinity purified antisera seen in Western blots and Ouchterlony, is not relevant in immunohistochemistry.

Further analysis with specific antibodies to SCP isoform I and II showed that SCP I-like material is present in high concentration in neurons, whereas a SCP II-like protein is exclusively present in various muscle cells (buccal mass, body wall, heart), where it is in close association with the contractile machinery as determined by polarization microscopy.[54] The expression of calcium-binding proteins in a cell-type specific fashion may indicate that their presence is not a transient response to an external stimulus but may serve a persistent functional property of these cells, such as a certain electrical behavior. In neurons, SCP I appeared to be present in high concentrations in electrically silent cells. It was suggested, therefore, that SCP may act as a calcium buffer in these cells which prevents the activation of calcium-dependent ion conductances and hence, prevents spontaneous oscillatory discharge activity.[24] The notion of SCP being a calcium buffer is also consistent with a high calcium-affinity of SCP with an apparent K_D in the range of 10^{-7}-10^{-8} M.[60]

A similar tissue specific distribution of SCP I and SCP II-like material was found in *Helix* nerve and muscle cells.[61] SCP II-like material had a Mr of ~20,000 whereas SCP I-like material had a Mr of ~10,000. Electrical recordings from SCP positive neurons showed that these cells were electrically silent, supporting the notion that SCP may play a role in the expression of this electrical behavior. SCP and parvalbumin have been suggested to exhibit similar functions in different animal phyla and it was postulated, therefore, that parvalbumin evolved from SCP.[62] SCP and parvalbumin immunoreactive material was, however, not colocalized in the same neurons (*Fig. 5c, 5d*) and differs in its topography and molecular weight suggesting that SCP is not an evolutionary precursor of parvalbumin and that parvalbumin-like material already exists in invertebrates. A further difference of SCP and PV containing cells is that PV is expressed in GABAergic neurons with high electrical discharge activity[63] whereas SCP-like protein is found in electrically silent neurons containing serotonin as transmitter.[61]

Using immunocytochemical techniques (PAP), S-100-like proteins were detected in the *Aplysia* nervous system. A positive reaction-product was exclusively found in the cytoplasm of neuronal somata and neurites but not in the nucleus. S-100 positive material was usually found in clusters of small to medium sized neurons (diameter below 100 μm). None of the identified giant neurons in the abdominal ganglion nor any other type of cell (glial, muscle or endocrine) was stained (unpublished observations).

In *Helix*, a protein immunologically related to *S-100* could also be localized in identified nerve cells including some giant neurons (LPa3 and LPa4) in the circumesophageal ganglia (*Fig. 5a*). S-100 was present exclusively in the cytoplasm of the soma and neurites.[64,65] In some cases, the cytoplasm close to the soma periphery is more intensively stained. S-100 was not contained in glial cells, the trophospongium or cells within the connective tissue enveloping the ganglia. Biochemical investigation (SDS-Gel-electrophoresis; Western blot) showed that neural ganglia contain a protein of Mr 12,000 which cross-reacts with antibodies to S-100. This is relatively similar to the vertebrate S-100 subunit with a Mr of 10,500. In vertebrates, S-100 exhibits a strong preservation of the amino acid sequence (~5% difference between mouse and men). The protein in *Helix* neurons that crossreacts with S-100 antibodies is in its Mr close to vertebrate S-100 and shares a similar epitope.

S-100 containing cells exhibit spontaneous electrical discharge activity (intracellular recordings),[64,65] which is characterized either by a "beating" mode where single action potentials are generated at a regular low frequency, or by a "bursting" mode where high burst-discharge activity of action potentials occurs in regular oscillatory cycles interspaced by silent periods. S-100 was not found in electrically silent cells. Furthermore, S-100 immunoreactive *Helix* neurons exhibit a "N-shaped current-voltage (I-V) relationship" indicative of a well developed calcium-activated potassium conductance in these cells. A N-shaped I-V relationship is characterized by a pronounced current dip at positive potentials. It was previously shown that this type of I-V relation is caused by superimposed voltage- and calcium-dependent potassium currents.[66] The more prominent the calcium-activated potassium current, the more pronounced is the N-shape of the outward currents. *Aplysia* burster R-15 cells, for example, have a pronounced N-shaped I-V relationship, with ~80% of the total potassium outward current being calcium-activated and ~20% voltage-activated.[22] Beater L-11 cells on the other hand, show the reverse quantitative distribution of outward currents. The correlation between S-100 immunoreactivity and calcium-activated potassium current may, however, also serve other functions in these cells, such as frequency adaptation or post-tetanic hyperpolarization. A more detailed characterization of the protein in identified neurons, its expression and its intra- and extracellular functions appears to be of major future interest.

Calcium currents in various cells are controlled by internal calcium ions.[67,68] In particular, the inactivation of the calcium current appears to be dependent on calcium-activated processes such as channel phosphorylation/dephosphorylation which may be mediated by calcium-binding proteins.[69,70] We further investigated the effects of intracellular S-100 on calcium currents (voltage clamp). In these experiments, calcium currents were separated from other membrane currents as indicated in "Materials and Methods" and S-100 was injected ionophoretically or by pressure. Injection of S-100 into the cells causes an increase in the inactivation of the calcium-inward current compared to controls without altering the peak calcium current.[64,65] S-100 appears not to act as a

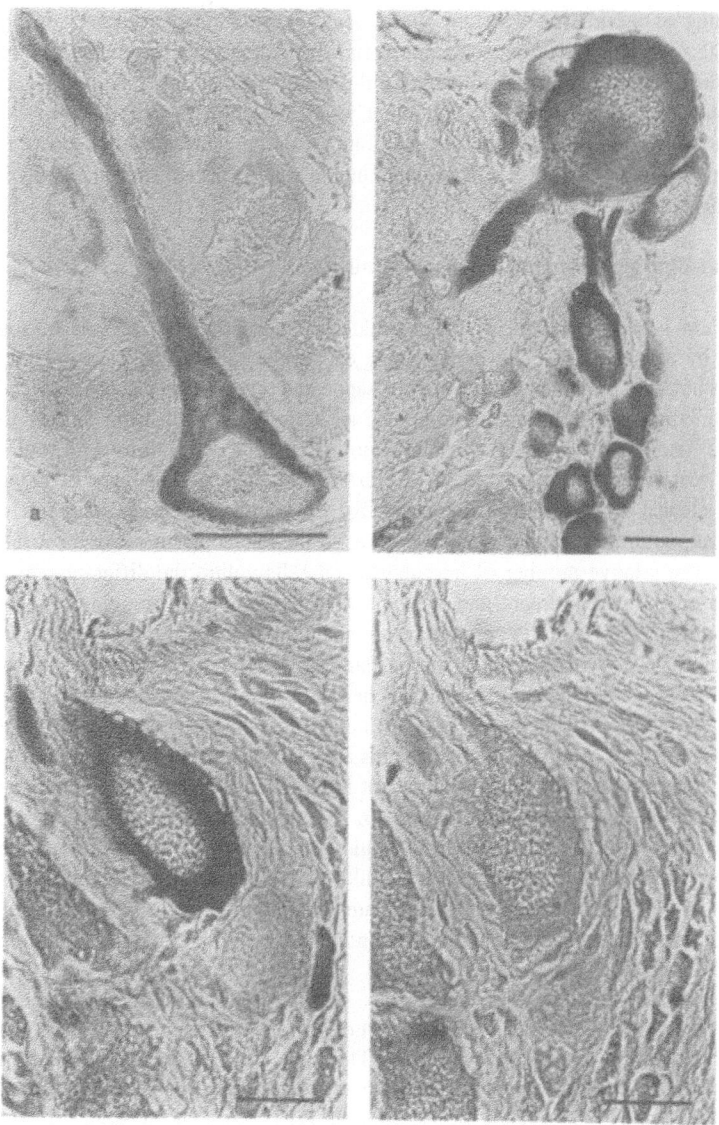

Figure 5. Calcium-binding proteins in neurons of the central nervous system of *Helix pomatia.*
a) S-100-immunoreactivity. Bar: 100 μm; b) various somata showing parvalbumin-immuno-reactivity[79] c)
SCP-immunoreactive material adjacent section to (d) treated with antisera against parvalbumin indicating
that SCP and PV are not colocalized.[61] Bar in b-d: 25 μm.

calcium buffer in this case since calcium buffering would cause a slowing of the
calcium-current inactivation phase.[67] The mechanism of the calcium-dependent
inactivation of the calcium current has been attributed to activation of calcineurin, a
calcium/calmodulin dependent phosphatase which causes dephosphorylation of the
calcium channel protein.[70] Since calcineurin is present in mollusc neurons[71] and S-100
has been found to bind to calcineurin,[72] it appears possible that S-100 is a mediator of
calcium channel dephosphorylation. The physiological effects of a S-100 dependent

shortening of the internal calcium signal could cause a faster repolarization of action potentials, allow for a faster action potential firing frequency or cause less adaptation (since the internal calcium accumulation will be less and therefore calcium-dependent ion conductances will be less). On the other hand, injection of *parvalbumin* into *Helix* neurons caused an increase of the maximum calcium current and a decrease of its inactivation phase.[64] These effects of parvalbumin were expected from the known calcium buffering function of this protein.

Calcium-Binding Proteins in Sensory Hair Cells

Calcium is also involved in the transduction of mechanical stimuli into electrical signals by sensory hair cells in the <u>inner ear</u> or the <u>lateral line organ</u>. The study of calcium-binding proteins in this context appeared interesting since these cells show oscillatory behavior upon stimulation with an underlying mechanism similar to that of mollusc oscillatory neurons.[73] Parvalbumin, S-100, calbindin and calmodulin-immunoreactive material was found exclusively in sensory hair cells of tadpoles and mature clawed frog, *Xenopus laevis*[74] (*Fig. 6*). Immunoreactivity was found in the cytoplasm, with the cuticular plate as well as in the nucleoplasm being most intensively stained. The relative number of immunostained cells within the different sensory areas depends on the compartment and the type of calcium-binding protein. In the auditory organs, both S-100 and PV are more abundant in the amphibian papilla (frequency response up to ~400 Hz) than in the basilar papilla (frequency response up to ~1,500 Hz), whereas calbindin-immunoreactive hair cells are most abundant in the basilar papilla. In the compartment serving vestibular functions (lagena, saccule, utricle and cristae ampullaris), calcium-binding proteins are expressed in high amounts and relatively equally distributed. Calmodulin is abundant in all compartments. The staining pattern is rather stable during development (stages 49, 53 and adult) and is only slightly increased in some cases during the transition to adulthood. The high concentration of calcium-binding proteins in the cuticular plate could well explain the high calcium buffering capacity in this part of the cell.[75] The calcium-binding proteins in sensory hair cells due to their properties as channel modulators and calcium buffers, may be involved in their resonant electrical behavior.

Sensory hair cells of the lateral line system are stimulated by water motion. Neuromasts in the lateral line system consist of sensory cells, supporting cells and mantle cells which are all covered by a cupula. The lateral line system of the body remains unchanged during metamorphosis whereas neuromasts of the head are reorganized. These changes are probably correlated with a different locomotory and feeding behavior of the tadpole and adult animals. As a result, the electrical behavior of sensory cells may be altered during the course of ontogeny which in turn may be reflected in the expression of CaBPs. The following study was designed to give us some indication if major changes in the expression of CaBPs occur during the time of innervation of the hair cells.

The animals used for experiments were derived from a single breed to reduce the possibility of variation between individuals. Stages 37, 47, 49, 53 , 59 (climax stage of metamorphosis) and adult animals were investigated. Four different CaBPs, parvalbumin, calbindin D-28K, calmodulin and S-100, were studied and immunoreactive material was detected in sensory cells of the stages 47 to adult. At the early stage 37 before innervation of hair cells, no immunoreactive material was detected. The number of sensory cells containing CaBPs usually first increased after innervation, obtained a maximum at about stage 53 and than decreased towards adulthood. In the case of calmodulin, immunoreactive cells decreased continously from stage 47 to adult. The results suggest that the expression of CaBPs may be involved in the development of these cells and that the advent of electrical functioning of the cells plays an important role. A similar dependence of the expression of CaBPs on innervation has been found for muscle parvalbumin.[76] After differentiation of the cells, CaBPs decline and further on, may be involved in various metabolic maintenance functions.

The sensory cells of the *Xenopus* lateral line organ are activated by calcium influx which can be blocked by external application of cobalt ions.[77] This preparation is advantageous in further studying the question of whether the expression of CaBPs may

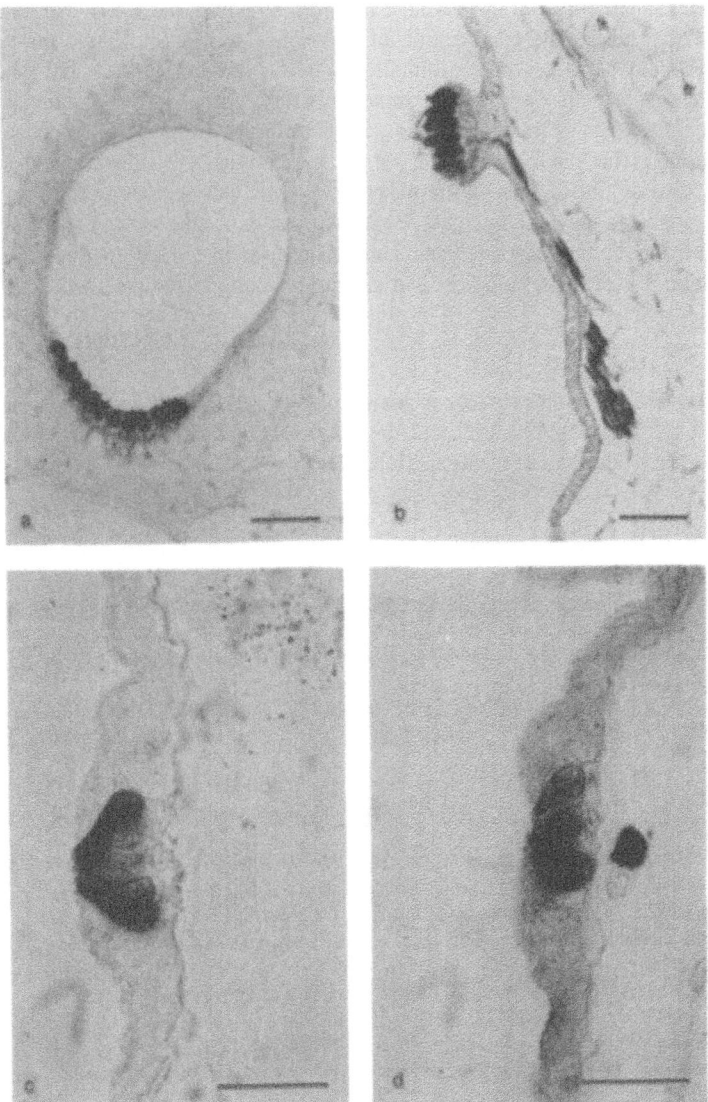

Figure 6. Calcium-binding proteins in sensory hair cells of the inner ear and lateral line system of *Xenopus laevis* a) Sensory cells in the basal papilla containing calmodulin-immunoreactive material b) Calmodulin-immunoreactive hair cells of the ampulla.[74] Bar in a-b: 50 μm. c) S-100 positive sensory cells and d) Parvalbumin-immunoreactive sensory cells of the lateral line organ. Bar in c-d: 25 μm.

be dependent on calcium activation. Tadpoles of stage 53 were kept either in artificial tap water or in tap water containing 0.1 or 1.0 mM cobalt. After 4 days in the appropriate solution, animals were anesthestized and fixed. The results show that the number of CaBP positive cells is suppressed by 60-70% after the calcium influx into the cells has been blocked. The total number of cells remained unaltered. Increased concentrations of calcium or sodium, or stimulating the cells by stirring the water did not alter the number of immunoreactive cells or the intensity of staining compared to controls. The experiments suggest that the expression of calcium-binding proteins is dependent on calcium entry into the cells.

Further investigation in this field will benefit from a more integrative approach using different techniques which brings together the working power and the exchange of ideas of various specialists. The combination of techniques such as molecular genetics and electrophysiology has been very fruitful already in studying the structure and function of ion channels. It may also help to surmount some of the problems in studying calcium-binding proteins. One of the major tasks will be to further investigate the physiological role of these proteins and try to find answers to questions such as, why are there so many different types of CaBP, what is the meaning of their cell specific distribution in the nervous system and how did they evolve? Particularly molecular genetic approaches will be helpful for the further characterization and quantification of CaBP and in the study of their evolution. Electrophysiological and calcium imaging techniques will be powerful tools to study function.

Acknowledgement. Supported by a grant of the Austrian Science Foundation, FWF-Fonds P8050-MED, P9247-MOB, and the Janssen-Foundation. For excellent technical assistance, we would like to thank K. Holzinger.

References

1. L. Lechleiter, S. Girard, E. Peralta, and D. Clapham, Spiral wave propagation and annihilation in Xenopus laevis oocytes, *Science* **252**:123 (1991).
2. A.K. Campbell, "Intracellular Calcium: Its universal role as regulator" J. Wiley & Sons, New York (1983).
3. P.G. Kostyuk. "Calcium Ions in Nerve Cells", Oxford Univ. Press, Oxford (1992).
4. R.P. Rubin, G.B. Weiss, and J.W. Putney. "Calcium in Biological Systems". Plenum Press, New York (1985).
5. T.G. Spiro. Calcium in Biology, J. Wiley & Sons, New York (1983).
6. R.H. Kretsinger, Structure and evolution of calcium-modulated proteins, *C R C Crit Rev Biochem* **8**:119 (1980).
7. N.D. Moncrief, R.H. Kretsinger, and M. Goodman, Evolution of EF-hand calcium-modulated proteins. I. Relationships based on amino acid sequences, *J Mol Evol* **30**:522 (1990).
8. S. Nakayama, N.D. Moncrief, and R.H. Kretsinger, Evolution of EF-hand calcium-modulated proteins. II. Domains of several subfamilies have diverse evolutionary histories, *J Mol Evol* **34**:416(1992).
9. K.G. Baimbridge, M.R. Celio, and J.H. Rogers, Calcium-binding proteins in the nervous system, *Trends in Neurosci* **15**:303 (1992).
10. R. Donato, Perspectives in S-100 protein biology, *Cell Calcium* **12**:713 (1991).
11. C.W. Heizmann. "Novel Calcium-Binding Proteins: Fundamentals and Clinical Implications", Springer Verlag, Berlin (1991).
12. C.W. Heizmann, and W. Hunziker, Intracellular calcium-binding proteins: more sites than insights, *Trends in Biochem Sci* **16**:98 (1991).
13. R.D. Burgoyne and M.J. Geisow, The annexin family of calcium-binding proteins, *Cell Calcium* **10**:1 (1989).
14. S.E. Moss. "The Annexins - A Novel Class of Calcium-Binding Proteins", Portland Press, London (1992).

15. S. Christakos, C. Gabrielides, and W.B. Rhoten, Vitamin D-dependent calcium binding proteins: chemistry, distribution, functional considerations, and molecular biology, *Endocrine Rev* **10**:3 (1989).

16. D.H. Lowenstein, M.F. Miles, F. Hatam, and T. McCabe, Up regulation of calbindin-D28K mRNA in the rat hippocampus following focal stimulation of the perforant path, Neuron 6:627 (1991).

17. C.W. Heizmann and K. Braun, Calcium binding proteins - molecular and functional aspects, in: "The Role of Calcium in Biological Systems", L.J. Anghileri, ed., CRC Press, Boca Raton, Fl (1990).

18. W.T. Frazier, E.R. Kandel, I. Kupfermann, R. Waziri, and R.E. Coggeshall, Morphological and functional properties of identified cells in the abdominal ganglion of Aplysia californica, *J Neurophysiol* **30**:1288 (1967).

19. E. Kandel. "Cellular Basis of Behavior. An Introduction to Behavioral Neurobiology", W.H. Freemann, San Francisco (1976).

20. W.B. Adams and J.A. Benson, The generation and modulation of endogeneous rhythmicity in the *Aplysia* bursting pacemaker neurone R15, *Prog Biophys Mol Biol* **46**:1 (1985).

21. A.L.F. Gorman, A. Hermann, and M.V. Thomas, Intracellular calcium and the control of neuronal pacemaker activity, *Fed Proc* **40**:2233 (1981).

22. A.L.F. Gorman, and A. Hermann, Quantitative differences in the currents of bursting and beating molluscan pacemaker neurones, *J Physiol* **333**:681 (1982).

23. A. Hermann, Neuronale Oszillationen, in "Die Anwendung von GnRH und GnRH-Analoga in der Urologie", J. FRICK, ed., Blackwell-MZV, Wien (1992).

24. A. Hermann, T. Pauls, and C.W. Heizmann, Calcium-binding proteins in Aplysia neurons, *Cell and Molec Neurobiol* **11**:371 (1991).

25. G.W. Hacker, L. Grimelius, G. Danscher, G. Bernatzky, W. Muss, H. Adam, and J. Thurner, Silver acetate autometallography: an alternative enhancement technique for immunogold-silver staining (IGSS) and silver amplification of gold, silver, mercury and zinc in tissues, J. Histotechnol. 11:213 (1988).

26. C.S. Holgate, P. Jackson, P.N. Cowen, and C.C. Bird, Immunogold-silver staining: new method of immunostaining with enhanced sensitivity, *J Histochem Cytochem* **31**:938 (1983).

27. D.R. Springall, G.W. Hacker, L. Grimelius, and J.M. Polak, The potential of immunogold-silver staining methode for paraffin sections, *Histochem* **81**:603 (1984).

28. L.A. Sternberger. Immunocytochemistry, 2nd ed., Wiley, New York (1979).

29. H. Schaegger, and G. Jagow, Tricine-sodium dodecyl sulfate-polyacrylamide gel electrophoresis for the separation of proteins in the range from 1 to 100 kDa, *Anal Biochem* **166**:368 (1987).

30. P.H. O'Farrell, High resolution two-dimensional electrophoresis of proteins, *J Biol Chem* 250:4007 (1975).

31. N. Jentoft, and D.G. Dearborn, Labeling of protein by reductive methylation using sodium cyanoborohydride, *J Biol Chem* 254:4359 (1977).

32. R. Billeter, C.W. Heizmann, H. Howald, and E. Jenny, Analysis of myosin light and heavy chain types in single human skeletal-muscle fibres, *Eur J Biochem* **116**:389 (1981).

33. O. Kuhn, and F.H. Wilt, Double labeling of chromatin proteins, *in vivo* and *in vitro*, and their 2-dimensional electrophoretic resolution, *Anal Biochem* **105**:274 (1980).

34. K. Maruyama, T. Mikawa, and S. Ebashi, Detection of calcium binding proteins by ^{45}Ca autoradiography on nitrocellulose membrane after sodium dodecyl sulfate gel electrophoresis, *J Biochem* **95**:511 (1984).

35. H. Towbin, T. Staehelin, and J. Gordon, Electrophoretic transfer of proteins from polyacrylamide gels to nitrocellulose sheets - procedure and some applications, *Proc Natl Acad Sci, USA* **76**:4350 (1979).

36. J. Johansen, L.H. Jensen, and C. Holm, Morphological and electrophysiological mapping of giant neurons in the subesophageal ganglia of Helix pomatia, *Comp Biochem Physiol* **71A**:283 (1982).

37. P.M. Conn. "Methods in Neurosciences: Electrophysiology and Microinjection", Academic Press, London (1991).

38. A. Hermann and A.L.F. Gorman, Effects of 4-aminopyridine on potassium currents in a molluscan neuron, *J Gen Physiol* **78**:63 (1981).

39. A. Hermann and K. Hartung, Specificity of the calcium activated potassium conductance in Helix neurones, *Pflüg Arch* **393**:248 (1982).

40. T.G. Smith, H. Lecar, S.J. Redman, and P.W. Gage. Voltage and Patch Clamping with Microelectrodes, Williams and Wilkens, Baltimore (1985).

41. N.B. Standen, P.T.A. Gray, and M.J. Whitaker. Microelectrode Techniques. The Plymouth Workshop Handbook, The Company of Biologists, Cambridge (1987).

42. R. Sherman-Gold. The Axon Guide, Axon Instruments (1993).

43. A.L.F. Gorman, and A. Hermann, Internal effects of divalent cations on potassium permeability in molluscan neurones, *J Physiol* **296**:393 (1979).

44. H. Drouin, On microelectrode ionophoresis, *Biophys J* **46**:579 (1984).

45. T.W. Stone. Microiontophoresis and Pressure Injection, J. Wiley & Sons, New York (1985).

46. A. Fine, W.B. Amos, R.M. Durbin, and P.A. McNaughton, Confocal microscopy: applications in neurobiology, *TINS* **11**:346 (1988).

47. J.B. Pawley. "Handbook of Biological Confocal Microscopy", Plenum Press, New York (1990).

48. D.M. Shotton, and N. White, Confocal Scanning Microscopy: threedimensional Biological imaging, *TIBS* **14**:435 (1989).

49. D.M. Shotton, Confocal scanning optical microscopy and its application for biological specimens, *J Cell Sci* **94**:175 (1989).

50. T. Wilson. Confocal Microscopy, Academic Press, New York (1990).

51. J.G. McCormack and P.H. Cobbold. "Cellular Calcium: A Practical Approach". Oxford Univ. Press, Oxford (1991).

52. J.P.Y. Kao, A.T. Harootunian, and R.Y Tsien, Photochemically generated cytosolic calcium pulses and their detection by Fluo-3, *J Biol Chem* **264**:8179 (1989).

53. K.H. Hilber and A. Hermann, Intracellular calcium measurements using confocal laser microscopy, *Eur J Neurosci* **5**:59 (1992).

54. T.L. Pauls, J.A. Cox, C.W. Heizmann, and A. Hermann, Sarcoplasmic calcium-binding protein in Aplysia nerve and muscle cells, *Eur J Neurosci* **5**:549 (1993).

55. R. Reifengerste, S. Grimm, S. Albert, N. Lipski, G. Heimbeck, A. Kofbauer, G.O. Pfugfelder, D. Quack, R. Reichmuth, B. Schug, K.E.. Zinsmaier, C.W. Heizmann, S. Buchner, and E. Buchner, An invertebrate calbindin sub-family: protein structure, genomic organization, and expression pattern of the calbindin-32 gene of Drosophila, *J Neurosci* **13**:2186 (1993).

56. J.A. Cox, Calcium vector protein and sarcoplasmic calcium binding proteins from invertebrate muscle, *in*: "Stimulus-Response Coupling: the Role of Intracellular Calcium", J.R. Dedman, and V.L. Smith, eds., CRC Press, Boca Raton, Ann Arbor, Boston (1990).

57. M.R. Celio and C.W. Heizmann, Calcium-binding protein parvalbumin is associated with fast contracting muscle fibres, *Nature* **297**:504 (1982).

58. J.A. Cox, W. Wnuk, and E.A. Stein, Isolation and properties of a sarcoplasmic calcium-binding protein from crayfish, *Biochem* **15**:2613 (1976).

59. J.F. Pechère, J. Demaille, J.P. Capony, E. Dutruge, F. Baron, and C. Pina, Muscular parvalbumins: Some explorations into their possible biological significance, *in*: "Calcium Transport in Contraction and Secretion", E. Carafoli, F. Clementi, W. Drabikowski, and A. Margreth, eds., North-Holland Publ, Amsterdam (1975).

60. W. Wnuk and J. Jauregui-Adell, Polymorphism in high-affinity calcium-binding proteins from crustacean sarcoplasm, *Eur J Biochem* **131**:177 (1983).

61. H.H. Kerschbaum, V. Kainz, and A. Hermann, Sarcoplasmic calcium-binding protein-immunoreactive material in the central nervous system of the snail, Helix pomatia, *Brain Res* **597**:339 (1992).

62. W. Wnuk, J.A. Cox, and E.A. Stein, Parvalbumins and other soluble high-affinity calcium-binding proteins from muscle, *in*: Calcium and Cell Function, W.Y. Cheung, ed., Academic Press, New York (1982).

63. M.R. Celio, Parvalbumin in most gamma-aminobutyric acid containing neurons of the rat cerebral cortex, *Science* **231**:995 (1986).

64. H. Kubista, Lokalisation und Funktion von S-100-immunreaktivem Material im Zentralnervensystem von Helix pomatia, Diploma work (1992).

65. H. Kubista, H.H. Kerschbaum, and A. Hermann, S-100 immuno-reactive material in the central nervous system of the snail, Helix pomatia, *Eur J Neurosci* **5**:23 (1992).

66. R.W. Meech and N.B. Standen, Potassium activation in Helix aspersa neurons under voltage clamp a component mediated by calcium influx, *J Physiol* **249**:211 (1975).

67. R. Eckert, and J.E. Chad, Inactivation of calcium channels, *Prog Biophys Molec Biol* **44**:215 (1984).

68. N.B. Standen, and P.R. Stanfield, A binding-site model for calcium channel inactivation that depends on calcium entry, *Proc Roy Soc* **217**:101 (1982).

69. D.M. Armstrong, Calcium channel regulation by calcineurin, a Ca^{2+}-activated phosphatase in mamalian brain, *TINS* **12**:117 (1989).

70. J.E. Chad and R. Eckert, An enzymatic mechanism for calcium current inactivation in dialyzed Helix neurons, *J Physiol* **378**:31 (1986).

71. T. Saitoh, and J.H. Schwartz, Phosphorylation dependent subcellular translocation of a Ca/calmodulin dependent protein kinase produces an autonomous enzyme in Aplysia neurons, *J Cell Biol* **100**:835 (1985).

72. R. Gapalakrishna, S.H. Barsky, and W.B. Anderson, Isolation od S-100 binding proteins from brain by affinity chromatography, *Biochem Biophys Res Commun* **128**:1118 (1985).

73. A.J. Hudspeth, The cellular basis of hearing: The biophysics of hair cells, Science 230:745 (1985).

74. H.H. Kerschbaum and A. Hermann, Calcium-binding proteins in the inner ear of <u>Xenopus laevis</u> DAUDIN, *Brain Res* **617**:43 (1993).

75. H. Ohmori, Mechanical stimulation and fura-2-fluorescence in the hair bundle of dissociated hair cells of the chick, *J Physiol* **399**:115 (1988).

76. K. Gundersen, E. Leberer, T. Lomo, D. Pette, and R.S. Staron, Fibre types, calcium-sequestering proteins and metabolic enzymes in denervated and chronically stimulated muscles of the rat. *J Physiol* **398**:177 (1988).

77. H.E. Karlsen and O. Sand, Selective and reversible blocking of the lateral line in freshwater fish, *J Exp Biol* **133**:249 (1987).

78. C.W. Heizmann and K. Braun, Changes in Ca^{2+}-binding proteins in human neurodegenerative disorders, *Trends in Neurosci* **15**:259 (1992).

79. H.K. Kerschbau, A. Hutticher, and A. Hermann, Parvalbumin-immunoreactive neurons in the brain of <u>Helix pomatia</u>, *Cell Tissue Res* **272**:109 (1993).

CHAPTER 12

PRE-EMBEDDING IMMUNOCYTOCHEMISTRY IN TRANSMISSION ELECTRON MICROSCOPY

Wolfgang Kummer,[1] Cornelia Hauser-Kronberger,[2]
Wolfgang H. Muss[2]

[1]Institute for Anatomy and Cell Biology, Philipps-University, Marburg, FRG (WK)
[2]Department of Pathological Anatomy, Immunohistochemistry and Biochemistry Unit, Salzburg General Hospital, Salzburg, Austria (CHK, WHM)

SUMMARY

This chapter focuses upon pre-embedding immunocytochemistry using peroxidase and 3,3'-diaminobenzidine (DAB) as the detection system. Since an ultrastructural study is attempted, particular emphasis is given to those parts of the technique which are specifically related to electron microscopy. This includes a detailed description of glutaraldehyde-free fixation media, regimes on how to facilitate penetration of immunoreagents without using detergents, the benefit of uranyl acetate *en bloc* staining in maleate buffer, and the choice of operational mode of the electron microscope. Even if optimally performed, the method is not suitable for every kind of ultrastructural immunocytochemical analysis. Its most evident limitation is the diffusion of the DAB reaction product which renders exact subcellular localization of the antigen very difficult, if not impossible. On the other hand, the positive aspects of this method are the use of large tissue areas, the possibility of a subsequent light and electron microscopical investigation of the same cell, the option to select distinct immunoreactive elements by light microscopy, very good ultrastructural tissue preservation, and the convenience of

[1] Address for Correspondence: Professor Wolfgang Kummer, M.D., Institut für Anatomie und Zellbiologie, Philipps-Universität, Robert-Koch-Str. 6, W-3550 Marburg, F.R.G., Phone: 06421/284035; Fax: 06421/285783

recognizing the immunolabel at low magnification. All these qualities have to be considered when choosing the appropriate method for a specific biological problem.

INTRODUCTION

Immunological techniques allow the specific histo- and cytochemical demonstration of many diverse classes of organic substances including free amino acids such as glycine. The broad spectrum of these techniques makes them interesting to various fields. In view of their specific requirements, several variations have been developed. Immunocyto-chemical techniques employed for transmission electron microscopy can be divided into: 1) pre-embedding (immunoreaction first, then tissue processing for electron microscopy),[1,2] 2) post-embedding (immunoreaction performed directly at ultrathin sections cut from embedded specimens),[3,4] and 3) non-embedding methods (immunoreaction performed at ultrathin frozen sections).[5,6] All these techniques require a label indicating the site of immunoreaction. These labels are usually electron dense and can be easily visualized by routine electron microscopy. It is also possible to use other labels, such as boronated protein A, which can be detected with electron spectroscopic imaging (ESI) and electron energy loss spectroscopy (EELS).[7] Commonly used electron dense labels are either particulate (ferritin, colloidal gold) or precipitates of osmicated 3,3'-diaminobenzidine (DAB) reaction product. The latter is mostly generated enzymatically by peroxidase in the presence of H_2O_2.[8] Alternatively, benzidine dihydrochloride (BDHC) might serve as substrate for peroxidase.[9] In general, each of these labels can be used for either technique. Colloidal gold conjugates dominate in post- and non-embedding methods, whereas peroxidase/DAB is commonly applied to pre-embedding immunohistochemistry. The present chapter will focus on pre-embedding immunocytochemistry using peroxidase/DAB, provide a detailed, basic procedure which is almost generally applicable, and highlight recent developments in this field.

THEORETICAL AND PRACTICAL CONSIDERATIONS

Protocols for pre-embedding immunocytochemistry can be divided into three main steps:
  1) Tissue processing
  2) Immunolabelling
  3) Processing for electron microscopy

The importance of points one and three cannot be overemphasized, since ultrastructural analysis will be attempted. The information provided by electron microscopical evaluation of poorly preserved tissue is often less than what can be achieved by appropriate light microscopical methods.

Tissue Processing

Tissue processing usually starts with *fixation* of the tissue. The fixation has to fulfill three requirements in pre-embedding immunocytochemistry:
  1)   It should provide an acceptable ultrastructural preservation of the tissue.
  2)   The antigen must be retained within the tissue.
  3)   The antigen must be recognized by the antibody after fixation ("maintenance of antigenicity").

Point no. 1 is the only one which is entirely independent of the antigen that will be studied. Only the cross-linking aldehydes, glutaraldehyde and paraformaldehyde, should be considered, either alone or in combination with each other and picric acid,[10,11] periodate-lysine,[12] and tannin.[13] Some of the most widely used fixative compositions and their advantages are as follows (each fixative is made up in a buffered solution, e.g. 0.1 M phosphate buffer, pH 7.4).

4% Paraformaldehyde

Most proteins and peptide antigens retain antigenicity when using this mild fixative, which, also provides very good ultrastructural preservation. It can be recommended as a starting point for most studies of animal material to be fixed by vascular perfusion, except when antigens of low molecular weight (e.g. GABA and dopamine) are to be detected. Such antigens shall be properly immobilized with glutaraldehyde.

Fixing with Paraformaldehyde and Glutaraldehyde (e.g. 4%/0.5%)

The ultrastructure of many tissues (e.g. kidney) can be improved several fold by even small amounts of glutaraldehyde. However, this might led to attenuation or even loss of antigenicity, especially in case of proteins and large peptide antigens.

Mixtures of Picric Acid and (para)Formaldehyde

A buffered mixture of 2% paraformaldehyde and 15% saturated picric acid was originally introduced by Stefanini et al.[10], and was commonly referred to as Zamboni's fixative. Surprisingly, we observed good ultrastructural preservation, even when the fixative was made up from formalin stock solution (37% formalin, acid-free) instead of freshly prepared paraformaldehyde (*Fig. 1*). This fixative can be stored for months. It has proved to be particularly useful for immersion fixation. These features make it very attractive for use in routine autopsies and biopsies. The spectrum of use with regard to retention of antigens and maintenance of antigenicity appears to be very similar to that of 4% paraformaldehyde.

4% Paraformaldehyde, 0.05-0.1% Glutaraldehyde, 0.2% Picric Acid

BGPA consists of 1% buffered glutaraldehyde and 15% saturated picric acid.[11] This was recommended by Somogyi and Takagi[14] for enzymes and neuropeptides. This mixture has been used in our laboratory for a number of peptide antigens (e.g. calcitonin gene-related peptide; substance P, vasoactive intestinal peptide) but it attenuates considerably the antigenicity of most protein antigens.

2.5 - 5% Glutaraldehyde

This very efficient cross-linking fixative is the first choice if neurotransmitters of small molecular weight (e.g. glutamate, glycine, serotonin, etc.) are to be reliably retained within the tissue. On the contrary, in our experience, it ruins antigenicity of almost all peptide and protein antigens. It should be mentioned, however, that Kosaka et al.[15] were well able to demonstrate transmitter-synthesizing enzymes (tyrosine hydroxylase, glutamate decarboxylase) in the rat brain after fixation with 5% glutaraldehyde if the sections were treated with 1% sodium borohydride ($NaBH_4$) for 30 min prior to application of the antisera.

In most cases, fixed specimens are too large to allow efficient *in toto* incubation with antisera, and, thus, have to be brought into appropriate size. This is usually achieved by sectioning, but there are also several ways of obtaining whole-mount preparations. For example, the gut can be separated by fine forceps into its different layers which are well suitable for immunocytochemistry both at light and electron microscopical level.[16,17] Sections can be obtained by vibratome or by cryostat. We prefer cryosectioning because the procedure for freezing and thawing enhances penetration of immunoreagents into the tissue section during the different steps of the immunocytochemical incubation. In contrast, vibratome sections allow only very limited penetration of antisera unless detergents are used. To avoid freezing artifacts, the specimens have to be cryoprotected, and then rapidly frozen. We use phosphate-buffered sucrose (18-30% sucrose, 0.1 M phosphate buffer, pH 7.4) for cryoprotection,[18,19] since other protectants such as dimethylsulfoxide (DMSO) result in less clear ultrastructural cell membrane preservation. For most tissues, 40 μm is a convenient section thickness: 1) Such sections are stable enough to survive the extensive handling during the mostly free-floating incubation procedure; 2) this thickness is at about the upper limit of penetration capacity of antisera into the tissue. In practice, the innermost region of 40 μm thick cryostat sections mostly will not contain detectable immunoreaction. Therefore, the use of even thicker sections will mostly increase the "dead space", but not the amount of tissue which can be evaluated.

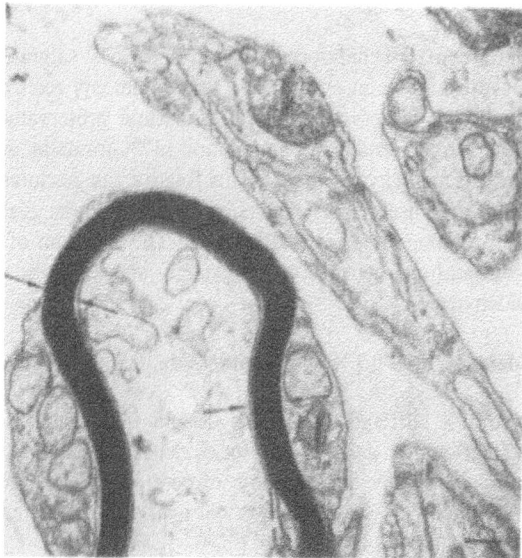

Figure 1. Nerve fibers in the human carotid body obtained at autopsy (4 hrs postmortem) and fixed by immersion with buffered formaldehyde/picric acid solution which had been stored for weeks at 4°C. Pre-embedding protocol as described under "optimal procedures" utilizing a rabbit calcitonin gene-related peptide antiserum. An immunoreactive unmyelinated axon is indicated by the asterisk; all other axons including the myelinated one are immunonegative. Note the good overall preservation of the tissue including the clearly resolved myelin lamellae (arrow). Bar = 0.5 μm.

Immunolabelling

If possible, the sections should be processed free-floating, because this exposes both surfaces to the immunoreagents and, together with continuous agitation, improves diffusion of reagents into the tissue section. In practice, sections might be collected in handled in any other convenient way. It is more common that free-floating cryostat sections subjected to immunocytochemistry and flat embedding in epoxy resin will rupture or fold. This will severely affect subsequent ultrastructural analysis only in very rare circumstances. We have successfully applied this procedure to such diverse organs as the central nervous system, peripheral ganglia, lymph nodes, lung, esophagus, adrenal gland, heart, blood vessels, and kidney. Nevertheless, it is not universally applicable as investigations of skeletal muscle and pancreas revealed that longitudinal sections of skeletal muscle entirely lost their structural integrity during incubation, and the islets of Langerhans were lost from otherwise intact sections of the pancreas. Such problems can be overcome if the sections are provided by some additional support. For this purpose, we have mounted them on either chrom-alum coated plastic foil or on celluloid (ordinary photographic film) with poly-L-lysine. As outlined above, this support reduced penetration of immunoreagents and, hence, the depth of tissue containing sufficient amounts of immunoreaction.

Prior to exposure of the sections to the primary antiserum, non-specific protein binding sites of the tissue should be occupied with a so-called "blocking solution." There are many different suggestions for the composition of this blocking solution, all of which contain a high amount of protein. Most commonly used ingredients are bovine serum albumin and normal serum from a non-immune animal. Of course, the blocking solution should *not* contain proteins that can be recognized by the detection system used after the primary antiserum. Thus, normal serum from that species in which the primary antibody was raised is prohibited. One should also carefully check cross-reactivities of the secondary antibody with immunoglobulins from other species. To be on the safe side, one can choose normal serum from that species in which the secondary antibody was raised. In case of problems with background labeling, it often proved useful to change the composition of the blocking solution (e. g. from normal swine serum to normal horse serum), but general recommendations for the optimal composition cannot be given.

For light microscopical investigations, the blocking solution mostly contains, in addition to the proteins, some detergent to enhance penetration of the immunoreagents. Commonly used detergents are Triton X-100 and Tween 20, each applied at a concentration of about 0.2-0.5%. Although some investigators also use these detergents for pre-embedding immunocytochemistry at ultrastructural level (especially when vibratome sections are utilized), we always try to omit it because of the fuzzy ultrastructural appearance of cell membranes treated with detergents (*Fig. 2*). Instead, digitonin might be considered for use at concentration of 0.005-0.015%, as recommended by Stirling.[20]

Following incubation with the primary antiserum and several buffer washes, an appropriate detection system must be chosen. This chapter will only consider peroxidase methods although other labels such as ferritin and colloidal gold have also been successfully applied.[20,21] However, these particulate labels penetrate much less into the tissue section than peroxidase conjugated reagents. This difference in penetration depth is not simply a matter of molecular size since a conjugate of an immunoglobulin and 1 nm colloidal gold particles is smaller than a peroxidase-anti-peroxidase complex. Instead, it might be related to the distribution and number of positive and negative charges.

As its name says, the detection system must be able to detect the unlabeled primary antibody. Proteins which fulfil this requirement are 1) antibodies raised against immunoglobulins of that species from which the primary antibody was obtained and 2) bacterial proteins with "pseudo-immune" properties.[22] These bacterial proteins (protein A and G) bind with high affinity to the Fc-fragment of immunoglobulin G although marked species differences in binding affinity occur.[23,24] Since one molecule of primary antibody can bind to only one molecule of protein A, this protein is often used in combination with

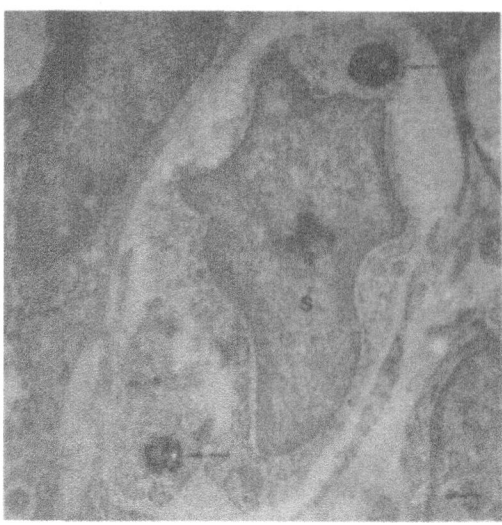

Figure 2. Effect of detergent treatment. Substance P-immunoreactive nerve fibers (arrows) in the wall of the guinea-pig carotid artery. This section was treated with 0.2% Triton X-100 added to the blocking solution prior to incubation with the primary antiserum. Note the fuzzy appearance of cell membranes, which often are not clearly delineated! Fixed by perfusion with 4% paraformaldehyde; section is counterstained with uranyl acetate and lead citrate. S = nucleus of a Schwann cell. Bar = 0.5 μm

wells of cell culture plates, placed in drops of antisera on Parafilm[R] or Nescofilm[R], or a particulate marker for post-embedding immunocytochemistry, especially if a quantitative evaluation is intended.[25]

In contrast, secondary antisera offer the advantage that several IgG molecules can bind to the primary antibody providing a multiplying effect.[25] This enhancing effect is warranted in pre-embedding immunocytochemistry where the sections are investigated light microscopically before ultrastructural analysis. Consequently, secondary antibodies instead of protein A or G are the tool of choice. As for light microscopy, these secondary antibodies either are directly conjugated to peroxidase or may serve as a bridge between the primary antibody and a peroxidase conjugated third reagent.

Among the peroxidase-labelled reagents, we observed a relationship between molecular size and penetration depth. In many instances, sufficient labeling was obtained with rather large detection reagents such as peroxidase-anti-peroxidase complexes[26] and streptavidin-biotinylated peroxidase complexes[27] which were used in combination with appropriate secondary antibodies as bridging reagents.[28,29] In some applications, only very limited penetration could be reached with these reagents. Penetration could be improved by use of peroxidase conjugated $F(ab)_2$-fragments as secondary antibodies.[30]

DAB is the most commonly used peroxidase substrate for ultrastructural immunocytochemistry. Since DAB is considered to be carcinogenic, handling with DAB powder should be absolutely minimized. This can be achieved either by preparation of a concentrated stock solution (100 mg/ml) of DAB in 0.05 M TRIS-HCl buffer and storing small aliquots in a freezer, or, even better, by the use of commercially available preparations which allow dissolving of DAB without any contact with the powder (e.g. ISOPAC, Sigma, St. Louis, MN, USA). In our laboratory, all solutions containing DAB (including the washing buffers following the DAB incubation) are handled in separate containers under a hood that is cleaned with bleach afterwards. For pre-embedding immunocytochemistry, it is unnecessary to monitor the development of the DAB reaction product under the microscope. This will avoid contamination of the microscope with DAB solution. We usually wait until brown background staining can be easily recognized macroscopically. It should be recalled that the 40 μm thick sections are cut into 0.08 μm thick sections for electron microscopical investigation. Thus, the amount of background labelling will be reduced to 0.2%.

Processing for Electron Microscopy

The general principles of embedding of free floating sections in any desired resin are the same as for small tissue samples routinely used in ultrastructural analysis. We were very satisfied with epoxy resins and always use epon but other resins might work as well. One of the advantages of pre-embedding compared to post-embedding techniques is that there is no need to take care of loss of antigenicity during processing for electron microscopy because the immunoreaction has already been performed. Therefore, the sections can be routinely osmicated. This will greatly help to preserve the lipids of the cell membranes in the course of dehydration. Moreover, DAB is osmiophilic so the contrast of the reaction product will be enhanced.

The times for each step during the following dehydration in a graded series of alcohol should be kept short since we are dealing with very thin samples (40 μm) which are rapidly dehydrated. Even after osmication, the samples may also be dilapidated to some extent if too long intervals are chosen. For a further stabilization of membranes, the specimens are subjected to *en bloc* staining with uranyl acetate.[31] This *en bloc* staining may be performed for 1 hr in 70% ethanol during dehydration. This will give excellent preservation of membranes but will extract many components of the cytoplasm (*Fig. 3*). Therefore, we stain the sections with uranyl acetate in maleate buffer after osmication but before dehydration.[32] In our material, this procedure gave by far the best results (*Figs. 1, 4*). It should be stressed that the differences in general ultrastructural appearance between the different protocols described above are much more pronounced when dealing with 40 μm sections than with tissue specimens of usual size!

It is obvious that the specimens should be embedded as flat as possible to facilitate subsequent ultrathin sectioning. In addition, flat embedded sections should be accessible to light microscopical investigation so that appropriate regions for ultrastructural analysis can be selected. These two goals can be achieved by embedding either between sheets of transparent plastic foil or between a slide and coverglass (e.g. conventional light microscopy). In the latter case, precautions have to be taken to allow removal of the coverglass, and the slide to have access to the resin-embedded section. For this purpose, we use teflonated slides and coverglasses. These can be easily prepared using commercially available teflon sprays. Alternatively, a plastic coverslip may be used.[33]

Regions of interest are cut out with a razor blade and mounted with cyanoacrylate or epon onto the top of prefabricated epon pyramids. Semithin and thin sections cut from such specimens will not reveal a coherent tissue section from the beginning on, since the surface of a 40 μm thick cryostat section is not ideally smooth. The rough outermost surface of this section will be 1) disrupted, and 2) covered by non-specific deposits of DAB reaction product. Thus, it is not suitable for ultrastructural investigation. Cutting thin sections slightly below this zone will give the best results in terms of immunocytochemical labeling intensity and structural integrity; in deeper regions, the structural preservation will still be optimal. Labeling intensity declines (*Fig. 5*) due to limited penetration of the immunoreagents.[34]

The cryostat section has been exposed to the antisera from both surfaces, so that proceeding through the often worthless center of the section will lead the investigator to the optimal region of the opposite side. The depth of the zone of interest varies largely from one application to another and is dependent on 1) the tissue (central nervous system is much worse than lymph nodes), 2) the fixation (glutaraldehyde is worse than formaldehyde/picric acid), 3) the amount of antigen in the tissue, and 4) the primary and secondary immunoreagents, as described above. In some fortunate cases, e.g. using monoclonal antibodies for detection of substance P in nerve fibers within lymph nodes fixed by 4% paraformaldehyde,[29] we obtained immunolabelling throughout the entire thickness of the cryostat section. On the other hand, dopamine-immunolabelling extended less than 1 μm into rat carotid body tissue fixed by 5% glutaraldehyde (own unpublished data).

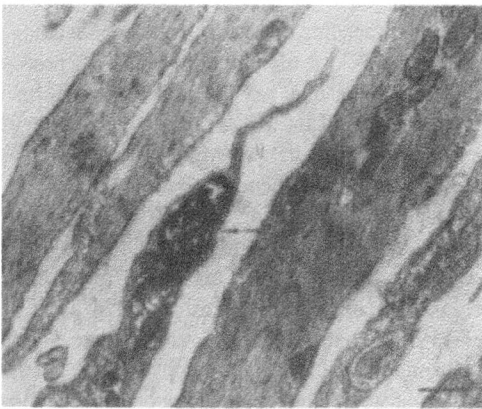

Figure 3. Effect of *en bloc* staining with uranyl acetate in 70% ethanol. Material and antisera are identical to Fig. 2, but in this case Triton X-100 was omitted from the blocking solution, and *en bloc* staining with uranyl acetate was performed during the dehydration. Cell membranes are now clearly preserved, but most of the cytoplasm is extracted. Arrow = substance P-immunoreactive axon, arrowhead = non-immunoreactive axon, S = Schwann cell processes. Bar = 0.5 μm

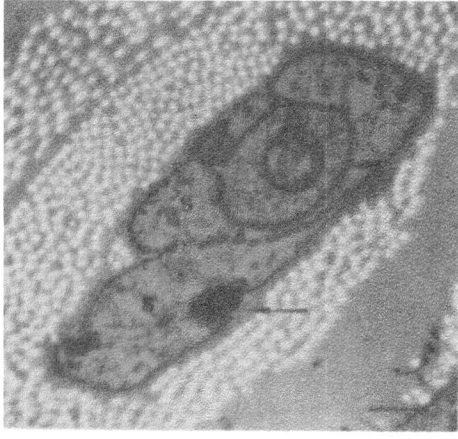

Figure 4. Protocol as described under "optimal procedures"; guinea pig fixed by perfusion with 4% paraformaldehyde; nitric oxide synthase-immunoreactive nerve fiber (arrow) in the bronchial smooth muscle. Both membranes and cytoplasmatic components are well preserved. Bar = 0.5 μm

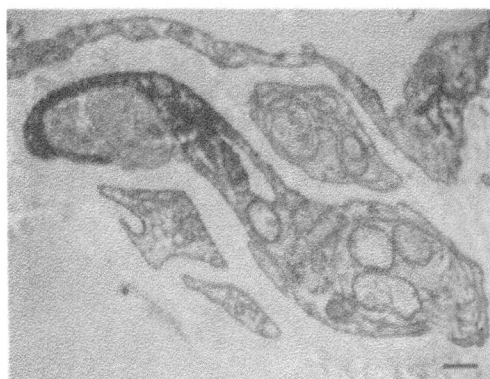

Figure 5. Demonstration of limited penetration of immunoreagents. Nerve fiber bundle in the human carotid body; autopsy material fixed by immersion in buffered formaldehyde/picric acid; protocol as described under "optimal procedures" using a rabbit glial fibrillary acidic protein antiserum. Parts of the same Schwann cell are intensely labeled at the left side, moderately labeled in the midportion, and non-labeled on the right hand side of the figure. Bar = 0.5 μm

Electron Microscopical Evaluation

Decrease of immunolabelling towards the center of the section occurs gradually. This implies that there is a region of low intensity labeling which cannot be evaluated unequivocally by routine electron microscopy. It is possible, however, to enhance the contrast of the osmicated DAB reaction product by appropriate operation of the electron microscope. A rather simple modification is to operate the electron microscope at lower accelerating voltage. In this respect, good results have been obtained at 60 kV and even 40 kV in a study of the innervation of the guinea-pig carotid sinus.[28] The ultrastructural analysis of thicker sections proved to be much more efficient, however. Calculated per surface area to be transmitted by the electron beam, a section 0.3 μm thick contains 5 times more reaction product than a section 60 nm thick. Modern electron microscopes (Philips, Zeiss) are able to handle sections of a thickness of up to 1 μm well. Investigating nitric oxide synthase-immunoreactive axons innervating the guinea-pig lingual artery by pre-embedding immunocytochemistry, 60-90 nm, 150 nm, 300 nm, and 500 nm thick sections were evaluated with a Zeiss EM 902 at elastic brightfield imaging mode.[30] Sections thicker than 100 nm were not counterstained with lead citrate. This study showed increasing contrast with increasing section thickness, but at a section thickness of 500 nm the geometry of the specimen rendered the interpretation difficult. Thus, in this application, 300 nm proved to be a favorable section thickness. Nevertheless, thin (60-90 nm) sections should also be included in the evaluation, since they revealed some subcellular details which were otherwise invisible.[30]

Finally, it should be mentioned that not all of the DAB reaction product in the tissue section reflects sites of immunoreaction. First, endogenous peroxidase will also convert the DAB in the presence of H_2O_2, leading to a reaction product identical to that generated by the peroxidase introduced by the immunoreagents. Such endogenous peroxidase is present in high amounts, for example, in the matrix of granula of eosinophils (*Fig. 6*). Moreover, erythrocytes posses pseudo-peroxidase activity leading to a precipitation of DAB reaction products at their surface. Endogenous peroxidase can be blocked prior to incubation with primary antisera by rinsing the sections in H_2O_2 in methanol, but this alcoholic treatment cannot be recommended for electron microscopy. Hydrogen peroxide in buffer does not fully block endogenous peroxidase activity. Thus, we do not attempt to block endogenous peroxidase in our pre-embedding protocol. Besides endogenous peroxidase, other non-immunological reactions may also cause labeling. If avidin or streptavidin conjugates are used, one must be aware of the possibility of endogenous biotin. Finally, some currently unexplained mechanisms may also occur. For instance, it has been reported that cells of the pig hypothalamus bind peroxidase-anti-peroxidase complex prepared in rabbit.[35]

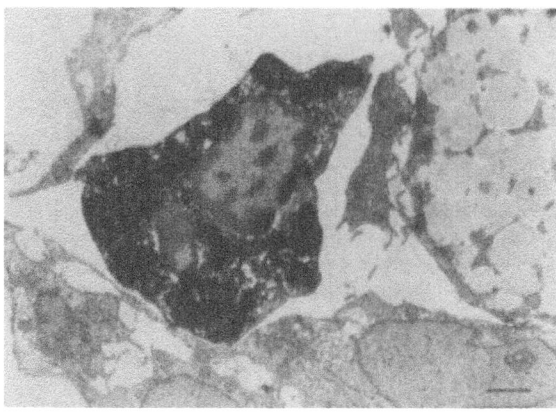

Figure 6. Endogenous peroxidase as a cause of electron dense reaction product. The granular matrix of an eosinophil granulocyte in the respiratory epithelium is heavily stained due to its content of eosinophil peroxidase. This micrograph is taken from the same specimen as in Fig. 4. Bar = 1 μm

OPTIMAL PROCEDURE

Tissue Processing

The different fixatives and their advantages and limitations are listed above. The most commonly used buffered 4% paraformaldehyde solution is prepared as follows:

Stock solution A: 27.6 g/l NaH_2PO_4 x $1H_2O$
Stock solution B: 35.6 g/l Na_2HPO_4 x $2H_2O$

115 ml solution A plus 385 ml solution B will give 500 ml 0.2 M phosphate buffer, pH 7.4. Add 40 g paraformaldehyde to 500 ml distilled water and heat on magnetic stirrer, then add 1 N NaOH until the solution turns clear; this will give 500 ml 8% paraformaldehyde. Now add to this solution 500 ml of 0.2 M phosphate buffer (see above). The fixative should be prepared on the day of usage and filtered twice (pore size: 3 μm and 0.65 μm).

The optimal duration of the fixation may vary between 10 min and 18 hrs; if no data are available for a certain antigen, we start with 15 min of perfusion fixation, and store the fixed animal for another 2 - 4 hrs at 4°C before dissecting tissue specimens. Specimens are then washed several times in 0.1 M phosphate buffer (pH 7.4) and stored overnight in the same buffer containing 18% (w/v) sucrose. Buffer is removed from the specimens by short drying on filter paper. They are mounted with OCT compound (Miles, Elkhart, IN, USA) on filter paper and snap frozen in liquid nitrogen. Tissues are stored sealed at -75°C.

Incubation and Embedding Protocol

All incubations are performed at room temperature, unless otherwise indicated.

- Collect 40 μm thick sections cut at a cryostat either in PBS or in blocking solution in reaction tubes of appropriate size.
- Incubate for 60 min with blocking solution (10% normal swine serum, 0.1% bovine serum albumin in PBS).
- Incubate overnight under continuous agitation with the primary antibody. The optimal dilution has to be determined empirically in each case; reasonable dilutions for crude rabbit sera are 1:1000 to 1:8000 in PBS.
- Wash at least 3 x 5 min in PBS.
- Incubate for 2 hrs with peroxidase-conjugated F(ab)$_2$-fragments of a goat-anti-rabbit-IgG (Zymed, South San Francisco, CA, USA) or any other appropriate secondary antiserum diluted 1:50 to 1:200 in PBS.
- Wash at least 3 x 5 min in PBS.
- Wash 5 min in 0.05 M TRIS-HCl, pH 7.6.
- React the section with DAB (see appendix) until it turns brown (10 - 40 min).
- Wash 5 min in 0.05 M TRIS-HCl, pH 7.6.
- Wash 5 min in PBS.
- Osmicate for 1 hr at 4°C (1% OsO$_4$ in 0.1 M phosphate buffer, pH 7.4).
- Wash at least 3 x 5 min in 0.05 M maleate buffer, pH 6.0.
- Stain *en bloc* with 1% uranyl acetate in 0.05 M maleate buffer, pH 5.2.
- Wash at least 3 x 5 min in 0.05 M maleate buffer, pH 6.0.
- Pass through a graded series of ethanol (30%, 50%, 70%, 90%, 96%), 5 min each.
- Rinse 2 x 5 min in 100% ethanol.
- Rinse 5 min in ethanol/propylene oxide 1:1.
- Rinse 2 x 5 min in propylene oxide.
- Immerse 30 min in propylene oxide/epon 1:1.
- immerse overnight in epon in a closed vessel.
- Flat-embed the sections in fresh epon between a teflon coated coverglass and teflon coated slide (easily prepared using commercially available teflon spray). Use only small amounts of epon; it should just spread as far as the coverslip reaches. This will ensure a really flat position of the section and facilitate subsequent semithin and ultrathin sectioning. Polymerize overnight at 60°C.

The sections are then ready for light microscopical investigation. The coverglass and slide can be removed with a razor blade. Regions of interest will now be cut out with a razor blade and mounted with epon (may be stored in small aliquots at -20°C) onto the top of preformed epon pyramids.

Appendix

1) PBS = phosphate buffered saline:

   Stock solution A: 27.6 g/l NaH$_2$PO$_4$ x 1H$_2$O (0.2 M)
   Stock solution B: 35.6 g/l Na$_2$HPO$_4$ x 2H$_2$O (0.2 M)

   28.75 ml solution A
   +96.20 ml solution B
   +45.00 g NaCl
   add to 5 l with distilled water

2)   0.05 M TRIS-HCl, pH 7.6

        Stock solution A: 121.1 g/l TRIS
        Stock solution B: 1 N   HCl

        75 ml solution A
        +60 ml solution B
        add to 1.5 l with distilled water

3)   0.05 M maleate buffer

        Stock solution A: 23.2 g maleic acid *or* 19.6 g maleic anhydride
        200 ml 1 N NaOH
        Add to 1 l with distilled water (0.2 M)

        Stock solution B: 0.2 N NaOH

        50.0 ml solution A
        + 7.2 ml solution B
        add to 200 ml with distilled water   (0.05 M, pH 5.2)

        50.0 ml solution A
        +26.9 ml solution B
        add to 200 ml with distilled water   (0.05 M, pH 6.0)

4)   DAB reaction medium:

Dissolve 12.5 mg 3,3'-diaminobenzidine-tetrahydrochloride in 100 ml 0.05 M TRIS-HCl, pH 7.6, filter. Incubate the sections in this solution for 10 min (preincubation medium; the peroxidase will be loaded with DAB); then take fresh solution and add immediately before use 7.5 µl 30% H_2O_2 (this will start the reaction; incubating medium)

DISCUSSION

The technical problems associated with the method have already been addressed above. Thus, the advantages and disadvantages of the technique shall be considered with respect to its contribution to the solution of specific biological problems. In other words, when shall this method be used, when are other techniques considered? From a user's point of view, there are five major differences between pre-embedding/DAB and post- and non-embedding/colloidal gold. Pre-embedding/DAB treated sections: 1) are much thicker, 2) cover a much greater area, 3) can be evaluated light microscopically, 4) are osmicated, and 5) contain a diffuse label instead a particulate marker.

Section Thickness

The use of a thick (40 µm) section which is labelled by an antibody in a single incubation is advantageous if only one antigen is to be detected in the specimen. In this case, many ultrathin sections can be obtained from a single incubation cycle, so it can be particularly easy to make serial reconstructions of certain cells. The disadvantage, however, is that serial ultrathin sections cannot be incubated with different antisera. Practically, it is impossible to re-identify single cells in adjacent 40 µm thick sections.

Section Area

With respect to this criterion, pre-embedding/DAB has some very clear advantages. Even from small specimens such as the human carotid body, we have been able to dissect seven different regions from a single cryostat section for further ultrastructural analysis. Thus, this technique is highly economical and helps to preserve precious tissues and antisera.

Light Microscopical Evaluation

Again, pre-embedding/DAB is highly advantageous in the following respects. First, the quality of the immunolabelling can be easily checked before the ultrastructural investigation. Second, labeled structures can be analyzed in their three-dimensional structure prior to electron microscopy. Third, regions of interest can be accurately selected. This is of utmost importance, if the tissue under investigation contains only a few immunoreactive structures. For example, we have studied the innervation of lymph nodes by substance P-immunoreactive nerve fibers.[9] Lymph nodes, in general, are not densely innervated, and substance P-immunoreactive axons comprise only a sub-population of nerve fibers innervating a lymph node. It is obvious that in most cases, randomly taken ultrathin sections will not contain even a single substance P-immuno-reactive axon! Therefore, regions of interest must be selected by light microscopy.

Osmication

Osmium tetroxide treatment will destroy antigenicity of most antigens and, therefore, is usually omitted in post-/non-embedding protocols. As a consequence, membrane preservation is rather poor, especially in the nervous system. Pre-embedding protocols, however, do include osmication which is performed *after* the immunolabelling. Hence, it usually provides much better tissue preservation.

Diffusible Versus Particulate Marker

This point covers the main disadvantage and limitation of the pre-embedding/DAB technique. Particulate markers clearly indicate the site of immunoreaction and are excellent for subcellular localization of the antigen. DAB reaction product, however, diffuses from the site of its generation by peroxidase;[36] the radius of diffusion is possibly smaller using BDHC instead of DAB.[9,20] The closer to the surface of the cryostat section the ultrathin sections are cut, the more pronounced the diffusion of the DAB reaction product. The reaction product tends to aggregate at intracellular membranes (e.g. outer membrane of mitochondria) and at the inner surface of the cell membrane. It is obvious that precise conclusions as to the intracellular localization of the antigen cannot be drawn. Therefore, if the study is intended mainly to clarify the subcellular distribution of an antigen, a particulate marker such as colloidal gold is preferred. On the other hand, if the study is intended to ask different questions - e. g. to identify the type of immunoreactive synapses and postsynaptic elements - the diffusible DAB reaction product is well suited. In addition, the diffusion of the DAB reaction product also offers an advantage: immunoreactive structures can be recognized at the electron microscopic level by screening the section at low magnification, whereas colloidal gold particles - which are usually 5 - 20 nm in size - require high magnification to be detected.

It is obvious that pre-embedding immunocytochemistry, using DAB as electron dense marker, offers a series of advantages that are inherently connected with disadvantages. The success of a study will not only depend on the quality of the pre-embedding immunolabelling, but also on the general suitability of the technique with regard to the specific probe to be investigated.

ACKNOWLEDGEMENTS

The authors' studies were supported by the DFG.

REFERENCES

1. V.M. Pickel, T.H. Joh, and D.J. Reis, Ultrastructural localization of tyrosine hydroxylase in noradrenergic neurons of brain. *Proc Natl Acad Sci USA* **72**:659 (1975).
2. V.M. Pickel, T.H. Joh, and D.J. Reis, Monoamine-synthesizing enzymes in central dopaminergic, noradrenergic and serotonergic neurons. Immunocytochemical localization by light and electron microscopy, *J Histochem Cytochem* **24**:792 (1976).
3. G. Pelletier, F. Labrie, A. Arimura, and A.V. Schally, Electron microscopic immunohistochemical localization of growth hormone-release inhibiting hormone (somatostatin) in the rat median eminence, *Am J Anat* **140**:445 (1974).
4. L. Probert, J. De Mey, and J.M. Polak, Distinct sub-populations of enteric p-type neurones contain substance P and vasoactive intestinal polypeptide, *Nature* **294**:470 (1981).
5. A. Dutton, K.T. Tokuyasu, and S.J. Singer, Iron-dextran antibody conjugates. General method for simultaneous staining of two components in high resolution immunoelectron microscopy, *Proc Natl Acad Sci USA* **76**:3392 (1979).
6. J.W. Slot and H.J. Geuze, Sizing of protein A-colloidal gold probes for immunoelectron microscopy, *J Cell Biol* **90**:533 (1981).
7. M. Bendayan, G.F. Bath, D. Gingras, I. Londono, P.T. Robinson, F. Alam, D.M. Adams, and L. Mattiazi, Electron spectroscopic imaging for high-resolution immunocytochemistry: use of boronated protein A, *J Histochem Cytochem* **37**:573 (1989).
8. R.C. Graham and M.J. Karnovsky, The early stages of absorption of injected horseradish peroxidase in the proximal tubules of mouse kidney: ultrastructural cytochemistry by a new technique, *J Histochem Cytochem* **14**:291 (1966).
9. S. Lakos and A.I. Basbaum, Benzidine dihydrochloride as a chromogen for single- and double-label light and electron microscopic immunocytochemical studies, *J Histochem Cytochem* **34**:1047 (1986).
10. M. Stefanini, C. de Martino, and L. Zamboni, Fixation of ejaculated spermatozoa for electron microscopy, *Nature* **216**:173 (1967).
11. G.R. Newman, B. Jasani, and E.D. Williams, A simple post-embedding system for the rapid demonstration of tissue antigens under the electron microscope, *Histochem J* **15**:543 (1983).
12. D.C. Hixson, J.M. Yep, J.R. Glenney, T. Hayes, and E.F. Walborg, Evaluation of periodate/lysine/paraformaldehyde fixation as a method for cross-linking plasma membrane glycoproteins, *J Histochem Cytochem* **29**:561 (1981).
13. J.W. Stirling, Ultrastructural localization of lysozyme in human colon eosinophils using the protein A-gold technique: effects of tissue processing on probe distribution, *J Histochem Cytochem* **37**:709 (1989).
14. P. Somogyi and H. Takagi, A note on the use of picric acid-paraformaldehyde-glutaraldehyde fixative for correlated light and electron microscopic immunocytochemistry, *Neuroscience* **7**:1779 (1982).
15. T. Kosaka, I. Nagatsu, J.Y. Wu, and K. Hama, Use of high concentrations of glutaraldehyde for immunocytochemistry of transmitter-synthesizing enzymes in the central nervous system, *Neuroscience* **18**:975 (1986).
16. M. Costa, R. Buffa, J.B. Furness, and E. Solcia, Immunohistochemical localization of polypeptides in peripheral autonomic nerves using whole mount preparations, *Histochemistry* **65**:157 (1980).
17. J.B. Furness, I.J. Llewellyn-Smith, J.C. Bornstein, and M. Costa, Chemical neuroanatomy and the analysis of neuronal circuitry in the enteric nervous system, *in*: "The peripheral nervous system," A. Björklund, T. Hökfelt, and C. Owman, eds., Elsevier, Amsterdam - New York - Oxford (1988).
18. K.T. Tokuyasu, A technique for ultramicrotomy of cell suspensions and tissues, *J Cell Biol* **57**:551 (1973).

19. H.J. Geuze and J.W. Slot, Disproportional immunostaining patterns of two secretory proteins in guinea pig and rat exocrine pancreatic cells. An immunoferritin and fluorescence study, *Eur J Cell Biol* **21**:93 (1980).

20. J.W. Stirling, Immuno- and affinity probes for electron microscopy: a review of labeling and preparation techniques, *J Histochem Cytochem* **38**:145 (1990).

21. J. Chan, C. Aoki, and V.M. Pickel, Optimization of differential immunogold-silver and peroxidase labeling with maintenance of ultrastructure in brain sections before plastic embedding, *J Neurosci Methods* **33**:113 (1990).

22. A. Forsgren and J. Sjoquist, "Protein A" from S. aureus. I. Pseudo-immune reaction with human gamma-globulin, *J Immunol* **97**:822 (1966).

23. R. Lindmark, K. Thoren-Tolling, and J. Sjoquist, Binding of immunoglobulins to protein A and immunoglobulin levels in mammalian sera, *J Immunol Methods* **62**:1 (1983).

24. B. Åkerström, T. Brodin, K. Reis, and L. Björck, Protein G: a powerful tool for binding and detection of monoclonal and polyclonal antibodies, *J Immunol* **135**:2589 (1985).

25. J.W. Slot, Methods in immunogold labelling of ultrathin frozen sections, *in*: "Molecular Neuroanatomy", F.W. van Leeuwen, R.M. Buijs, and C.W. Pool, eds., Netherlands Institute for Brain Research, Amsterdam (1987).

26. L.A. Sternberger, P.H. Hardy, J.J. Curculis, and H.G. Meyer, The unlabelled antibody enzyme method of immunohistochemistry, *J Histochem Cytochem* **18**:315 (1970).

27. C. Bonnard, D.S. Papermaster, and J.P. Kraehenbuhl, The streptavidin-biotin bridge technique: application in light and electron microscope immunocytochemistry, *in*: "Immunolabelling for electron microscopy", J.M. Polak, and I.M. Verndell, eds., Elsevier, Amsterdam (1984).

28. W. Kummer, M. Reinecke, and C. Heym, Neurotensin-like immunoreactivity in presumptive baroreceptor neurons innervating the guinea pig carotid sinus, *J Autonom Nerv Syst* **35**:107 (1991).

29. R. Kurkowski, W. Kummer, and C. Heym, Substance P-immunoreactive nerve fibers in tracheobronchial lymph nodes of the guinea pig: origin, ultrastructure and coexistence with other peptides, *Peptides* **11**:13 (1990).

30. W. Kummer and B. Mayer, Nitric oxide synthase-immunoreactive axons innervating the guinea-pig lingual artery: an ultrastructural immunohistochemical study using elastic brightfield imaging, Histochemistry, (in press).

31. M.W. Brightman and T.S. Reese, Junctions between intimately apposed cell membranes in the vertebrate brain, *J Cell Biol* **40**:648 (1969).

32. M.J. Karnovsky, The ultrastructural basis of capillary permeability studied with peroxidase as tracer, *J Cell Biol* **35**:213 (1967).

33. A. Hendrickson, Technical modificatin to facilitate tracing synapses by electron microscopic autoradiography, *Brain Res* **85**:241 (1975).

34. V.M. Pickel, Immunocytochemical methods, *in*: "Neuroanatomical Tract-Tracing Methods", L. Heimer, and M.J. Robards, eds., Plenum Press, New York - London (1981).

35. J. Meijer, P. Poot, G. Molenaar, and R. Goede, PAP complex: a pitfall in immunocytochemistry of the pig hypothalamus, *J Neurosci Methods* **17**:269 (1986).

36. P.J. Courtoy, D.H. Picton, and M.G. Farquhar, Resolution and limitation of the immunoperoxidase procedure in the localization of extracellular matrix antigens, *J Histochem Cytochem* **31**:945 (1983).

CHAPTER 13

POST-EMBEDDING LOCALIZATION METHODS FOR ELECTRON MICROSCOPY: BASIC APPROACH AND PROTOCOLS

Peter M. Lackie[1] and Jürgen Roth[2]

[1]Southampton University Medicine
Centre Block, Level D
Southampton General Hospital
Southampton SO9 4XY, England (PML)
[2]Division of Cell and Molecular Pathology
Department of Pathology
University of Zürich
Schmelzbergstr 12
CH8091 Zürich, Switzerland (JR)

SUMMARY

Immunocytochemistry and other affinity labelling methods for electron microscopy provide high resolution information about the subcellular localization of specific molecules which is not available by other means. We review here the basic strategies for localization and indicate the approaches most likely to be successful within the constraints of the structural resolution required and available resources. These techniques have been the subject of several good and extensive reviews. The emphasis of this chapter is to provide a practical introductory guide rather than to detail all the available methods. We hope to provide the information needed to devise a rational and-- so far as possible-- workable protocol for localizing a molecule at the ultra-structural level. The most widely used methods, from tissue collection visualization in the microscope are reviewed together

[1]Address for correspondence: Dr. Peter Lackie, Southampton University Medicine, Centre Block, Level D, Southampton General Hospital, Southampton SO9 4XY, England. Tel: + 44 703 794146 (direct); Fax: + 44 703 701771

with the rationale of each technique and an indication of conditions for which it is useful. Basic key protocols are provided for direct labelling using immunogold and lectin probes, and indirect labelling with immunogold and protein A. Colloidal gold is our recommended label of choice. A protocol for preparing gold labelled probes is given. Alternative methods that may be suitable for specific localization problems are outlined together with sources of more detailed information.

INTRODUCTION

Light microscopy has revealed little of the fine structure of cells because of the physical limitations of the resolution of light. Details of cellular sub-structure only became clear with the development of electron microscopy which, because of the shorter wavelength of the radiation used, allowed these structures to be resolved. Recent advances in cell and molecular biology have provided great insights into the molecular and genetic basis of cellular and developmental processes. In eukaryotes, these processes occur within and between the multi compartmental structure of cells and cell groups. Although as a result of basic electron microscopic studies the structure of these compartments is well defined, this is not true of all their functions. Immunocytochemistry and other affinity localization techniques for EM allow us to establish the distribution of the constituent molecules and, hence, the functions of the compartments.

The use of electron dense markers to label proteins, such as antibodies for localization by electron microscopy, was first introduced by Singer in 1959, using ferritin as a label.[1] Horseradish peroxidase labelling of antibodies, subsequently visualized by diaminobenzidine, was also used from 1966 after its introduction by Nakane and Pierce.[2] A further landmark was the introduction of colloidal gold as a marker for antisera[3] and specific antibodies[4] for electron microscopy. Localization of antibodies on sections using protein A-gold marked an important change in the approach to post-embedding labelling for electron microscopy.[5] Since that time, colloidal gold has become accepted as the best general marker for post-embedding electron microscopic labelling. Colloidal gold has also been used for double labelling (i.e. the simultaneous localization of two tissue components) in conjunction with peroxidase[6,7] and with combinations of different sized gold particles, either with protein A[8,9] or immunoglobin probes.[10] Single and double gold labelling procedures have subsequently been very widely used and have been the subject of many reviews[11-14] and multi-chapter volumes.[15,16] Gold labels have now largely replaced peroxidase and ferritin labels for post-embedding immunocytochemistry and other affinity-cytochemical methods for electron microscopy because of their high electron density, ease of labelling and, due to the discrete sizes in which gold particles can be produced, their applicability to double labelling. Those interested in further details of the background and methodology of these techniques refer to references 12 and 16. Only methods of electron immunocytochemistry and other affinity methods as applied to ultrathin sections on EM grids will be considered here (i.e. post-embedding staining). Pre-embedding EM methods are dealt with elsewhere in this volume. Affinity- cytochemistry is used in its widest meaning to include all those methods based upon the specificity of an interaction between a probe (e.g. lectin or antibody) and a corresponding cellular component.

Affinity-cytochemistry for transmission EM is only one of several complimentary high-resolution methods of localization applicable to biological problems. Video microscopy[17,18] and confocal microscopy[19,20] can provide high resolution localization data particularly valuable for studies of dynamic processes in living cells. Replica methods such as freeze fracture,[21] surface replicas[22] and scanning EM[23,24] combined with labelling, can provide additional information about cell surfaces and intracellular membrane structures. All of these methods can provide valuable additional information, but do not replace post-embedding labelling for EM. The latter remains the only appropriate method for detailed subcellular localization of molecules and for providing information about the cell at high resolution, like that relating to molecular trafficking, cell-cell apposition and adhesion mechanisms.

GENERAL CONSIDERATIONS AND APPROACH

In biological terms, we are interested in the location of target molecules within the context of tissue and cell structure in vivo. The major advantage of EM, its resolution, is accompanied by various technical problems when used in conjunction with affinity cytochemistry, specifically: (1) preserving tissue structure, (2) maintaining target structure for affinity binding and (3) localization with appropriate high resolution labels. This chapter will deal with some of the methods used to overcome these problems, indicating the relative merits of each approach.

A localization study will generally center around a target molecule of interest, knowledge of the specific information required, and the resolution needed to obtain this. The characteristics of the available probes are crucial and may limit the technical means by which these aims can be achieved. The affinity probe determines the specificity of the localization of the target molecule. Target molecule detection requires (1) target recognition and (2) the accessibility of a sufficient number or concentration of target sites to distinguish specific from non-specific binding. Target recognition is affected by preparation procedures, especially fixation, which alter the target and hence its interaction with the probe. Accessibility depends largely upon the cellular location of the target molecule and the properties of the constituent matrix, in turn affected by fixation and embedding. It is usually most productive to first screen all available probes, since the greater the ability of the probe to recognize the target molecules after fixation and embedding, the easier the localization studies will be.

Affinity probes require a detection procedure to render them visible in sections by EM. Detection may be either direct, when the probe itself is labelled, or indirect when a further (secondary) labelled affinity step is used to localize the probe. Labels must be bound with high avidity, without affecting the affinity of the probe (primary or secondary) for its target. It must also be easily visible against the tissue structure, yet small enough to provide suitable resolution. The label of choice used by most groups in recent years has been and is likely to remain, colloidal gold. This is clearly identifiable by EM, can be easily used to label a wide range of biologically relevant probes and can be prepared in different homogeneous sizes which are easily distinguished for use in double labelling.

Having addressed the problem of the probe to be used and the labelling method to visualize it, the appropriate protocol(s) can be chosen. The various aspects of the choices which go to make up a successful protocol are dealt with in more detail below. Some examples of protocols which have been found to be optimal for our studies are given.

INITIAL INVESTIGATIONS

At this early stage, a localization system will be required and the simplest possible method compatible with EM localization should be used. An indirect gold labelling protocol is recommended where possible, using either protein A-gold or immunogold. For lectin affinity cytochemistry, direct gold labelling or digoxigenin (DIG) labelled lectins are recommended. To check that the localization system is working, a model system such as dot blots of purified target molecule on nitro-cellulose, or a tissue known to contain large amounts of the target molecule, may be used. This provides a means to compare different methods. Next, tissue collection should be optimized as far as possible, reducing the delay in collection and ensuring good sampling and optimal structure. It is vital to use appropriate controls at each stage to ensure the results do represent the true localization of the target and these should be planned at an early stage. For reasons of economy and the availability of purified target molecule, it may be necessary to delay absorption and competition controls until the optimal conditions for localization are established.

TISSUE PREPARATION

Fixation. Methods of tissue fixation which provide optimal structural preservation for conventional EM are well established[25,26,27] and commonly based on a cross-linking fixative (e.g. glutaraldehyde) as the primary fixative, often followed by osmium tetroxide used as both a fixative (particularly of lipids) and as a contrasting agent. Material prepared in this way can be successfully used for some immunocytochemical studies[28] and allows use of archive material. Unfortunately, these fixatives are often incompatible with antigen localization by affinity cytochemistry and, in this case, a compromise between structure and antigen detectability must be sought.

Glutaraldehyde provides good structural preservation, but often affects recognition by affinity probes. For affinity cytochemistry (*Fig. 1*), only ultra pure grades of glutaraldehyde should be used. Formaldehyde, which should be freshly prepared from paraformaldehyde, causes fewer problems, although it does not preserve ultrastructure so well. The combination of formaldehyde (2-8%) with low concentrations of glutaraldehyde (down to 0.01%) can significantly enhance structural preservation without preventing affinity binding.

As a starting point, freshly prepared paraformaldehyde (2-4%) with a low concentration of glutaraldehyde (0.5-0.01%) in an isotonic phosphate buffer (e.g. 0.01M phosphate buffer pH 7.4, containing 0.15M sodium chloride-PBS) is recommended. Ammonium chloride (50mM in buffer) should be used immediately after fixation to quench free aldehyde groups and remove residual fixative activity.[29] If this protocol proves ineffective, paraformaldehyde alone can be used. If none of these approaches are successful, special fixative procedures such as Periodate-Lysine-Paraformaldehyde[30] and dimindoester cross-linking[31] may be useful for some antigens.

To determine the fixation sensitivity of the probe's binding to its target molecule, and hence an optimal fixation and preparation regime, a model system can be used. For example, 10 µl dots from a doubling dilution series of the purified target molecule on nitrocellulose may be incubated with various fixatives and then with the affinity probe at appropriate dilution. Binding is then visualized using an immunoblotting detection system. Probe penetration is also affected by fixation, but for this parameter, there is no alternative to testing preparation regimes on tissue samples.

Embedding and Sectioning

Resins

Embedding should produce a block of tissue sufficiently hard and uniform to allow thin sections to be cut from it. Conventionally, tissues are dehydrated, equilibrated with a suitable organic solvent and then infiltrated with the embedding material. This is later hardened by crosslinking, usually promoted by heat or light. Each of these steps can cause problems for tissues to be used for affinity cytochemistry. The properties of the embedding agent affect both sectioning and the accessibility of the target within the section, unless the embedding medium is partly or completely removed. Sections of resin embedded material[32] using epoxy resins such as "Araldite" and "Epon" can be used successfully, after hydrogen peroxide treatment, for some affinity studies (*Fig. 1*). Archive material embedded in these resins is often available. Hydrophilic acrylic resins, especially when cross-linked at low temperature using visible or UV light, have been found to give especially good results for immunocytochemistry and lectin staining (*Figs. 2-5*).

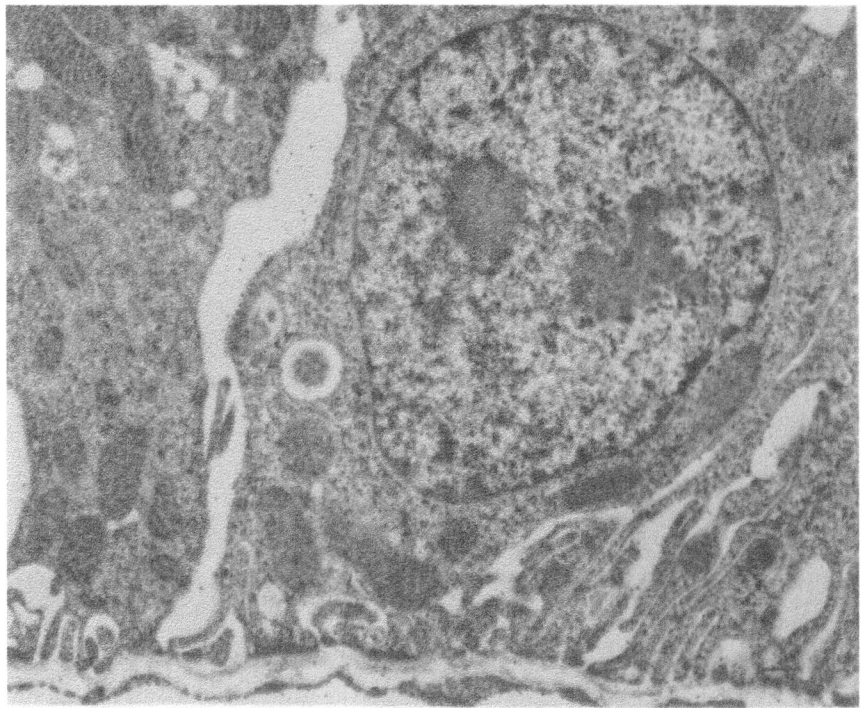

Figure 1. Chicken kidney, glutaraldehyde fixed and embedded in Epon. Detection of immunoreactivity for vitamin D-dependent calcium binding protein with the protein A-gold technique. Gold particle labelling (15 nm) is present over the cytosol and euchromatin of a principle cell of a convoluted distal tubule (right). No specific labelling is detectable over the cytoplasm of an intercalated cell (left). Full details of procedure and results have been published.[63,64] Scale bar = 250 nm.

These resins, commercially available in the Lowicryl[33] and LR-White and LR-Gold series,[34] have good penetration due to low viscosity, even at low temperature. They will tolerate the presence of some water during cross-linking, permitting less rigorous processing steps. Hydrophilic resins do not require etching treatments because of their hydrophilicity. Sectioning tissues embedded in these resins produces sections with a contoured surface with exposed target molecules for labelling.[35] This characteristic also affects the result of counterstaining tissue components in sections for EM.[36]

Cryo-Sectioning

Cryo-sectioning represents a major departure from the conventional EM approach in that no organic solvent and no cross-linking is involved, the hardness and strength of the embedding medium is attained by freezing. Since the tissue does not need to be dehydrated, heated, or exposed to harsh chemicals for embedding, a relatively mild fixation can be used for studies with cryosections without compromising structure.

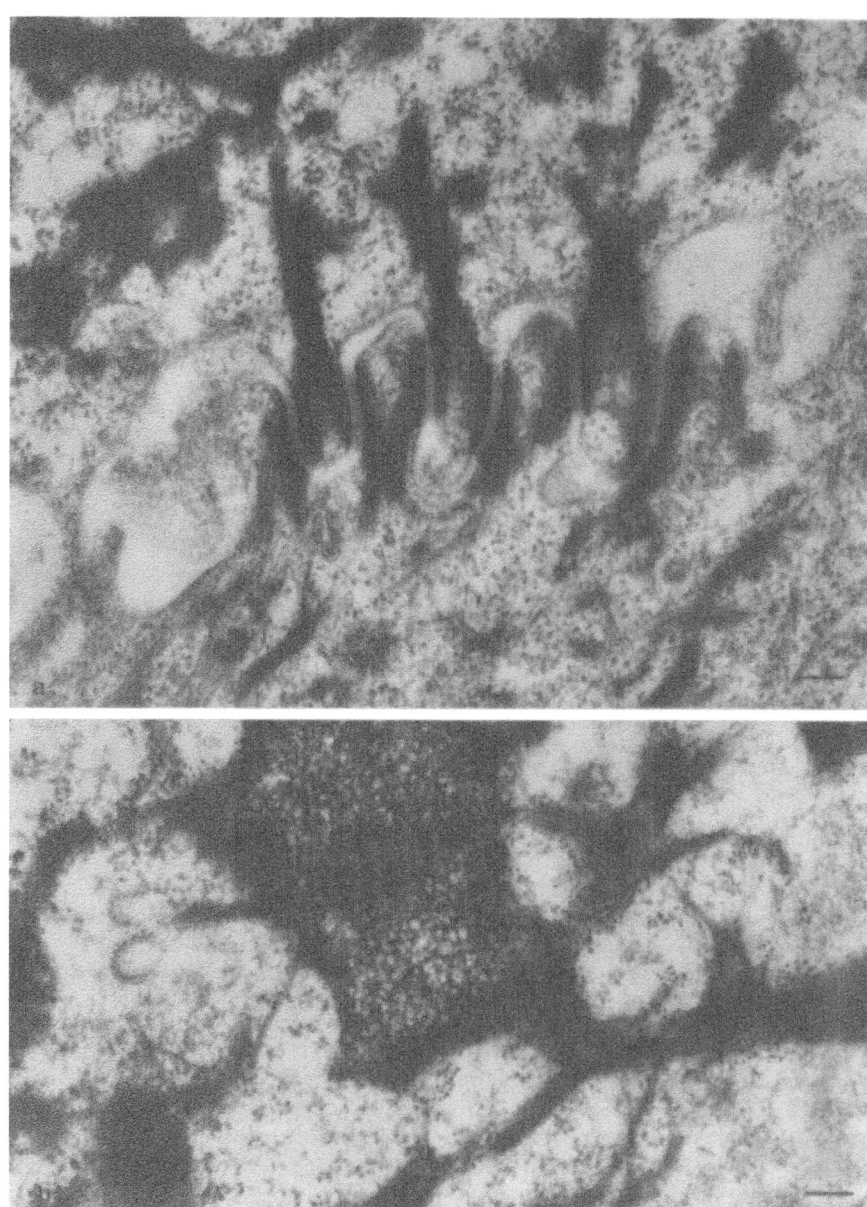

Figure 2. Human skin, glutaraldehyde fixed and low temperature embedded in Lowicryl K4M. Immunolocalization of cytokeratin with monoclonal antibody KL 1 and protein A-gold (8 nm).[29] Note the abrupt change in labelling with the cells in the stratum basale being unlabelled (a). All cells in the stratum granulosum (b) show gold particle labelling over the cytokeratin filament bundles. However, regions of the filament bundles apparently not exposed on the surface of the thin section are unlabelled. Neither keratohyalin nor the surrounding cytoplasm show any gold particles. Scale bar = 150 nm.

Although it is possible to use tissue frozen in physiological salt solutions, very rapid freezing is needed to prevent ice crystal damage. The technique developed by Tokuyasu and coworkers[37,38] avoids this by infiltrating tissues with a cryoprotectant such as 2.3M sucrose. This prevents ice crystal formation and the associated damage to the tissue, even with relatively slow rates of freezing.[37] This method greatly increases the depth of well preserved tissue in the block and plunge freezing in liquid nitrogen or nitrogen slush produces good tissue preservation. When PVP is added to the infiltrating sucrose solution,[39] the PVP does not enter cells. In this way, it helps to produce a more uniformly plastic block which can be sectioned more easily. The development of ionizing systems,[40] the routine use of cryo-diamond knives and improvements in ultracryomicrotome design[40] have greatly improved the ease with which semi-thin (*Fig. 6*) or ultrathin (*Fig. 7*) frozen sections can be produced.

After cutting, sections are picked up on a drop of sucrose, transferred to carbon coated grids with Formvar or Parlodion films, and stored on gelatin plates until use. Before use, the gelatin is liquified by warming to 37°C and grids recovered using a loop. The sucrose embedding medium is thus removed, while ensuring that the sections are never allowed to dry out. The gelatin also provides a blocking step which reduces non-specific background. The whole localization procedure can then be carried out with the sections fully hydrated and floating on drops of incubation and washing media. The final steps to prepare sections for EM is to post-fix in glutaraldehyde, and re-embed them on-grid, in methyl cellulose, PVA or a low viscosity resin. Uranyl acetate counterstaining is included in the final methyl cellulose or PVA embedding step[38] while osmium tetroxide and uranyl acetate counterstaining can be combined with resin post-embedding after dehydration.[41] The critical step in each of these processes is to leave a very thin embedding layer while ensuring that all the tissue is protected by the embedding medium against air drying.

General Comments

To obtain optimal resolution for electron microscopy, sections should be as thin as possible, although there is a trade off in terms of quantity and size limitation of the structural components in very thin sections. Since little or no penetration of the antibody occurs with most embedding media, only binding sites exposed on the surface will be labelled, although the structure may be visible in the transmission image (*Figs. 2,6*). If sections are cut onto coated grids or stained floating on the surface of the labelling solutions, only one of the surfaces will be labelled, again reducing the apparent labelling of structures visible within the sections.

PRIMARY AFFINITY STEP

At this stage it may be useful to apply the proposed protocol to semi-thin sections (resin or cryo-sections) of tissue as a rapid screening process. For immunogold probes, we have found that the results from gold labelled reagents, enhanced with a photographic physical silver development method,[35] are transferable to the EM level (omitting the silver enhancement).

Immunocytochemistry

Antibodies have proved to be the most versatile affinity labels yet exploited with high specificity and affinity. Specific antibodies against a very wide range of macromolecules can be produced, almost irrespective of their structure, without needing a pure preparation

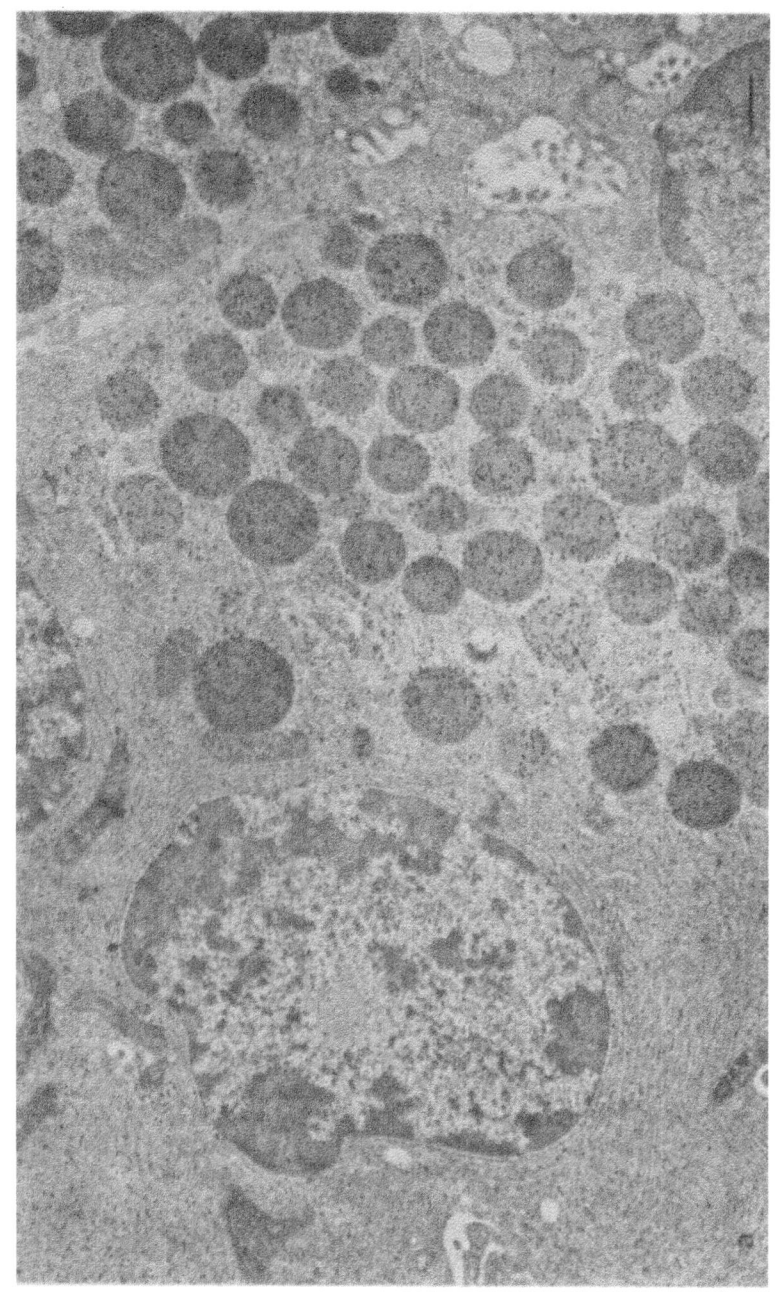

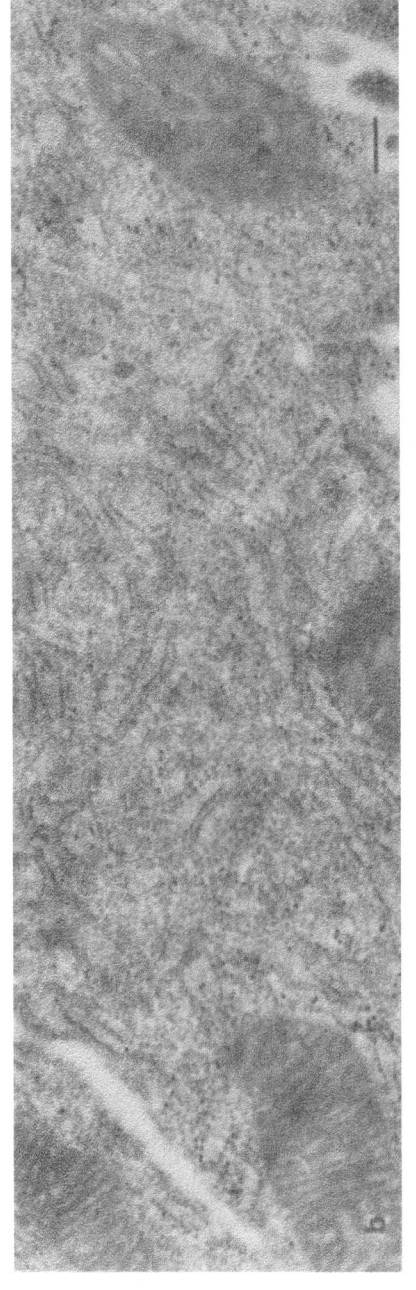

Figure 3. Rat exocrine pancreas, fixed in glutaraldehyde and embedded at low temperature in Lowicryl K4M.[65] Amylase immunoreactivity is detected with the protein A-gold technique. Exocrine acinar cells shown at low magnification exhibit gold particle (15 nm) labelling over the rough endoplasmic reticulum, Golgi apparatus, zymogen granules and alcinar lumen (a). Immunolabelling for amylase over the lumen of the rough endoplasmic reticulum is shown at higher magnification in (b). Scale bar = 300 nm (a), 150 nm (b).

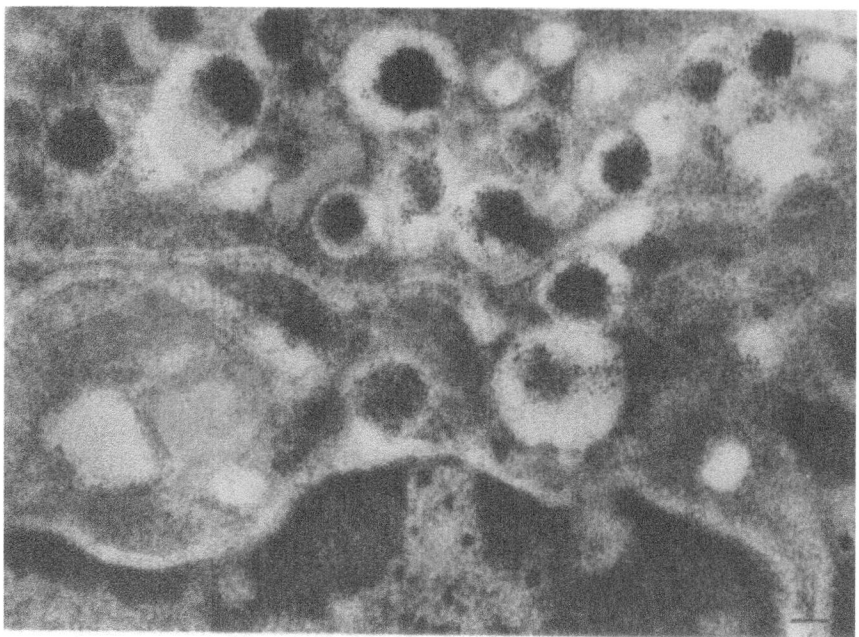

Figure 4. Human insulinoma fixed in 3% formaldehyde, 0.1% glutaraldehyde and embedded at low temperature in Lowicryl K4M resin. Insulin immunoreactivity is detected in secretory granules with a monoclonal anti-insulin antibody and protein A-gold (8 nm). The gold particle labelling is principally confined to the secretory granule cores. Scale bar = 150 nm.

of the target molecule. Both polyclonal and monoclonal antibodies are used widely and successfully for immunocytochemistry at the electron microscope level (Figs. 1-4, 6, 7). Monoclonal antibodies are homogeneous preparations which recognize a single epitope and are easy to purify and characterize. Polyclonal antibodies, on the other hand, may recognize more than one epitope which increases the probability of recognizing antigens after fixation, and of having an antibody of higher specificity. Further details and reviews of the preparation of antibodies may be found elsewhere.[42,43] The initial method of choice for immunocytochemistry should be a simple two step method using labelled immunogold or protein A-gold. Good quality gold reagents are available commercially for both procedures or can be prepared in the lab (see protocols). Direct gold labelling of an antibody may help if problems are encountered with the indirect system, or for double labelling using two antibodies raised in the same species.

Gold Labelled Antigen Detection (GLAD)

The gold labelled antigen technique,[44,45] is based upon the use of antisera in conjunction with gold labelled purified antigen. The antigen-gold will bind specifically only to divalent antibodies which recognize the antigen in the tissue and hence form a bridge to the gold labelled antigen. The gold labelled-antigen may be formed directly with large antigens while small antigens are covalently linked to a carrier such as BSA or ovalbumin before gold labelling. The main advantage, and disadvantage, of the method for antibody detection is that the specificity of labelling resides primarily in the quality of the purified antigen. This means that broad spectrum polyclonal antisera, which may recognize more than one antigen specifically can nonetheless be used for specific localization of molecules.

The same antiserum can be used for region specific localization of molecules,[44] and even for double labelling. However, the availability of purified antigen may be severely limiting, although the advent of modern molecular biology techniques may reduce the magnitude of this problem in future, at least for peptide antigens. The method also requires antigen-gold complexes to be produced for each antigen of interest and is therefore most appropriate for use with difficult antigens and for double labelling. It is recommended that antigen is labelled with gold using the protocol given later in the chapter.

Lectins

Lectins are a heterogeneous class of compounds, of plant or animal origin, characterized by their affinity for carbohydrate structures. Lectins are purified by biochemical means and are often poorly characterized in terms of their binding specificity. Recent improvements in purification and characterization procedures have resulted in the availability of lectins which are known to be highly specific for certain sugar residues in particular linkages. For example, the lectin from Maackia amurensis, recognizes only 2-3 linked sialic acids[46,47] while that from Sambucus nigra recognizes only 2,6-linked sialic acids.[48,49] Lectin-substrate affinity is often very high, even compared to antibodies, providing ideal affinity labels. Since the structure of lectins themselves is heterogeneous, localization of lectin binding generally relies on directly labelling the lectin with gold (*Fig. 5*) or haptens such as digoxigenin (DIG)[50] which subsequently can be localized by immunocytochemistry. For lectins with multivalent binding sites, labelled carbohydrate structures such as fetuin-gold[51] can be used for localization, analogous to the GLAD technique. Since lectins differ in their optimal pH for binding, appropriately buffered solutions should be chosen for each lectin and care should be taken not to include glycosylated proteins or sugar residues in blocking steps or buffers. Direct gold labelling of lectins will often be found to be the most effective and appropriate method and has been reviewed by Roth.[52] DIG labelled probes[50] are more versatile than many other hapten labels because there are no endogenous DIG-like binding sites. Non-specific interactions between lectins and hydrophobic resin can cause problems and the hydrophilic resin K4M is recommended.[52,33]

LOCALIZATION

Colloidal gold labelled probes are the de facto standard for electron microscopy, they are also versatile and can be used for parallel studies using light microscopic affinity cytochemistry and immunoblotting. The concentration of gold label is best expressed as the optical density (O.D. 525: absorbance) using a 525 nm wavelength light source. Gold probes are generally stable[13] but the O.D. may vary significantly during storage and whenever possible, dilutions should be checked immediately before use. The O.D. will need to be significantly increased for equivalent labelling with larger gold particles. The higher resolution of electron microscopy largely negates the need for signal amplification since it is possible to detect binding of individual probes by using gold probes. However, label visibility and ease of interpretation can be enhanced by amplification (see below). Recently, silver development of small gold probes has been used to enhance their visibility in the EM; details will be found elsewhere in this volume.

Direct Labelling

Directly labelled affinity probes provide a simple and sensitive method of localization.

They are applicable to both lectins (*Fig. 5*) and antibodies (*Fig. 7*) but must be individually prepared for each probe. For this reason, direct labelling methods are not ideal for the novel localization of tissue components, although once established, they are simple to use and easy to apply to double labelling (*Fig. 7*). Direct gold labelling of a probe requires a minimum of 100 mg of purified probe but is economical once produced. Details are given later in this chapter. Initial affinity labelling tests should use gold diluted in PBS + 1% BSA and including detergents if they do not inhibit affinity binding (see protocols for examples). An O.D. 525 = 0.1 is a good starting point for immunogold and for most procedures. O.D.'s in the range of 0.05 to 0.5 for 8 nm gold should be effective.

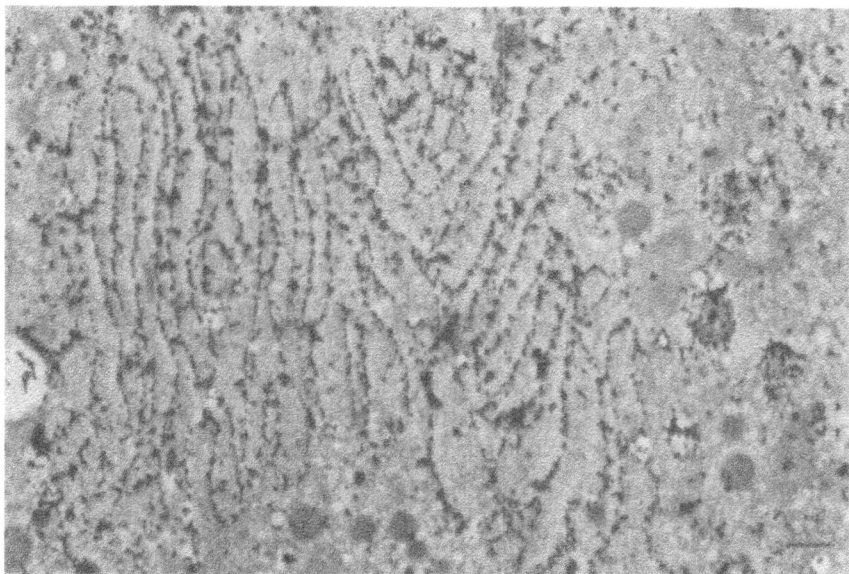

Figure 5. Chicken duodenum, fixed in glutaraldehyde and embedded in Lowicryl K4M resin at low temperature. Directly gold-labelled Ricinus communis lectin I detects galactose-like residues.[52] Gold particle (15nm) labelling is found along the lateral plasma membrane interdigitations of two absorptive enterocytes and over electron dense bodies in the cytoplasm. Scale bar = 300 nm.

Indirect Labelling- (1) Immunogold Label

Gold labelled secondary antibodies (*Fig. 6,7*) provide a versatile and specific method to localize primary affinity probes (both antibodies and hapten labelled probes such as biotin or DIG). Bridge methods are also applicable but not widely used for EM. Immunogold secondary labels will give some amplification which may be increased by introducing extra layers, i.e. applying a second unlabelled anti-immunoglobulin layer before the labelled layer. The comments for direct labelled immunogold are equally applicable to visualization with secondary immunogold labels. Starting O.D.'s of 0.1 (8 nm gold) are recommended, details can be found in the protocols later in the chapter.

Indirect Labelling- (2) Protein A-Gold

The most important and widely used indirect non-immune affinity labelling method for EM is the protein A-gold method[12] (*Figs. 1-4*) which takes advantage of the affinity of protein A for the Fc region of a range of antibodies. When used as a single secondary step, protein A-gold binding is proportional to primary antibody binding and gives little or no amplification, a valuable property for quantitative studies. For use with mouse monoclonal IgG antibodies, an intermediate rabbit anti-mouse immuno-globin layer should be applied, which will result in some amplification. This can be further enhanced by additional immunoglobulin layers as required. Starting O.D.'s of 0.06 (8 nm gold) or 0.4 (14 nm gold) in PBS containing 1% BSA and up to 0.2% Triton X-100 and Tween 20 are recommended. Details can be found in the protocols later in the chapter.

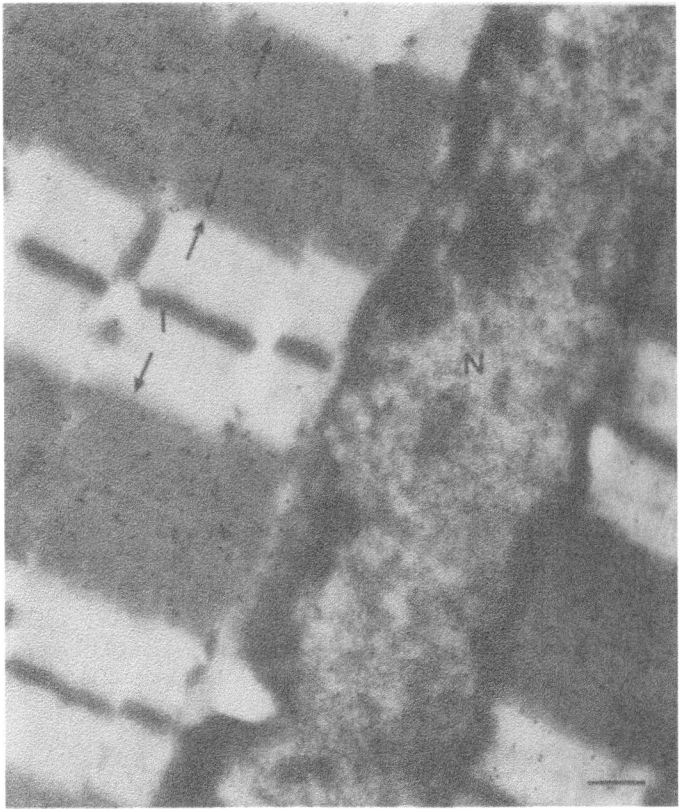

Figure 6. A semi-thin (approximately 300nm) cryosection of human gluteus maximus muscle immunolabelled with an antibody (N0Q7.5.4D) specific for the slow myosin heavy chain isoform of human skeletal muscle. Indirect immunogold detection (15nm gold) gives specific labelling of type 1 fibers, showing preferential localization within the A band areas (A) and negligible background staining of I bands (I)[66] In some regions, labelling apparently extends over the nucleus (N), but this is in fact associated with thick filaments overlying the nucleus at the surface in this section. Scale bar = 300 nm.

Double Labelling

The wide range of different gold sizes available are ideal for double labelling studies. In addition to the normal problems of affinity localization, double labelling techniques must avoid cross-reaction between the two probes. This is achieved in two different ways: (1) avoidance of cross-reaction, e.g. using different sides of the section (8), masking binding sites by fixation, silver enhancement, addition of excess free probe, etc.; (2) ensuring that cross interaction does not occur, by using directly labelled reagents for example, primary antibodies from different species, using different hapten labels or using the GLAD method. Detailed consideration of all the possible methods is outside the scope of this chapter. However, combination of two non-cross reactive probes or using at least one direct label is recommended (see *Fig. 7* and the associated protocols) and will generally not present particular problems. The smaller gold particles generally should be used to visualize the probe which gives the weakest labelling since larger gold particles result in less labelling. Double labelling with Protein A-gold of two different sizes has also been found to be very effective when appropriately applied.[9,53] The use of good control procedures to detect cross-reactivity is essential and should quickly show if the procedure is effective.

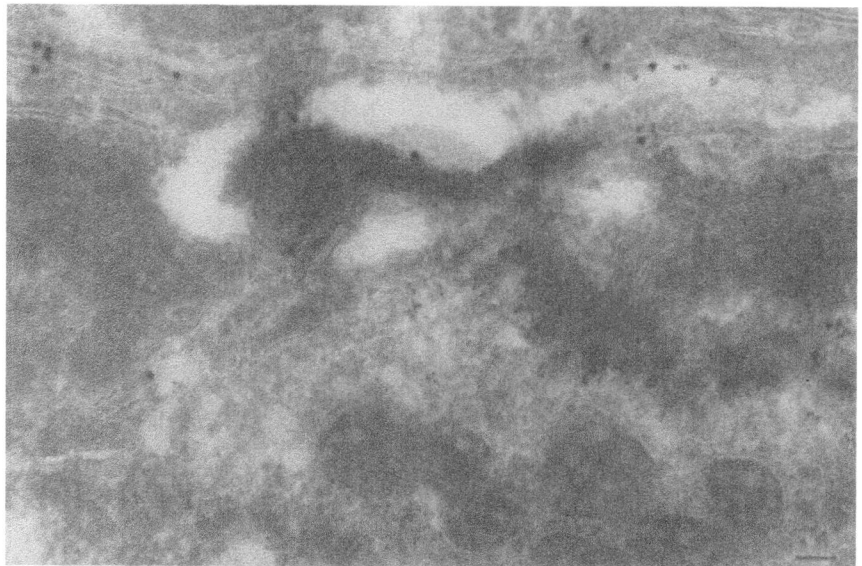

Figure 7. Ultrathin-frozen section of rat brain fixed in 3% paraformaldehyde and 0.1% glutaraldehyde. Double immunolabelling for polyscelic acid (8 nm gold) and N-CAM (14 nm gold) in the olfactory lobe.[60] Scale bar = 100 nm.

Blocking Steps And Reducing Background

Affinity probes often bind to other sites in the section either because of simple electrostatic interactions, or due to incidental affinity binding. Conditions should therefore be chosen under which the affinity binding of the probe to the target molecule is maximal compared to binding to other sites in the section.[29] For EM, as for other systems, the probe should be diluted to give maximal labelling and an acceptable level of background binding. Further dilution risks the loss of specific labelling.

Before the primary affinity step, non-specific binding sites can be blocked with protein or serum which can also be included in the probe diluent. Frequently used blocking agents include diluted normal sera (1-10%), fat-free milk powder (1-5%) and proteins such as BSA (0.1-5%). Detergents, typically Tween 20 and Triton X-100 (both at 0.01-0.2%) also reduce non-specific interactions. Blocking agents and detergents can be used alone or in combination and should be selected to suit the target, the primary label and the method of localization. Some antibodies are sensitive to detergents while gold labelled probes generally benefit from their addition to the gold diluent. One to two percent fat-free dried milk powder in PBS is recommended as a pre-blocking step, but not for diluting gold reagents. A combination of 1% BSA, 0.05% Triton X-100 and 0.05% Tween 20 in PBS (PBS+++) has been found to be useful for antibody and gold diluents. If dilution and blocking steps do not remove the problem of non-specific binding, then purification of the probe or cross absorption with soluble purified preparations of the cross-reactive component should be considered.

Problems of Detection-- Accessibility

In many embedding media, penetration of the probe into the section may not be possible, and labelling depends primarily on the sites available at the section surface. Detergent treatment can improve access and reduce non-specific surface interactions. For epoxy resin sections, hydrogen peroxide etching will enhance penetration (see protocols). For semi-thin sections, saturated sodium hydroxide in ethanol or methanol can be used to remove the resin although tissue preservation is adversely affected. The size and characteristics of the binding site required for the probe, the size of the probe, and the label are also relevant in this context. Gold probes do not generally penetrate sections well, although small gold particle sizes down to 1 nm significantly enhance penetration and labelling of cryosections for electron microscopy. Such gold labels usually require silver enhancement for convenient visualization (see other chapters in this volume for details).

CONTROLS

Controls should show that the detection system is working, do not result in unacceptable non-specific background label, and that none of the detection steps in themselves produce a signal. Omission of the probe and the substitution of a similar one for a component not found in the tissue should be used. Specificity of the probe may be more difficult to establish but is clearly essential in any study. Pre-absorption with purified target molecule preparations should abolish specific staining but not background, in a concentration dependent manner. Treatment of the sections with enzymes or antagonists can be valuable controls if their specificity is well established, particularly for directly labelled probes. Specificity of binding in other assays (e.g. immunoblotting, radioimmunoassay, ELISA) also helps to establish specificity although care should be take when interpreting results from assays which use substantially different probe concentrations or methods. For example, competitive radioimmunoassay does not provide useful evidence of exclusive specificity for immunocytochemistry since any number of non-specific antibodies will not affect the specific competitive binding.

OPTIMAL PROCEDURES AND PROTOCOLS

Preparing gold probes: gold sol preparation
Tannic acid and sodium citrate method after Slot and Geuze[53]

All solutions must be prepared with clean, siliconized glassware and using high quality double distilled water.

Solution A	79 ml water
	1 ml of 1% tetrachloroauric acid
Solution B	15 ml water
	4 ml of 1% tri-sodium citrate
	20-3000 ml of 1% tannic acid

Solutions A and B are heated to 60°C in separate siliconized glass beakers, and B is rapidly added to A while stirring vigorously. The solution should change color immediately. Boil the solution until the color changes to orange-red. After cooling, add 0.7 ml of hydrogen peroxide (30% volume), mix well, leave for 10-60 minutes and then boil for 10 minutes to remove excess hydrogen peroxide. The final volume of the prepared sol should then be adjusted to 100 ml by addition of water. Do not use if the final sol shows any blue/grey color. When necessary, especially for lectin preparations, the pH of the sol can be adjusted using minimal volumes of potassium carbonate or hydrochloric acid. The pH should only be tested using a gel-filled pH electrode.

The following volumes of tannic acid will yield these approximate sizes of gold:

volume of tannic acid (ml)	approximate mean gold size (nm)
20	14
100	10
200	8
1000	5
3000	3*

* for 3nm gold particles, the solutions should be heated to 80°C and the pH of solution B adjusted by the addition of 375 ml of 0.2 M potassium carbonate.

Other Methods

The sols produced with the tannic acid/sodium citrate method after hydrogen peroxide treatment should be applicable to the majority of labelling applications without loss of probe affinity. If, however, probe affinity is lost, it may be helpful to use a gold sol produced by another method. We have found the methods of Frens,[54] using sodium citrate alone (produces approximately 14 nm monodispersed sol) or Stathis[55] using ascorbic acid (produces polydispersed sol approximately 6-12 nm) to be most convenient and useful. Other methods are detailed and reviewed elsewhere.[13,56]

Gold Labelling Affinity Probes

The basic steps in the procedure are (1) to establish the minimum amount of the molecule to be labelled which stabilizes the gold sol; (2) mix the gold and molecule to be labelled at this concentration or slightly above, subsequently adding additional stabilizers; (3) spin the gold sol to sediment the gold particles and discard unlabelled probe in the supernatant.

Establishing The Minimum Stabilizing Amount of Probe

Titration of the amount of probe needed to stabilize a gold sol is carried out by mixing different amounts of the probe molecule with constant amounts of gold and then adding destabilizing amounts of salt. If flocculation occurs, as detected by a color change to blue/grey, or by EM, the amount of probe added was not stabilizing. A simple and economical way to do this is to take 20 ml droplets of gold on a hydro-phobic film such as parafilm over a white surface, add 20 ml of appropriately diluted probe, leave for a few minutes and then add 10 ml of 10% sodium chloride. After 30 minutes, the color of the drops is checked and the appropriate concentration of probe selected by floating poly-L-lysine coated EM grids on a parallel series of drops without added salt.[57] It is also possible to check by EM examination to see that aggregates do not form.[57,58] Probes to be labelled should be checked individually for the stabilizing amount for each gold particle size, however some guidance as to the range of concentrations can be given. We have found when using 8nm gold sols, that approximately 40-60 mg of polyclonal and monoclonal antibodies are needed for each 1 ml of gold sol whereas approximately 12.5 mg/ml protein A is needed. Other probes, particularly lectins, may require significantly different amounts for stabilization and may need to be labelled at different pH values.[13]

Gold Labelling and Storage of Labelled Probes

Labelling is achieved simply by mixing gold sol and probe (approximately 1.1 x the minimum stabilizing amount). It is crucial that the two are well mixed; rapid addition of the gold to an equal volume of diluted probe and brief vortex mixing will achieve this. The probe can be diluted in buffer providing the mixing is very rapid since the probe will stabilize the sol before the salts in the buffer induce flocculation.[58] Immediately after mixing, it is useful to add stabilizing amounts of protein (e.g. BSA) or Carbowax 20 (PEG) and then concentrated (e.g. 10X) buffer such that the probe is kept in optimal pH and salt conditions. Gold preparations should then be spun at 63,000 x gav for 45 minutes in a Kontron 70.5Ti rotor (or similar). The gold will then form a loose "pellet"; the supernatant containing unlabelled probe is discarded. Gold labelled probes should be stored away from strong reducing agents. Stabilizing proteins or Carbowax and addition of azide will increase storage life which is several months or more at 4°C.

Direct Labelling: Antibody-Gold

Tissue fixed in 3% para-formaldehyde, 0.1% glutaraldehyde in PBS for 2 hours and subsequently quenched in 50mM ammonium chloride in PBS (2 changes, 30 minutes total). This procedure is applicable to a wide range of embedding materials including cryosectioning.

1.   Blocking step: 2% defatted milk in PBS, 5-10 minutes.
2.   Direct labelling step: Gold labelled antibody (8 nm) O.D. 525 = 0.2-0.05 in PBS + 1% BSA, 0.05-0.01% Triton X-100 and 0.05-0.01% Tween 20, 60 minutes.
3.   Wash 6 x PBS (total of 15 minutes) by floating on drops of PBS (cryosections) OR 3 x PBS by mild jet washing (resin sections)
4.   Post-fix 1% glutaraldehyde in PBS 20 minutes (may be omitted for resin sections)
5.   Wash 3 x in distilled water
6.   Counterstain as appropriate

Specific Example: Direct Labelling of Ultrathin Cryosections

We have used this direct labelling method with a mouse monoclonal antibody against a 2, 8-linked polyscelic acid residues, characteristically found on N-CAM peptides.[59,60,61] Protein A affinity purified antibody mAb 73562, (from D Bitter- Suermann, Hannover, Germany [a mouse IgG_1 monoclonal antibody]) was directly labelled with 8 nm gold (see protocol). Approximately 45 mg of purified protein was used for each 1 ml of 8 nm gold sol and was used at a working O.D. 525 = 0.05.

Ultrathin frozen sections from rat brain were used in this procedure. After fixation, tissue was infiltrated with 2.3 M sucrose containing 10% w/v PVP for 4 hours, snap frozen in liquid nitrogen slush and ultrathin frozen sections were cut. These were stored overnight on 2% gelatin plates at 4°C, thawed at 37°C and labelled by floating on drops of incubation and washing solutions.

Post-embedding in methyl cellulose/uranyl acetate was performed after labelling. Briefly, grids were transferred to drops of this mixture (see below) at 4°C, rapidly transferred to a second and third drop (to ensure residual water was removed) and then left on the third drop for a minimum of 10 minutes. The grids were then removed and the excess embedding medium removed using filter paper while holding the grid in a fine wire loop slightly larger than the diameter of the grid. The grids were air dried, resulting in a thin protective layer which was electron-lucent. Methyl cellulose/ uranyl acetate solution was made as follows: 2% w/v methyl-cellulose was dissolved in distilled water by stirring for 2-3 days at 4°C and spun at 230,000 x g$_{av}$ for 1 hour in a Kontron 70.5Ti rotor (or similar). The solution was best if made up a few weeks in advance. Tubes were stored at 4°C without disturbing the pellet and could be used for several months. Saturated aqueous uranyl acetate was mixed with the 2% methyl cellulose in the ratio 1:9 immediately before use.

Direct Labelling: Lectin-Gold

Tissue Fixed as for Direct Antibody Labelling. The detailed protocols for lectin affinity localizations vary more widely than those for antibodies and reference should be made to the literature for specific conditions.[52,13]
1. Blocking step: 1% BSA in PBS, 5 minutes
2. Gold-lectin complexes O.D.525 = 0.1-0.5 in appropriate diluent, 45 minutes
3. Wash 6 x PBS (total of 15 minutes) by floating on drops of PBS (cryosections) OR 3 x PBS by mild jet washing (resin sections)
4. Post fix in 1% glutaraldehyde in PBS

Specific example: Tissue was embedded in Lowicryl K4M at low temperature. The lectin isolated from Maackia amurensis, recognizes specifically terminal a2, 3-linked and not a2, 6-linked, sialic acid residues.[47] Gold labelled lectin (8 nm gold) was made using the method detailed with hydrogen peroxide treated gold and lectin dissolved in 0.01 M phosphate buffer pH 7.4[47]. In this case, the gold was diluted to O.D.525 = 0.5 in PBS + 1% BSA, 0.05% Triton X-100 and 0.05% Tween 20. Sections were incubated for 45 minutes.

Indirect Labelling– Immunogold

Tissue fixed and prepared as for direct antibody labelling protocol (above).
1. Blocking step: 1% defatted milk in PBS, 5-10 minutes
2. Primary labelling step: Rabbit polyclonal antibody diluted 1:50 to 1:5000, 1 hour
3. Wash 6 x PBS (total of 15 minutes) by floating on drops of PBS (cryosections) OR 3 x PBS by mild jet washing (resin sections)
4. Secondary labelling step: gold labelled goat anti-rabbit antibody O.D.525 = 0.1 in PBS + 1% BSA, 0.05% Triton X-100 and 0.05% Tween 20, 1 hour
5. Repeat step 3.
6. Post-fix 1% glutaraldehyde in PBS
7. Wash 3 x in distilled water

Specific example: We have used this method with ultrathin cryosections for the localization of N-CAM during developmental processes (see also 9.6). A rabbit polyclonal antibody (from C Goridis, Marseille, France) was diluted 1:500 in 1% defatted milk in PBS. Cryosections were post-embedded in methyl cellulose/uranyl acetate as detailed above.

Indirect Labelling– Protein A-Gold

Tissue fixed as for direct antibody labelling, embedded in Lowicryl K4M at low temperature.

1. Blocking step: 2% defatted milk in PBS 5-10 minutes
2. Primary antibody step: appropriately diluted polyclonal rabbit antiserum   1-2 hours
3. Wash 4 x in PBS, total 10 minutes
4. Secondary labelling step protein A-gold (8 nm) O.D.525 = 0.06 in PBS + 1% BSA, 0.05% Triton X-100 and 0.05% Tween 20
5. Wash 3 x PBS, total 10 minutes
6. Wash 3 x in distilled water, air dry
7. Store as required, counterstain

Double Labelling

Tissue fixed and prepared as for direct antibody labelling protocol (above). This method is a combination of the direct immunolabel method and the indirect immunolabelling methods given above. The specific details are given in this protocol as an example of how two single labelling protocols can be combined.[60]
1. Blocking step: 1% defatted milk in PBS, 5-10 minutes
2. Primary labelling step: Rabbit polyclonal antibody against N-CAM diluted 1:500, 1 hour
3. Wash 6 x PBS (total of 15 minutes) by floating on drops of PBS (cryosections) OR 3 x PBS by mild jet washing (resin sections)
4. Secondary labelling step & direct labelling step: (A) gold labelled rabbit anti-mouse antibody (14 nm gold) O.D.525 = 0.1 and
(B) directly (8 nm) gold labelled polyscelic acid antibody O.D. 525 = 0.06 both in PBS + 1% BSA,   0.05% Triton X-100 and 0.05% Tween 20, 1 hour (double strength dilutions of (A) and (B) are made and mixed 1:1 to give the final dilutions required)
5. Repeat step 3
6. Post-fix 1% glutaraldehyde in PBS
7. Wash 3 x in distilled water

The same pattern of staining was seen when the gold sizes were reversed. It is also possible to apply the direct label during the first labelling step, but the milk must be omitted from the antibody diluent which increases background staining from the N-CAM antiserum. Combining the two primary labels also results in somewhat reduced labelling for N-CAM, probably because of steric hindrance from the direct PSA label.

DISCUSSION

The quality of information about the localization of molecules which has been obtained by electron microscopy is a tribute to the inventiveness and technical achievements of scientists over the past 50 years. Affinity probes and methods of applying them have been developed that solve many technical difficulties and retain structure closely related to that found *in vivo*. Affinity labelling of specific cellular components and their visualization at sub-cellular resolution is likely to find increasing application as we try to relate the discoveries of molecular biology to the basic structure of cells. Many of the methods which will be required are available and will be gradually improved. For instance, the use of smaller gold labels and other means to enhance penetration into sections will improve the quality of information about some difficult antigens. The latest generation of microtomes and attachments for frozen thin sectioning will facilitate more routine use of frozen sections. Equally, hydrophilic resins are likely to become more widely needed and used. Post-embedding electron microscopy is still the best way of obtaining high resolution localization data, and increasingly, it will be complimented by data from high resolution light microscopic methods such as video microscopy and confocal scanning microscopy, providing information about fully hydrated and living cells.

In conclusion, we stress that affinity cytochemistry methods are not difficult to apply. Almost every university or hospital biology or pathology department around the world has access to an electron microscope and microtome -- little extra in the way of equipment or facilities is needed. There are an increasing number of research areas where ultrastructural localization studies using tried and tested methods are the only way to solve important problems.

ACKNOWLEDGEMENTS

We would like to thank Mr. D. Wey for his help preparing the photographs and Dr. A. Semper for critical reading of the manuscript and for permission to use figure 6.

REFERENCES

1. S. J. Singer, Preparation of an electron dense antibody conjugate, *Nature* 183:1523 (1959).
2. P. K. Nakane and G. B. Pierce, Enzyme-labeled antibodies: preparation and application for the localization of antigens, *J Histochem Cytochem* 14:929 (1966).
3. W. P. Faulk and G. M. Taylor, An immunocolloid method for the electron microscope, *Immunochemistry* 8:1081 (1971).
4. E. L. Romano, C. Stolinski and N. C. Hughes-Jones, An antiglobulin reagent labelled with colloidal gold for use in electron microscopy, *Immunochemistry* 11:521 (1974).
5. J. Roth, M. Bendayan and L. Orci, Ultrastructural localization of intracellular antigens by the use of protein A-gold complex, *J Histochem Cytochem* 26:1074 (1978).
6. J. Roth and M. Wagner, Peroxidase and gold complexes of lectins for double labelling of cell-surface binding sites by electron microscopy, *J Histochem Cytochem* 25:1181 (1977).
7. J. Roth and M. Binder, Colloidal gold, ferritin and peroxidase as markers for electron microscopic double labelling lectin, *J Histochem Cytochem* 26:163 (1978).
8. M. Bendayan, Double immunocytochemical labeling applying the protein A-gold technique, *J Histochem Cytochem* 30:81 (1982).
9. H. J. Geuze, J. W. Slot, P. A. van der Ley and R. C. T. Scheffer, Use of colloidal gold particles in double labelling immunoelectron microscopy of ultrathin frozen sections, *J Cell Biol* 89:653 (1981).
10. F. J. Tapia, I. M. Varndell, L. Probert, J. De Mey and J. M. Polak, Double immunogold staining method for the simultaneous ultrastructural localisation of regulatory peptides, *J Histochem Cytochem* 31:977 (1983).
11. M. Horisberger, Evaluation of colloidal gold as a cytochemical marker for transmission and scanning electron microscope, *Biol Cell* 36:253 (1979).
12. J. Roth, The protein A-gold (pAg) technique - a qualitative and quantitative approach for antigen localization on thin sections, *in*: "Techniques in Immunocytochemistry," Vol I. G.R. Bullock, P. Petrutsz, eds., Academic Press, London (1982).
13. J. Roth, The colloidal gold marker system for light and electron microscopic cytochemistry, in: "Techniques in Immunocytochemistry," Vol II. G.R. Bullock, P. Petrutsz, eds., Academic Press, London p217 (1983).
14. I. M. Varndell and J. M. Polak, Electron microscopical immunocytochemistry, in: "Immunocytochemistry: Modern Methods and Applications," 2nd Edition. J.M. Polak and S. Van Noorden, eds., John Wright & Sons, Bristol, U.K., p146 (1986).
15. J. M. Polak and I. M. Varndell, "Immunolabelling for Electron Microscopy," Elsevier Science Publishers, Amsterdam. (1984).
16. A. J. Verkleij and J. L. M. Leunissen, "Immuno-gold Labelling in Cell Biology," CRC Press, Inc., Boca Raton, USA. (1989).
17. D. M. Shotton, Review: Video-enhanced light microscopy and its applications in cell biology, *J Cell Sci* 89:129 (1988).

18. M. De Brabander, H. Geerts, R. Nuydens, S. Geuens, M. Moermans and J. R. De Mey, Detection and use of gold probes with video-enhanced contrast light microscopy, *in*: "Immuno-gold Labelling in Cell Biology," A. J. Verkleij and J. L. M. Leunissen, eds., CRC Press, Inc., Boca Raton, USA. p217 (1989).

19. D. M. Shotton, Confocal scanning optical microscopy and its applications for biological specimens, *J Cell Sci* **94**:175 (1989).

20. W. A. M. Linnemans, G. J. Brakenhoff, E. A. Van Spronsen and N. Nanninga, Immuno-gold labelling and confocal scanning electron microscopy, *in*: "Immuno-gold Labelling in Cell Biology," A. J. Verkleij and J. L. M. Leunissen, eds., CRC Press, Inc., Boca Raton, USA. p233 (1989).

21. N. J. Severs, Freeze-fracture cytochemistry: Review of methods, *J Electron Microsc Tech* **13**:175 (1989).

22. H. Hohenberg, Replica preparation techniques in immuno-gold cytochemistry, *in*: "Immuno-gold Labelling in Cell Biology," A. J. Verkleij and J. L. M. Leunissen, eds., CRC Press, Inc., Boca Raton, USA. p157 (1989).

23. M. Horisberger, J. Rosset and H. Bauer, Collodidal gold granules as markers for cell surface receptors in the scanning electron microscope, *Experientia* **31**:1147 (1975).

24. G. M. Hodges, M. A. Smolira and D. C. Livingston, Scanning electron microscope immunocytochemistry in practice, *in*: "Immunolabelling for Electron Microscopy," J.M. Polak and I.M. Varndell, eds., Elsevier Science Publishers B.V., Amsterdam. p189 (1984).

25. A. M. Glauert, Fixation, dehydration and embeddding of biological specimens, in: "Practical Methods in Electron Microscopy," A.M. Glauert, ed., North Holland, Amsterdam (1974).

26. M. A. Hayat, "Fixation for Electron Microscopy" Academic Press, London. (1981).

27. G. R. Bullock, The current status of fixation for electron microscopy: a review, *J Microsc*, Oxford **133**:1 (1984).

28. M. Bendayan and M. Zollinger, Ultrastructural localization of antigenic sites on osmium-fixed tissues applying the protein A-gold technique, *J Histochem Cytochem* **31**:101 (1983).

29. J. Roth, D. Taatjes and M. J. Warhol, Prevention of non-specific interactions of gold-labeled reagents on tissue sections, *Histochemistry* **92**:47 (1989).

30. I. W. McLean and P. K. Nakane, Periodate-lysine-paraformaldehyde fixative a new fixative for immunoelectron microscopy, *J Histochem Cytochem* **22**:1077 (1974).

31. J. Hassell and A. R. Hand, Tisue fixation with diimidoesters as an alternative to aldehydes. I. Comparison of cross-linking and ultrastructure obtained with dimethylsuberimididate and glutaraldehyde, *J Histochem Cytochem* **22**:223 (1974).

32. B. E. Causton, The choice of resins for electron immunocytochemistry, *in*: "Immunolabelling for Electron Microscopy," J.M. Polak and I.M. Varndell, eds., Elsevier Science Publishers B.V., Amsterdam. p29 (1984).

33. E. Carlemalm, R. M. Graavito and W. Villiger, Resin development for electron microscopy and an analysis of embedding at low temperature, *J Microsc Oxford* **126**:123 (1982).

34. G. R. Newman, B. Jasani and E. D. Williams, A simple post-embedding system for rapid demonstration of tissue antigens under the electron microscope, *Histochem J* **15**:543 (1983).

35. J. Roth, Postembedding labeling on lowicryl K4M tissue sections: Detection and modification of cellular components, *Methods in Cell Biology* **31**:513 (1989).

36. J. Roth, D. J. Taatjes and K. T. Tokuyasu, Contrasting of Lowicryl K4M thin sections, *Histochemistry* **95**:123 (1990).

37. K. T. Tokuyasu, Application of cryoultramicrotomy to immunocytochemistry, *J Microsc- Oxford* **143**:139 (1986).

38. G. Griffiths, K. Simons, G. Warren and K. T. Tokuyasu, Immunoelectron microscopy using thin, frozen sections: Application to studies of the intracellular transport of semliki forest virus spike glycoproteins, *Methods in Enzymology* **96**:466 (1983).

39. K. T. Tokuyasu, Use of poly(vinylpyrrolidone) and poly(vinyl alcohol) for cryoultramicrotomy, *Histochem J* **21**:163 (1989).

40. M. Michel, H. Gnegi and M. Müller, Diamonds are a cryosectioner's best friend, *J Microsc - Oxford* **166**:43 (1991).

41. G.A. Keller, K. T. Tokuyasu, A. H. Dutton and S. J. Singer, An improved procedure for immunoelectron microscopy: ultrathin plastic embedding of immunolabelled ultrathin frozen sections, *Proc Natl Acad Sci USA* **81**:5744 (1984).

42. J. De Mey and M. Moeremans, Raising and testing polyclonal antibodies for immunocytochemistry, *in*: "Immunocytochemistry: Modern Methods and Applications," 2nd Edition. J.M. Polak and S. Van Noorden, eds., John Wright & Sons, Bristol, U.K., p3 (1986).

43. M. A. Ritter, Raising and testing monoclonal antibodies for immunocytochemistry, *in*:

"Immunocytochemistry: Modern Methods and Applications," 2nd Edition. J.M. Polak and S. Van Noorden, eds., John Wright & Sons, Bristol, U.K., p13 (1986).

44. L.I. Larsson, Simultaneous ultrastructural demonstration of multiple peptides in endocrine cells by a novel immunocytochemical method, *Nature* **282**:743 (1979).

45. L.I. Larsson, Peptide immunocytochemistry, *Prog Histochem Cytochem* **13**:1 (1981).

46. W. C. Wang and R. D. Cummings, The immobilized leukoagglutinin from the seeds of Maackia amurensis binds with high affinity to complex-type asn-linked oligosaccarides containing terminal sialic acid-linked a 2,3 to penultimate galactose residues, *J Biol Chem* **263**:4576 (1988).

47. T. Sata, P. M. Lackie, D. J. Taatjes, W. Peumans and J. Roth, Detection of the neu5Ac (a2,3) gal (b1,4) GalNAc sequence with the leukoagglutinin from Maackia amurensis: Light and electron microscopic demonstration of differential tissue expression of terminal sialic acid in a2,3- and a2,6-linkage, *J Histochem Cytochem* **37**:1577 (1989).

48. N. Shibuya, I. J. Goldstein, W. F. Broekaert, M. Nsimba-Lubaki, B. Peeters and W. J. Peuman, The elderberry (Sambucus nigra L.) bark lectin recognizes the Neu5Ac (a2,6)Gal/GalNAc sequence, *J Biol Chem* **262**:1596 (1987).

49. D. J. Taatjes, J. Roth, W. Peumans and I. J. Goldstein, Elderberry bark lectin-gold techniques for the detection of NeuNAc (a2,6) Gal/GalNAc sequences: Applications and limitations, *Histochem J* **20**:478 (1988).

50. T. Sata, C. Zuber and J. Roth, Lectin-Digoxigenin conjugates - A new hapten system for glycoconjugate cytochemistry, *Histochemistry* **94**:1 (1990).

51. J. Roth, J. M. Lucocq and P. M. Charest, Light and electron microscopic demonstration of sialic acid residues with the lectin from Limax flavus: A cytochemical affinity technique with the use of fetuin-gold complexes, *J Histochem Cytochem* **32**:1167 (1984).

52. J. Roth, Application of lectin-gold complexes for electron microscopic localization of glycoconjugates on thin sections, *J Histochem Cytochem* **31**:987 (1983).

53. J. W. Slot and H. J. Geuze, A new method of preparing gold probes for multiple labeling cytochemistry, *Eur J Cell Biol* **38**:87 (1985).

54. G. Frens, Controlled nucleation for the regulation of the particle size in monodisperse gold suspensions, *Nature Phys Sci* **241**:20 (1983).

55. F. C. Stathis and A. Fabrikanos, Preparation of colloidal gold, *Chem Ind London* **27**:860 (1958).

56. J. L. M. Leunissen and J. R. De Mey, Preparation of gold probes, *in*: "Immuno-gold Labelling in Cell Biology," A. J. Verkleij and J. L. M. Leunissen, eds., CRC Press, Inc., Boca Raton, USA. p4 (1989).

57. W. Baschong and N. G. Wrigley, Small colloidal gold conjugated to Fab fragments or to immunoglobulin G as high-resolution labels for electron microscopy: A technical overview, *J Electron Microsc Tech* **14**:313 (1990).

58. J. M. Lucocq and W. Baschong, Preparation of protein colloidal gold complexes in the presence of commonly used buffers, *Eur J Cell Biol* **42**:332 (1986).

59. P. M. Lackie, C. Zuber and J. Roth, Polysialic acid and N-CAM in embryonic rat kidney: mesenchymal and epithelial elements show different patterns of expression, *Development* **110**:933 (1990).

60. C. Zuber, P. M. Lackie, W. A. Catterall and J. Roth, Polysialic acid is associated with sodium channels and the neural cell adhesion molecule (N-CAM) in adult rat brain, *J Biol Chem* **267**:9965 (1992).

61. P. M. Lackie, C. Zuber and J. Roth, Expression of polysialylated N-CAM during rat heart development, *Differentiation* **47**:85 (1991).

62. M. Frosch, J. Gergen, G. J. Bulnois, K. M. Timmis and D. Bitter-Suermann, NZB mouse system for production of monoclonal antibodies to weak bacterial antigens - Isolation of an IgG antibody to the polysaccharide capsules of Escherichia coli K1 and group B meningococci, *Proc Natl Acad Sci USA* **82**:1194 (1985).

63. J. Roth, B. Thorens, W. Hunziker, A. W. Norman and L. Orci, Vitamin D-depen-dent calcium binding protein: immunocytochemical localization in chick kidney, *Science* **214**:197 (1981).

64. J. Roth, D. Brown, A. W. Norman and L. Orci, Localization of the vitamin D-dependent calcium binding protein in mammalian kidney, *Amer J Physiol* **243**:F243 (1982).

65. J. Roth, M. Bendayan, E. Carlemalm, W. Villiger and M. Garavito, The enhancement of structural preservation and immunocytochemical staining in low temperature embedded pancreatic tissue, *J Histochem Cytochem* **29**:663 (1981).

66. A. E. Semper, R. B. Fitzsimons and D. M. Shotton, Ultrastructural identification of type 1 fibres in human skeletal muscle: Immunogold labelling of thin cryo-sections with a monoclonal antibody against slow myosin, *J Neurol Sci* **83**:93 (1988).

CHAPTER 14

IMMUNOGOLD-SILVER STAINING (IGSS) FOR IMMUNOELECTRON MICROSCOPY AND IN MULTIPLE DETECTION AFFINITY CYTOCHEMISTRY

Tibor Krenács, DMB, PhD;[1] Lászlá Krenács, M.D.

Department of Pathology, Albert Szent-Györgyi University
of Medicine, Szeged, Hungary

SUMMARY

Applications and optimal conditions of immunogold-silver staining (IGSS) as utilized in post-embedding ultrastructural single and double immunolabelling are reviewed. Light microscopic multiple antigen detection and in non-radioactive *in situ* hybridization, are discussed. In IGSS, the silver shell formed around the gold particles by an autometallographic process results in an electron-dense product with a particle size proportional to the duration of silver enhancement. Application of gold conjugates with the smallest particle size results in the highest reaction density, but electron microscopic visualization of non-amplified gold particles requires high instrumental magnification and weak tissue contrasting. Amplification of the gold particle diameters by silver precipitation can improve both the signal and, by permitting almost normal counterstaining, structural recognition. This may be of particular advantage when the antigens of interest are situated in electron-dense structures, such as dense-core granules of neuroendocrine cells, immunoglobulin deposits, or cytoskeletal filament clusters. At the light microscopic level, the dark-brown to jet-black product of IGSS contrasts well with the colors of most of the chromogens generally used in immunoenzyme techniques, and IGSS is, therefore, particularly useful in multiple immunostaining combinations. By covering the immunological sequences completely, the silver precipitate can prevent unwanted side-bindings. This offers the possibility of using two primary antibodies of the same animal species for double labeling. The principle of IGSS can be exploited to reveal gold particles adsorbed, not only to immunoglobulins, but also to protein A, lectins,

[1]Address for Correspondence: Tibor Krenács, Ph.D., Albert Szent-Györgyi University of Medicine, Kossuth L. sgt. 40, H-6701 Szeged, P.O.Box 401, Hungary. Telephone: 36-62 14 522/14; Fax: 36-62 314 156

RNAase or streptavidin, providing a general and more sensitive marker in affinity cytochemistry.

INTRODUCTION

In 1983, Holgate[1] and co-workers described a sensitive light microscopic antigen detection method, utilizing immunoglobulin-gold conjugates in combination with selective intensification of the gold signal in tissues by means of physical silver development.[2] This method has been known since then as immunogold-silver staining (IGSS), and has proven to offer great potential in several fields of immunocytochemistry. IGSS has already been exploited for light microscopic antigen detection in cell cultures and smears, in cryostat, paraffin and resin-embedded sections, pre-and post-embedding immunoelectron microscopy and scanning electron microscopy.[3-11]

The use of colloidal gold adsorbed to protein A, RNA-ase, lectins and streptavidin has further widened the scope of the silver-amplified gold marker.[12-15] The term IGSS should, therefore, have a wider meaning. It should include all reactions based on autometallographic silver amplification of gold label coupled to specific tissue targets by high affinity bindings. Reactions utilizing these special bindings, like those formed between antigen and antibody, enzyme and substrate, lectin and sugar, protein A and the Fc part of immunoglobulin, avidin and biotin, and DNA or RNA and complementary nucleic acid sequences, can be collected under the term affinity cytochemistry. Utilization of IGSS for immunoelectron microscopy, light microscopic multiple antigen detection, and the detection of other affinity cytochemical reactions, as combined with immunoenzyme stainings for double labeling, has recently become more and more popular.[14-22]

However, technologies involved are highly heterogeneous, and special demands have to be considered. Therefore, it would be worthwhile to find the optimal conditions for each of these methodologies. The aim of this chapter is to review the utilization of IGSS and to discuss some of the technical aspects of the method based on recent experiments, and to describe possible applications of gold/silver markers. This will include single and double antigen detection at the light and electron microscopic levels, and *in situ* hybridization. For this, selected histopathological tissues from our routine practice have been chosen.

Principle of IGSS and Its Reaction Product Formation

Proteins, at their isoelectric point, can be attached by non-covalent forces to the surface of colloidal gold particles and prevent their aggregation.[16] Immunoglobulins, protein A, lectins, RNA-ase and streptavidin can take part in this action. This renders gold particles able to bind to specific targets, such as antigens, immunoglobulins, and sugars. Specific DNA and RNA sequences can also be detected by this means through biotin, digoxigenin or fluorescein-isothiocyanate (FITC) labeled complementary probes. Gold particles on these specific sites can catalyze the electron transport from an electron donor (i.e. hydroquinone) to an acceptor (ionic silver), resulting in the formation of a metallic silver shell around the gold particle (autometallographic reaction).[1,13] The precipitated silver catalyses further reduction, making the process autocatalytic. This may result in at least two important advantages of the application of the gold/silver marker: (1) it results in a theoretically endless intensification, which explains its high sensitivity, and (2) it gives a compact, jet-black, electron-dense reaction product with adjustable particle size. The shielding effect of the silver shell on the immunological sequences permits sequential use of another primary antibody, in some cases, even from the same animal species, for double labeling both light and electron microscopically.[1,22] In practice, autometallographic solutions are somewhat light-sensitive. Silver grains in the solution are captured and non-specific argyrophilic sites may appear. This causes background staining through self-nucleation, leading to a time limit of specific intensification.

The light sensitivity is primarily dependent on type and concentration of the silver source, the amount of reducing agent, the presence of protecting colloid, (e.g. gum acacia), and the temperature.[5,6] Specific and non-specific reactions appear fastest in silver nitrate and more slowly in the lactate. The use of silver acetate, as the most moderately dissociating silver salt, provides an easily controlled reaction. Unlike with silver lactates, and in particular with the nitrate, there is no dire need for light protection during development when the acetate is used.[23] The protecting colloid can strongly moderate the speed of development, but in our studies, its effect in preventing background staining was not sufficient to justify its utilization.[9,10] To attain the highest quality IGSS signal, the following steps are important: (1) application of an oxidizing agent such as Lugol's iodine, before the immunoreaction, (2) blocking of the non-specific protein binding sites of the tissues between the incubation steps with normal serum from the species providing the secondary antibody, (3) use of a high salt concentration buffer for washing, and (4) coating the sections with 0.1-0.5% gelatin before silver amplification may have significance in attainment of the highest quality IGSS signal.[5,6] Moderate non-specific background staining can be eliminated against the strong specific signal by immersion of the IGSS-treated sections in a mixture of 0.3 ml of 7.5% potassium ferricyanide and 1.2 ml of 20% sodium thiosulfate diluted to 60 ml with distilled water (Farmer's solution).[5]

IGSS in Ultrastructural Antigen Detection

Soon after its description for light microscopy, IGSS was also introduced for ultrastructural antigen detection by Van Den Pol,[8] who applied it in free-floating vibratome sections and for double labeling.[24] Subsequently, Lackie et al.[7] compared the sensitivities and studied the kinetics of silver precipitation of the reactions made with anti-rabbit immunoglobulins adsorbed on 5, 15 and 40 nm colloidal gold. Probably due to its better penetration and lower steric hindrance on the antigenic sites, 5 nm gold immunoconjugate was found to give the highest labelling density, an observation later confirmed by other laboratories.[16,25,26] Amplification of the 5 nm particles, either with a silver nitrate or silver lactate-containing developer, resulted in strong reactions of sharp contrast, but heterodispersed particles. Recently, ultra small gold particles (0.8-1.4 nm diameter) adsorbed to streptavidin, immunoglobulin or protein A have become commercially available from several companies. Such reagents have been successfully applied for the detection of intracellular antigens even in pre-embedding methods, due to the good penetration of the conjugates.[18] Since the direct visualization of ultrasmall particles is almost impossible even at the ultrastructural level, silver amplification is inevitable. One nm gold-streptavidin has been utilized with a novel neutral pH silver developer for the detection of synaptic vesicle and cytoplasmic proteins in cultured neuron cells.[18] In the hands of Lah et al., the neutral pH developer gave more reliable results and improved ultrastructural morphology than the methods of Danscher[2] and that involving the IntenSE M (Kit from Amersham, see Appendix). One nm gold adsorbed goat-anti-mouse immunoglobulin, but none of those reagents with larger diameters, were found to penetrate even through the nuclear envelope. It provided a strong labeling of nuclear matrix proteins with heterogeneous particle size when a silver lactate amplification step was additionally used.[27] A sensitive localization of anti-gamma-glutamyltranspeptidase with a 1 nm gold adsorbed goat-anti-rabbit immuno-globulin conjugate was performed in rat lung epithelial cells by the application of another commercial silver enhancement kit (Biocell, see Appendix), although the final particle size was also rather heterogeneous with that kit.[28]

There are fields in ultrastructural pre-embedding immunolabelling where penetration problems do not seriously affect the quality of the immunolabelling. Despite this, immunogold reagents with small gold particle sizes (1 nm and 5 nm) are preferred to achieve the strongest possible immunolabelling under the given conditions. Antigens on ultrathin cryosections are readily available - at least at the surface - even when using conjugates with larger gold particle diameters.[29,30] Furthermore, most of the antigens residing in the cytoplasmic membrane (e.g., lymphocyte differentiation antigens), are accessible for large gold conjugates. However, if the antigenic determinant is located within the internal part of the membrane, 5 nm or even 1 nm gold conjugates can be used with the greatest of success.[17]

For post-embedding "on grid" application, antibodies recognizing their epitopes after routine tissue fixation and embedding may be of potential value. Tissues embedded in epoxide, (e.g. Epon 812, Dnoupan ACM) or acrylate (e.g. LR-White, LR-Gold, Lowicryl K4M and K11M) resins have been subjected to silver-enhanced immunogold detection.[9,10,31-33] Hydrophobic epoxy resin sections have to be etched to allow the access of antibodies and gold conjugates to the antigen in the surface region. In a study where we used 5 nm gold immunoconjugate, longer silver enhancement of the SR-Ca^{2+} ATPase reaction revealed the entire SR-membrane system in a rat muscle as observed at low magnification, while short amplification allowed for a better correlation of the enzyme localization between the myofibrils with the delicate ultrastructure.[9] In our laboratory, relatively mild fixation and silver amplification of the 5 nm gold signal allowed the sensitive detection of several intracellular, extracellular and bacterial antigens in tissues embedded in epoxy or LR-White resins.[10] Most recently, IGSS with 1 nm gold-streptavidin was found to be very effective for the demonstration of intra- and extracellular antigens in cryofixed, freeze-substituted and lowycril K11M embedded skin tissues.[33]

IGSS in Ultrastructural Double Labelling

In the early stages, IGSS was combined with a DAB/immunoperoxidase method for ultrastructural pre-embedding double labeling of two neurotransmitters in the rat hypothalamus, with primary antibodies of different animal origins.[8,16,24] Since silver precipitation completely hides the protein reagents applied in its sequence, double labeling with the same immunoconjugate without unwanted side-bindings can also be performed. Bienz et al.[22] utilized this principle first for the "on grid" detection of different poliovirus proteins in infected cells. The use of fine grain photographic developer (Agfa-Gevaert) and photoemulsion (Ilford L4) resulted in very uniform particles, allowing easy differentiation between the two particle sizes. Since then, several other protocols have been applied for double labeling on the same basis, to analyze questions of protein co-localization in the same cells or compartments. Pre-embedding double-labeling with monoclonal antibodies revealed the progesterone receptor in the cell nuclei and heat-shock protein associated with the unoccupied steroid receptors in the cytoplasm of chick oviduct cells.[34] Five nm gold adsorbed to streptavidin was silver-amplified and the same conjugate was subsequently used without enhancement.

An alkaline developer applied for 5-10 min resulted in particles of irregular shape and heterogeneous size of about 30-50 nm.[35] In a combined pre- and post-embedding technique, protein A coupled with 15 nm colloidal gold was used to reveal two antigens in the same cells.[36] A cell surface antigen was detected on intact cells by means of silver amplification. This was followed by the embedding, cutting and immunostaining of the same cells for an intracellular antigen with the same protein A conjugate, but without silver enhancement.

Another method used biotinylated lectins and antibodies for post-embedding double detection in LR-Gold resin-embedded human gastric parietal cells, using streptavidin-15 nm colloidal gold conjugate.[15] The gold-silver signal of the method carried out first, enhanced with the silver lactate method[2] without protecting colloid, was well distinguishable from the unamplified signal. Biotinylated lectins were suggested to be versatile markers when detected with streptavidin-15 nm colloidal gold. They were found to provide better results than the use of lectin-gold or antibody-gold conjugates. The same principle could be used in the combined pre- and post-embedding lectin-antibody double labeling of murine fibroblasts, but permeabilization with 0.1% saponin was necessary for a better conjugate penetration.

Two-sided ultrastructural double labeling with the same protein A-10 nm gold conjugate and silver amplification with the IntenSE M on one side, was performed in LR-White-embedded human pituitary sections.[32] In this study, TSH and FSH hormones could be detected within the same dense core endocrine granules.

The double labeling methods noted above may help to reduce the artefactual variations resulting from different labeling characteristics of gold conjugates with different sizes, and may, therefore, permit a better comparative quantification of the antigens.

IGSS in Scanning and Photoelectron Microscopy

Ultrastructurally, immunostaining results achieved with the IGSS methods on cell surfaces can also be analyzed after UV stimulation because of the strong electron emission of the particles (photoelectron microscopy),[37] and in the backscattered electron imaging (BEI) mode of the scanning electron microscope.[11,38,39] In photoelectron microscopy, autometallographic amplification of the 6 nm gold label dramatically improved visualization of the fibronectin reaction on human fibroblast cell surfaces.[37] This effect was probably due to better access of smaller particles to the antigen, and an improved electron emission of the silver coating in comparison with the protein coating of the unenhanced label.

The same advantage could be utilized in the scanning electron microscopic analysis of the IgE receptor distribution on mast cells, when the hapten-gold signal was silver-enhanced.[38] The application of 15 nm and 40 nm immunogold reagents with a subsequent silver amplification for the double surface labeling of T and B lymphocyte-associated antibodies (CD3, and CD20), allowed the simultaneous observation of the surface characteristics and antigen profiles of the different lymphocyte types in the same microscopic field.[39]

Multiple Immunolabelling with IGSS and Immunoenzyme Methods at the Light Microscope Level

Detection of Antigens Situated in Different Cell Types

At the beginning of the eighties, some attempts were made to use immunocolloids with the peroxidase-antiperoxidase method (PAP), or protein A-gold with protein A-silver without signal post-intensification for sequential double antigen detection of endocrine cells.[40,41] These methods did not attain widespread acceptance. Despite the use of high concentrations of the colloids, only low signal levels were obtained, and unwanted bindings between the different sequences could not be fully excluded.

Intensification of the gold signal by autometallography provided the key for resolution of these problems, as already demonstrated in the original description of the IGSS method.[1] There, IgA was detected by means of IGSS, with a primary antibody concentration well beyond the sensitivity of the subsequent PAP method revealing IgG-producing plasma cells in human tonsil tissue.

When advantage was taken of the sheltering effect of the gold-silver signal, the B and T lymphocytes in acetone-fixed frozen sections of normal lymph nodes[42] and in methacarn-fixed paraffin sections of tumorous lymph nodes[20] could be clearly detected in one and the same section. The avidin-biotin-peroxidase complex method (ABC) was used to detect the second antigen. Also, T lymphocyte rosettes around Hodgkin's cells with the tumor-specific intracellular and membrane-bound hapten X antigen - previously never shown simultaneously in tissue - could be demonstrated in parallel on sections of Hodgkin's disease.[20]

Since the non-tumorous endocrine pancreas produces the four main hormones (insulin, glucagon, somatostatin and pancreatic polypeptide) in separate cell types, a useful organ was found for studies of sequential multiple immunolabelling where primary antibodies of the same animal source could be combined.[19] Triple, and even quadruple immunostaining combinations could be devised without disturbing color mixing, despite three of the four primary antibodies used were of the same species origin.[19,43] Multistained sections offered the possibility of rapid visual checking of cell types in normal Langerhans islets, in ductoendocrine proliferation and in other unusually arranged endocrine cell accumulations associated with chronic pancreatitis.[43]

The IGSS method has also been utilized for the detection of carbohydrates by biotinylated lectin in combination with immunoenzyme methods for the simultaneous identification of different nephron segments.[14] Sequential use of the same protein A-gold conjugate and IGSS methods may sometimes give distinguishable black and brown signals, provided the second development is not complete.[44] The combination of IGSS with an immunofluorescence method permitted a fine correlation of vasoactive intestinal polypeptide (VIP)-reactive nerves with VIP receptors in the human colonic mucosa.[45]

Besides the brown-black color, the IGSS reaction product can be developed to be cyan- or magenta-like colors, by borrowing a developer from color photography.[46] These products are also applicable in combinations with immunoenzyme methods, although this approach does not appear to have too much merit in comparison with the others above.

Detection of Antigens within the Same Cells

In limited cases, when two antigens occupy separate parts of the same cells, such as the nucleus, the cytoplasm, or the cell membrane, careful application of two antibodies from the same animal species with separate detection systems can be used sequentially for double labelling, particularly when IGSS is in the pair. An example is the identification of the virus-harboring cell type by the detection of intracellular herpes virus antigen with the ABC-peroxidase-DAB method and cell surface IgG with the IGSS method.[47]

In all other cases, the possibility of unwanted side-bindings between antibody layers is to be eliminated. To date, two useful ways have been found to do this. The first is the choice of non-crossreacting antibodies obtained from different species. In this way, intermediate filament co-expression, such as desmin and vimentin in muscle tumors, or GFAP and vimentin in glial tumors, could be directly established with a combination of the IGSS and ABC methods.[20,48] The other principle involves the use of two directly-labeled antibodies from the same species, or the combination of an unlabeled and a labeled antibody from the same species. Gillitzer *et al.*[49] used unlabeled and FITC-conjugated mouse monoclonal antibodies specific for lymphocyte differentiation antigens in skin sections of patients with diseases involving an immunological background. The unlabeled mouse monoclonal antibody was detected with biotinylated second layer antibodies against mouse immunoglobulins followed by streptavidin-gold.

The remaining free binding sites of the biotinylated second layer were saturated with 10% normal mouse serum. This was followed by a second immunolabelling with FITC-conjugated mouse monoclonal primary antibody, rabbit antibodies against FITC and peroxidase- or alkaline phosphatase-conjugated antibodies against rabbit immunoglobulins. The double labeling was most successful when the reaction products were developed after the full treatment with both immunological sequences. Silver enhancement preceded the enzyme developments. Discernible mixed cell-surface labeling of lymphocyte subpopulation-restricted antigens (CD4+ and CD8+ T-cells) and activation antigens (HLA-DR) were obtained in some cells. The same approach proved to be generally applicable, e.g. for the detection of vimentin and cytokeratin intermediate filaments within the same adenocarcinoma cells in a Grawitz tumor.[21]

Combination of *In Situ* Hybridization and Immunocytochemistry at the Light Microscope Level

The most recent field where the silver-amplified gold label seems to be of promise is non-radioactive *in situ* hybridization. This allows the detection of specific RNA or DNA sequences in tissue sections without the hazard of radioactive contamination.[50] *In situ* hybridization, combined with immunocytochemistry, may allow the correlation of the genetic information or the presence of viral infection with the translated protein products in the cells. Hybridization probes labeled with biotin, digoxigenin, or FITC, hybridizing to specific tissue-bound or viral DNA or RNA regions, can be the targets for immunoglobulin-gold or streptavidin-gold labeling with subsequent silver amplification. Because of the predominantly nuclear localization of most DNA sequences to be detected, immunogold reagents with gold particles of 1 to 5 nm in diameter are preferred. These may penetrate through both the cytoplasmic membrane and the nuclear envelope.[51,52] In some cases, even 10 nm gold containing conjugates can also be used as shown later in this chapter.

IGSS for *in situ* hybridization has been utilized to reveal human cytomegalovirus (CMV) DNA. Combined with the immuno-beta-galactosidase method, it has been applied to demonstrate either an immediate early antigen of CMV, or the vimentin expression of virus-infected cells and inclusion bodies in the adrenal gland of an AIDS patient.[21] In another case with a triple staining method, Jacob-Creutzfeldt virus (JCV) DNA was established to be contained in the carbonic anhydrase-immuno-reactive oligodendrocytes, but not in GFAP-immunoreactive astrocytes in an AIDS case.[53] In that study, visualization of the gold-silver precipitate was facilitated with bright-field epipolarization microscopy.

OPTIMAL PROCEDURE - TISSUE HANDLING

For ultrastructural immunostaining, IGSS can be used as a pre-embedding procedure in unfixed, or weakly fixed cell suspensions,[17] in ultrathin cryosections[29,30] and in fixed and cryoprotected vibratome sections.[8,16,18,24] It can also be used as a post-embedding procedure in thin sections collected on grids[9,10,22,33] or in re-embedded immunostained semithin sections.[7] In all of these cases, glutaraldehyde, (para)formaldehyde (or mixtures of both), or periodate-lysine-paraformaldehyde (PLP) can be used as fixative. In pre-embedding procedures, in order to detect intracellular antigens, the use of detergents such as Tween-20 or -80, or Triton-X-100, or freezing and thawing of the sections are necessary to improve the poor penetration of the immunogold reagents through the plasmalemma and intracellular membranes.[16,18] IGSS is particularly useful for the detection of cell membrane antigens, where penetration is not required. By following the method described by Otsuki *et al.*,[17] we have used IGSS for the detection of a lymphocytic cell membrane antigen in a peripheral mononuclear cell suspension from mouse to demonstrate this ability of IGSS (*Fig. 1*).

As a routine post-embedding immunocytochemical procedure, the use of a mixture of 4% paraformaldehyde and 0.1% glutaraldehyde assures immunostaining of a variety of antigens while providing an acceptable preservation of ultrastructure. It is, therefore, recommended for general use.[10] In our laboratory, we usually take tissue specimens measuring 1 mm[3,] fixed as noted above and embed them routinely in plastic resins by polymerizing Durcupan ACM resin at 56°C or LR-White at 52°C for 24 hours. We cut ultrathin sections with a yellow to gold interference color from resin-embedded blocks and place them on Formvar-coated Ni or Au grids. Copper grids are not suitable for IGSS, since they are corroded by the immunogold labelling and acidic silver developer. Etching of the epoxy sections with an oxidizing agent (e.g. hydrogen peroxide, sodium methoxide, or sodium ethoxide) may increase the hydrophilic character in the surface region.[54] Since tissue-bound osmium can reduce silver ions, osmicated sections are not applicable for IGSS reaction, even after partial elimination of osmium by sodium metaperiodate oxidation.[55] The permeabilization of the epoxy sections by etching, the long incubation, the frequent handling of the grids with forceps, and the acidic pH of the developer can explain the need for Formvar support for the sections and the application of not less than 100 mesh grids (in our work 200 mesh is the most frequent).

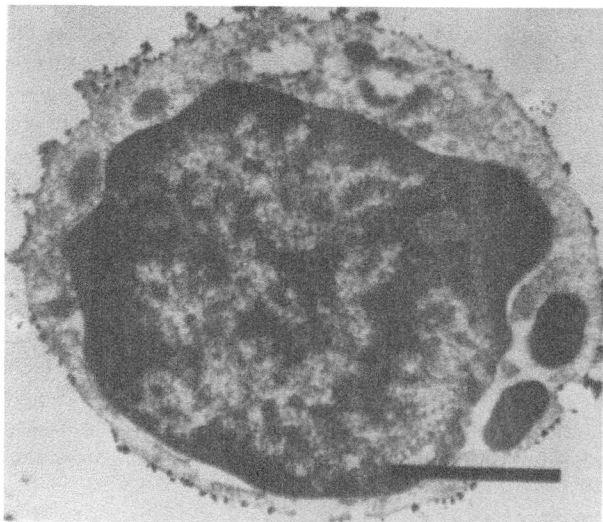

Figure 1. Mouse peripheral blood mononuclear cell immunostained for immune-associated (Ia) antigen with GAM-G5 immunogold reagent and silver enhancement for 3 min with the nitrate developer. Note the high density of heterodisperse particles on the cell surface. Bar: 2.0 μm

In light microscopic multiple labeling, the conditions appropriate for the IGSS method usually allow any immunoenzyme techniques to be used. IGSS has been used with success in tissues fixed by means of a wide variety of fixatives including 4% formaldehyde, mixtures of paraformaldehyde and a low concentration of glutaraldehyde, Bouin's solution, formol sublimate and methacarn.[1,20,23] Pretreatment of the sections with Lugol's iodine and sodium thiosulfate is inevitable for elimination of the remaining mercury pigments from sublimate-containing fixatives. Disregarding the fixation used, this treatment has also been very useful in facilitating immunogold binding. The basis of this effect is not yet exactly understood.[1,5,6] The effect must be connected with the improved penetration of the gold conjugate by "loosening" the cell membranes probably as a result of some splicing of alkanes in phospholipids.[54] Lugol's iodine treatment has no significant effect in immunoenzyme methods where penetration is usually not a problem. Tissues fixed in methacarn, a precipitating fixative which eliminates most of the lipids from the membranes, usually permits a stronger reaction with IGSS than do additive aldehydes. This occurs even when the reaction intensity with immunoenzyme labeling is the same (T. Krenács, unpublished observation).

Post-embedding Immunoelectron Microscopy with IGSS

From tests under different conditions, the protocol below is recommended as that providing the best results. All steps can be performed in a Petri-dish covered with Parafilm, containing a small pot filled with distilled water. The grids are placed with the sections facing down on a drop of reagent (i.e., floating).

Protocol for IGSS on Grids

(1) Put the grids on TBS (0.1 M Tris-buffered saline, pH 7.6) and allow to hydrate for 1 hour.

(2) Perform etching with 5% hydrogen peroxide for 10 min or with a mixture of equal parts of 1 M sodium hydroxide and concentrated methanol for 5 min. (Harsh etching usually has no significant benefit in sections embedded in LR-White, but it may destroy the structure.)

(3) Wash the grids by immersing them in one drop of distilled water (immersion) for 3x1 min, and then repeat this with TBS for 3x1 min.

(4) Saturate non-specific protein binding sites of the sections with TBS containing 10% normal goat serum (NGS), 1% bovine serum albumin (BSA) and 0.1% sodium azide for 30 min.

(5) Put the grids directly onto the drops of the appropriate primary antibodies. Incubate at 4°C overnight by maintaining the Petri dish at room temperature for 30 min before and after the incubation period. Rabbit primary antibodies and the following tissues were used in this study: anti-SR-Ca^{2+} ATPase (1:100) made as described elsewhere,[9] in rat fast muscle; anti-glial fibrillary acidic protein (GFAP, 1:300) and anti-ubiquitin (1:100) in astrocytomas; anti-IgA (1:800) in IgA nephritis; and anti-glucagon (1:200) in human endocrine pancreas. As a negative control, a section was treated with normal rabbit serum diluted 1:100 instead of specific primary antibody.

(6) Wash in TBS for 6x1 min. as in step 3.

(7) Incubate the grids again on drops of normal serum as in step 4 for 15 min.

(8) Put the grids directly on drops of goat anti-rabbit Ig adsorbed to 10 nm colloidal gold (GAR-G10, 1:20), or on rat-anti-rabbit Ig adsorbed on 5 nm gold as described elsewhere[9] (RAR-G5, 1:10), and incubate for 1 hour.

(9) Wash by immersing the grids in TBS for 4x1 min, and fix the reaction product on the specific binding sites with 2% glutaraldehyde for 5 min.

(10) Wash by immersion in TBS for 3x1 min. and in double distilled water for 6x1 min.

(11) Silver acetate autometallography according to Hacker et al.:[23] Dissolve: (a) 50 mg hydroquinone in 50 ml 0.1 M citrate buffer, pH 3.8; and (b) 100 mg silver acetate in 50 ml double distilled water. Use a magnetic stirrer for about 15 min in both solutions. Mix equal amounts of solutions (a) and (b) just before use (Prepare a proportionally smaller amount for electron microscopic IGSS). Put the grids on one drop of the developer.

233

After 2-4 min, stop the reaction by putting the grids on a drop of a photographic fixer diluted 1:10 (e.g. Agefix) or of 5% sodium thiosulfate for 2 min. (Citrate buffer: dissolve 2.56 g citric acid and 2.36 g trisodium citrate in 50 ml double distilled water. Adjust pH to 3.8 just before use. The buffer can be stored at 4°C for about one month.) In our laboratory, the developer used formerly contained 20 mg silver nitrate in 10 ml double distilled water instead of silver acetate.

(12)   Rinse gently by jet-washing and wash with double distilled water for 3x1 min.

(13)   In the case of double labeling, an additional immunoreaction can be carried out, e.g., for ubiquitin (1:200) by repeating steps 4-10 without silver amplification of the gold label of the second immunostaining.

(14)   Counterstain the grids with 1% uranyl acetate and lead citrate for about 80% of the duration used for routine transmission electron microscopy.

Simultaneous Double Labeling with the IGSS and Indirect Immunoalkaline Phosphatase (IAP) and Peroxidase (IPO) Methods at Light Microscopic Level

(1)   Cut 5 um thick paraffin sections (in our study, all the tissues were fixed in 4% formaldehyde). Dewax in xylene and rehydrate in a descending series of ethanol and distilled water.

(2)   Immerse the sections in Lugol's iodine (1% iodine and 2% potassium iodide in distilled water) for 5 min, wash in water (1 min) and then immerse in 2.5% aqueous sodium thiosulfate until decoloration. Wash in tap water.

(3)   Block endogenous peroxidase if peroxidase is to be used as a label in the pair, with methanol containing 0.5% hydrogen peroxide for 30 min, and wash in tap water.

(4)   Wipe and label the glass around the sections with a wax pen (DAKO pen).

(5)   Treat the sections with TBS/NGS/BSA/azide for 30 min as in step 4 of the electron microscopic protocol from this chapter.

(6)   After shaking off normal serum, incubate with mixtures of non-crossreactive primary antibodies from different species, diluted in TBS/BSA/NGS (see step 5) for at least 60 min at room temperature, or overnight at 4°C.

Our study involved the application of mixtures of: (A) rabbit anti-somatostatin (1:400) with guinea-pig anti-insulin (1:400) in sections of human pancreas; (B) monoclonal mouse anti-UCHL-1 (1:100) with rabbit anti-anti-cyto-keratin (1:100), in sections of a thymoma; and (C) monoclonal mouse anti-vimentin (1:50), or anti-p53 (1:50) with polyclonal rabbit anti-desmin (1:100), in sections of a pleomorph rhabdomyosarcoma.

(7)   Wash with TBS containing 0.05% Tween 20 for 3x5 min.

(8)   Treat the sections with the labeled second layer antibodies diluted in 0.1 M Tris-HCl (pH 8.2) containing 1% BSA, for 60 min. We have used mixtures of: (A) GAR-G10 (1:20) and peroxidase-conjugated rabbit anti-guinea pig Ig (RAGP-PO, 1:100); (B) GAR-G10 (1:20) and alkaline phosphatase-conjugated goat anti-mouse Ig (GAM-AP, 1:100); (C) goat anti-mouse Ig-coated 10 nm colloidal gold (GAM-G10) and peroxidase-conjugated goat anti-rabbit Ig (GAR-PO, 1:100).

(9)   Wash with TBS containing 0.05% Tween 20 for 3x5min.

(10)   Postfixation with 1% glutaraldehyde in TBS for 1 min. This step somewhat decreases peroxidase and alkaline phosphatase activities. It should be used only after antibodies with low affinities and is not necessary for all batches of immunogold-reagents.

(11) Wash twice with TBS, rinse with double distilled water for 3x30 sec., and also with double distilled water for 3x3 min.

(12) Immerse sections in the same silver acetate developer[23] as in step 11 of the immunoelectron microscopic protocol. Use infrared safelight or shield the staining-jar containing the sections from daylight by putting it into a cupboard, or by covering. After 5-6 min, check the development intensity in a microscope.

13) Wash with TBS for 3x3 min, and immerse the sections into the same buffer as to be used for the enzyme development.

(14) Develop the enzyme markers by immersing the sections in one of the following incubation media.

(I) Peroxidase: (for A and C) Dissolve 20 mg 3,3'-diaminobenzidine tetrahydrochloride (DAB) in 100 ml 0.05 M Tris-HCl buffer, pH 7.6, and add 100 µl hydrogen peroxide just before use. A brown product is formed.

(II) Alkaline phosphatase:(for B)(a) Dissolve 2 mg naphthol AS-MX phosphate (SIGMA) in 200 µl, N-dimethylformamide, add 9.8 ml of 0.1 M Tris-HCl buffer (pH 8.2) and 1 drop of 1 M levamisole. This stock solution, without levamisole, can be stored frozen in ten times concentrated portions for some months. Mix 1 drop of 4% New Fuchsin with 1 drop of 4% sodium nitrite; after 2 min, mix this with 10 ml of the above stock solution. Filter the mixture onto the sections.

(b) Dissolve 10 mg Fast Red TR in 10 ml stock solution and filter the mixture onto the sections. Ready to use substrate/chromogen systems of the same kind (produced by DAKO) can shorten the performance and help standardization.

(15) Wash the sections with the same buffer as used for the enzyme development and then with distilled water.

(16) Counterstaining is possible with Mayer's Hematoxylin. As a negative control, the cross-reactivities of the secondary antibodies were checked by exchanging them in single staining methods. No cross-reaction was found, either between anti-rabbit Ig and mouse monoclonal antibodies or between anti-mouse Igs and rabbit primary antibodies.

Following the immunostaining for insulin and somatostatin, pancreas sections can be further treated with rabbit anti-glucagon (1:400) and alkaline phosphatase conjugated goat anti-rabbit Ig (GAR-AP, 1:100) according to steps 6-14, but excluding step 12, resulting in a triple immunolabelling. Additional double immunostainings made in two more cases, also follow the above protocol, except that the methods are compiled sequentially. In pancreas sections, glucagon (IPO/DAB) with somatostatin (IGSS), and in a B cell immunoblastoma, kappa (IGSS) and lambda (IAP/Fast Red TR) immunoglobulin light chains can be detected one by one in this sequence. Rabbit anti-kappa and lambda light chain Ig are diluted 1:1000.

In situ Hybridization with a FITC-labeled DNA Oligoprobe and the IGSS Method, Combined with Immunocytochemistry at Light Microscopic Level

Use 3-aminopropyltriethoxy-silane(APES)-coated glass slides for mounting the sections (see Appendix). (APES coating: clean the glass slides by immersion in ethanol for 5 min. After drying, immerse the slides in 2% APES made with acetone for 5 min. Rinse the sections briefly in distilled water and dry at 60°C overnight). Follow steps 1, 2 and 4 of the simultaneous immuno-labelling protocol above. (Endogenous peroxidase blocking can be omitted.)

(5) After washing with 0.05 M Tris-HCl buffer, pH 7.6, pretreat the sections with 0.1% Protease XIV for 10-20 min.

(6) Wash again with Tris-HCl containing 0.1% Tween-20, and then with Tris-HCl containing 30% ion-free formamide.

(7) Treat with a prehybridization buffer containing 30% formamide, 10% dextran sulfate, 0.1% sodium pyrophosphate, 0.2% polyvinylpyrrolidone and 5 mM disodium-EDTA in 0.05 M Tris-HCl, pH 7.6, for 30 min.

(8) Hybridization with 10-20 μl of the concentrated FITC-labeled EBV (EBER) DNA oligoprobe. Cover the sections with silicone-coated coverslips, and incubate overnight at 37°C. (Preparation of coverslips: clean coverslips in 96% ethanol, then immerse them in 2% silicone for 5 min. Rinse briefly in distilled water and dry at 60°C overnight.) In our study, as a negative control, one section was treated with the prehybridization buffer alone under the same conditions as those incubated with the probe.

(9) Soak the coverslips off by washing with Tris-HCl containing 30% ion-free formamide and immerse the sections in the same mixture for 2x5 min. Wash with Tris/Tween-20 for 3x5 min.

(10) Incubate the sections with mouse anti-FITC (1:50) diluted in TBS/Tween containing 1% BSA and 10% NGS for 1 hour.

(11) Wash with TBS/Tween for 3x5 min.

(12) Incubate with GAM-G10 (1:20) diluted in 0.05 M Tris-HCl, pH 8.2, containing 1% BSA for 1 hour.

(13) Wash with TBS for 3x3 min. (Fixation of the reaction product with 1% glutaraldehyde is possible, but not always necessary). Rinse the sections twice in double distilled water and repeat this for 3x3 min.

(14) Develop the sections in the same way as in step 11 of the electron microscopic protocol above.

(15) After washing with distilled water and TBS, perform the immunoreaction, e.g. by using monoclonal mouse anti-LMP1 Ig ("latent membrane protein", 1:100), or monoclonal mouse anti-L-26 (1:100) for 1 hour, then treat the sections with GAM-AP (1:100) for 30 min. Follow the guidance of the simultaneous immunolabelling protocol steps 6-16 by omitting steps 10-12. Develop the alkaline phosphatase reaction with the New Fuchsin chromogen system as described above.

RESULTS AND DISCUSSION

IGSS for Immunoelectron Microscopy

The historical review in the first half of this chapter mentioned some technical aspects at the relevant place. Observations made with the optimal protocols will now be presented and discussed in comparison with the published results of other laboratories.

Although several papers have appeared in which silver intensification was applied ultrastructurally for the better visualization of the immunogold signal, few paid special attention to the silver development step.[7-10,15-18,22,24,27-34,36] A wide range of home-made and commercially available developing media (see the Appendix) have been used with different results. Concerning the homogeneity of the particles, in post-embedding methods, the best results can be expected when a fine grain photoemulsion and developer is applied at photographic safelight illumination.[22] In pre-embedding methods, because of the need for penetration, a photoemulsion can not be applied. Therefore, it must be substituted by one of the media below. The IntenSE series of Amersham Int. is the most well-known of the commercial silver amplification kits. These kits usually work at a neutral or mildly alkaline pH and have the advantage of relative stability for some months at 4°C. However, they provide more heterodisperse particles and sometimes more background than the original silver lactate developer of Danscher.[2,18,28,33]

A developing medium containing silver nitrate instead of lactate also gave a heterogeneous amplification pattern, particularly when used for more than 3 min.[8-10,16,24] Interestingly, with this particular application, none of the developers showed a significant difference in reaction product specificity and appearance if gum acacia as protecting colloid was omitted from the solutions. The only difference which we could observe was the roughly 5 times longer amplification time needed when a solution containing the protecting colloid is applied. Without the protecting colloid, the enhancement is so quick that no light protection is required. The neutral pH developer of Lah et al.[18] may give more reliable results than those above, particularly in pre-embedding methods. Postfixation of the gold label particles onto the sections in 2% glutaraldehyde before silver amplification, may not only improve the immunostaining density by preventing detachment of the gold particles during the development, but sometimes also has a beneficial effect on the ultrastructural morphology as well, particularly when etching is made before the immunostaining (T. Krenács, unpublished observation).

Recent advances in autometallography following the application of silver acetate, have resulted in the impressive detection of heavy metals, and immunohistochemical and *in situ* hybridization signals at a light microscopic level.[23] Silver acetate developer was also successfully used in this study for ultrastructural antigen detection. This provided more uniform particles in kidney biopsy specimens from IgA nephritis than the developer using silver nitrate (*Fig. 2*).

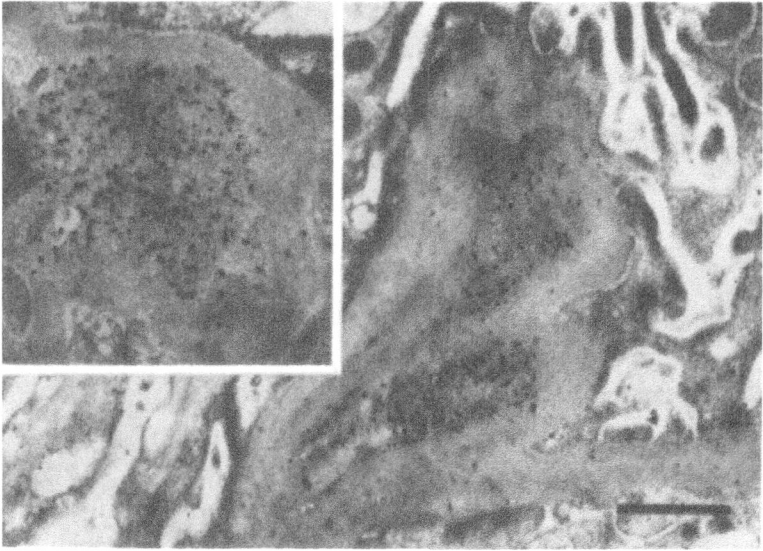

Figure 2. Immunodeposits along the basal lamina in a kidney glomerulus with IgA nephritis. Immunostaining for IgA alpha heavy chains with GAR-G10 immunogold reagent and silver enhancement for 2-3 min. The silver nitrate developer gave more heterogeneous particles (in the middle), than the acetate developer did (inset) showing evenly amplified gold-silver particles. Durcupan embedding. Bar: 1.0 μm

The examination of the reaction with 5 nm gold immunoconjugate was very difficult, even at high instrumental magnification (*Fig. 3*). This was not much better with 10 nm particles. However, the amplification made with the acetate developer substantially improved the recognition of the particles (*Fig. 4*).

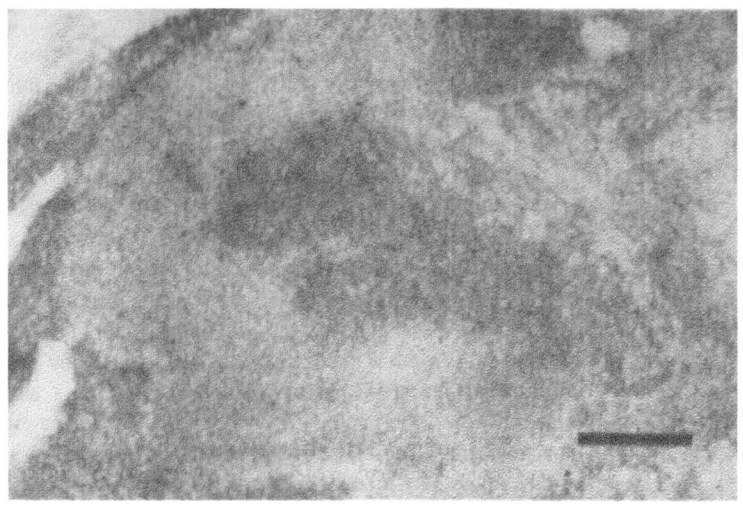

Figure 3. The same IgA nephritis as in *Fig. 2* immunostained for IgA with RAR-G5 immunogold reagent without silver amplification. Despite the high magnification, 5 nm gold particles are hardly seen. Durcupan embedding. Bar: 0.25 µm

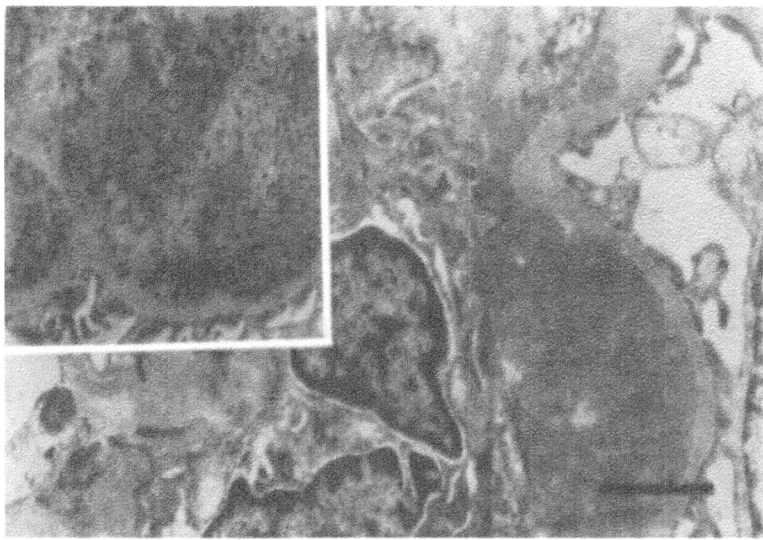

Figure 4. The same IgA nephritis as in Fig. 2 immunostained for IgA with GAR-G10 conjugate without (right) and with 2-3 min. of silver acetate enhancement (inset). Note the improvement in visualization and the uniform particle sizes obtained with the acetate developer. Durcupan embedding. Bar: 2.0 µm

We have studied the GFAP expression characteristics of several glial tumors.[48] Astrocytomas containing abundant bundles of intermediate filaments and electron-dense bodies, "Rosenthal fibers", were selected for further study. When the effect of enhancement duration with the silver acetate amplifier was analyzed,[23] we found quite homogeneous final particle sizes after 2-3 min. These were about 2-3 times the size of the original particles (*Fig. 5*).

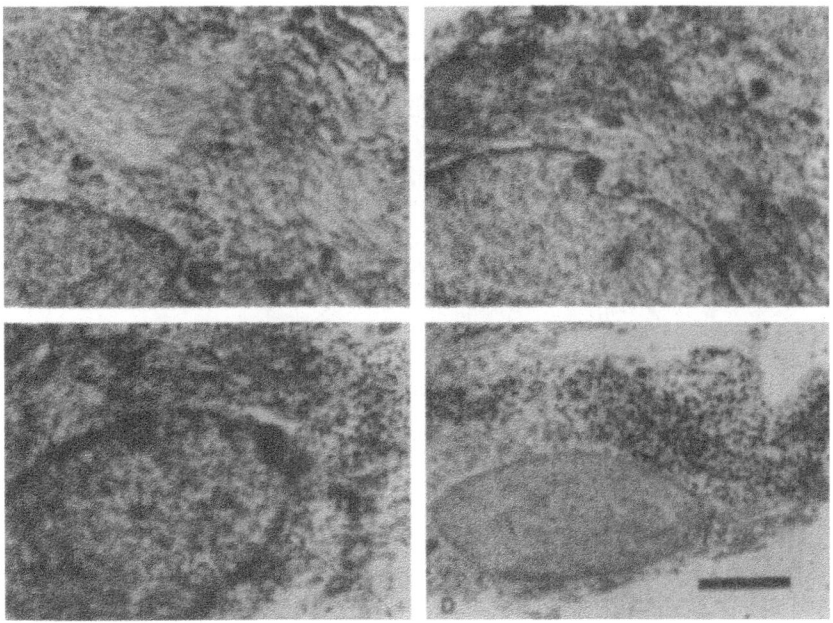

Figure 5. Astrocytoma cells immunostained for GFAP with GAR-G10 immunogold reagent without (A) and with (B-D) silver amplification for 2 (B), 3 (C), and 4 (D) min with the acetate developer. Despite the same particle densities on the intermediate filament bundles, signal recognition is proportional to the duration of the enhancement. Bar: 2.0 μm

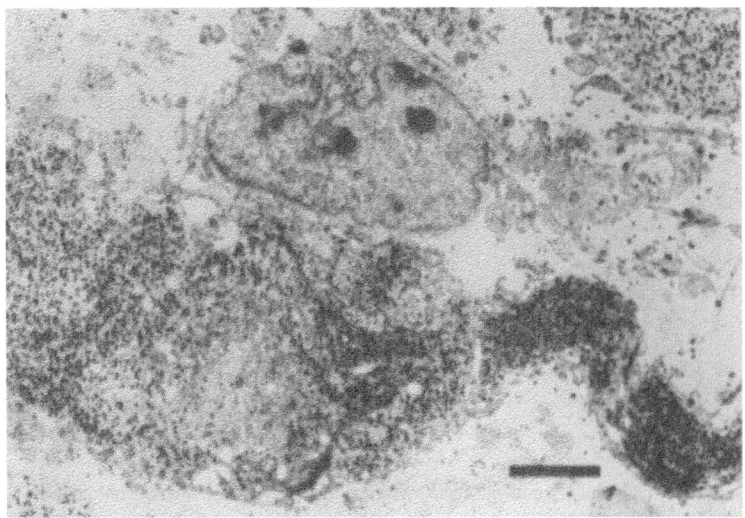

Figure 6. Pilocytic astrocytoma immunostained for GFAP with GAR-G10 immunogold reagent and 4 min of silver enhancement with the acetate developer. The reaction density is proportional to the intermediate filament content within the cell. Although the shape and size of the labeling particles are somewhat irregular, the specificity of the reaction is without any doubt. Bar: 2.5 μm

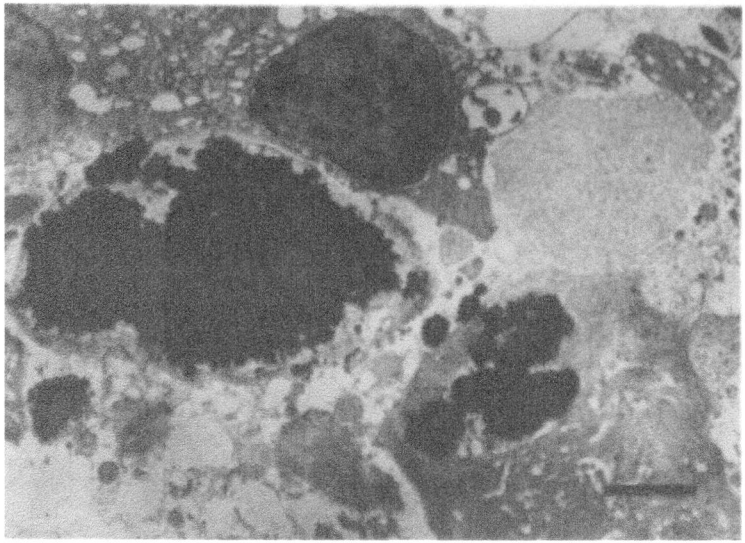

Figure 7. Astrocytoma with electrondense corpuscles called Rosenthal-bodies, immunostained for ubiquitin with GAR-G10 immunogold reagent and 4 min of silver enhancement with the acetate developer. Note the concentration of the particles at the periphery of the larger corpuscle (left). Bar: 2.5 µm

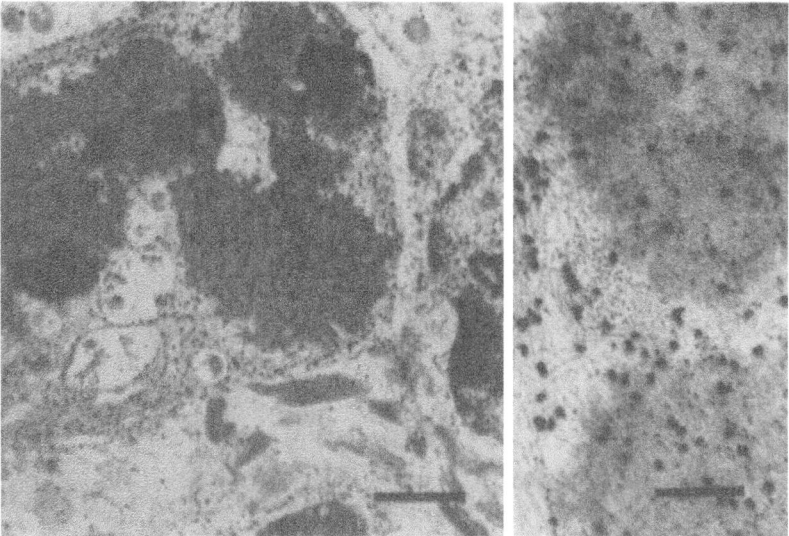

Figure 8. Astrocytoma doubly immunostained with the same GAR-G10 immunogold reagent. GFAP was detected with silver enhancement for 3 min. with the acetate developer (larger particles), followed by the ubiquitin reaction, but without signal amplification (smaller, 10 nm particles). The electrondense corpuscles, Rosenthal bodies express both antigens, while the surrounding intermediate filaments show mainly GFAP reactivity. Note the uniform particle sizes of about 30 nm of the amplified reaction, representing the GFAP (inset). Bar: 1.5 µm; inset: 0.4 µm

Intensification for 4 min made the particles quite irregular in shape and size. This must be due in part to the aggregation of neighboring particles because of the high density of the original signal. Under optimal conditions, intensification of the signals for 3-4 min gave the strongest reaction in astrocytic tumor cells, while the labeled ultrastructure remained discernible (*Fig. 6*). The sharp contrast and the relatively large size of the individual particles allowed an easy assessment of the antigen distribution for ubiquitin within the electrondense structures of labeled cells (*Fig. 7*).

Ubiquitination, which plays a crucial role in the formation of intermediate filament inclusion bodies,[56] was studied in an astrocytic tumor, with double labeling for the cell-type specific GFAP reactivity and ubiquitin (*Fig. 8*). The gold signal of the GFAP reaction was silver-nitrate enhanced for 2-3 min, resulting in a homogeneous particle size of about 30 nm, which could be easily differentiated from the original 10 nm particles used in a subsequent step for the detection of ubiquitin. Since both antigens are detected with the same gold conjugate, this kind of double labeling may allow an adequate quantitative assessment, provided that the first antibodies are of the same affinities and their optimal dilutions used. Also, the better visualization of the larger particles may facilitate computerized image analysis.

When the ultrastructural localization of SR-Ca^{2+} ATPase, an enzyme responsible for muscle relaxation, was studied,[9] we found that gold particles labeling antigens in the hydrophilic LR-White resin needed a longer duration of amplification than those used in epoxy media (*Fig. 9*).[10] Although a comparison has not been made, a similar situation can be supposed for the case of the other hydrophilic embedding resins, LR-Gold, Lowycril K4M and K11M, since better penetration of the gold conjugate into their surface region may result in a somewhat delayed access of the silver ions in comparison with the free particles on the resin surface.

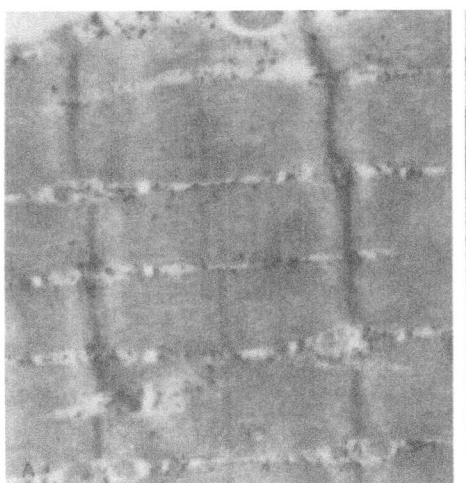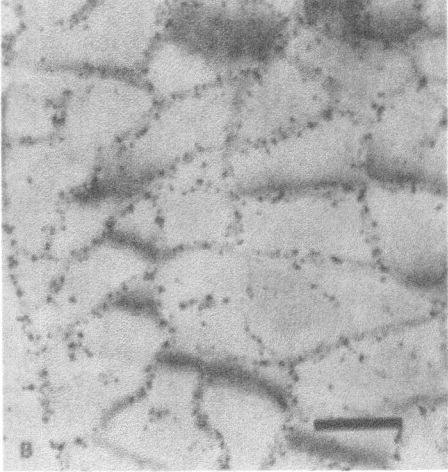

Figure 9. Sarcoplasmic reticulum (SR)-Ca^{2+} ATPase reaction with the RAR-G5 immunogold reagent and silver enhancement in rat gastrocnemius muscle. Application of the nitrate developer for 3 min resulted in smaller and more uniform particles in the section embedded in LR-White (A) than in that embedded in Durcupan ACM (B). Note the dense labeling of the SR-membrane system among the myofibrils in both the longitudinal (A) and cross (B)-sections. Bar: 0.5 μm

Silver enhancement can also be of particular advantage in the detection of endocrine hormones situated in "dense-core" granules.[10] A high reaction product density, detected with 5 nm immunogold marker could only be visualized after silver enhancement, when normal counterstaining with uranyl acetate and lead citrate was applied (*Fig. 10*).

For pre-embedding ultrastructural immunocytochemistry, immunoperoxidase methods, with DAB development alone or with combined DAB/silver development, have been preferred to date.[57] For a selective DAB/silver reaction, the endogenous tissue argyrophilia has to be suppressed, whereas there is no need to do this in the case of gold signal amplification, because of the differences in kinetics of the two types of catalysis.[57] In addition, the diffuse DAB/silver signal may hide labeled structures. A family of gold conjugates with 0.8-1.4 nm particles, available from several companies provides reliable results in pre-embedding detection of intracellular and intranuclear antigens,[18,27,28,33,34] allowing the general acceptance of IGSS in this field too. For this, the recently produced 1.4 nm gold covalently bound to IgG Fab' fragments can be a valuable addition, due to its better stability and electron microscopic resolution, than those of non-covalently bound conjugates.[58]

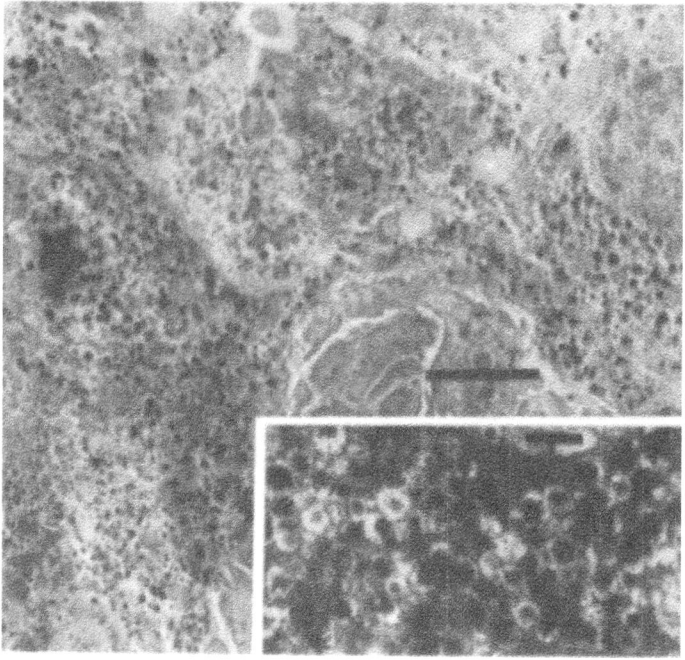

Figure 10. Human pancreatic Langerhans islet embedded in LR-White and Epon 812 (inset), and immunostained for glucagon with RAR G5 immunogold reagent and silver enhancement with the nitrate developer. Despite the dark contrasts of the dense-core granules, the reaction products can be well recognized within them. Bar: 3.0 μm; inset: 0.5 μm

IGSS in Multiple Detection Combinations for Light Microscopy

The IGSS reaction allows facile discrimination of tissue morphology, since routine histological stains or enzyme histochemical reactions can be applied consecutively. Furthermore, the black color and the sharp contrast of the IGSS reaction product provides an excellent color contrast with most of the chromogens usually applied in immunoenzymatic methods, including DAB (brown) and AEC (red-brown) used for peroxidase, Fast Blue BB (blue), Fast Red TR (red) and New Fuchsin (crimson red) used for alkaline phosphatase, and potassium ferro/ferricyanide (blue-green) used for beta-galactosidase development.[14,19-21,42,43,47-49] Most of the combinations can be used with hematoxylin counterstaining. With selected combinations of primary antibodies, detection methods, enzyme labels and chromogens, and with application of the methods in the optimal sequence under the given conditions, reliable triple and even quadruple immunostaining combinations can be devised.[19,20,43,53]

For combined antigen detection, two basic situations must be considered. If the antigens of interest have been established as residing in different cells or in separate parts of the same cells, e.g. the cell membrane, cytoplasm or nucleus, primary antibodies from the same species can be used sequentially.[20] In these cases, the sheltering effect of the silver precipitate of IGSS or that of DAB in immunoperoxidase methods can be exploited.[19,20,43] The methods combined sequentially should be used in the sequence of their sensitivities (IGSS, ABC, PAP or APAAP, indirect PO/AP) to exclude the amplification effect of a sensitive method on the still-forming side-bindings.[43] Chromogens used in immunoperoxidase methods, particularly DAB, can also reduce ionic silver from the IGSS developer[16,35,57] (*Fig. 11A*).

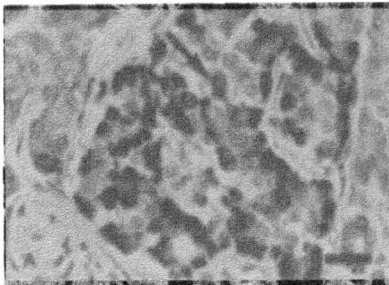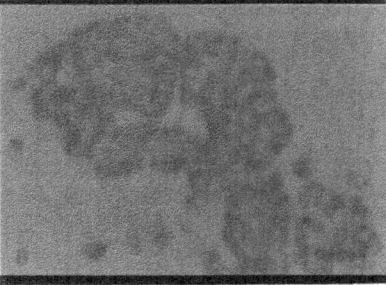

Figure 11. Human pancreatic Langerhans islet cells. (A) Immunostaining for glucagon with IPO (DAB), followed by the IGSS reaction for somatostatin. Note that DAB precipitate reduced also silver ions already at the beginning of the silver enhancement, resulting in grayish shade on brown glucagon cells. Single grayish signal in a cell (arrow) probably indicates somatostatin production. Hematoxylin. Bar: 25 μm (B) Simultaneous immunostaining for somatostatin (IGSS, black), and insulin IPO (DAB, brown), followed by the detection of glucagon (IAP/Fast Red TR, purple). There is no color mixing. Bar: 40 μm

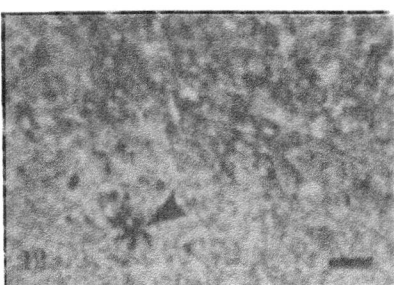

Figure 12. Lymph node with B-cell immunoblastoma sequentially immunostained for kappa (IGSS, black) and lambda (IAP, purple) immunoglobulin light chains. Tumor cells express exclusively kappa light chains in their cytoplasm (top), while a group of reactive plasma cells (arrowhead) of both types are also present. Hematoxylin. Bar: 60 μm

Immunoalkaline phosphatase or beta-galactosidase methods have not been found to interfere with IGSS if used before it.[19,20,49] Nevertheless, in our experience, IGSS as the first method in combinations is to be preferred. In our hands, the sequential use of IGSS and immunoalkaline phosphatase or immunoperoxidase methods provided unequivocal evidence of tumorous transformation, e.g. by establishing monoclonality in a B cell immunoblastoma[59] (*Fig. 12*), or p53 oncoprotein overexpression[60] with the phenotypic antigen expression (*Fig. 14B*) in pathological alterations. If non-crossreacting antibodies are available, they can be used simultaneously, followed by a sequential marker development, the silver amplification being achieved before the enzyme reaction, to shorten the combined staining procedure. Somatostatin and insulin-producing cells in a pancreatic Langerhans islet (*Fig. 11B*) and cytokeratin plus T lymphocyte marker expression in a thymic tumor probably of epithelial origin[61] (*Fig. 13*), were revealed in this way. In *Fig. 11B*, an additional immunostaining for glucagon with immunoalkaline phosphatase method resulted in a selective triple immunolabelling.

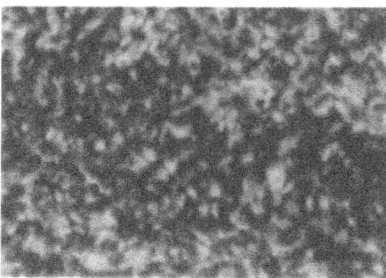

Figure 13. Thymoma, simultaneously immunostained with polyclonal vs - keratin (IGSS, black) for the tumorous epithelial component (larger cells), and with monoclonal UCHL-1 (IAP, purple) for T lymphocytes. Hematoxylin. Bar: 40 μm

When the antigens to be detected are confined to the same structures or cells, or their location is unsure and is to be investigated, non-crossreacting antibody layers originating from different species can be applied simultaneously.[20]

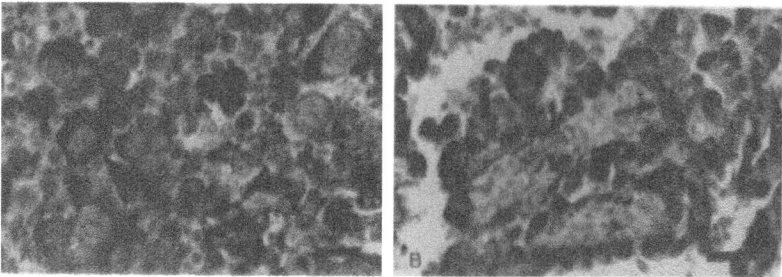

Figure 14. Pleomorph rhabdomyosarcoma. (A) Simultaneous immunostaining for vimentin (IGSS, granular black) and desmin (IPO/DAB,brown) intermediate filaments, and (B) sequential double immunostaining for p53 oncoprotein (IGSS, black) in the cell nuclei and desmin (IPO/DAB, brown), the phenotypic marker of the tumor. Note that even the weak reactions in both cases (arrows) can be easily recognized and their quantity assessed due to the deep color and sharp contrast of the silver intensified gold marker. The vimentin and desmin reaction can clearly be separated in some cells (arrowhead). Hematoxylin. Bars, A: 25 μm; B: 40 μm

Immunogold silver staining can be very useful in both basic situations, but during the development, care should be taken that the strong signal of IGSS not to obscure the next marker if they are not separated within the same cells. When one of the antigens is present in a trace amount, IGSS can be the method of choice for detection due to its high sensitivity (*Fig. 14*). The uneven distribution of vimentin and desmin intermediate filaments co-expressed in the same rhabdomyosarcoma cells[62] can be assessed by this means. Appropriate controls, with secondary antibodies exchanged in single stainings, established the specificity of all simultaneous labelling in this study.

In our experience with IGSS in frozen sections, the non-specific background staining is more difficult to overcome than in embedded tissues. This is probably due to the poorer penetration of the gold conjugates and the higher level of tissue argyrophilia in unembedded versus embedded chemically more uniform tissues.[20] Since anti-immunoglobulin antibodies of high specificities and avidities coupled to enzymes or colloidal gold have recently been available, there is no need to apply three-step PAP or ABC methods to achieve high detection sensitivity. This allows simplification of the double and multiple labeling protocols.

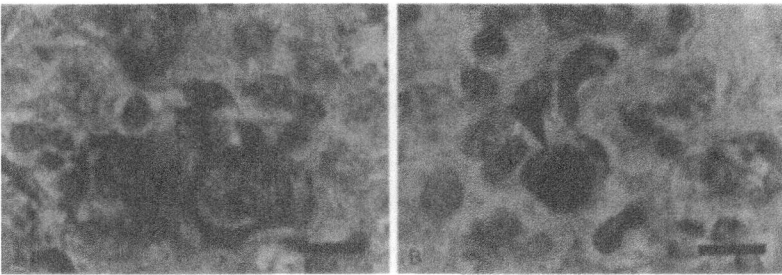

Figure 15. Hodgkin's disease, lymphocytic depletion type. Combination of *in situ* hybridization with EBER 1/DNA oligoprobe for the EBV "early nuclear RNA", detected with IGSS (black); with immunocytochemical reactions for a B lymphocyte-associated antigen (L-26/CD20) (A), and the "latent membrane protein" (LMP) of the virus (B), detected with an IAP method (New Fuchsin, purple). The *in situ* hybridization signals as granular grayish-black masses in some tumor cell nuclei (arrowheads), and the L-26 (A) or LMP (B) reactions are co-localized in some tumor cells. Bars: 15 μm

The utilization of IGSS in non-radioactive *in situ* hybridization with 5 nm and 1 nm conjugates has only recently been described,[51,52] and the combinations of this approach with immunocytochemistry have not been frequently used yet.[21,53] While investigating the role of Epstein-Barr virus (EBV) in the pathogenesis of lymphomas,[63] we have studied the EBV gene expression characteristic for viral latency with a FITC-labeled DNA oligoprobe combined with immunocytochemistry, for identification of the immunophenotype of the cells containing the virus. In Hodgkin's disease, some of the tumor cells expressing the early viral RNA proved to be of B lymphocytic origin (Figure 15A) on the basis of dual labeling, which is supported by the recent literature.[64] In addition, the viral latency was established through the simultaneous presence of the *in situ* hybridization signal in the nuclei and the expression of the "latent membrane protein" (LMP[65]) attached to the cell membrane (*Fig. 15B*) in some Hodgkin's tumor cells. The grayish-black signal of IGSS contrasts well with the purple color of the immunoalkaline phosphatase method. It can be clearly distinguished from the hematoxylin counterstain of the unhybridized cells.

Although 10 nm gold-absorbed anti-mouse Ig antibodies allowed clear visualization of the hybridization signal in some cells, conjugates with 5 nm or 1 nm colloidal gold can be expected to reveal more cells with less RNA copies, because of their easier access to the antigens through the cellular membranes.[7,18,27,34]

CONCLUSION

As demonstrated in this study and several others that have appeared since the introduction of IGSS into affinity cytochemistry, silver-enhanced gold seem to be one of the most versatile markers. It is applicable at both light microscopic and ultrastructural level. Gold markers with small (1-10 nm) particle diameters provide the best results. The particulate nature and possibility of the adjustment of the final particle size of the IGSS reaction can allow the most sensitive visualization of the antigenic sites without obscuring ultrastructural details. Ultrastructurally, the silver amplification step must be stopped just before the neighboring particles join, probably the point at which the reaction becomes visible light microscopically. For light microscopic observations, the enhancement must be continued to obtain compact aggregates.

In post-embedding electron microscopy, the most homogeneous particle size of the amplified reaction, which may be important in quantitative antigen analysis and in double labeling, can be achieved by using either a photoemulsion or silver acetate as a silver source. In ultrastructural pre-embedding studies, the use of 1 nm gold conjugates with a following silver amplification has proved to be a step forward in intracellular antigen detection. The inherent limitations of the post-embedding approach, i.e. the sensitivity of the antigens to the fixation and resin embedding, and also the impeded access of the antibodies and gold conjugates into the resin, can hopefully be limited by the introduction of more and more monoclonal antibodies that recognize aldehyde-fixed epitopes, and of such hydrophilic embedding media, LR-White, LR-Gold and Lowycrils, into the routine practice of immunohistochemical laboratories.

In light microscopic utilization of the technique where deeper penetration through both the sarcolemma and the nuclear envelope (i.e. detection of nuclear antigens, or *in situ* hybridization) is sometimes necessary, penetration is facilitated by alcoholic fixatives versus aldehydes, or in embedded versus frozen tissue sections. Pretreatment of the aldehyde-fixed sections with Lugol's iodine may improve penetration without hindering the subsequent immunolabelling. Light microscopically, the combinations of the IGSS technique with immunoperoxidase, immunoalkaline phosphatase or immuno-beta-galactosidase methods are well established. Multiple labeling of specific tissue targets by means of IGSS and different immunoenzyme methods may reveal topomorphological and morphometric features and may establish paired targets of structural or functional interest in normal and diseased tissues or cells.

In the future, properly used IGSS may be of great potential in the simultaneous examination of oncogene or viral gene transcription in correlation with the presence of translated proteins. This would allow a better understanding of structure-function relationships even at a single cell level. Although the protocols described have been found to be applicable in several other cases, further technical improvements and the standardization of silver autometallography in different embedding media can be expected.

Acknowledgements

We are greatly indebted to Jenö Ormos for advice and to Béla Bozóky, Hirotsugu Uda and Sumiko Tanaka, for collaboration. We are also grateful to DAKO (Glostrup, Denmark) for providing some antibodies and support for this study. Likewise to the SOROS Foundation of Hungary for special support to attend the First International Workshop of Modern Analytical Methods in Histochemistry, Salzburg, Sept. 14-18, 1992, where the results were presented in part. This work was supported by grant OTKA-F5256 from the Hungarian Academy of Sciences.

Appendix: Sources of Antibodies, Reagents and Chemicals Used

		Code
1.	DAKO a/s, Glostrup, Denmark:	

Mouse monoclonal antibodies to:

EBV/latent membrane protein (LMP)	M 897
FITC	M 878
L-26 (CD20)	M 755
p53 protein (DO7)	M 7001
UCHL-1 (CD45R0)	M 742
vimentin (V9)	M 725

Rabbit polyclonal antibodies against:

desmin	A 611
glial fibrillary acidic protein (GFAP)	Z 334
glucagon	A 565
IgA (alpha heavy chains)	A 262
kappa light chains	A 191
ws-keratin	Z 622
lambda light chains	A 193
ubiquitin	Z 458
somatostatin	A 566

Guinea pig polyclonal antibodies to insulin	A 564
FITC-conjugated EBV/EBER oligonucleotides	Y 017

Alkaline phosphatase conjugated-

goat anti-mouse Igs (GAM-AP)	D 486
goat anti-rabbit Igs (GAR-AP)	D 487

10 nm gold adsorbed to:

goat anti-mouse Ig (GAM-G10)	G 441
goat anti-rabbit Ig (GAR-G10)	G 442

Horseradish peroxidase conjugated-

goat anti-rabbit Ig (GAR-PO)	P 448
rabbit anti-guinea pig Ig (RAGu-PO)	P 141

DAB tablets	S 3000
New Fuchsin chromogen/substrate kit	K 698
Fast Red chromogen/substrate kit	K 699

2.	Amersham Int. Aylesbury, Buckinghamshire, England:	

goat anti-mouse immunoglobulins adsorbed to
5 nm gold particles (GAM-G5)   RPN 420

3. Merck, Darmstadt, Germany:

citric acid	247
trisodium citrate-dihydrate	6448
silver nitrate	1512
Lugol's solution	9261
sodium nitrite	6549
dextrane sulfate sodium salt	3093
polyvinylpyrrolidone	7370
hydroquinone	4610

4. Sigma Co., St. Louis, Mo., USA:

bovine serum albumin (BSA)	A 4503
protease XIV (from Streptomyces griseus)	P 5147
naphthol-AS-MX-phosphate sodium salt	N 5000
Fast Red TR salt	F 8764
3-amino-9-ethylcarbazole (AEC)	A 5754
3-aminopropyltriethoxy-silane (APES)	A 3648

5. Fluka, Buchs, Switzerland:

Durcupan ACM	44 610
silver acetate	85 140
glutaraldehyde 25%	49 626

6. Polysciences Inc., Warrington, Pa., USA:
    LR-White (London Res. Co.)   17 411

7. Others
    Agefix (Agfa-Gevaert, Germany)
    Parafilm "M" (American Natl. Can, Grewnwich, Ct.,USA)

Commercial kits for silver enhancement: IntenSE M (Amersham)-Silver enhancement kit (Biocell Research Lab., Cardiff, UK)-Aurion R-Gent (Aurion, Wareningen, The Netherlands)-HQ Silver (Nanoprobes, Inc., Stony Brook, NY, USA)

Ig = Immunoglobulins

References

1. C.S. Holgate, P. Jackson, P. Cowen, C., and Bird, Immunogold-silver staining: new method of immunostaining with enhanced sensitivity *J Histochem Cytochem* **31**:938 (1983).
2. G. Danscher: Histochemical demonstration of heavy metals. A revised version of the sulfide silver method suitable for both light and electron microscopy, *Histochemistry* **71**:1 (1981).
3. M. De Waele, J. De Mey, W. Renmans, C. Labeur, PH. Reynaert, and B. Van Camp, An immunogold-silver staining method for the detection of cell surface antigens in light microscopy. *J Histochem Cytochem* **34**:935 (1986).
4. M. De Waele, J. De Mey, P. Reynaert, M.F. Dehou, W. Gepts, and B. Van Camp, Detection of cell surface antigens in vryostat sections with immunogold-silver staining, *Am J Clin Pathol* **85**:573 (1986).
5. D.R. Springall, G.W. Hacker, L. Grimelius, and J.M. Polak, The potential of the immunogold-silver staining method for paraffin sections, *Histochemistry* **81**:603 (1984).
6. G.W. Hacker, D.R. Springall, S. Van Noorden, A.E. Bishop, L. Grimelius, and J.M. Polak, The

immunogold-silver staining method. A powerful tool in histopathology, *Virchows Arch (Anat Pathol)* **406**:449 (1986).

7. P.M. Lackie, R.J. Hennessy, G.W. Hacker, and J.M. Polak, Investigation of immunogold-silver staining by electron microscopy, *Histochemistry* **83**:545 (1985).

8. A.N. Van Den Pol, Double pre-embedding ultrastructural immunocytochemistry with intensified colloidal gold and peroxidase: Dopamine neurons in the dorsomedial hypothalamus receive GABA synapses, *Soc Neurosci Abstr* **10**:440 (1984).

9. T. Krenács, E. Molnár, E. Dobó, and L. Dux, Fibre typing using sarcoplasmic reticulum Ca-ATP^{2+}ase and myoglobin immunohistochemistry in rat gastrocnemius muscle, *Histochem J* **21**:145 (1989).

10. T. Krenács, B. Iványi, B. Bozóky, Z. Lászik, L. Krenács, Zs. Rázga, and J. Ormos, Postembedding immunoelectron microscopy with immunogold-silver staining (IGSS) in Epon 812, Durcupan ACM and LR-White resin embedded tissues, *J Histotechnol* **14**:75 (1991).

11. L. Scopsi, L.I. Larsson, L. Bastholm, and M. H. Nielsen, Silver-enhanced colloidal gold-probes as markers for scanning electron microscopy. *Histochemistry* **86**:35 (1986).

12. J. Roth and P.U. Heitz, Immunolabeling with the Protein A-gold technique: An overview, *Ultrastruct Pathol* **13**:467 (1989).

13. G. Danscher and J.O.R. Nörgaard, Light microscopic visualization of colloidal gold on resin embedded tissue, *J Histochem Cytochem* **31**:1394 (1983).

14. B. Iványi and T.S. Olsen, Immunohistochemical identification of tubular segments in percutaneous renal biopsies, *Histochemistry* **95**:351 (1991).

15. J.M. Pettitt and D.C. Humphris, Double lectin and immunolabelling for transmission electron microscopy: Pre- and post-embedding application using the biotin-streptavidin system and colloidal gold-silver staining, *Histochem J* **23**:29 (1991).

16. A.N. Van Den Pol, Thyrosine hydroxylase immunoreactive neurons throughout the hypothalamus receive glutamate decarboxylase immunoreactive synapses: A double pre-embedding immunocytochemical study with particulate silver and HRP, *J Neurosci* **6**:877 (1986).

17. Y. Otsuki, L.E. Maxwell, S. Magari, and H. Kubo, Immunogold-silver staining method for light and electron microscopic detection of lymphocyte cell surface antigens with monoclonal antibodies, *J Histochem Cytochem* **38**:1215 (1990).

18. J.J. Lah, D.M. Hayes, and R.W. Burry, A neutral pH silver development method for the visualization of 1 nanometer gold particles in pre-embedding electron microscopic immunocytochemistry, *J Histochem Cytochem* **38**:503 (1990).

19. T. Krenács, Z. Lászik, and E. Dobó, Application of immunogold-silver staining and immunoenzymatic methods in multiple labelling of panreatic Langerhans islet cells. *Acta Histochem* **85**:79 (1989).

20. T. Krenács, L. Krenács, B. Bozóky, and B.Iványi, Double and triple immunocytochemical labelling at the light microscope level in histopathology, *Histochem J* **22**:530 (1990).

21. W. Van Den Brink, C. Van der Loos, H. Volkers, R. Lauwen, F. Van Den Berg, H.J. Houthoff, and P.K. Das, Combined beta-galactosidase and immunogold/silver staining for immunocytochemistry and DNA in situ hybridization, *J Histochem Cytochem* **38**:325 (1990).

22. K. Bienz, D. Egger, and L. Pasamontes, Electron microscopic immunocytochemistry. Silver enhancement of colloidal gold marker allows double labeling with the same primary antibody, *J Histochem Cytochem* **34**:1337 (1986).

23. G.W. Hacker, L. Grimelius, G. Danscher, G. Bernatzky, W. Muss, H. Adam, and J. Thurner, Silver acetate autometallography: An alternative enhancement technique for immunogold-silver staining (IGSS) and silver amplification of gold, silver, mercury and zink in tissues, *J Histotechnol* **11**:213 (1988).

24. A.N. Van Den Pol, Silver-intensified gold and peroxidase as dual ultrastructural immunolabels for pre- and postsynaptic neurotransmitters, *Science (NY)* **228**:332 (1985).

25. J. Gu and M. D'Andrea, Comparison of detecting sensitivities of different sizes of gold particles with electron microscopic immunogold staining using atrial natriuretic peptide in rat atria as a model, *Am J Anat* **185**:264 (1989).

26. S. Yokota, Effect of particle size on labeling density for catalase in protein A-Gold immunocytochemistry, *J Histochem Cytochem* **36**:107 (1988).

27. A.D. Graaf, P.M.P. Van Bergen en Henegouwen, AM.L. Meijne R. Van Driel, and A.J. Verkleij, Ultrastructural localization of nuclear matrix proteins in HeLa cells using silver enhanced ultrasmall gold probes, *J Histochem Cytochem* **39**:1035 (1991).

28. D. Dinsdale, J.A. Green, M.M. Manson, and M. J. Lee, The ultrastructural immunolocalization of

gamma-glutamyltranspeptidase in rat lung in correlation with the histochemical demonstration of enzyme activity, *Histochem J* **24**:144 (1992).

29. A.J. Atherton, P. Monaghan, M.J. Warburton, and B.A. Gusterson, Immunocytochemical localization of the ectoenzyme aminopeptidase N in the human breast, *J Histochem Cytochem* **40**:705 (1992).

30. P. Monaghan and A. Atherton, Immunocytochemistry of cryosections, *in*: Electron Microscopic Immunocytochemistry. Principles and Practice, J.M. Polak and J.V. Priestley eds., Oxford Univ. Press, Oxford p. 123 (1992).

31. M. Slater, Differential silver enhanced double labeling in immunoelectron microscopy, *Biotech Histochem* **66**:153 (1991).

32. I.C. Velde and F.A. Prins, New sensitive light microscopical detection of colloidal gold on ultrathin sections by reflection contrast microscopy. Combination of reflection contrast and electron microscopy in post-embedding immunogold histochemistry, *Histochemistry* **94**:61 (1990).

33. H. Shimizu, A. Ishida-Yamamoto, and R.A.J. Eady, The use of silver-enhanced 1-nm gold probes for light and electron microscopic localization of intra- and extracellular antigens in skin, *J Histochem Cytochem* **40**:883 (1992).

34. A.K. Pekki, Different immunoelectron microscopic locations of progesterone receptor and HSP90 in chick oviduct epithelial cells, *J Histochem Cytochem* **39**:1095 (1991).

35. Zs. Liposits, C. Phelix, and W.K. Paull, Adrenergic innervation of corticotropin releasing factor (CRF)-synthesizing neurons in the hypothalamic paraventricular nucleus of rat, *Histochemistry* **84**:201 (1986).

36. M.G. Zelechowska and R. Mandeville, Immunogold and immunogold/silver staining in the ultrastructural localization of target molecules identified by monoclonal antibodies, *Anticancer Res* **9**:53 (1989).

37. G.B. Birrell, D.L. Habliston, K.K. Hedberg, and O.H. Griffith, Silver enhanced colloidal gold as a cell surface marker for photoelectron microscopy, *J Histochem Cytochem* **34**:339 (1986).

38. R.F. Stump, J.R. Pfeiffer, J. Seagrave, and J.M. Oliver, Mapping of gold-labeled IgE receptors on mast cells by scanning electron microscopy: Receptor distributions revealed by silver enhancement, backscattered electron imaging, and digital image analysis, *J Histochem Cytochem* **36**:493 (1988).

39. E. De Harven, D. Soligo and H. Christensen, Double labelling of cell surface antigens with colloidal gold markers, *Histochem J* **22**:18 (1990).

40. J. Gu, J. De Mey, M. Moeremans and J.M. Polak, Sequential use of the PAP and immunogold staining methods for light microscopical double staining of tissue antigens, *Regul Pept* **11**:365 (1981).

41. J. Roth, Applications of immunocolloids in light microscopy. Preparation of protein A-silver and protein A-gold complexes and their application for localization of single and multiple antigens in paraffin sections, *J Histochem Cytochem* **30**:691 (1982).

42. H. Sako, Y. Nakane, K. Okino, K. Nishihara, M. Kodama, M. Kawata, and H. Yamada, Simultaneous detection of B-cells and T-cells by a double immunohistochemical technique using immunogold-silver staining and avidin-biotin-peroxidase complex method, *Histochemistry* **86**:1 (1986).

43. T. Krenács, E. Dobó, and Z. Lászik, Characteristics of endocrine pancreas in chronic pancreatitis, as revealed by simultaneous immunocytochemical demonstration of hormone production, *J Histotechnol* **13**:213 (1990).

44. O. Fujimori, A double protein A-gold-silver staining method for tissue antigens in light microscopy, *Histochem J* **24**:61 (1992).

45. W. Kummer, Simultaneous immunohistochemical demonstration of vasoactive intestinal polypeptide and its receptor in human colon, *Histochem J* **22**:249 (1990).

46. J.A. Bertsch, V. Bialecki, R. Emmons and L. Korytko, Colored silver-intensified gold technique for light microscopy, *BioTechniques* **6**:448 (1988).

47. Z. Nagy-Oltvai, T. Brady, and G.D. Hsiung: Double labeling of SIG+ cells Harboring a lymphothropic herpesvirus by avidin-biotin complex immunoperoxidase and immunogold-silver techniques, *J Histochem Cytochem* **36**:1187 (1988).

48. B. Bozóky, T. Krenács, Zs. Rázga, and A. Erdôs, Ultrastructural characteristics of glial fibrillary acidic protein expression in epoxy embedded human brain tumors, *Acta Neuropath* **86**:295 (1993).

49. R. Gillitzer, R. Berger, and H. Moll, A reliable method for simultaneous demonstration of two antigens using a novel combination of immunogold-silver staining and immunoenzymatic labeling, *J Histochem Cytochem* **38**:307 (1990).

50. V. T.V. Chan and O'D. McGee, Non-radioactive probes: preparation, characterization, and detection. *in*: In Situ Hybridization. Principles and Practice, J.M. Polak and J.O'D. McGee eds., Oxford Sci. Publ., Oxford, p.59 (1990).

51. P. Jackson, F.A. Lewis, M. Wells, *In situ* hybridization technique using an immunogold-silver staining system. *Histochem J* **21**:425 (1989).

52. P. Jackson, D.A. Dockey, F.A. Lewis, and M. Wells, Application of 1 nm gold probes on paraffin wax sections for *in situ* hybridization histochemistry, *J Clin Pathol* **43**:810 (1990).

53. H.H. Volkers, R. Rock, F.M. Van den Berg, Immunogold-silver staining for *in situ* hybridization, *Life Science (Amersham)* **4**:9 (1990).

54. B.E. Causton, The choice of resins for electron immunocytochemistry, *in*: Immunolabelling for Electron Microscopy, J.M. Polak and I.M. Varndell, eds., Elsevier Sci. Publ., Amsterdam, p.29 (1984).

55. M. Bendayan, and M. Zollinger, Ultrastructural localization of antigenic sites on osmium-fixed tissues applying the protein A-gold technique, *J Histochem Cytochem* **31**:101 (1983).

56. N. Tomokane, T. Iwaki, A. Takeishi, and J.E. Goldmann, Rosenthal fibers share epitopes with alfa-B-crystalline, glial fibrillary acidic protein, and ubiquitin, but not with vimentin, *Am J Pathol* **138**:875 (1991).

57. I. Merchenthaler, F. Gallyas, and Zs. Liposits, Silver intensification in immunocytochemistry, *in*: Techniques in Immunocytochemistry, Vol. 4, G.R. Bullock and P. Petrusz eds., Academic Press, London, p.217 (1989).

58. J.F. Hainfeld and F.R. Furuya, a 1.4 nm gold cluster covalently attached to antibodies improves immunolabeling, *J Histochem Cytochem* **40**:177 (1992).

59. S.H. Swerdlow, Small cell lymphoid neoplasms of B cell origin, *in*: Interpretation of Lymph Nodes, Raven Press, New York, p. 143. (1992).

60. J.M. Nigro, S.J. Baker, A.C. Preisinger, J.M. Jessup, R. Hostetter, K. Ceary, S.H. Binger, N. Davidson, S. Baylin, and P. Devilee, Mutations in the p53 gene occur in diverse human tumor types. Nature **342**:705 (1989).

61. G. Janossy, M. Bofill, L.K. Trejdosiewicz, and H.N.A. Willcox, Cellular differentiation of lymphoid subpopulations and their microenvironment in the human thymus, *in*: The Human Thymus, H.K. Müller-Hermelink, ed. Springer-Verlag, Berlin, p.89 (1986).

62. W.M. Molenaar, J.W. Oosterhuis, and F.C.S. Ramaekers, Mesenchymal and musclespecific intermediate filaments (vimentin and desmin) in relation to differentiation in childhood rhabdomyosarcomas, *Human Pathol* **16**:838 (1985).

63. K. Sandvej, L. Krenács, S.J. Hamilton-Dutoit, J.L. Rindum, J.J. Pindborg, and G. Pallesen, Epstein-Barr virus latent and replicative gene expression in oral hairy leukoplakia. Histopathol **20**:387 (1992).

64. C. Schmid, L. Pan, T. Diss, and P.G. Isaacson, Expression of B-cell antigens by Hodgkin's and Reed-Sternberg cells, *Am J Pathol* **139**:701 (1991).

65. G. Pallesen, S.J. Hamilton-Dutoit, M. Rowe, and L.S. Young, Expression of Epstein-Barr virus latent gene products in tumour cells of Hodgkin's disease, *Lancet* **337**:320 (1991).

CHAPTER 15

POSTEMBEDDING LIGHT AND ELECTRON MICROSCOPIC IMMUNOCYTOCHEMISTRY IN PINEAL PHOTONEUROENDOCRINOLOGY

Ingeborg Vigh-Teichmann[1] and Bela Vigh

Neuroendocrine Section (I.V.T.) of the Hungarian Academy of
Sciences - Semmelweis Medical University Joint Research Organization
(EKSZ) at the 2nd Department of Anatomy (B.V.)
Semmelweis Medical University

SUMMARY

Postembedding immunocytochemistry (ICC) is a reliable tool for demonstrating both structural and molecular components in tissue section Opsin, vitamin A, S-antigen and gamma-aminobutyric acid (GABA)ICC were used to identify immunopositive and immunonegative cellular elements in the pineal organ of various vertebrates. The peroxidase and the gold methods were used at the light microscopic (LM) and the electron microscopic (EM) levels, respectively. Fixation and protocols in postembedding ICC were dealt with. Problems with background staining were discussed in connection with vitamin A and GABA ICC. Differences were found in vitamin A immunoreactivity between photoreceptor types dependent on the light conditions used before fixation. Three to four types of pinealocytes could be distinguished by their ultrastructure, opsin or S-antigen immunoreactivity. S-antigen-immunoreactive axons of pinealocytes formed synapses on the dendrites of pineal neurons and neurohormonal terminals on the basal lamina of the organ. GABA-immunoreactive axons synapsed with pinealocytes and GABA-immunonegative as well as GABA-immunopositive neurons. The synaptic complexes of the mammalian pineal resembled those of the retina and submammalian

[1]Address for correspondence to: Doz. Ingeborg Vigh-Teichmann, 2nd Department of Anatomy, Semmelweis University Medical School, Tüzoltó utca 58, H-1094 Budapest IX, Hungary. Tel: 36-1-215-6920; Fax: 36-1-215-3064

pineal. They represent a structural basis for the integration of perceived light information with various afferents from the central nervous system. They are presumed to guide the neural and neurohormonal output of the pineal organ.

INTRODUCTION

Postembedding ICC provides a valuable tool for identifying well preserved immunopositive and immunonegative cellular elements and their associated antigenic molecules in fixed or unfixed resin-embedded tissues at the LM and the EM levels. For a better appreciation of the principles of this technique, the reader is referred to the excellent reviews in several Chapters of this volume, particularly those on postembedding ICC by Peter Lackie and T. Krénács. For visualization of antigenic sites, most researchers employ conjugates of horseradish peroxidase (HRP), colloidal gold with immunoglobulins (IgG), protein A, G, or streptavidin. Less frequently used are ferritin conjugates which need special counterstaining with bismuth chelatum.[1]

Every laboratory has its own set of antigens,[2] antibodies and optimal protocols. We too, have our special conditions of performing immunocytochemical reactions. Our methods may not be suitable for other research needs. In this chapter, practical considerations of HRP-anti-HRP (PAP), immuno-HRP, avidin biotin-HRP (ABC), protein A-gold, immunogold and streptavidin-gold postembedding ICC are discussed in connection with studies of the pineal organ and retina. Both tissues are indispensable components of the photoneuroendocrine system.

The pineal organ is known to play an important role in the regulation of circadian and circannual rhythms of the organism, color change of the skin, locomotor activity, reproduction and thermoregulation.[3-9] When we started morphological studies of the pineal organ, the view of several authors suggested that during evolution the photosensitive pineal complex of lower vertebrates developed into a gland-like structure in reptiles, birds and mammals.[10-12] a.o. However, more recent ultrastructural, immunocytochemical, and electrophysiological data indicate that the pineal organ has a secretory as well as a photoperceptive function from cyclostomes to birds.[13-16] A similar question arises concerning the mammalian pineal - whether it functions exclusively as an endocrine gland, or during phylogeny, it retained structures and components of an apparatus dealing with direct light perception as known for the retina and submammalian pineal.

In a series of studies, we dealt with the ultrastructure of the pineal organ in various vertebrate classes.[13-15,17-25] The pineal organ consists of glial cells, intrinsic neurons and pinealocytes (*Fig. 1*).

We found that the pinealocytes are endowed with inner segments and outer segments morphologically resembling retinal cones, not only in lower vertebrates but also in reptiles and birds (*Fig. 2*). The mammalian pinealocytes also give rise to 9x2+0 smooth (*Fig. 27*) or bulbous vesiculated cilia (*Fig. 3*). Such vesiculated cilia resemble differentiating photoreceptors.[14,18,26] Their presence appears to depend on species, age and environmental light conditions. The perikarya of the pinealocytes are characterized by a cytoplasm of either strong, medium or weak electron density to which the amount of nuclear heterochromatin appears to be correlated (*Fig. 1, 4, 25*).

The photoreceptor pinealocytes of several submammalian species transmit perceived direct light information synaptically via their synaptic ribbon- or spherules-containing axons.[24,27,28] These terminate on the dendrites and perikarya of intrinsic pineal nerve cells (*Fig. 5*) already known in monkeys since 1929.[29,30] Such large secondary neurons send their axons to centers of the brain stem via the pineal tract,[4,31-33] or in mammals and humans via fiber bundles into the habenula and posterior commissures.[24,34,35] Efferent nerve fibers of the lizard parietal eye terminate in structures such as hypothalamic nuclei[36] which also play a role in photoneuroendocrine hormonal regulation.

The morphological basis of a pineal hormone-secretory function shown by us,[14,18,20,22,37] is represented by neurohormonal terminals formed by the synaptic ribbon- and/or spherule-containing axons of pinealocytes on the pineal basal lamina (*Fig. 6*) opposite to the pial capillaries. This pattern is quite similar to that of other neurohormonal areas such as the median eminence, neurohypophysis or urophysis.[38] We could show, for example, in the pigeon[14,37] and rabbit pineal,[39] that such neurohormonal terminals were

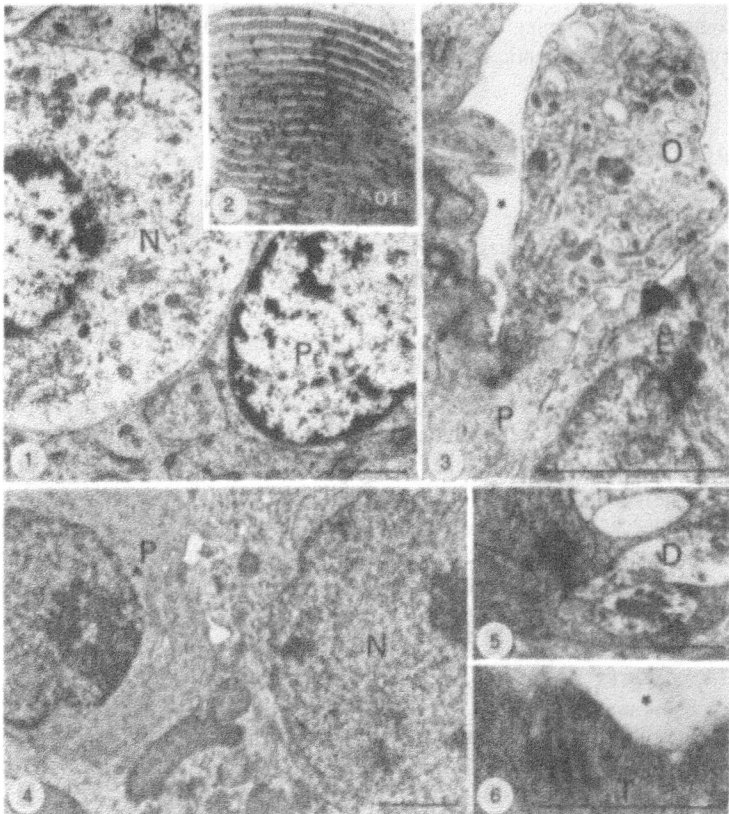

Figures 1-6. Ultrastructural details of the pineal organ of various vertebrates. 1) Electron-lucent intrinsic pineal neuron (N) with axo-somatic synapse (arrow) and pinealocyte (P) displaying few nuclear heterochromatin. Hedgehog, x10,400. 2) Portion of a pineal cone-type outer segment of Lacerta viridis. COS-1/immunogold reaction (black dots), x100,000. 3) Outer segment-like (o) vesiculated cilium of the ferret pinealocyte (P) Asterisk: lumen of follicular cavity. E: ependyma, x44,000. 4) Electron-dense pinealocyte (P) of cat. N: nucleus of neuron, x16,000. 5) Synaptic spherules-containing axon (A) of pinealocyte synapses (arrow) with neuronal dendrite (D). Guinea pig, x22,500. 6) Synaptic ribbon-containing neurohormonal terminal (T) of rat pinealocyte faces the pial space (asterisk). Black dots: vitamin A/immunogold reaction, x41,800.

serotonin-immunoreactive (IR). This raises the question that still needs to be clarified, i.e., whether serotonin is released in accordance with photoperception during the day in addition to melatonin[40] which is known to be secreted at night.[16,41]

Perception of light and integration of this information only takes place when photopigments, the necessary proteins of the photochemical transduction cascade[42-44] and transmitters are present. Animal photopigments consist of the transmembrane receptor glycoprotein opsin[43] and the chromophore 11-cis vitamin A aldehyde (retinal)[45,46] that serves to catch the photons. One of the quenching proteins of the cascade is the S-antigen.[22,47] For the pineal organ, the study of its photopigments, transduction enzymes[48] and transmitters is still in a preliminary stage, especially concerning EM-ICC.

In the following, we summarize our results and practical experience in the postembedding ICC of the pineal organ of different vertebrates after having applied antibodies raised against rhodopsin, green/red and green/blue pigments, vitamin A, S-antigen and GABA. Technical specialties of postembedding ICC were dealt with in connection with HRP and gold methods applied to the pineal organ and retina, the latter being used as a control tissue. It was shown that postembedding ICC was sometimes troublesome, time-consuming but a valuable tool in the elucidation of different types of pinealocytes and their synaptic connections, especially in mammals.

MATERIAL AND METHODS

The animals were kept under a natural light cycle and anesthetized either with 1% to 3% MS222 (MABA for fishes, Sandoz Ltd., Basle, Switzerland), urethane, ether, chloroform or sodium hexabarbituate between 9-22 hr. Some animals were dark-adapted for at least 2 hours and fixed under special light (twilight, dim red or yellow light), within 1 to 15 min after light exposure. Other material was fixed completely dark-adapted, or after several hours of light adaptation.[34]

Animal species used were: lamprey (Lampetra planeri), ratfish (Chimaera monstrosa), hatchet fish (Argyropelecus hemigymnus), eel (Anguilla anguilla, A. rostrata), creek chub (Semotilus atromaculatus), goldfish (Carassius auratus), frog (Rana esculenta), turtle (Pseudemys scripta elegans), lizard (Lacerta agilis, L. viridis), pigeon (Columba livia), hedgehog (Erinaceus roumanicus), ferret (Putorius furo), bat (Myotis myotis, M. blythi oxygnathus), mink (Mustela vison), stone martin (Martes foina), cat (Felis domestica), guinea pig (Cavia cobaya), cattle (Bos taurus), Mongolian gerbil (Meriones unguiculatus), white laboratory rat.

Optimal Procedures

Fixation. The pineal organs and the corresponding retinas were fixed with 1% to 3% glutaraldehyde (GA) or 0.1% to 4% paraformaldehyde (PF) containing up to 2% GA, when the primary antibody was raised against an unconjugated antigen. The fixative was solved in 0.1M Sörensen phosphate, Millonig phosphate or cacodylate buffers. The material, postfixed in the same fixative or in 4% PF for up to 4 hr at 4°C, was thoroughly washed in buffer. In order to reduce background staining, we also started to wash in 50 mM solution of ammonium chloride in isotonic 0.02 M phosphate buffered saline[49] containing 2.5% sucrose for osmotic protection. Some of the material was cut frozen in a cryostat in order to preserve vitamin A.

After dehydration in ethanol, the tissue was embedded in Araldite (Fluka, Buchs, Switzerland), Poly Bed 812-Araldite 6005 mixture, LR White or LR Gold (Polysciences, St. Goar, FRG). The hydrophilic, acrylic based LR White and LR Gold resins allowed embedding from 70% and 90% alcohol directly, i.e. lipids were better preserved. Further, the LR sections were found to be more permeable for antibodies at the EM level than the hydrophobic epoxy-based Poly Bed 812 and Araldite resins that need etching. There was no difference in the antigenicity depending on which of the plastics was applied. Semi-thin and ultrathin sections were collected on very clean slides and grids, respectively. Nickel or gold grids were used for ICC instead of copper grids, since they may be necessarily exposed to reducing chemicals.

Immunocytochemistry

Each antigen has its own physiological and immunocytochemical conditions under which it could be optimally demonstrated. First of all, the appropriate dilution of each primary, secondary, and tertiary antiserum had to be established by a series of dilutions starting with a dilution of at least 1:100. It was advisable to include positive control sections, which contained the antigen in consistent and relatively large amounts, to be sure that the method was performed correctly and all chemicals worked satisfactorily. If the intensity of an immunoreaction decreased considerably, the expiration date of the secondary antibody was checked first. A decreasing activity of a secondary antibody could be compensated by using a more concentrated dilution of the primary antibody.

Primary antibodies used: a) Antibovine rhodopsin antiserum from sheep[50] or rat[51] and the monoclonal mouse antichicken cone antibodies (Abs OS-2 (detecting green-, blue- and greenblue-sensitive pigments)[23,25] and COS-1[51] (for green- and red-sensitive pigments[23,25,52]). b) The Vitamin A antiserum was raised against retinoic acid[53] conjugated to human serum albumin (HSA) and crossreacted with both the cis-and trans-forms of retinol and retinal[54]. c) Antibodies against HPLC-purified bovine S-antigen were raised in rabbit (I. Gery, Bethesda, MD, USA) or as monoclonal antibodies in mouse (L.A. Donoso, Philadelphia, PA, USA). d) Antibodies against GABA-BSA-GA (H. Petter, Leipzig, Germany) were raised in rabbit and diluted in Coons buffer,[55] containing Triton X 100.

The schedules of the PAP,[56] ABC,[57] anti-IgG-HRP and anti-IgG fragment-HRP (a-IgG-F/a,b'/2-HRP) reactions are given below.[4, 23,24] The antibodies were applied as drops on top of the semi-thin epoxy sections in a moist chamber, followed by washing in buffer in staining jars.

Light-Microscopic Immunoreaction

1. Etching of semi-thin Araldite and Poly Bed 812 sections with sodium methoxide,[58] 2% NaOH in absolute ethanol[59] 1:2 in distilled water for 15 min, sodium periodate or H_2O_2 to unmask antigenic sites. Wash sections in absolute methanol I, II, acetone I, II and water for 3-5 min each.
2. Blocking of endogenous HRP by 1% H_2O_2 in HPBS.
3. Blocking of non-specific reactive sites (mainly aldehyde groups) by 0,1% sodium borohydride in 1% Na_2HPO_4,[60] or ice cold distilled water for 5 min (stop when hydrogen bubbles develop).
4. Wash thoroughly 3 x 5 min in the appropriate solvent used.
5. Blocking of non-specific binding sites by 1% to 10% normal serum of the species from which the secondary antibody was obtained, in 0.05 M histochemical phosphate buffered saline, pH 7.4-7.6 (HPBS). Gently shake off. If protein A-gold will be applied as secondary antibody, use 1-5% bovine serum albumin or ovalbumin, since protein A specifically binds to the Fc region of IgG, also present in normal serum.
6. Primary antibody (e.g. from rabbit) appropriately diluted in HPBS for 1-2 h, or overnight in a moist chamber, at room temperature.
7. Wash in HPBS 3x3-5 min.
8. Secondary antibody (goat antirabbit-HRP, or goat antirabbit IgG for PAP, biotinylated (B) antirabbit IgG from goat for ABC) diluted 1:100 to 1:400 in HPBS. Incubation time: 60 min.
9. Wash in HPBS, 3x3-5 min.
10. HRP conjugate (PAP, or the avidin biotin-HRP complex=ABC) usually diluted 1:100 to 1:800 in HPBS, 1 hr, depending on source, mg of protein content and age of the sample.

11. Wash in HPBS 2x3-5 min followed by 0.05 M Tris-HCl, pH 7.6.

12. Development of the HRP reaction product by a freshly prepared solution of 0.03% 3,3'-diaminobenzidine tetrahydrochloride (DAB) and 0.0013% hydrogen peroxide in 0.05 M Tris-HCL buffer, pH 7.6 in the dark, for 10 min.

13. Wash in distilled water and cover with glycerine-gelatin.

The HPBS buffer may also contain 0.25% BSA, small ions like 2.7 mM KCl,[61] 0.02 M glycine or lysine to quench non-specific binding sites. Further, 0.05-0.1% Tween 20 or 0.1-0.25% Triton X-100 may be added to reduce background staining, but this may cause considerable elution of protein.[62]

Electron Microscopic Immunoreactions with Gold Conjugates

Grids were immersed into or on top of incubation drops. The diluted antibodies were centrifuged at 12,000 g for 5-10 min to eliminate precipitates.

1. Etch ultrathin sections in sodium methoxide 1:2, 5-10 min depending on hardness of the resin.

2. Quench free aldehyde groups with 0.1% sodium borohydride.

3. Block reactive tissue groups by 1-5% normal serum (or 0.25% BSA, ovalbumin for protein A-gold conjugates), 0.5% carrageenan type IV (Sigma, Munich, FRG) or freshly prepared 0.01-0.1% gelatin (No. 4070, Merck, Darmstadt, FRG) in 0.05 M TBS or HPBS, pH 7.4; 15 min or more, shake off.

4. Primary antibody appropriately diluted in HPBS, 1 to 18 hr.

5. Wash with HPBS.

6. Appropriately diluted secondary antibody in HPBS
   a) Secondary Biotinylated anti-IgG of the primary antibody species, diluted 1:25-1:400, for 30 min, followed by streptavidin-gold for 1 hr.
   b) A secondary IgG link antibody may be used 1:100-1:1000 for 15 min before the protein A-gold step to increase the intensity of the reaction.
   c) Gold-conjugated secondary antibody diluted 1:5 to 1:200 depending on duration of incubation (1h and more). A self prepared and diluted protein A-gold or IgG gold-conjugate is usually faintly reddish colored, at O.D. 525 nm[49] if the gold particle size is above 5 nm in diameter.

To reduce non-immunological binding of stabilized gold particles to mitochondria, nuclei and connective tissue fibers 0.01-0.1% gelatin was recommended as a competitive protein due to its high affinity for gold over a wide pH range.[63] In addition, 0.05-0.2% Triton X-100 and/or 0.05%-0.1% Tween 20 may be added to increase permeability[49] and reduce background staining. 0.05% gelatin in 0.1% Tween 20-PBS buffer pH 7.6 was sufficient to reduce streptavidin gold background staining of the GABA immunoreaction. In the case of the vitamin A reaction, we applied a Tris-saline buffer of pH 8.0 containing 0.1-0.2% Triton X-100 to get rid of non-specific anti-IgG-gold binding to collagen fibers, tubules, secretory granules of glands and myofibrils of muscle.

Specifically, tests of the method usually involved leaving off one of the reaction steps like the secondary antibody. For testing of the antiserum specificity, the primary antiserum was substituted for pre-immune serum, or the antiserum was absorbed with increasing ng (1-100 ng) of antigen given to 1 ml of the working dilution of the antibody. With either of these procedures all gold labelling should disappear in the sections, if the antiserum is monospecific.[64]

RESULTS

Opsin Immunoreactivity

Most of the outer segments of the photoreceptor cells of the pineal organ of lampreys, freshwater fish and amphibians were OS-2- and rhodopsin-immunoreactive indicating the presence of rhodopsin or crossreacting porphyropsin.[4,15,21,22,25,31,34,46,52,65] Therefore, these photoreceptors were called rod-like in contrast to rhodopsin-immunonegative cone-like photoreceptors.[15,17]

At the EM level, gold particles of protein A, anti-IgG or streptavidin conjugates labeled the binding sites of the opsin antibodies onto the photoreceptor membranes of the pineal and retinal outer segments (Fig. 2). Gold particles of 5-15 nm in diameter were only visible at relatively high magnification. Under such optimal conditions, also unlabeled outer segments were easily recognized.

In this connection, it was of interest that in reptiles and birds, the number of rhodopsin-IR pineal photoreceptors was relatively low - similar to the pattern in the corresponding retinas. There were some photoreceptors that reacted with the antibodies COS-1 (lizard) and OS-2 (lizard, pigeon) binding to red or green, and green or blue pigments, respectively (*Fig. 2, 7*). However, as already observed in lamprey, fish and frog by us,[23,46,52] certain photoreceptors were always unstained. These immunonegative photoreceptors were supposed to represent ultraviolet-sensitive elements.[34,52,78] Our hypothesis is supported by electrophysiological and behavioral data of different vertebrate classes,[66-70] which indicated the presence of UV-sensitive photoreceptors in the retina as well as in the pineal organ.[71] On the basis of the deep sea light conditions, we draw a similar conclusion for some of the opsin-immunonegative pineal photoreceptors of the fish Chimaera monstrosa.[23]

There were mainly three to four types of pinealocytes in the pineal organ from cyclostomes to birds.[4,19,23,34,46,72] Their differences in size, ultrastructure, and opsin immunoreactivity in lamprey and frog suggested the presence of green-, blue-, red- and UV-sensitive pineal photoreceptors.[23,34,46] Obviously, the submammalian pineal is involved in measuring direct light intensity and light quality according to the environmental light conditions under which the animal lives.

The question of a direct light sensitivity also arises in connection with the mammalian pineal organ. In certain species like hedgehog and ferret, pinealocytes were found to give rise to smooth and/or bulbous vesiculated 9x2+0 cilia (*Figs. 3, 27*).[14,26,35] Unfortunately, the four photopigment antibodies used in our study, did not bind to these bulbous outer segment-like cilia. This was probably due to a loss of the antigen during fixation, to the masking of the antigenic sites, further to the particular properties of the primary antibodies applied. Recently, some authors[73-76] found some rhodopsin-IR pinealocytes and picograms of rhodopsin in the pineal organ of mouse, cat, hamster, rat and sheep by LM ICC, immunoblotting and/or ELISA, mainly with the help of antibodies raised against abs COS-1 that dilapidated rhodopsin.

Further, we noticed[15] that sodium borohydride applied for blocking reactive groups of the fixative, destroyed an epitope usually detected by the mouse, binds to red and green photopigments. Since this agent first of all reduced aldehyde groups[60] and was able to induce dark isomerization of the vitamin A chromophore,[77] the epitope detected by COS-1 might be associated with the binding site of the vitamin A ligand in the opsin molecule.[43] Therefore, when one applies antibodies directed against uncharacterized epitopes, the sequential steps of the ICC reaction have to be adjusted cautiously. This experience has lead us to reinvestigate the pineal organ under modified ICC conditions and with antibody CERN 886 raised against detergent-treated light-exposed rhodopsin. This antiserum-LgGgold reaction labelled rhodopsin-IR sites onto the cell membrane of an electron-dense

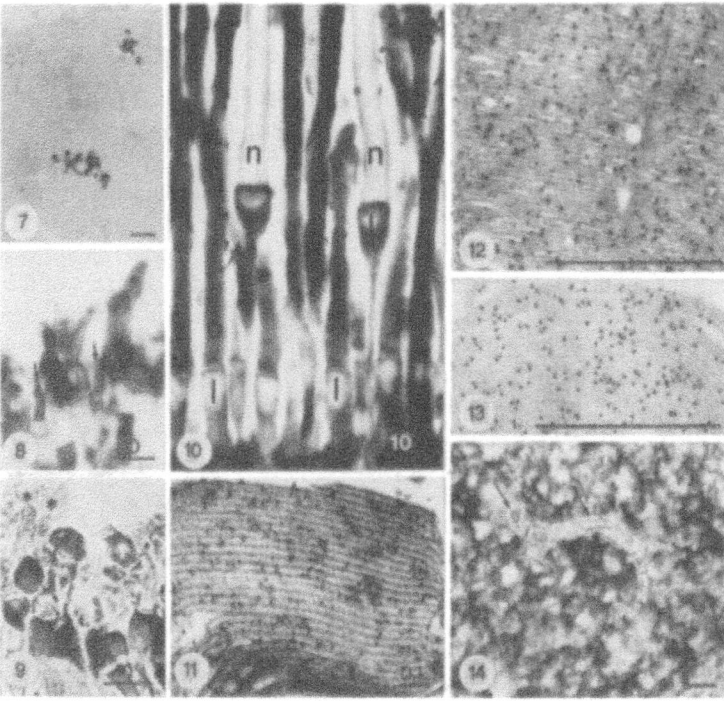

Figures 7-14. OS-2-opsin and vitamin A ICC in retina and pineal organ. 7) Few OS-2-IR outer segments of pinealocytes in the nestling pigeon pineal. ABC reaction. 8) Vitamin A-IR rod (asterisk) and cones (arrows) of dark-adapted chicken retina. PAP reaction x1,100. 9) Vitamin A-IR cones (arrows) and negative rods (asterisks) of chicken retina fixed under yellow light. PAP reaction x1,100. 10) Vitamin A-IR long green-sensitive rods (o) and immunonegative short rod outer segments (n). I: inner segment. Frog retina. ABC reaction, x 1,100. 11) Portion of a vitamin A-IgG gold-labeled outer segment of frog pineal organ, x51,500. 12) and 13) Double labeling of rhodopsin- (15 nm gold particles) and vitamin A-binding sites (10 nm gold particles) in retinal rod, and pineal rod-type outer segment, respectively, of frog, x40,0000, x43,800. 14) Vitamin A immunoreactivity in rat pineal. Arrows: capillary with immunonegative erythrocytes. PAP reaction, x450.

small-sized pinealocyte (a fourth type).[95-96] Therefore, we believe that the mammalian pineal organ -- mainly composed of pinealocytes being rhodopsin- immunonegative, i.e. cone-like[34] -- may still have retained a certain capacity to react to dim light directly.

Vitamin A Immunoreactivity

As already mentioned, photopigments consist not only of the receptor protein opsin coupled to G-protein,[43] but also of the vitamin A chromophore retinal. This is an 11-cis retinal or 11-cis 3-dehydroretinal in the dark-adapted state. With retinoic antibodies crossreacting with retinal and retinol, we found immunoreactivity in the rod- and cone-type outer and inner segments of retina and pineal[39,45,46,78] fixed under dim red or infrared light (*Fig. 8, 10, 11*). After short exposure to room light or to yellow light, the rod photoreceptor was immunonegative (*Fig. 9*) indicating that the bleached chromophore[80] was not accessible to the antibody and had probably entered the visual cycle outside of the outer segment. In contrast, cones were still immunoreactive probably due to their longer time of light adaptation[34,46,78] (Fig. 9). Under dim red light fixation, short rod photoreceptors of the frog retina known to be sensitive to a more blue-shifted green light, exhibited immunoreactive inner segments, but unstained outer segments (*Fig. 10*). Furthermore, we were surprised that material fixed in complete darkness did not bind the vitamin A antibodies[34]. Under similar dark conditions, vitamin A could not be demonstrated in rat retina by chromatography.[81] This ICC pattern appears to indicate that certain conformational changes take place under dim red light. These changes are probably necessary to make accessible the epitope of the cyclohexene ring of the retinoids[54] recognized by the vitamin A antibodies in the dark-adapted photopigment.

Under regular conditions, vitamin A immunoreactivity was observed in preembedding frozen as well as in postembedding L. R. White sections by us (*Figs. 8-14*).[34,45,46,78] The immunogold technique of L.R. White sections proved to be valuable not only for detecting stained and unstained outer segments but also for demonstrating different antigens in one and the same profile. We performed sequential double immunogold staining reactions that labeled binding sites of rat antibovine rhodopsin with 15 nm gold particles and of rabbit antivitamin A with 10 nm gold particles in pineal and retinal outer segments[78] (*Figs. 12,13*). Our results prove that not only the retinal but also the pineal photoreceptors contain vitamin A-associated photopigments.

We also found Vitamin A in the mammalian pineal (*Figs. 6, 14*).[34,39,78,79] This immunoreactivity may indicate the presence of retinal, retinol and retinoic acid, since all these retinoids react with the vitamin A antibodies used. Retinol and retinoic acid occur in very low amounts in every cell type and play a role in gene expression,[82] development, growth, and maintenance of cell differentiation.[45] Different quantities of retinol -- known to also participate in the visual cycle of the retina -- and of retinyl ester have recently been demonstrated in bovine pineals adapted either to undescribed light conditions[83] or to yellow light[84] by high performance liquid chromatograph (HPLC). HPLC studies of retinal were performed successfully in pineals[85-87] displaying many rhodopsin- and/or porphyropsin-containing pinealocytes.

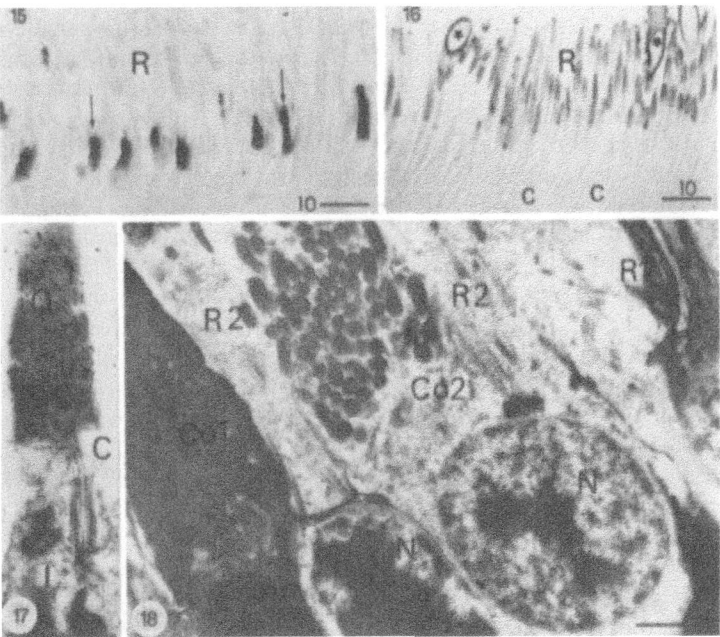

Figures 15-18. Details of the bovine retina. 15) and 16) COS-1-IR (arrows) and OS-2-IR cones (asterisks), respectively. c: immunonegative cones, R: rods, x1,100, x1,100. 17) Rhodopsin-IR outer segment (o) of electron-lucent rod. C: connecting cilium, I: inner segment, x24,650. 18) Four types of retinal photoreceptors: electron-dense rod (R1), electron-lucent rod (R2), electron-dense cone (Co1) and electron-lucent cone (Co2). N: nuclei of photoreceptors. S-antigen/immunogold, x9,000.

The presence of rhodopsin[73-76] in picogram quantities, of a particular rod-like pinealocyte[95,96] and of cellular retinol- and retinal-binding proteins[88] in the mammalian pineal suggests the occurrence of a detectable amount of the chromophore of the photopigment. Further studies of pineals prepared under various light conditions are needed to contribute to our present knowledge of the pineal non-image visual cycle.

S-Antigen Immunoreactivity

S-antigen, also called arrestin, is a 46-48 kD polypeptide of the photoreceptor transduction cascade that induces autoimmune uveitis and pinealitis[89,90] and may be expressed in tumors[91]. Normally, this protein binds to and quenches phosphorylated, light-activated rhodopsin. Thus, S-antigen is involved in finishing the transduction cascade by preventing rhodopsin from further activating the G-protein transducin.[43]

The polyclonal and monoclonal S-antigen antibodies used[22,34] were raised against S-antigen of the bovine retina. The latter was known for its numerous electron-dense green-sensitive rods, but we found[14,34] that it also contains electron-lucent rods (blue-shifted sensitive ones) and two kinds of cones, electron-dense (red-sensitive ones) and electron-lucent (blue-sensitive) ones (Fig. 15,16,17,18).

In the fish (except hatchet fish and ratfish) and amphibians studied, the S-antigen antibodies bound to the many rod-type photoreceptors of the retina and pineal organ (Figs. 14,22), while cones were unstained. The association of S-antigen with the rod photoreceptors was also proved by sequential double labeling with rat antibovine rhodopsin visualized by 15 nm IgG-gold particles, and with rabbit antibovine S-antigen bound to 10 nm IgG-gold particles in frog pineal organ and hedgehog retina.[22] However, we also found S-antigen immunoreactivity in photoreceptor nuclei and retinal and pineal nerve cells (Figs 20, 21). This latter reaction could not be eliminated by any procedure applied to get rid of background staining.[2,64] Thus, polyclonal as well as monoclonal S-antigen antibodies may also recognize non-photoreceptor-specific members of the S-antigen superfamily like the beta-arrestins.[92]

In the mammalian retina (cattle, cat, ferret), the electron-dense and electron lucent rods as well as electron-dense cones exhibited S-antigen immunoreactivity (Fig. 18), while the electron-lucent cone was immunonegative. This indicates that the S-antigen of the immunoreactive rods and cones shares common epitopes and that -- contrary to general belief -- the cattle, cat and ferret retinas have the structural capacity for chromatic vision. In the mammalian pineal, various numbers of pinealocytes were S-antigen-IR (Figs. 23,24), many in the cat. Differences in the intensity of their immunoreactivity speak in favor of different stages of S-antigen gene expression, but also of different types of pinealocytes.[22,34] In the ferret, only small numbers of pinealocytes were S-antigen-IR (Fig. 24). These pinealocytes displayed electron-dense cytoplasm and nuclei with spotted chromatin like those of the immunopositive electron-lucent rods of the retina (Figs. 25, 26). Furthermore, a few medium electron-dense pinealocytes were immunolabelled and many were unstained exhibiting nuclei with relatively poor chromatin.[34] This ICC pattern in cattle, cat and ferret is in accord with our previous results on the presence of at least three types of pinealocytes in bat and hedgehog.[22] This also furnishes new cytological details for their characterization.

It should also be emphasized that the S-antigen not only occurs in the perikarya of the pinealocytes, but also in their synaptic ribbon-containing axons, forming neurohormonal terminals on the pineal basal lamina[22] and synapses on dendrites of secondary neurons. However, the cilia originating from the immunopositive perikarya of the pinealocytes appeared to be almost free of immunoreactivity (Fig. 27). Thus, although retinal and pineal S-antigen have identical amino acid sequences[93] -- a fact favoring identical function -- it has still to be clarified, whether and under which circumstances the S-antigen of the

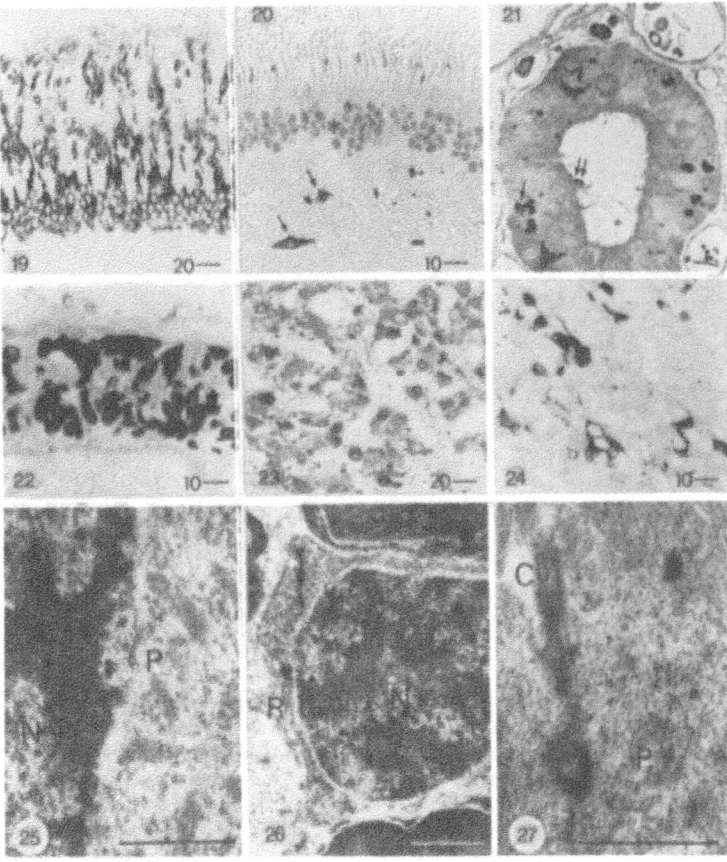

Figures 19-27. S-antigen ICC in retina and pineal organs. 19) Only rod photoreceptors are immunoreactive in goldfish retina. ABC reaction, x450. 20) and 21) Note positive nuclei of photoreceptor layer (asterisk) and positive ganglion cells (arrows) of retina and pineal. Double arrow: CSF-contacting neuronal dendrite. Ratfish, x450. 22), 23), and 24) Immunoreactive pinealocytes in goldfish, cattle and ferret pineal, x450, x280, x450. 25) and 26) S-antigen-IgG-gold-labeled pinealocyte (P) and electron-lucent retinal rod (R) perikaryon. N: nuclei rich in heterochromatin. Asterisk: synaptic ribbon of rod, x26,000, x15,600. 27) Weak IgG-gold labeling in cilium (C) of an S-antigen-immunoreactive pinealocyte (P) x32,000.

mammalian pinealocyte plays a role in quenching synaptic and/or secretory as well as perceptive activities.

GABA Immunoreactivity

The immunoreaction for the demonstration of the small amino acid transmitter GABA -- a hapten not sufficiently antigenic by itself to elicit antibody production -- can easily be performed on Poly Bed 812, Araldite or LR sections (*Fig. 28-32*). However, some background problems have to be overcome like the peroxidase and gold-labeling of glial fibrils and collagen tissue elements by applying Triton X-100 or Tween 20, a buffer of double ionic strength, further the anti-IgG F(a,b')2-HRP complex instead of PAP or ABC. The preparations layered with the HRP-conjugated fragment of IgG had a very low background. This even led to rather faint staining intensity that could be compensated by osmification of the semi-thin sections in 0.01-1% OsO_4 in distilled water (*Figs. 28,29*).

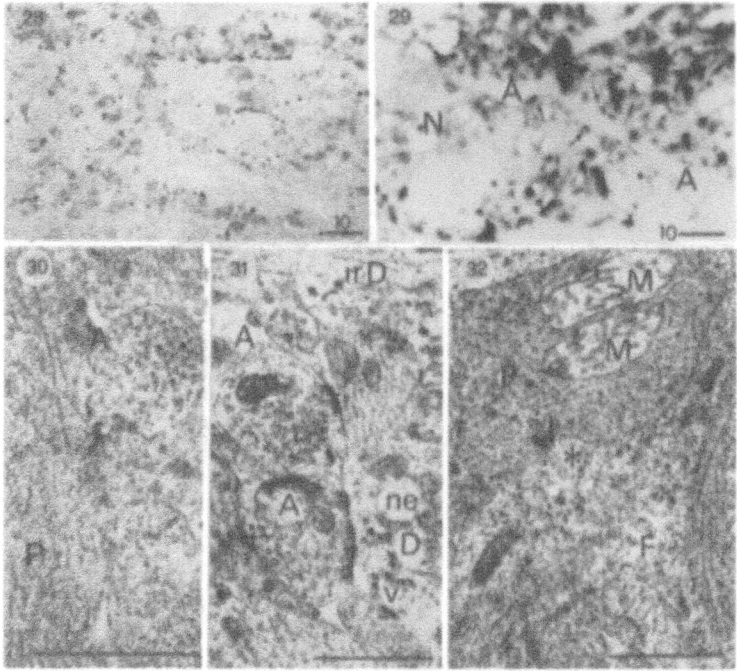

Figures 28-32. GABA ICC in the mammalian pineal organ. 28) Beaded GABA-IR nerve fibers of cattle pineal. Osmicated ABC reaction, x2,100. 29) The apparent axon (A) of the GABA-immunonegative large secondary neuron (N) receives GABA-IR axon terminals (arrows). Ferret x1.100. 30) Synaptic triad between synaptic ribbon-containing axon of pinealocyte (P), GABA-IR axon (A) and GABA-immunonegative process (asterisk). Arrow: synaptic membrane density, x38,300. 31) GABA-IR axons (A) form synapses on GABA-IR (ir) and GABA-immunonegative (ne) dendrites (D). v: granulated vesicles, x25,000. 32) Synaptic ribbon-containing axon of pinealocyte (P) synapses (asterisk) with GABA-IR (black dots) nerve fiber (F). M: mitochondria x24,500.

Under the conditions of ICC processing used, there were small GABA-immunoreactive and large GABA-immunonegative neurons in the pineal organ of the mammals[24] studied. Axons of the GABA-IR nerve cells formed rare synapses with some pinealocytes (*Fig. 30*), but more frequently with the large non-immunoreactive secondary neurons and immunoreactive interneurons (*Figs. 31*).[34] Additional myelinated, apparently afferent GABA-IR axons lost their myelin sheath and terminated on secondary neurons. These -- and some GABA-immunoreactive dendrites -- also received synapses from synaptic ribbon-containing axons of pinealocytes (*Fig. 32*). The latter and the glial cells were non-immunoreactive. Thus, postembedding ICC was time-consuming when one had to struggle with background problems, but useful for the elucidation of synaptic circuits in the pineal organ (*Fig. 33*).

DISCUSSION

The advantage of the postembedding light and electron microscopic methods -- employing peroxidase or gold conjugates of anti-IgGs, protein A and streptavidin on resin-embedded sections -- lies in the following simple facts: these techniques are convenient and enable the researcher to clearly distinguish immunoreactive profiles from non-immunoreactive ones; this can be done repeatedly from one and the same block of material. A precondition is an acceptable ultrastructural preservation and embedding into a resin, chosen appropriately with respect to the maintenance of the antigen, to be demonstrated by immunoreaction. In the postembedding method, the antigen is detected on a plane cut surface: the antibody does not need to penetrate layers of the tissue, and this excludes the penetration problems with which one has to contend in the preembedding techniques. The masking of the antigen by the epoxy resins can easily be overcome by the etching of the sections prior to the reaction.

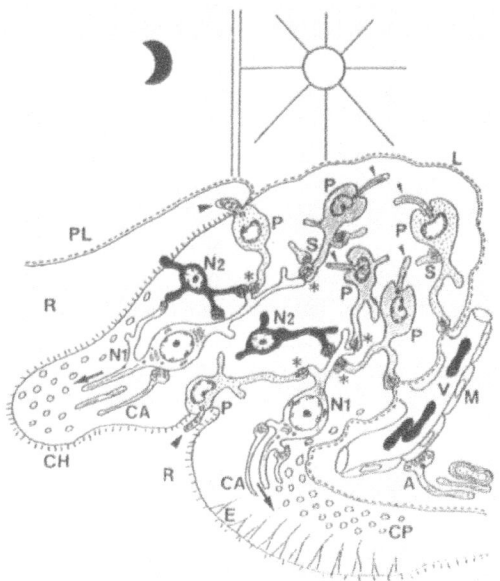

Figure 33. Scheme on the cellular and synaptic organization of the mammalian pineal organ. Symbol of moon and sun: supposed influence of circadian and circannual changes of environmental light on pineal activity. Arrowheads: sensory 9x2+0 cilia of three types of pinealocytes (P), asterisks: the synaptic ribbon/spherule-containing axons of pinealocytes terminate on large secondary neurons (N1) and synapse rarely with small GABA-immunoreactive interneurons (N2). Note the neurohormonal axon terminals of pinealocytes abutting the pineal basal lamina (L) opposite to fenestrated capillaries. The secondary neurons send their efferent axons (arrows) into the habenular (CH) and posterior commissures (CP). A: peripheral vegetative axons innervating smooth muscle cells (M) of vessel (V), CA: central afferents, E: ependyma of 3rd ventricle, PL: choroid plexus, R: suprapineal and pineal recesses, S: GABA-immunoreactive and GABA-immunonegative synapses on pinealocytes.

The postembedding technique proved to be more reliable in our hands, especially for the detection of the opsin of the photopigments. This was preferable to the method of reembedding and cutting of PAP immunoreacted and osmicated semi-thin sections, a technique we started with before the era of the immunogold techniques.[26] Similarly, results obtained by preembedding HRP-ICC on cryo cut ultrathin sections were insufficient perhaps due to inappropriate penetration of the antibodies into the tissue and the problem of appropriate storage of the frozen blocks for repeated sectioning. The opsin immunoreaction on frozen ultrathin sections was faint with the 10 nm gold conjugates available to us at that time. When penetration was enhanced by a detergent, the delicate photoreceptor and synaptic membranes were damaged; it was difficult to evaluate the structures. All these problems were avoided by the postembedding method. As in every technique, here also profiles may be non-immunoreactive, not only because of the apparent absence of crossreactivity, but also due to the loss of the antigen during fixation, loss of antigenicity by masking of the antigenic sites during embedding or processing.

Based on these technical, practical considerations, we can state that the postembedding techniques used by us were rather helpful for elucidating structural and molecular components of the photosensitive submammalian pineal organ.[19,52] Furthermore, we could show together with other authors[22,89,94] that the mammalian pineal organ possesses the S-antigen component of the photochemical transduction cascade. In mammals, several authors observed rhodopsin,[95,96] rhodopsin kinase, retinal-binding protein and intercellular retinol-binding protein, substances typical of photoreceptor cells.[39,73-76,88] Our studies revealed the presence of at least three cytologically and immunocytochemically different types of pinealocytes and of a neurohormonal, as well as a synaptic apparatus including GABA-immunoreactive elements.[14,20,24,34,37,39] In our opinion, all these elements represent a structural and molecular basis that allows the pineal organ to integrate information on directly perceived light.

The different types of pinealocytes and the synaptic circuits observed by common electron microscopy and GABA-ICC speak (*Fig. 33*) in favor of a considerable integrative activity performed in addition to the perception of light. The synapses found on the secondary pineal neurons suggest an integration of perceived light information with that coming from nuclei of the central nervous system by afferent pinealopetal nerve fibers. Our observations of neurohormonal terminals and efferent pinealofugal nerve fibers underline the existence of both an endocrine and a neural efferentation of the pineal organ. This morphological pattern may represent the basis of the regulation of the behavior in animals according to circadian and circannual changes of the environmental light.

ACKNOWLEDGEMENT

This work was supported by Hungarian OTKA grants 1679 and 1109.

REFERENCES

1. S.K. Ainsworth and M.J. Karnovsky, An ultrastructural staining method for enhancing the size and electron opacity of ferritin in thin sections, *J Histochem Cytochem* **20**:225-229 (1972).
2. G.V. Childs, G. Unabia, adf D. Ellison, Immunocytochemical studies of pituitary hormones with PAP, ABC, and immunogold techniques: Evolution of technology to best fit the antigen. *Am J Anat* **175**:307-330 (1986).
3. J. Herbert, The role of the pineal gland in the control by light of the reproductive cycle of the ferret. *In*: The pineal gland, G.E.W. Wolstenholme, J Knight (eds) Churchill Livingstone, Edinburgh, London, p:303-327 (1971).
4. I. Vigh-Teichmann, H.W. Korf, F. Nürnberger, A. Oksche, B. Vigh, and R. Olsson, Opsin-immunoreactive outer segments in the pineal and parapineal organs of the lamprey (Lampetra fluviatilis), the eel (Anguilla anguilla), and the rainbow trout (Salmo gairdneri), *Cell Tissue Res* **230**:289-307 (1983).
5. B.D. Goldman and J.M. Darrow, The pineal gland and mammalian photoperiodism, *Neuroendocrinology* **37**:386-396 (1983).
6. H. Illnerova, Entrainment of mammalian circadian rhythms in melatonin production by light, *Pineal Res Rev* **6**:173-217 (1988).
7. J.D. Glass, Neuroendocrine regulation of seasonal reproduction by the pineal gland and melatonin, *Adv Pineal Res* **6**:219-259 (1988).
8. V.M. Cassone, The pineal gland influences rat circadian activity rhythms in constant light, *J Biol Rhythms* **7**:27-40 (1992).
9. M. Max and M. Menacker, (1992) Regulation of melatonin production by light, dark, and temperature in the trout pineal, *J Comp Physiol* **A170**:479-490 (1992).
10. J.K. Kappers, The morphological and functional evolution of the pineal organ during its phylogenetic development, *Excerpta Med Int Congr Ser* **184**:619-626 (1968).

11. J.P. Collin, Differentiation and regression of the cells of the sensory line in the epiphysis cerebri, *in*: "The Pineal Gland", G.E.W. Wolstenholme, J. Knight (eds. Churchill Livingstone, Edinburgh, London (1971) p. 79-125.

12. J.P. Collin, P. Brisson, J. Falcon, and P. Voisin, Multiple cell types in the pineal, Functional aspects, *in*: "Pineal and Retinal Relationships", P.J. O'Brien and D.C. Klein, eds. Academic Press, New York (1986) p:15-32.

13. B. Vigh, I. Vigh-Teichmann, and B. Aros, Comparative ultrastructure of cerebrospinal fluid-contacting neurons and pinealocytes, *Cell Tissue Res* **158**:409-424 (1975).

14. B. Vigh and I. Vigh-Teichmann, Comparative neurohistology and immunocytochemistry of the pineal complex with special reference to CSF-contacting neuronal structures, *Pineal Res Rev* **6**:1-65 (1988).

15. I. Vigh-Teichmann, M.A. Ali, and B. Vigh, Comparative ultrastructure and opsin immunocytochemistry of the retina and pineal organ in fish, *Progr Brain Res* **91**:307-313 (1992).

16. A. Zachmannn, J. Falcon, S.C.M. Knift, V. Bolliet, and M.A. Ali, Effects of photoperiod and temperature on rhythmic melatonin secretion from the pineal organ of the white sucker (Calostomus-Commersoni) invitro, *Gen Comp Endocrinol* **86**:26-33 (1992).

17. B. Vigh, I. Vigh-Teichmann, B. Aros, and A. Oksche, Sensory cells of the "rod-" and "cone-type" in the pineal organ of Rana esculenta, as revealed by immunoreaction against opsin and by presence of an oil (lipid) droplet, *Cell Tissue Res* **240**:143-148 (1985).

18. B. Vigh, I. Vigh-Teichmann, and B. Aros, Neurohemal areas bordering the internal cerebral veins in the pineal organ of the bat (Myotis blythi oxygnathus), *Z Mikrosk-Anat Forsch* **100**:745-758 (1986).

19. B. Vigh and I. Vigh-Teichmann, Three types of photoreceptors in the pineal and frontal organs of frogs: Ultrastructure and opsin immunoreactivities, *Arch Histol Jpn* **49**:391-414 (1986).

20. B. Vigh and I. Vigh-Teichmann, The pinealocyte forming receptor and effector endings: Immunoelectron microscopy and calcium histochemistry, *Arch Histol Cytol* **52**:433-440 (1989).

21. I. Vigh-Teichmann and B. Vigh, The pinealocyte: Its ultrastructure and opsin immunocytochemistry, *Adv Pineal Res* **1**:31-40 (1986).

22. I. Vigh-Teichmann, B. Vigh, I. Gery, and Th. van Veen, Different types of pinealocytes as revealed by immunoelectron microscopy of anti-S-antigen and antiopsin binding sites in the pineal organ of toad, frog, hedgehog and bat, *Exp Biol* **45**:27-43 (1986).

23. I. Vigh-Teichmann, A. Szél, P. Röhlich, and B. Vigh, A comparison of the ultrastructure and opsin immunocytochemistry of the pineal organ and retina of the deep-sea fish Chimaera monstrosa, *Exp Biol* **48**:361-371 (1990).

24. I. Vigh-Teichmann, H. Petter, and B. Vigh, GABA-immunoreactive intrinsic and GABA-immunonegative secondary neurons in the cat pineal organ, *J Pineal Res* **10**:18-29 (1991a).

25. I. Vigh-Teichmann, M.A. Ali, and B. Vigh, Ultrastructure and opsin immunocytochemistry of the pineal complex of the larval Arctic charr Salvelinus alpinus: A comparison with the retina, *J Pineal Res* **10**:196-209 (1991b).

26. B. Vigh, and I. Vigh-Teichmann, Light- and electron-microscopic demonstration of immunoreactive opsin in the pinealocytes of various vertebrates, *Cell Tissue Res* **221**:451-463 (1981).

27. L. Vollrath, (1981) The pineal organ, *in*:"Handbuch der mikroskopischen Anatomie des Menschen", Vol 6/7, A. Oksche, and L. Vollrath, eds, Springer, New York (1981).

28. J.A. McNulty, and L.M. Fox, Pinealocyte synaptic ribbons and neuroendocrine function, *Microsc Res Techn* **21**:175-187 (1992).

29. W. Kolmer, Ganglienzellen als konstanter Bestandteil der Zirbel von Affen, *Z ges Neurol Psychiat* **121**:423-428(1929).

30. P.M. Levin, A nervous structure in the pineal body of the monkey, *J Comp Neurol* **68**:405-409 (1938).

31. I. Vigh-Teichmann, H.W. Korf, A. Oksche, and B. Vigh, Opsin-immunoreactive outer segments and acetylcholinesterase-positive neurons in the pineal complex of Phoxinus phoxinus (Teleostei, Cyprinidae), *Cell Tissue Res* **227**:351-369 (1982).

32. P. Ekström and Th. van Veen, Pineal connections with the brain in two teleosts, the crucial carp and the European eel, *J Pineal Res* **1**:245-262 (1984).

33. R.L. Puzdrowski, and R.G. Northcutt, Central projections of the pineal complex in the silver lamprey Ichthyomyzon unicuspis, *Cell Tissue Res* **255**:269-274 (1989).

34. I. Vigh-Teichmann, Immunocytochemistry of photoneuroendocrine structures, (In Hungarian), Dr. Sci. Thesis, Hung. Acad. Sci. Budapest, pp 1-135 (1991).

35. B. Vigh and I. Vigh-Teichmann, Pinealofugal pathways, Int Symp Pineal Hormones, Bowral, Australia, July 21-24, 1991, Abstr. p.70 (1991).

36. H.W. Korf and U. Wagner, Nervous connections of the parietal eye in adult Lacerta s. sicula Rafinesque as demonstrated by anterograde and retrograde transport of horseradish peroxidase, *Cell Tissue Res* **219**:567-583 (1981).

37. B. Vigh, Comparative cytomorphology of pineal organ with special regard to CSF-contacting neuronal elements, (In Hungarian), Dr. Sci. thesis, Budapest 1987, p. 1-231 (1987).

38. I. Vigh-Teichmann, and B. Vigh, Comparison of epithalamic, hypothalamic and spinal neurosecretory terminals, *Acta Biol Acad Sci Hung* **30**:1-39 (1979).

39. I. Vigh-Teichmann, and B. Vigh, Immunocytochemistry and calcium cytochemistry of the mammalian pineal organ: A comparison with retina and submammalian pineal organs, *Microsc Res Techn* **21**:227-241 (1992).

40. A.B. Lerner, J.D, Case, Y. Takahashi, T.H, Lee, and W. Mori, Isolation, of melatonin, the pineal gland factor that lightens melanocytes, *J Amer Chem Soc* **80**:2587 (1958).

41. M. Saboureau, B. Vivien-Roels, and P. Pévet, Pineal melatonin concentrations during day and night in the adult hedgehog: Effect of a light pulse at night and superior cervical ganglionectomy, *J Pineal Res* **11**:92-98 (1991).

42. B. Rayer, M. Naynert, and H. Stieve, Phototransduction: different mechanisms in vertebrate and invertebrates, *J Photochem Photobiol, B Biol* **7**:107-146 (1990).

43. P.A. Hargrave, and J.H. McDowell, Rhodopsin and phototransduction: a model system for G protein-linked receptors, *FASEB J* **6**:2323-2331 (1992).

44. J.B. Hurley, Signal transduction enzymes of rhodopsin of vertebrate photoreceptors, *J Bioenergetic Biomembranes* **24**:219-226 (1992).

45. I. Vigh-Teichmann, B. Vigh, A. Szél, P. Röhlich, and G.H. Wirtz, Immunocytochemical localization of vitamim A in the retina and pineal organ of the frog, Rana esculenta, *Histochemistry* **88**:533-543 (1988).

46. I. Vigh-Teichmann, B. Vigh, and G.H. Wirtz, Immunoelectron microscopy of rhodopsin and vitamin A in the pineal organ and lateral eye of the lamprey, *Exp Biol* **48**:203-213 (1989).

47. L.A. Donoso, D.S. Gregerson, L. Smith, S. Robertson, V. Knospe, T. Vrabec, and C.M. Kalsow, S-antigen - Preparation and characterization of site-specific monoclonal antibodies, *Current Eye Res* **9**:343-355 (1990).

48. R.N. Lolley, C.M. Craft, and R.H. Lee, Photoreceptors of the retina and pinealocytes of the pineal gland share common components of signal transduction, *Neurochem Res* **17**:81-89 (1992).

49. J. Roth, D.J. Taatjes, and M.J. Warhol, Prevention of non-specific interactions of gold-labeled reagents of tissue sections, *Histochemistry* **92**:47-56 (1989).

50. D.S. Papermaster, B.G. Schneider, M.A. Zorn, and J.P. Kraehenbuhl, Immunocytochemical localization of opsin in outer segments and Golgi zones of frog photoreceptor cells. An electron microscope analysis of cross-linked albumin embedded retinas, *J Cell Biol* **77**:196-210

51. A Szél, L. Takács, E. Monostori, T. Diamantstein, I. Vigh-Teichmann, and P. Röhlich, Monoclonal antibody recognizing cone visual pigment, *Exp Eye Res* **43**:871-883 (1986).

52. I. Vigh-Teichmann and B. Vigh, Opsin immunocytochemical characterization of different types of photoreceptors in the frog pineal organ, *J Pineal Res* **8**:323-333 (1990).

53. D.H. Conrad, and G.H. Wirtz, Characterization of antibodies to vitamin A, *Immunochemistry* **10**:273-275 (1973).

54. G.H. Wirtz and S.S. Westfall, Reactivity of vitamin A derivatives and analogues with vitamin A antibodies, *J Lipid Res* **22**:869-871 (1981).

55. A.H. Coons, Fluorescent antibody methods, in: "General Cytochemical Methods," J.F. Danielli, ed., Academic Press, New York (1958) pp 399-422.

56. L.A. Sternberger, Immunocytochemistry, John Wiley, New York (1979).

57. S.M. Hsu, L. Raine and H. Fanger, The use of avidin-biotin-peroxidase complex (ABC) in immunoperoxidase techniques: A comparison between ABC and unlabeled antibody (PAP) procedures, *J Histochem Cytochem* **29**:577-580 (1981).

58. H.D. Mayor, J.C. Hampton, and B. Rosario, (1961) A simple method for removing the resin from epoxy-embedded tissue, *J Biophys Biochem Cytol* **9**:909-910 (1961).

59. B.P. Lane and D.L. Europa, Differential staining of ultrathin sections of Epon-embedded tissues for light microscopy, *J Histochem Cytochem* **13**:579-582 (1965).

60. R.D. Lillie and P. Pizzolato, Histochemical use of borohydrides as aldehyde blocking reagents, *Stain Technol* **47**:13-16 (1972).

61. J.C. Blanks, and J. Stone, Selective lectin binding of the developing mouse retina, *J Comp Neurol* **221**:31-41 (1983).

62. D.I. Stott, Immunoblotting and dot blotting, *J Immunol Methods* **119**:153-187 (1989).

63. O. Behnke, T. Ammitzboll, H. Jessen, M. Klokker, K. Nilausen, J. Tranum-Jensen, and L. Olsson, Non-specific binding of protein-stabilizer gold sols as a source of error in immunocytochemistry, *Eur J Cell Biol* **41**:326-338 (1986).

64. F. van Leeuwen, Pitfalls in immunocytochemistry with special reference to the specificity problems in the localization of neuropeptides, *Amer J Anat* **175**:363-377 (1986).

65. I. Vigh-Teichmann, P. Röhlich, B. Vigh, and B. Aros, Comparison of the pineal complex, retina and cerebrospinal fluid contacting neurons by immunocytochemical antirhodopsin reaction, *Z Mikrosk-Anat Forsch* **94**:623-640 (1980).

66. D.M. Chen, J.S. Collins, and T.H. Goldsmith, The ultraviolet receptor of bird retinas, *Science* **225**:337-339 (1984).

67. J.E.G. Downing, M.B.A. Djamgoz, and J.K. Bowmaker, Photoreceptors of a cyprinid fish, the roach: morphological and spectral characteristics, *J Comp Physiol A* **159**:859-868 (1986).

68. D. Burkhardt and E. Maier, The spectral sensitivity of a passerine bird is highest in the UV, *Naturwissenschaften* **76(1)**:82-83 (1989).

69. G.H. Jacobs, J. Neitz, and J.F. Deegan II, Retinal receptors in rodents maximally sensitive to ultraviolet light, *Nature* **353**:655-656 (1991).

70. H. Pohl, Ultraviolet radiation: A zeitgeber for the circadian clock in birds, *Naturwissenschaften* **79**:227-229 (1992).

71. K. Uchida and Y. Morita, Intracellular responses from UV-sensitive cells in the photosensory pineal organ, *Brain Res* **534**:237-242 (1990).

72. I. Vigh-Teichmann and B. Vigh, CSF-contacting neurons and pinealocytes, *in*: "The Pineal Gland. Current State of Pineal Research", B. Mess, S. Ruzsás, L. Tima, and P. Pévet, eds., Akadémiai Kiadó, Budapest 1985, pp 71-88.

73. H.W. Korf, R.G. Foster, P. Ekström, and J.J. Schalken, Opsin-like immunoreaction in the retinae and pineal organs of four mammalian species, *Cell Tissue Res* **242**:645-648 (1985).

74. R.G. Foster, A.M. Timmers, J.J. Schalken, and W.J. DeGrip, A comparison of some photoreceptor characteristics in the pineal and retina. II. The Djungarian hamster (Phodophus sungorus), *Comp Physiol A* **165**:565-572 (1989).

75. M. Araki, K. Watanabe, F. Tokunaga, and T. Nonaka, Phenotypic expression of photoreceptor and endocrine cell properties by cultured pineal cells of the newborn rat, *Cell Differ Dev* **25**:155-164 (1982).

76. K. Palczewski, M.E. Carruth, G. Adamus, J.H. McDowell, and P.A. Hargrave, Molecular, enzymatic and functional properties of rhodopsin kinase from rat pineal gland, *Vision Res* **30**:1129-1138 (1990).

77. D. Bownds, and G. Wald, Reaction of the rhodopsin chromophore with sodium borohydride, *Nature* **205**:254-257 (1965).

78. I. Vigh-Teichmann, B. Vigh, and G.H. Wirtz, Vitamin A immunocytochemistry of the retina and pineal complex in various vertebrates, *Serono Symp* **44**:61-64 (1987).

79. I. Vigh-Teichmann and B. Vigh, The cerebrospinal fluid-contacting neuron: A peculiar cell type of the central nervous system, Immunocytochemical aspects, *Arch Histol Cytol* **52 Suppl**:195-207 (1989).

80. G.S. Brindley."Physiology of the Retina and Visual Pathway", Arnold, London (1970).

81. W.F Zimmermann, The distribution and proportions of vitamin A compounds during the visual cycle in the rat, *Vision Res* **14**:795-802 (1974).

82. F. Chytil and D.E. Ong, Intracellular vitamin A-binding protein, *Annu Rev Nutr* **7**:321-335 (1987).

83. A.T.C. Tsin, T.S. Phillips, and R.J. Reiter, An evaluation of the level of retinoids in the bovine pineal body, *Adv Pineal Res* **3**:147-150 (1989).

84. H.L. Shi, H.C. Furr, and J.A. Olson, Retinoids and carotenoids in bovine pineal gland, *Brain Res Bull* **26**:235-239 (1991).

85. M. Tabata, T. Suzuki, and H. Niwa, Chromophores in the extraretinal photoreceptor (pineal organ) of teleosts, *Brain Res* **338**:173-176 (1985).

86. S. Tamotsu and Y. Morita, Blue sensitive visual pigment and photoregeneration in pineal photoreceptors measured by high performance liquid chromatography, *Comp Biochem Physiol PT B* **96**:487-490 (1990).

87. J.H. Sun, R.J. Reiter, N.L. Mata, and A.T.C. Tsin, Identification of 11-cis-retinal and demonstration of its light-induced isomerization in the chicken pineal gland, *Neurosci Lett* **133**:97-99 (1991).

88. C.D.B. Bridges, R.G. Foster, R.A. Landers, and S.L. Fong, Interstitial retinol-binding protein and cellular retinal-binding protein in the mammalian pineal, *Vision Res* **27**:2040-2060 (1987).

89. C.M. Kalsow, and W.B. Wacker, Localization of a uveitogenic soluble retinal antigen in the normal guinea pig eye by an indirect fluorescent antibody technique, *Int Arch Allergy* **44**:11-20 (1973).

90. H.S. Dua, P. Hossain, P.A.J. Brown, A. McKinnon, J.V. Forrester, D.S. Gregerson, and L.A. Donoso, Structure-function studies of S-antigen - Use of proteases to reveal a dominant uveitogenic site, *Autoimmunity* **10**:153-164 (1991).

91. J.M. Bonnin, and R. Perentes, Retinal S-antigen imunoreactivity in medulloblastoma, *Acta Neuropathol* **76**:204-208 (1988).

92. U. Scheuring, M. Franco, B. Flievet, H. Guizouarn, M. Mirshahi, J. Faure, and R. Motais, Arrestin from nucleated red blood cells binds to bovine rhodopsin in a light-dependent manner, *FEBS Lett* **276**:192-196 (1990).

93. T. Abe, and T. Shinohara, S-antigen from the rat retina and pineal gland have identical sequences, *Exp Eye Res* **51**:111-112 (1990).

94. L.A. Donoso, C.F. Merryman, K.E. Edelberg, R. Naids, and C. Kalsow, S-antigen in the developing retina and pineal gland: A monoclonal antibody study, *Invest Ophthalmol Vis Sci* **26**:561-568 (1985).

95. I. Vigh-Teichmann, L.A. Donoso, W.T. de Grip, and B. Vigh, Opsin- and S-antigen-immunoreactive pinealocytes in mammals. 6th Coll. Europ. Pineal Soc., Copenhagen, July 23-27, 1993. Abstr. A5 (1993).

96. I. Vigh-Teichmann, W.T. de Grip, and B. Vigh, Immunocytochemistry of pinealocytes and synapses in the mammalian pineal organ. *Microscopia Elettronica* **14(2)Suppl**:387-388 (1993).

CHAPTER 16

IN SITU HYBRIDIZATION FOR PEPTIDE mRNA

Giorgio Terenghi and Julia M. Polak[1]

Department of Histochemistry, Royal Postgraduate Medical School
Hammersmith Hospital, London W12 ONN, UK

SUMMARY

The introduction of *in situ* hybridization for mRNA using labelled complementary nucleic acid probes has made possible the understanding of gene expression at the cellular level. Different types of probe have been used for *in situ* hybridization, including cDNA sequences, synthetic oligonucleotides, and complementary RNA (cRNA) probes. Hybridization can be carried out on sections of fixed tissue as well as on cell culture preparations. The approach to radioactive and non-radioactive *in situ* hybridization is similar, with some variations in the concentration of the solutions or the timing of different steps. The choice of radiolabel is generally determined by a balance between speed and resolution, and both ^{32}P and ^{35}S are suited for studies of the neuroendocrine system. The use of digoxigenin is recommended when working with non-radioactive probes, as it sometimes gives lower background than biotin. The detection of either non-isotopic reporter molecule can be carried out using a wide range of immunohisto-chemical methods. We have applied *in situ* hybridization for various studies of regulatory peptides, such as anatomical identification of peptide mRNA, detection of functional changes of peptide synthesis and secretion in different endocrine situations. *In situ* hybridization can offer a suitable complement or alternative to morphological investigation at light microscopical level by immunohistochemistry.

[1]Address for correspondence: Julia M. Polak, D.Sc., M.D., FRCPath, Professor, Department of Histochemistry, Royal Postgraduate Medical School, Hammersmith Hospital, Du Cane Road, London W12 ONN, UK. Tel: 081-743-2030 ext 3231; Fax: 081-743-5362.

INTRODUCTION

Currently, molecular biology is widely used in research and pathological investigation, either in the form of hybridization blot analysis or of *in situ* hybridization analysis. Southern and Northern blot hybridization analysis are useful techniques to confirm the presence of a specific nucleic acid sequence. However, both are time consuming procedures which require a considerable amount of expertise and facilities. Also, because of the nucleic acid extraction technique used in these methods, it is not possible to define precisely the cell type encoding the RNA or DNA sequence. Furthermore, if the sequence is only present in small amounts, the dilution effect due to the extraction technique might prevent its detection. By using *in situ* hybridization, it is possible to overcome some of these problems and, more importantly, to localize precisely nucleic acid sequence in whole cells or tissue sections, giving unique information which is not available from other techniques. The technique is simply based on the use of labelled nucleic acid probes which are able to hybridize with complementary RNA or DNA target sequence, but because of its combination of histological and molecular biology methods, sometimes it may appear daunting to the beginner. In this chapter, we will give an overview of different aspects of this methodology and the problems which might be encountered by the operator paying particular attention to the detection of target mRNA. More extensive discussion of different aspects of the *in situ* hybridization technique can be found in other publications.[1,2,3]

FIXATION AND PREPARATION OF THE TISSUE

A prerequisite for obtaining satisfactory results is the retention of the nucleic acid sequence under investigation and the preservation of the tissue morphology. These can be achieved by using a suitable fixative solution, which can also stop the degradation of the target molecule. Nucleic acid degradation is due to a variety of factors, of which endogenous nucleases play a major role.[4] Manipulation, dissection, and dehydration of unfixed samples are also responsible for an acceleration of nucleic acid degradation, possibly through the release of endonucleases and lysosomal products.[5] DNA is stable in the tissue for a relatively long time, as DNAses are not very resistant and may be denatured quickly. Hence DNA sequences can be identified even in tissue with poor preservation. Conversely, RNAse is an ubiquitous and very heat-stable enzyme, which can degrade target mRNA and RNA probes at any stage of the hybridization procedure. RNAse contamination should be carefully avoided. Protocols are available for the preparation of RNAse-free equipment and solutions.[6]

Different nucleic acid sequences show different degradation curves, and it is difficult to give a definite time limit for the fixation of the tissue. As target degradation could be very fast, it is recommended that the delay between tissue collection and fixation should be minimal, particularly for surgical specimens.[7,8] In case of post mortem material, good hybridization results have been obtained on samples collected up to 10 hours after death,[9,10] indicating that RNA degradation is slower in untouched cooled tissue, also described in studies of post mortem tissue by mRNA blot analysis.[5,11]

A variety of fixatives for *in situ* hybridization including precipitating and cross-linking solutions, have been used with different results.[3,7,9,12-17] Precipitating reagents are sometimes preferred as they provide a better penetration to the probes, but the preservation of the tissue morphology is often unsatisfactory and loss of target is a possibility.[14,15] For mRNA peptide studies, cross-linking fixation is preferred and a 4% paraformaldehyde solution is generally used. However, prolonged fixation might affect adversely the hybridization results.[18,19] Tissue permeabilization is often necessary to increase probe penetration.[7,15,18]

Choice of Probe and Labels

There are three different types of probes currently in use for *in situ* hybridization: DNA, RNA, and oligonucleotide probes. Complementary DNA probes (cDNA) were the first to be applied for *in situ* hybridization[20-22] and, although still used, they have been supplanted by the use of complementary RNA (cRNA) and oligonucleotide probes. cRNA probes are transcribed from a template sequence using an appropriate RNA polymerase enzyme in the presence of ribonucleotides.[23,24] They are considered more sensitive than cDNA probes,[25] as they are single stranded, can be uniformly labelled, and form stable RNA-RNA hybrids, thus allowing the use of higher hybridization and wash temperatures. Also, their size can be optimized for maximum tissue penetration (generally between 250 and 400 bases), either by restriction enzyme digestion of the template or by alkaline hydrolysis of the transcribed probe.[25] More recently, RNA probes have been prepared by using the polymerase chain reaction (PCR) methods with promoter sequence containing an RNA polymerase sequence.[26] This allows preparation of an RNA probe template from small amount of tissue.

Oligonucleotide probes can be easily synthesized from any given sequence.[27,28] Their small size, generally 30-50 bases, allows easy tissue penetration, although care should be taken in the choice of sequence and the setting of the hybridization conditions in order to avoid non-specific hybridization to homologous sequence.[29] However, the sensitivity of oligonucleotide probes is generally considered lower than that of RNA probes because of the limited amount of label that can be incorporated into these probes. The number of detectable labels can be increased by using a "cocktail" of different oligonucleotides recognizing different parts of the same target nucleic acid. Alternatively, it is possible to synthesize oligonucleotides which include a promoter sequence for RNA polymerase. This can be used as templates for the transcription of uniformly labelled short RNA probes.[30]

Isotopes or non-radioactive reporter molecules can be used to label the probes. The more commonly used isotopes are ^{32}P, ^{35}S and ^{3}H, although ^{125}I has also been used and ^{33}P was recently introduced on the market as an alternative. The choice between isotopes is dictated by the system under investigation and by the need for resolution, sensitivity and speed.[31] The detection of radiolabelled probes is possible by using either film or emulsion autoradiography. The book by Roger[32] offers a comprehensive guide to this technique. ^{32}P has very high specific activity allowing a very rapid detection by autoradiography. However, the wide spectrum of ß emission results in a wide scatter of silver grain and, consequently, a poor resolution. In contrast, resolution is very good with ^{3}H-labelled probes, but the exposure time for autoradiography is long, generally 3-4 weeks. ^{33}P and ^{35}S offer a relatively low range of ß particle emission (*Fig. 1*), with a relative short time of autoradiographic exposure (2-10 days). However, the former is particularly costly, while the latter might give high background signal because of the non-specific binding of the sulphur to the tissue. This problem can be partly overcome by including dithiothreitol (DTT, 100 mM) in the hybridization reaction and in the post-hybridization washes.[15]

In the past, the use of non-radioactive labels has been restricted by their limited sensitivity of detection. However, recent studies have shown that the sensitivity of biotin- and digoxigenin-labelled probes is comparable to that of radiolabelled ones.[33-37] Biotin can be use for labelling either cRNA or oligonucleotide probes,[16,38-40] although the use of this label has been restricted because of the high background signal that may result from the high levels of tissue endogenous biotin. Digoxigenin is a plant-derived molecule, recently introduced as probe marker.[41] Digoxigenin-labelled probes appear to have an increased sensitivity as compared to biotinylated one,[35,42] possibly due to the low levels of background noise. The detection of both biotin and digoxigenin is based on immunohisto-chemical methods, with either fluorescent or enzymatic reporter molecules (*Fig. 2*).

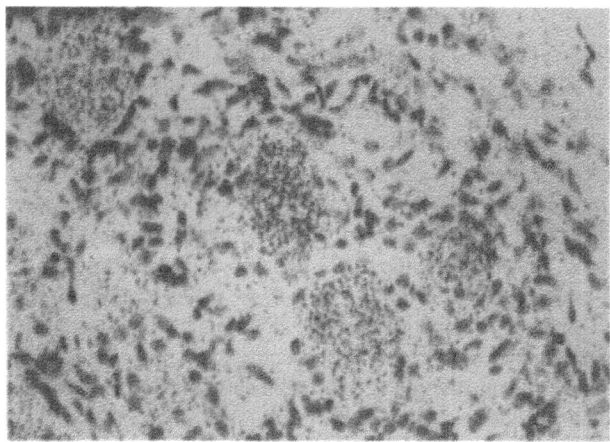

Figure 1. Calcitonin gene-related peptide (CGRP) mRNA localized to primary sensory neurons in rat dorsal root ganglia using ^{35}S-labelled cRNA probe. Emulsion autoradiography, 10 days exposure time, hematoxylin counterstaining.

Different immunohistochemical enhancing methods can also be applied, which can improve both resolution and sensitivity of detection considerably.[40] It is possible to use biotinylated and digoxigenin-labelled probes for the simultaneous identification of different target sequences within the same tissue.[43,44] Double *in situ* hybridization can also be carried out by using a combination of directly labelled probes and immunohisto-chemical detection.[45]

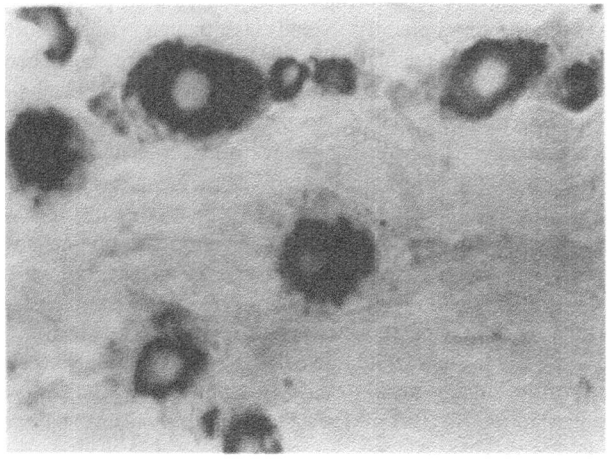

Figure 2. A section of rat dorsal root ganglia comparative to that in Figure 1 showing CGRP mRNA detected using a digoxigenin-labelled cRNA probe. The probe was detected using antibodies against digoxigenin conjugated with alkaline phosphatase, developed according to the NBT/BCIP method.

HYBRIDIZATION CONDITIONS

The hybridization procedure can be broadly divided into tissue pretreatment, hybridization of the probe and post-hybridization washes. The scope of the pre-hybridization treatments is to increase the accessibility of the target nucleic acid to the probe and to prevent non-specific binding of the probe. Different studies have tried to define an optimal probe size,[14,20-22,46] but there is a general agreement that for cRNA probes a size ranging between 200 and 500 bases may give the best results. This problem is not encountered with oligonucleotide probes, as their small size is ideal for easy tissue penetration.

Tissue permeabilization can be achieved by limited enzyme digestion with proteases, although it is necessary to determine empirically the optimal protease concentration and digestion time for each tissue type. Excessive permeabilization might cause deterioration of cell morphology and loss of target RNA.[15,22,47] Other permeabilization procedures can include the use of lipolytic agents (such as Triton X-100), dehydration and rehydration through graded alcohol and xylene, or soaking the sections in diluted acid to remove basic proteins. Acetylation of tissue protein by treatment with 0.25% acetic anhydride can decrease electrostatic interaction between the probe and basic protein within the cells.[48] Acetylation may also prevent electrostatic binding of the probe to poly-L-lysine, which is often used to coat the slides to prevent loss of section.[49]

Hybridization can be described as the formation of hydrogen bonds between nucleotides of complementary nucleic acid sequences. Different factors can influence the formation and stability of the hybrids, such as the length of the sequence, the frequency of G-C pairs, the salt concentration and the temperature of the reaction. The latter two are particularly important in defining the experimental stringency of the reaction, which determines the stability of specifically bound probes (see below). Theoretically, the maximum hybridization rate occurs at about 25-30°C below the melting temperature (Tm),[50] which is the temperature at which half the hybrids dissociate. The Tm of a probe can be calculated with a mathematical formula which is different for DNA and RNA probes.[51-53] However, these calculations derive from solution hybridization, which do not take into account the influence of the tissue components. Hence, the hybridization temperature should be determined experimentally for each probe, using a range of temperatures (see protocol for indicative values).

Following hybridization, the tissue undergoes serial washes with solutions of decreasing concentration of salt and with increasing temperature. In effect, these changes of salt and temperature establish an increase of the stringency of the washes in order to remove more unstable non-specifically bound probes. This should result in a decrease of background noise in the final preparation. When using RNA probes, a further wash containing RNAse helps in removing single stranded RNA while leaving the double stranded hybrid molecule intact.[25]

IN SITU HYBRIDIZATION OF PEPTIDE mRNA

In situ hybridization has been applied in the study of numerous tissues to assess the expression of peptide mRNA in both normal and pathological situations. In the gut during fetal development, peptide-containing neurons are not easily identified using only immunohistochemistry. For example, at 8 weeks gestation, antibodies to the pan-neuronal marker protein gene product 9.5 (PGP) immunostain the full network of neuronal cells and fibers, but VIP immunoreactivity is present in only few fibers in the myenteric plexus. VIP-immunoreactive neuronal cell bodies can be identified only at 18 weeks gestation, when the amount of stored peptide is sufficient to be detected by immunohistochemistry. However, VIP mRNA can be detected in the neuronal cells using in situ hybridization from 9 weeks gestation onwards (Fig. 3), thus allowing one to follow the pattern of neuronal colonization of the gut at an earlier gestational age than would be possible with immunohistochemistry.[54]

Endothelin-1 (ET-1) is a potent vasoconstrictor peptide which was first isolated from endothelial cells. Morphologically, the peptide has been localized to the endothelium in

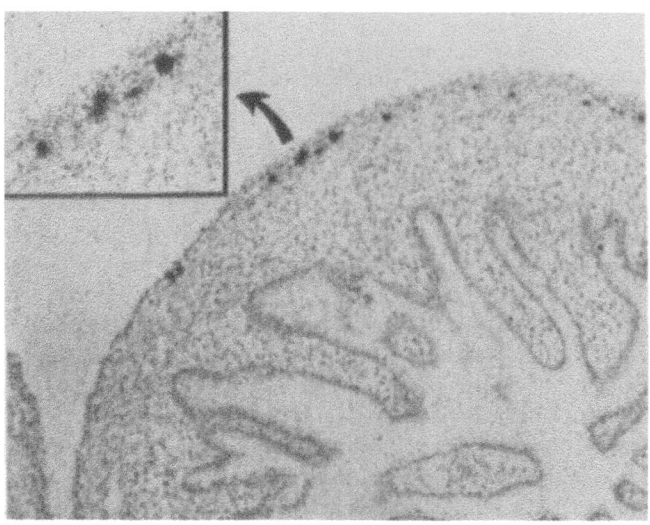

Figure 3. Vasoactive intestinal peptide (VIP) mRNA in ganglion cells of the myenteric plexus of human small intestine from 12 weeks gestation. Inset: details of some of the neuronal cells. ^{32}P-labelled cRNA probe, emulsion autoradiography after 5 days exposure. Hematoxylin counterstaining.

blood vessels using immunohistochemistry, and its synthesis in these cells has been confirmed using *in situ* hybridization[55] (*Fig. 4*). Electrophysiological studies have also shown a direct effect of endothelin on neuronal excitability, raising the possibility that ET-1 might also be localized to neuronal cells. Using a combination of immunohisto-chemistry and *in situ* hybridization, ET-1 peptide and its mRNA were localized to a large number of primary sensory neurons in dorsal root ganglia (*Fig. 5*) and in motoneurons of the spinal cords.[56] Comparison of sections also showed that ET-1 mRNA almost always co-exists with substance P and/or calcitonin gene-related peptide (CGRP) transcripts in neuronal cells, similar to the results obtained from immunostaining for all these peptides.

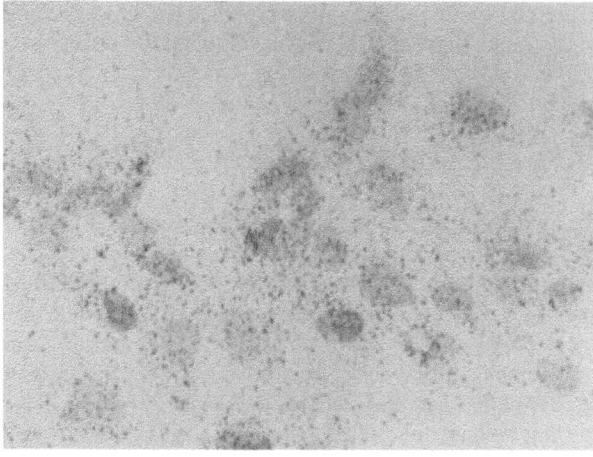

Figure 4. Endothelin-1 (ET-1) mRNA localized in cultured endothelial cells of human dermal microvessels using a ^{32}P-labelled cRNA probe. Emulsion autoradiography after 10 days exposure, hematoxylin counterstaining.

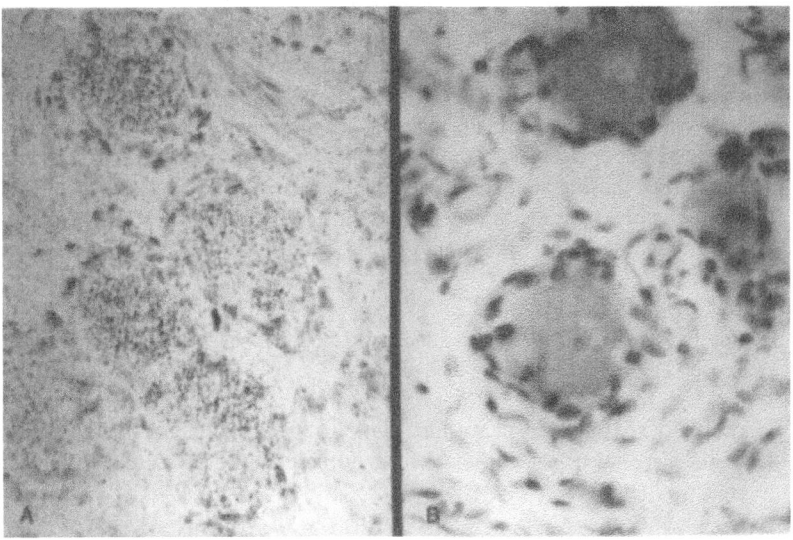

Figure 5. ET-1 mRNA **(A)** and immunoreactivity **(B)** localized in primary sensory neurons of rat dorsal root ganglion. *In situ* hybridization was carried out using a ^{35}S-labelled cRNA probe, detected by emulsion autoradiography.

Other immunohistochemical studies showed the localization of ET-1 in endocrine cells of the lung and in bronchial epithelium.[57] These results, and the finding that ET-1 has a trophic role, prompted further studies in order to identify the presence of ET-1 in different types of pulmonary tumors. *In situ* hybridization and immunohistochemistry showed that ET-1 is synthesized and stored in tumor cells of squamous cell carcinomas and adenocarcinomas, but not in small cell carcinomas.[58] This is in contrast to the results obtained with probes for gastrin-releasing peptide (GRP), a peptide with trophic effect on tumor cells, whose mRNA was identified in tumor cells of small cell carcinoma of the lung.[59] The differential distribution of GRP and ET-1 would suggest that each peptide has specific growth promoting effects on different tumor cell types.

These are only a few examples of a growing field of studies that use *in situ* hybridization in conjunction with other morphological and biochemical techniques, in order to more fully understand the role and functions of peptides and proteins in different tissues.

PROTOCOLS

Fixation

1. Cut tissue into small pieces (approx. 1 x 1 x 0.5 cm) using a sterile sharp blade, and fix them in freshly made 4% paraformaldehyde/0.1 M phosphate buffered saline (PBS) solution for 6 hours at room temperature. When mRNA is the target, special care should be taken. The specimen handling procedures should be carried out using clean disposable gloves and sterile instruments, in order to avoid RNAse contamination.

2. Animal tissue can be fixed *in situ* by perfusion with 4% paraformaldehyde/ PBS, followed by immersion fixation of the dissected tissue (1-4 hours, depending on

the tissue size and fixation obtained with perfusion). This method is strongly recommended if brain or spinal cord tissues are used. These tissues do not fix well by immersion only, due to the poor penetration of the fixative in the tissue matrix.

3.   Fixative other than paraformaldehyde may be more appropriate when investigating specific target molecules. Some indication of other possible fixatives can be gained from the literature. However, it is good practice to test more than one fixative in order to establish which gives the best hybridization results as indicated by the highest signal/background noise ratio with optimal preservation of morphology.

4.   After fixation, rinse the tissue blocks in 4-5 changes of PBS containing 15% sucrose (2 hours each change or overnight).

5.   Store the fixed material in washing buffer at 4°C for only a limited period (1 month). Cryostat or paraffin blocks should be prepared as soon as possible and stored ready to cut.

6.   Tissue sections should be mounted onto poly-L-lysine (PLL) coated slides and dried for at least 4 hours (or overnight) at 37°C to obtain maximum tissue adhesion. PLL-coated slides are best used immediately, but batches may be prepared and stored in racks, wrapped in the aluminum foil to protect from dust and RNAse contamination. PLL-coated slides can be stored at room temperature for up to 1 month.

Hybridization

All the steps up to hybridization (included) should be carried out in RNAse-free conditions. Solutions should be autoclaved or prepared with sterile ingredients using RNAse-free equipment. Equipment should be autoclaved, or baked at 250°C for 4 hours, as appropriate. Select the slides, number and mark them as necessary with pencil (not pen as ink may disappear during the various incubation steps).

1.   Rehydrate the sections by immersion in 0.2%Triton in PBS for 15 mins. Wash in PBS 2x3 mins.

2.   Carry out the tissue permeabilization by incubating the tissue in 0.1 M Tris/50 mM EDTA prewarmed at 37°C, containing 1 μg/ml proteinase K. Indicated incubation time is 15-20 min. However, the incubation time for proteinase K should be tested for each tissue type, as prolonged proteinase digestion could damage the tissue, with a loss of morphology and of target nucleic acid.

3.   Stop the proteinase K activity by immersion in 0.1 M glycine/PBS for 5 min. Then immerse the sections in 4% paraformaldehyde/PBS for 3 min to post-fix the target nucleic acid. Rinse the sections briefly in PBS, twice, to remove the paraformaldehyde.

4.   Place the slides in a staining jar containing 0.1M triethanolamine and, while stirring, add acetic anhydride to 0.25% (v/v) and incubate for 10 min. Rinse the slides briefly in double distilled water and dry them at 37°-40°C. This takes approximately 10 min.

5. Dissolve the probe in hybridization buffer at 50°C to a final concentration of 0.5 ng/μl (radiolabelled probes) or 2.5 ng/μl (non-radioactive probes).

6. Apply 10 μl of diluted probe per section to the dry slides. The volume of diluted probe can be increased for large sections.

7. Using fine forceps, gently place a siliconized coverslip onto the section to spread the probe solution. If there are any air bubbles, remove them by pressing gently on the coverslip with the forceps.

8. Hybridize the section for 16-20 hours at suitable temperature in a sealed humid chamber containing 5xSSC. It has been demonstrated that the hybridization reaction reaches an equilibrium after 4-6 hours incubation, when a maximum of hybrid has formed.[25] Good hybridization results can be obtained with short incubation times,[15,60] but generally incubation is carried out overnight for convenience.

9. A different hybridization temperature will be needed for various probes according to their Tm. The hybridization temperature is also dependent on the type of probes which have been used. With hybridization buffer containing 50% formamide, it is suggested that the following range of temperature should be tested initially: 42°-48°C for cDNA probes; 42°-55°C for cRNA probes; 37°-40°C for oligonucleotide probes.

10. Following hybridization, remove the coverslip by immersing the slide in 2xSSC containing 0.1% SDS. The coverslips will float off after a few minutes soaking. There is no need to autoclave the solutions used for post-hybridization washes.

11. Wash the sections in 2xSSC containing 0.1% SDS at room temperature, shaking gently, for four changes of 5 min. Then wash the slides in 0.1xSSC containing 0.1% SDS at the same temperature used for hybridization, shaking gently, for 2 changes of 10 min.

12. High background signal can be removed by prolonged washes of the slides at higher stringencies. Increase the number of washes in 0.1xSSC containing 0.1% SDS, at the same time, progressively increasing the temperature. Take care not to allow the sections to dry out in between any of the washes. Use the same container throughout, changing the solutions quickly.

13. If using cRNA probes, rinse the sections twice briefly in 2xSSC, then incubate 10 μg/ml RNAse A solution in 2xSSC at 37°C for 15 min. It is essential to remove any trace of SDS from the sections before the incubation with RNAse solution, as SDS would inhibit the action of the enzyme. It is desirable to use RNAse at this stage of the hybridization procedure if you have been using cRNA probes, as RNAse will denature the single stranded cRNA probe which is bound non-specifically to the section, hence decreasing the background staining. Double stranded RNA hybrids (cRNA-mRNA) are unaffected by the enzyme.

14. Briefly rinse sections in 2xSSC, then PBS before proceeding with either autoradiography (radiolabelled probes) or immunohistochemical detection (non-radioactive probes). For autoradiography see Roger.[32] For non-radioactive probes, detailed protocols are supplied with the kit by the manufacturer.

References

1. K. Valentino, J.H. Erberwine and J.D. Barchas (eds), *In situ* hybridization. Applications to neurobiology, Oxford University Press, Oxford (1987).
2. J.M. Polak and J.O.D. McGee (eds), *In situ* hybridization - Principles and practice, Oxford University Press, Oxford (1990).
3. G. Terenghi and R.A. Fallon, Techniques and applications of *in situ* hybridization, *in*: "Current topics in pathology - Pathology of the nucleus", Underwood JCE (ed), Springer Verlag, Berlin, p. 290-337 (1990).
4. D.A. Nielsen and D.J. Shapiro, Insight into hormonal control of messenger RNA stability, *Mol Endocrin* **4**:953-957 (1990).
5. S.A. Johnson, D.G. Morgan and C.E. Finch, Extensive post-mortem stability of RNA from rat and human brain, *J Neurosci Res* **16**:267-280 (1986).
6. J. Sambrook, E.F. Fritsch, and T. Maniatis, Molecular cloning - A laboratory manual. 2nd Edition, Laboratory Press Cold Spring Harbor, (1989).
7. H. Höfler, H. Childers, M.R. Montminy, R.M. Lechan, R.H. Goodman RH and H.J. Wolfe, *In situ* hybridization methods for the detection of somatostatin mRNA in tissue sections using antisense RNA probes, *Histochem J* **18**:597-604 (1986).
8. M. Asanuma, N. Ogawa, K. Mizukawa, K. Haba and A. Mori, Comparison of formaldehyde-preperfused frozen and freshly frozen tissue preparation for the *in situ* hybridization for α-tubulin mRNA in the rat brain, *Res Comm Chem Pathol Pharmacol* **70**:183-192 (1990).
9. G. Terenghi, J.M. Polak, Q. Hamid, E. O'Brien, P. Denny, S. Legon, J. Dixon, C.D. Minth, S.L. Palay, G. Yasargil and V. Chan-Palay, Localization of neuropeptide Y mRNA in neurons of human cerebral cortex by means of *in situ* hybridization with complementary RNA probes, *Proc Natl Acad Sci USA* **84**:7315-7318 (1987).
10. S.J. Gibson, J.M. Polak, A. Giaid, Q.A. Hamid, S. Kar, P.M. Jones PM, P. Denny, S. Legon, S.G. Amara, R.K. Craig, S.R. Bloom, R.J.A. Penketh, C. Rodek, N.B.N. Ibrahim and A. Dawson, Calcitonin gene-related peptide mRNA is expressed in sensory neurons of the dorsal root ganglia and also in spinal motoneurons in man and rat, *Neurosci Lett* **91**:283-288 (1988).
11. G.R. Taylor, G.I. Carter, T.J. Crow, J.A. Johnson, A.F. Fairbairn AF, E.K. Perry and R.H.Perry, Recovery and measurement of specific RNA species from postmortem brain tissue: a general reduction in Alzheimer's disease detected by molecular hybridization, *Exp Mol Pathol* **44**:111-116 (1986).
12. A.T. Haase, M. Brahic, and L. Stowring, Detection of viral nucleic acids by *in situ* hybridization, *in*: "Methods in virology, VII", Maramorosch K, Koprowski H (eds), Academic Press, New York, pp 189-226 (1984).
13. H.A. McAllister and D.L. Rock, Comparative usefulness of tissue fixatives for *in situ* viral nucleic acid hybridization, *J Histochem Cytochem* **33**:1026-1032 (1985).
14. T.R. Moench, H.E. Gendelman, J.E. Clements, O. Narayan, and D.E. Griffin, Efficiency of *in situ* hybridization as a function of probe size and fixation technique, *J Virol Methods* **11**: 119-130 (1985).
15. R.H. Singer, J.B. Lawrence, and C. Villnave, Optimization of *in situ* hybridization using isotopic and non-isotopic detection methods, *Biotechniques* **4**:230-250 (1986).
16. A.F. Guitteny, B. Fouque, C. Mongin, R. Teoule, and B. Boch, Histological detection of mRNA with biotinylated synthetic oligonucleotide probes, *J Histochem Cytochem* **36**:563-571 (1988).
17. I. Tournier, D. Bernau, A. Poliard, D. Schoevaret, and G. Fedman, Detection of albumin mRNAs in rat liver by *in situ* hybridization: usefulness of paraffin embedding and comparison of various fixation procedures, *J Histochem Cytochem* **35**:453-459 (1987).
18. D.J. Brigati, D. Myerson, J.J. Leary JJ *et al*, Detection of viral genomes in cultured cells and paraffin-embedded tissue sections using biotin labelled hybridization probes. *Virology* **126**:32-50 (1983).
19. J.N. Wilcox, C.E. Gee, and J.L. Roberts, *In situ* cDNA-mRNA hybridization: development of a technique to measure mRNA levels in individual cells, *Methods Enzymol* **124**:510-533 (1986).
20. M. Brahic and A.T. Haase, Detection of viral sequences of low reiteration frequency by *in situ* hybridization, *Proc Natl Acad Sci USA* **75**:6125-6129 (1978).
21. L.M. Angerer and R.C. Angerer, Detection of poly A⁺ RNA in sea urchin eggs and embryos by quantitative *in situ* hybridization, *Nucleic Acid Res* **9**:2819-2840 (1981).
22. J.B. Lawrence and R.H. Singer, Quantitative analysis of *in situ* hybridization methods for the detection of actin gene expression. *Nucleic Acid Res* **13**:1777-1799 (1985).

23. D. Melton, P. Kneg, M. Rabagliati, T. Maniatis, K. Zinn and M.R. Green, Efficient in vitro synthesis of biologically active RNA and RNA hybridization probes from plasmids containing a bacteriophage SP6 promoter, *Nucleic Acid Res* **12**:7035-7056 (1984).

24. R.C. Angerer, K.H. Cox and L.M. Angerer, *In situ* hybridization to cellular RNAs, *Genet Eng* **7**:43-65 (1985).

25. K.H. Cox, D.V. De Leon, L.M. Angerer and R.C. Angerer, Detection of mRNAs in sea urchin embryos by *in situ* hybridization using asymmetric RNA probes, *Dev Biol* **101**:485-502 (1984).

26. I.D. Young, L. Ailles, K. Deugau, and R. Kisilevsky. Transcription of cRNA for *in situ* hybridization from polymerase chain reaction-amplified DNA, *Lab Invest* **64**:709-712 (1991).

27. M.H. Caruthers, S.L. Beaucage, J.W. Efcavitch, E.F. Fisher, R.A. Goldman, P.L. de Haseth, W. Mandecki, M.D. Matteucci, M.S. Rosendahl and Y. Stabinski, Chemical synthesis and biological studies on mutated gen control regions, Cold Spring Harbor Symp Quant Biol 47: 411-418 (1982).

28. M.E. Lewis, T.G. Sherman, and S.J. Watson, *In situ* hybridization histochemistry with synthetic oligonucleotides; strategies and methods, *Peptides* **6(suppl 2)**:75-87 (1985).

29. Y. Kajimura, J. Krull, S. Miyakoshi, K. Itakura and H. Toyoda, Application of long synthetic oligonucleotides for gene analysis: Effect of probe length and stringency conditions on hybridization specificity, *GATA* **7**:71-79 (1990).

30. W. Brysch, G. Hagendorff and Schlingensiepen, RNA probes transcribed from synthetic DNA for *in situ* hybridization, *Nucleic Acid Res* **16**:2333 (1988).

31. M.A.W. Brady and F.M. Finlan, Radioactive labels: autoradiography and choice of emulsion for *in situ* hybridization, in *"In situ* hybridization - Principles and practice", Polak JM, McGee JOD (eds), Oxford University Press pp. 31-58 (1990).

32. A.W. Rogers, Technique of autoradiography, Elsevier, Amsterdam (1979).

33. B. Bhatt, J. Burns, D. Flamery and J.O.D. McGee, Direct visualization of single copy genes on banded metaphase chromosome by non-isotopic *in situ* hybridization, *Nucleic Acid Res* **16**:3951-3961 (1988).

34. J.B. Lawrence, C.A. Villnave and R.H. Singer, Sensitive, high-resolution chromatin and chromosome mapping *in situ*: presence and orientation of two closely integrated copies of EBV in a lymphoma line, *Cell* **52**:51-61 (1988).

35. Y. Furuta, T. Shinohara, K. Sano, M. Meguro and K. Nagashima, *In situ* hybridization with digoxigenin-labelled DNA probes for detection of viral genomes, *J Clin Pathol* **43**:806-809 (1990).

36. S. Podell, W. Maske, E. Ibanez and E. Jablonski, Comparison of solution hybridization efficiencies using alkaline phosphatase-labelled and ^{32}P-labelled oligodeoxynucleotide probes, *Mol Cell Probes* **5**:117-124 (1991).

37. E.R. Unger, M.I. Hammer, and M.L. Chenggis, Comparison of ^{25}S and biotin as labels for *in situ* hybridization: use of an HPV model system, *J Histochem Cytochem* **39**:145-150 (1991).

38. M. Zabel and H. Schafer, Localization of calcitonin and calcitonin gene-related peptide mRNA in rat parafollicular cells by hybridocytochemistry, *J Histochem Cytochem* **36**:543-546 (1988).

39. L-I. Larsson, T. Christensen, and H. Dalboge, Detection of POMC mRNA by *in situ* hybridization using biotinylated oligodeoxynucleotide probes and avidin-alkaline phosphatase histochemistry, *Histochemistry* **89**:109-116 (1988).

40. A. Giaid, Q. Hamid, C. Adams, D.R. Springall, G. Terenghi and J.M. Polak, Non-isotopic RNA probes. Comparison between different labels and detection systems, *Histochemistry* **93**:191-196 (1989).

41. C. Kessler, The digoxigenin anti-digoxigenin (DIG) technology - a survey on the concept and realization of a novel bioanalytical indicator system, *Mol Cell Probes* **5**:161-205 (1991).

42. R.G. Morris, M.J. Arends, P.E. Bishop, K. Sizer, E. Duvall, and C.C. Bird, Sensitivity of digoxigenin and biotin labelled probes for detection of human papillomavirus by *in situ* hybridization, *J Clin Pathol* **43**:800-805 (1990).

43. C.S. Herrington, J. Burns, A.K. Graham, B. Bhatt, and McGee, Interphase cytogenetics using biotin and digoxigenin labelled probes. II: simultaneous detection of two nucleic acid species in individual nuclei, *J Clin Pathol* **42**:601-606 (1989).

44. B.J. Trask, H. Massa, S. Kenwrick and J. Gitschier, Mapping of human chromosome Xq28 by two colour fluorescence *in situ* hybridization of DNA sequences to interphase cell nuclei, *Am J Hum Genet* **48**:1-15 (1991).

45. R.W. Dirks, R.P.M. van Gijlswijk, R.H. Tullis, A.B. Smit, J. van Minnen, M. van der Ploeg and

A.K. Raap, Simultaneous detection of different mRNA sequences coding for neuropeptide hormones by double *in situ* hybridization using FITC-and biotin-labelled oligonucleotides, *J Histochem Cytochem* **38**:467-473 (1990).

46. C.E. Gee and J.L. Roberts, A technique for the study of gene expression in single cells, *DNA* **2**:157-163 (1983).

47. B.D. Shivers, B.S. Schachter and D.W. Pfaff, *In situ* hybridization for the study of gene expression in the brain, *Methods Enzymol* **124**:497-510 (1986).

48. S. Hayashi, I.C. Gillam, A.B. Delaney and G.M. Tener, Acetylation of chromosome squashes of *Drosophila melanogaster* decreases the background in autoradiographs from hybridization with ^{125}I-labelled RNA, *J Histochem Cytochem* **26**:677-679 (1978).

49. W-M. Huang, S.J. Gibson, P. Facer, J. Gu, and J.M. Polak, Improved section adhesion for immunocytochemistry using high molecular weight polymers of L-lysine as a slide coating, *Histochemistry* **77**:275-279 (1983).

50. R.J. Britten, D.E. Graham and B.R. Neufeld, Analysis of repeating DNA sequences by reassociation, *Method Enzymol* **29**:363-418 (1974).

51. C.A. Thomas and B.M. Dancis, Ring stability, *J Mol Biol* **77**:44-55 (1973).

52. D.K. Bodkin and D.L. Knudson, Assessment of sequence relatedness of double stranded RNA genes by RNA-RNA blot hybridization, *J Virol Methods* **10**:45-52 (1985).

53. L.M. Angerer, M.H. Stoler and R.C. Angerer, *In situ* hybridization with RNA probes: an annotated recipe, in: "*In situ* hybridization. Applications to neurobiology", Valentino K, Eberwine JH, Barchas JD (eds), Oxford University Press. pp 42-47 (1987).

54. P. Facer, A.E. Bishop, G. Moscoso, G. Terenghi, Y.F. Liu, R.H. Goodman, S. Legon and J.M. Polak, Vasoactive intestinal polypeptide gene expression in the developing human gastrointestinal tract, *Gastroenterology* **102**:47-55 (1992).

55. H.A. Bull, C.B. Bunker, G. Terenghi, D.R. Springall, Y. Zhao, J.M. Polak and P.M. Dowd, Endothelin-1 in human skin: immunolocalization, receptor binding, mRNA expression and effects on cutaneous microvascular endothelial cells, *J Invest Dermatol* **97**:618-623 (1991).

56. A. Giaid, S.J. Gibson, N.B.N. Ibrahim, S. Legon, S.R. Bloom, M. Yanagisawa, T. Masaki, I.M. Varndell and J.M. Polak JM, Endothelin-1, an endothelium-derived peptide, is expressed in neurons of the human spinal cord and dorsal root ganglia, *Proc Natl Acad Sci USA* **86**:7634-7638 (1989).

57. D.R. Springall, P. Howarth, H. Counihan, R. Djukovic, S. Holgate and J.M. Polak, Endothelin immunoreactivity of airway epithelium in asthmatic patients. *The Lancet* **337**:697-701 (1991).

58. A. Giaid, Q. Hamid, D.R. Springall, M. Yanagisawa, O. Shinmi, T. Sawamura, T. Masaki, S. Kimura, B. Corrin and J.M. Polak, Detection of endothelin immunoreactivity and mRNA in pulmonary tumours, *J Pathol* **162**:15-22 (1990).

59. Q. Hamid, A.E. Bishop, D.R. Springall, C. Adams, A. Giaid, P. Denny, M. Ghatei, S. Legon, F. Cutitta, J. Rode, E. Spindel, S.R. Bloom, and J.M. Polak, Detection of human probombesin mRNA in neuroendocrine (small cell) carcinoma of the lung, *Cancer* **63**:266-271 (1989).

60. J. McCafferty, L. Cresswell, C. Alldus, G. Terenghi and R.A. Fallon. A shortened protocol for *in situ* hybridization to mRNA using radiolabelled RNA probes, *Techniques* **1**:171-182 (1989).

CHAPTER 17

PRINCIPLES OF THE POLYMERASE CHAIN REACTION

Erich Arrer,[1] Gerhard W. Hacker,[2] Angelika Schiechl,[2]
Jiang Gu[3]

[1]Central Laboratory and
[2]Institute of Pathological Anatomy, General Hospital
 Salzburg, Austria
[3]Deborah Research Institute, Browns Mills, NJ, USA

SUMMARY

The polymerase chain reaction (PCR) is a nucleotide sequence amplification procedure allowing the production of large amounts of a specific DNA or RNA sequence from a complex DNA or RNA template. It is based on oligonucleotide primer annealing onto complementary nucleic acid sequences followed by enzymatic DNA synthesis by application of a heat-stable DNA polymerase. These steps, together with melting of the double-stranded DNA to obtain single-stranded templates, are performed sequentially using a computerized thermocycler which allows the automatic control of melting, annealing, and synthesis temperature during some 25-35 cycles. From the starting amount of as low as one single DNA or RNA molecule, several hundred thousand or even millions of copies of the initial sequence may be obtained. During the last decade, the technique has been widely applied for *in vitro* studies, and, very recently, *in situ* PCR techniques have been developed. The following chapter will familiarize the readers with the basic principles of this exciting technique. As an example of application, our results on the diagnosis of cystic fibrosis are presented.

---

[1]Address for Correspondence: Dr. E. Arrer, Central Laboratory, Salzburg General Hospital, Muellner Hauptstrasse 48, A-5020 Salzburg, Austria. Tel. ++43-662-4482-3804; Fax.++43-662-4482-885

INTRODUCTION

In the history of molecular biology, the emergence of new techniques such as Southern blotting, DNA sequencing, pulse field gel electrophoresis, molecular cloning, and others has often transformed the way of thinking about how to approach fundamental biomedical problems. About 20 years ago, recombinant DNA technology was introduced as a tool for the biological sciences. Molecular cloning has allowed the study of the structure of individual genes from living organisms. This method depends on the replication of the DNA from plasmids or other vectors during cell division of microorganisms. In 1984, based on an *in vitro* process, a DNA amplification procedure known as the polymerase chain reaction (PCR) was developed.[1-4] With this method, large amounts of a specific DNA fragment from complex DNA template can be produced using a simple enzymatic reaction.

Gene amplification by PCR can simplify many of the standard procedures for cloning, analyzing, and modify nucleic acids. It may be regarded as a form of "cell-free molecular cloning." PCR is characterized by its high sensitivity, selective amplification of the sequence of interest, and speed. Virtually pure DNA fragments from complex DNA templates can be obtained in a matter of hours rather than the weeks or months that traditional cloning requires.

The logic of the reaction is simple in principle. It is characterized by annealing of oligonucleotide to complement DNA or RNA sequences followed by enzymatic DNA synthesis *in vitro* primed by these oligonucleotides. The amount of starting material needed for PCR can be as little as a single molecule rather than the usual millions of molecules required for standard cloning and molecular biological analysis. Although purified DNA is used in many applications, it is not required for PCR. Crude cell lysates may also provide excellent templates. In contrast to requirements of other standard molecular biological procedures, the DNA need not even be intact as long as some molecules exist that contain sequences complementary to both primers.

The speed and sensitivity of PCR have been widely acknowledged by scientists. Since its introduction, PCR has transformed the way in which DNA analysis is carried out in both medicine and basic research. Such methods have also been applied to problems that only a few years ago were thought to be inaccessible to molecular analysis. The amplified products are verified by a number of criteria. The first and most frequently used is the length of the product visualized in ethidium bromide stained agarose or polyacrylamide gels or, in several cases, by capillary electrophoresis. Further criteria are predictable restriction sites which are monitored directly[5] or by oligomer restriction.[6] Amplified products can also be blotted on filters after electrophoretic separation or dotted directly and hybridized with labeled sequence-specific probes.[7] More recently, probes have been converted to nonisotopic colorimetric systems by labeling them with an enzyme and bound to a membrane or to a well of a microtiter plate. One of the detection systems is a colorimetric oligonucleotide ligation assay[8] and another, a luminescent probe detection system, called hybridization protection assay.[9] Labeled primers were once used in place of probe-based detection systems. However, this requires perfect target specificity in the amplification reaction to discriminate between target sequences that differ by only one or a small number of nucleotides.[10] Finally, a modified PCR protocol leads to an accumulation of single-stranded DNA in the amplification assay and facilitates DNA sequencing.[11] Direct sequencing of PCR fragments is also an important tool for detection of mutations.

It is foreseeable that PCR will greatly enhance the power of diagnostic activities that depend on the analysis of DNA or RNA sequences. This is true in the diagnosis of inherited disorders, genetic counseling, infectious diseases, forensic investigations and other related topics. In addition, this new method turns out to be very useful in a number of basic research applications in molecular biology and genetics. Rarely, has a new technique been so successful within such a short time.

BASIC TECHNIQUES OF PCR

To amplify a specific DNA segment by PCR, it is not necessary to know the nucleotide sequence of the target DNA. A pair of oligonucleotide primers complementary to a short stretch of known unique sequence that flank the target region of interest, is used to direct DNA synthesis. Design of these primers is a key point in carrying out a successful PCR. The length of the primers (about 20 to 30 bases) must be sufficient to overcome the statistical likelihood that their sequences would occur randomly in the overwhelmingly large number of non target DNA sequences in the sample.

PCR is carried out by repeated cycles of enzymatic *primer extension* in opposite and overlapping directions (*Fig.1*). Each cycle is initiated by *melting* double-stranded DNA to obtain single-stranded templates. This step is followed first by *annealing* of the primer oligonucleotides which are added in large molar excess over template strands. The 3'-hydroxyl ends face the target. Finally, each primer is elongated by a brief pulse of DNA synthesis. This three-step cycle is repeated 25-35 times until a sufficient amount of product is yielded.

Since both strands of a given DNA segment serve as templates, the number of target sequences increases exponentially. After the first few cycles, the major product is a DNA fragment that is exactly equal in length to the sum of the length of the two primers and the intervening target DNA.

The original protocol made use of the Klenow fragment of Escherichia coli DNA polymerase I. The unmodified DNA polymerase I, the DNA polymerase of the phage T4 or the modified T7 DNA polymerase, can also be applied.[12] Critical disadvantages of these enzymes are their heat lability and the fact that the reaction at a temperature of 37°C often produces an incompletely pure target product. The isolation of a heat-resistent DNA polymerase from the hot spring archebacterium *Thermus aquaticus*, designated "Taq polymerase," allows primer annealing and extension to be carried out at an elevated temperature, thereby reducing primer annealing to nontarget sequences. Each primer acts independently in the first few cycles as a probe, screening all the targets. If a pair of these primers hybridize nearby in the correct orientation, they select a subset of targets. This primer pair subsequently acts as a vector to amplify the target in further cycles, resulting in the production of large amounts of virtually pure target DNA. Starting with only 1 μg of whole human genomic DNA containing about 300,000 copies of each unique sequence, 25 cycles can generate up to several μg of a specific product several hundred base pairs in length.

A very important advantage of the Taq polymerase is its heat-stability because it survives even extended denaturation steps. This offers a number of advantages. First, it does not have to be added in each cycle. Secondly, this has allowed automation of PCR-using machines that have controlled heating and cooling capability.[13] A number of such devices, called thermocyclers, are now commercially available at relatively low cost.

PCR is efficient, fast, and extremely sensitive. Regarding efficiency, the theoretical upper limit of the number of product molecules is 2^n, where n is the number of cycles. This means that every single target sequence present at the start point could, in only 20 cycles, give rise to about a million progeny molecules. An attractive feature in many cases, is that target sequences do not have to be purified extensively prior to amplification. PCR can be successfully applied to amplify sequences from genomic DNA as a whole or from crude mixtures of total cellular RNA. Serum samples can be taken directly as a source of infectious DNA or RNA. However, uncontrolled biochemical sample compositions and a high degree of DNA or RNA complexity may reduce the specificity of the reaction. Random primer-target interactions cannot be excluded. To enhance specificity, an optimization of the annealing temperature, a "hot start" technique, the use of "nested" sets of primers,[14] or partial fractionation of crude preparation may be helpful. Nested sets of primers involve one set of cycles of repeated synthesis with a pair of primers and a second set of cycles with primer sequences located between the primers used in the previous round.

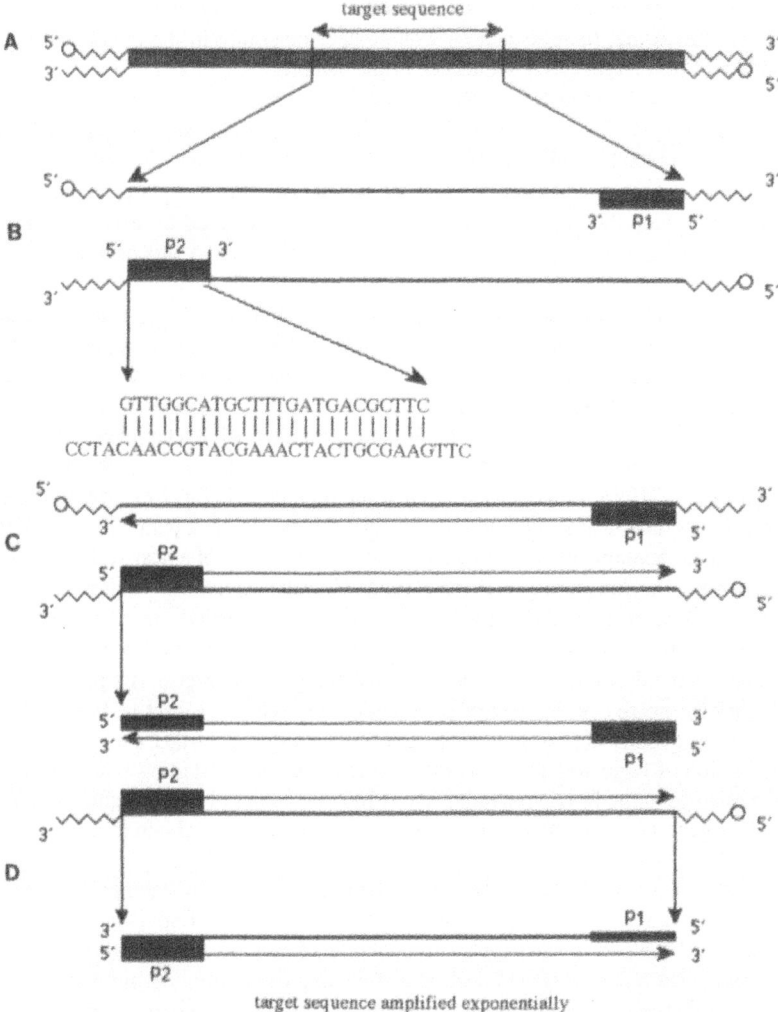

Figure 1. Simplified scheme of the polymerase chain reaction. A) The target sequence, limited by the arrows as shown, should be amplified. B) After separation of the double-stranded DNA, two primers P1 and P2 are annealed to their complementary sequences. C) Both primers are elongated at the 3'-end by DNA polymerase. D) In the second cycle the synthesized DNA sequence is limited in length by the 5'-ends of the primers, and only these sequences are amplified exponentially.

DNA or RNA as a template may be isolated from any biological source. It can be obtained from cells,[15] hair roots,[16] sperm,[17] and other body fluids. Even DNA extracted from embedded archival tissue[18] is suitable for the amplification reaction. The *in vitro* amplification of RNA sequences is also possible and is preceded by a reverse transcription step resulting in the generation of a single-stranded cDNA complementary to the original RNA. Then, in a second step, this cDNA is converted into double-stranded DNA by the action of the amplifying DNA polymerase.[19] Co-amplification of the mRNA with an internal standard may allow quantitation of the amount of specific mRNA.[20] A new development in PCR technology is *in situ* PCR which allows the amplification of DNA sequences in formalin-fixed cells of cytological paraffin-embedded materials.[21,22]

STANDARD PCR REACTION

The standard conditions[23] will amplify most target sequences and are presented here principally to provide starting conditions for designing new PCR applications. The reaction outlined below should be adequate for most *in vitro* PCR amplifications. It is suggested to optimize the protocol of each application especially for repetitive diagnostic or analytical procedures in which optimal performance is required.

Standard PCR is typically done in a 100 µl or 50 µl volume or even in smaller quantities.

1. A 100 µl reaction vial consists of:

Components	Volume	Final Concentration
Double distilled, sterile water	53.5 µl	{fill up to 100 µl volume}
[10x] reaction buffer*	10.0 µl	
dNTP's mix, 1.25 mM each dNTP	16.0 µl	{200 µM each dNTP}
primer I, 20 µM	5.0 µl	{1.0 µM}
primer II, 20 µM	5.0 µl	{1.0 µM}
template DNA	10.0 µl	{>10⁴ target copies}**
Taq polymerase	0.5 µl	{5 units/µl Taq polymerase}
mineral oil	2-3 drops***	

*)   The [10x] reaction buffer consists of 100 mM Tris-HCl (pH 8.3; 25°C), 500 mM KCl, 15 mM $MgCl_2$, 0.01% gelatine.
**)   1 µg of human single-copy genomic DNA equals 3×10^5 targets.
***)   Carefully overlay the reaction mix with 2-3 drops of mineral oil to prevent evaporation, or use AmpliWax®(Perkin Elmer).

2. As a guideline, the following temperature-profile may be used and applied for 30 to 35 cycles. The reaction is stopped by chilling to 4°C.

Denaturation:	95°C for ~1 min in the first cycle, then 30 sec.
Primer annealing:	55°C for 30 sec (depends on the G/C content of the primers).
Primer extension:	72°C for 1 min.
Final extension:	72°C for 5 min.

PRIMER SELECTION AND DESIGN

Unfortunately, the approach to selecting efficient and specific primers remains somewhat empirical. There is no set of rules that will ensure the synthesis of an effective primer pair. Yet it is the primers more then anything else that determine the success or failure of an amplification reaction. The following guidelines will help to find an efficient primer pair.[23]

1. The primers should be between 18-30 bases in length.
2. Select primers with random base distribution and a G/C content near 50%. Avoid significant secondary structures, particularly at the 3'-ends of the primers and stretches of polypurines and polypyrimidines.
3. Sequences should not complement within themselves or to each other, particularly at the 3'-ends, to reduce the incidence of "primer dimer" artefacts.
4. Sequences not complementary to the template can be added at the 5'-end of the primers to introduce restriction sites or regulatory elements in the PCR product.
5. For shorter primers or degenerate primers, the thermal profile of the reaction have to be adapted.
6. In general, concentrations ranging from 0.1 to 1.0 µM of each oligonucleotide are acceptable.

Finally, although most primers will work with varying degrees of success, occasionally, some primers may fail completely. In many of these cases, simply moving the primers by a few bases in either direction may solve the problem.

PCR BUFFER

The PCR buffer contains 10 mM Tris-HCl (pH 8.3 at room temperature), 50 mM KCl, and 1.5 mM $MgCl_2$. The stock solution [10x] should be aliquoted and frozen; it should not be refrozen for more than three times. The concentration of $MgCl_2$ in the reaction mix has a profound effect on the specificity and yield of an amplification. Concentrations of about 1.5 mM are usually optimal (with 200 µM each dNTP). As deoxynucletide triphosphates bind $MgCl_2$, the amount of dNTPs present in a reaction will determine the amount of free magnesium available.

The concentration of magnesium chloride influences:
* Primer annealing
* Temperature for denaturation of the DNA templates
* Specificity of the PCR
* Activity of the Taq polymerase
* Fidelity of the Taq polymerase

In some circumstances, different amounts of magnesium may provide better results. Generally however, excess of magnesium will result in the accumulation of non-specific products.

DEOXYNUCLEOTIDE TRIPHOSPHATES

Deoxynucleotide concentrations between 50 and 200 µM each result in the optimal balance among yield, specificity, and fidelity. The four dNTPs must be balanced to minimize misincorporation errors.

DNA POLYMERASE

Purified DNA polymerases to synthesize DNA *in vitro* have been available for a wide variety of molecular biological techniques. A critical disadvantage in the use of the Klenow fragment of E. coli DNA polymerase I, the DNA polymerase of the phage T4, or the modified T7 DNA polymerase is their heat-lability. When these DNA polymerases are used in the PCR, fresh DNA polymerase has to be added after each denaturation step.

DNA polymerase isolated from the archebacterium *Thermus aquaticus* (Taq polymerase)[24,25] survives even extended denaturation temperatures, does not have to be added in each cycle, and allows synthesis on elevated temperatures. Mismatch-priming is less likely to occur at higher than at lower annealing and DNA synthesis temperature.[5] *Thermus aquaticus* strain YT1 is a thermophilic, eubacterial microorganism capable of growth at 70-75°C. The native and recombinant form of the Taq polymerase[26] has a molecular weight of about 94 kD with a specific activity of 200,000 units/mg.

The extension rate depends on the nature of the DNA target and has its optimum approximately at 75°C. Lower temperatures lower the rate of extension:

* 150 nucleotides/sec at 75°C
* 60 nucleotides/sec at 70°C
* 24 nucleotides/sec at 55°C
* 1.5 nucleotides/sec at 37°C
* 0.25 nucleotides/sec at 22°C

Purified 94 kD Taq polymerase does not contain a 3',5'-exonuclease activity, and the rate of misincorporation was calculated to be about 1.1×10^{-4}. In PCR, the misincorporation rate of non-proofreading DNA polymerases is determined critically by the concentration and balance of nucleotide triphosphates. In contrast, the rate of misincorporation of the Klenow fragment DNA polymerase I is about 4-fold lower.[27] This disadvantage is of no consequence for the PCR for most purposes except cloning, because of the random distribution of the wrong bases. Mismatched bases, on the other hand, are inefficiently extended and promote chain termination. Because of its heat-stability, Taq polymerase retains 50% of its activity in a PCR-mix after heating 130 min at 92.5°C and 40 min at 95°C.

TEMPERATURE PROFILING

Denaturation Time and Temperature

Typically applied denaturation conditions are 95°C for 30 seconds; higher temperatures and extended denaturation are only required in certain cases, especially for G/C-rich targets. Incomplete denaturation reduces the product yield.

Primer Annealing

The temperature and time required for primer annealing depend on the base composition and on the length and concentrations of the primers. Starting with an annealing temperature 5°C below the calculated T_m (melting temperature) of the amplification primers is recommended. An increased annealing temperature decreases incorrect annealing of the primers, reduces misextension, and, therefore, will help to increase the specificity of the reaction. Some of the nonspecific amplification products can be eliminated under so-called "hot-start" conditions. This means that the reaction is started after initial heating the reaction mix over the calculated temperature for primer annealing by adding Taq polymerase or another reaction component to the PCR mix.

Primer Extension

Primer extension is performed at the optimum temperature for the Taq polymerase activity. An extension time of 1 minute a 72°C is sufficient for products up to 2 kb in length. For short sequences of less than 200 bp in length, two step cycles without an extension step are sufficient. However, longer elongation may be helpful in early cycles if the number of the target molecules is small.

Number of Cycles

The number of cycles needed to yield sufficient amplification product depends on the starting concentration of the target DNA when other parameters are optimized. Based on these conditions, approximately 30 cycles are necessary to amplify a single-copy sequence of 1 µg human genomic DNA to a detectable amount. Too many cycles increase nonspecific background products.

PLATEAU EFFECT

The theoretical upper limit of the number of product molecules is 2^n, where n is the number of cycles. Under normal conditions this value is not obtained, and an average efficiency of 60-85% per cycle is more realistic. This reduces the overall yield to a value of about:

$$T = t \, (1+E)^n$$

T is the final amount of product molecules
t is the starting amount of target
E is the efficiency of the amplification

The amplification reaction is limited. After a certain number of cycles, the desired amplification gradually stops accumulating exponentially and enters a linear or stationary phase (plateau). This phenomenon is influenced by many factors. The molar ratios of PCR reagents with respect to template are highest at the beginning of the amplification. This ratio is mainly reduced by the accumulation of target molecules but not by consumption of reagents. In a typical PCR, the initial excess of primers is 10^7 and the excess of deoxynucleotide triphosphates is 10^{10} with respect to DNA template. Assuming a 50% degradation of dNTP at the end of amplification, the final dNTP excess will be about 10^4 and 95% of the primers remain unconsumed after 30 cycles.

Taq DNA polymerase is in the lowest molar excess (10^5) of all reaction components. After a 10^6-fold amplification of the target DNA, there are more template molecules than enzyme molecules. In this stage, the enzyme becomes totally occupied and the ratio of primer to template decreases, promoting self-annealing of the strands. The reaction begins to saturate and shift from an exponential pattern to a more linear accumulation of product molecules.

APPLICATION OF PCR FOR DIAGNOSIS OF CYSTIC FIBROSIS

Cystic fibrosis (CF) is the most common fatal inherited disease in the Caucasian population, affecting one in 2.500 live births. The disease is devastating to both children and family alike. The increased viscosity of exocrine secretions, caused by disturbed permeability of the cellular membrane for chloride ions, leads to bronchopulmonary obstruction and pancreatic failure, followed by malabsorption, pneumonia, and bronchiectasia.

The recent isolation of cystic fibrosis transmembrane conductance regulator (CFTR) gene[28,29] and the discovery of the predominant mutation causing the disease,[30] a 3 bp-deletion in exon 10 (codon 508), have provided a new dimension in diagnosis of CF. Delta-F508 causes the loss of phenylalanine in a supposed CF transmembrane regulation protein, and this affects a prospective ATP-binding site. Consequently, the regular transport of chloride ions. Identification of Delta-F508 mutation using PCR allows direct identification of CF patients and carriers.

Patients and Methods

Blood samples collected in sterile K-EDTA tubes were obtained from children suffering from CF diagnosed according to clinical criteria. Rapid preparation of DNA for amplification by PCR[31-33] was done as described below:

1. An aliquot of 500 μl whole blood is mixed with 500 μl "lysis buffer" in a 1.5 ml Eppendorf microcentrifuge tube.
2. Centrifugation at 10,000 x g for 30 sec.
3. Supernatant is decanted and the pellet resuspended in 1.0 ml of "lysis buffer" and vortexed briefly.
4. Steps 2 and 3 are to be repeated twice.
5. Finally, the pellet is resuspended in 500 μl of "PCR buffer", and 3 μl of proteinase K (10 mg/ml) are added.
6. The sample is incubated at 55°C for 3 hrs.
7. To inactivate proteinase K, incubate for 5 min. at 95°C.
8. An aliquot of 10 μl lysate is introduced to PCR.

Lysis Buffer: 0.32 M sucrose, 10 mM Tris-HCl (pH 7.5), 5 mM $MgCl_2$, 1% Triton X-100

PCR Buffer: 10 mM Tris-HCl (pH 8.3), 50 mM KCl, 2.5 mM $MgCl_2$, 0.1 mg/ml gelatine, 0.45% NP40, 0.45% Tween 20. Aliquot, autoclave and store frozen.

The PCR was carried out according to the standard protocol described under *basic procedures* with denaturation of DNA template at 95°C for 30 seconds and the final elongation at 72°C for 5 minutes. The primers, C16B and C16D[30], were annealed at 59°C for 1 minute and elongated at 72°C for 1 minute throughout a total of 35 cycles. In the presence of the delta-F508 mutation, a 94 bp fragment was amplified, and in the absence of the mutation, a 97 bp fragment of the exon 10 of the CFTR gene was amplified. Detection was done after separating the fragments on a 12% polyacrylamide gel and ethidium bromide staining (*Fig. 2*).

Based on the analysis of 76 CF chromosomes from the CF population in Salzburg, we determined the frequency of the delta-F508 mutation to be 58% of the CF cases investigated (Arrer, unpublished data). As mentioned, the a priori risk of bearing a child who suffers from CF is 1:2.500. Testing for Delta-F508 reduces the risk enormously, if neither of the parents carry this mutation. In addition to the major mutation, more than 150 other mutations have been detected up to now. This fact makes CF carrier screening very difficult. However, more efficient strategies have been worked out recently.[34]

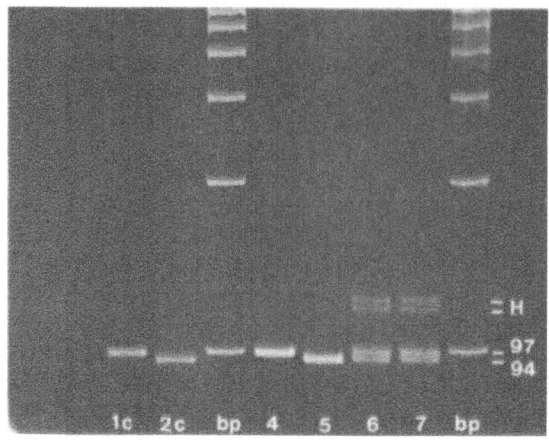

Figure 2. Detection of the Delta-F508 mutation of the CFTR gene. PCR was performed with 10 μl DNA prepared as described. The primers C16B and C16D[30] were used for amplification of a 94 bp fragment, in the presence of the Delta-F508 mutation. Without the mutation, a 97 bp fragment was amplified. The PCR product was separated on a 12% polyacrylamide gel and stained with ethidium bromide. Lines (1c) and (2c) show homocygot normal and homocygot Delta-F508 mutant controls; (bp) is a 100 base-pair ladder; (4-7) is a family with a homocygot normal daughter (4) and a homocygot Delta-F508 son (5); both, father and mother (6/7), are heterocygot for Delta-F508 and show heterodublices (H). The most left portion (without labelling) is a PCR reaction without target DNA, to detect carry over.

GOOD LABORATORY PRACTICE FOR PCR

Because of its extreme sensitivity, contamination is a serious problem in using PCR methods. The following guidelines may help to prevent carry over of PCR products from earlier amplifications:

* Always use gloves
* Physically separated areas/rooms for the different steps of PCR should be used (separate rooms for preparing molecular biological methods, for PCR setup, thermocycling and handling amplified PCR products)
* Use a set of positive-displacement pipettes only for PCR
* Use aliquoted master mixes even when possible
* Substitute dTTP by dUTP whenever necessary
* Use negative and positive controls
* Add DNA template to the PCR mix as the final step
* Be as carefully as when working with cell cultures

The yield and specificity of PCR amplification depends upon a number of physical and chemical conditions. Primer specificity, magnesium chloride and enzyme concentration, as well as the annealing and denaturation temperature also influence the success of PCR. The following operating practices are recommended to maintain optimum instrument performance:

*   Regular visual inspection of the thermocycler
*   Cleaning of the thermoblock after each amplification
*   Confirmation of the cycle time and temperature accuracy and reproducibility.

PERSPECTIVES

The tremendous number of PCR methods and applications published in the last three years indicate the enormous growth of the field of PCR technology (Table 1). The first highlight after establishing the basic concepts of PCR[2], was the application in human genetics and diagnostics.[4] Meanwhile, it has become an extraordinarily useful tool in many fields of medical research and molecular biology. It has revolutionized the way in which a wide variety of experiments and clinical molecular genetic tests are being performed. New knowledge will become available on the causes and pathogenetic mechanisms of multifactorial disorders. Also, some contributions of genetic polymorphisms to common diseases may be clarified and the concept of risk prediction may come one step closer to its realization.

Sequence-tagged sites (STSs) are short sequences of DNA that can be amplified using PCR. This enables us to construct a set of overlapping yeast artificial chromosome(YAC) clones and will enhance mapping strategy for the physical mapping of the human genome.[35,36] The knowledge of an estimated 100,000 genes will elevate PCR to one of the most widely used tools in this field.

Table 1. APPLICATIONS OF PCR

1. Molecular genetics:
    Detection of mutations (Screening)
    Carrier detection
    Parental diagnosis of inherited diseases
    Linkage analysis
    Population genetics
    Forensic applications
2. Clinical investigations:
    Pathogen detection (viral, bacterial, fungal, parasitic)
    Oncogene expression
    Minimal residual diseases
    Risk assessment
3. Molecular biology:
    Gene targeting
    DNA sequencing
    Gene expression
    Physical mapping of the human genome

PCR will help to overcome some difficulties in diagnosis of viral neonatal infections, early infections, and undetermined serological states; it will also support the identification of new agents. Pathogens that are dangerous, slow, or fastidious to grow, e.g. mycobacteria, will provide another potential opportunity for using PCR. Rapid detection and differentiation of fungal and parasitic infections will permit more accurate treatment and prevent extensive tissue destruction.

In situ PCR[21,22] is another step towards bringing morphological, immunological, and molecular characteristics together, possibly leading to a better understanding of cell growth and elucidation of tumor outcome. Such techniques will undoubtedly entrance the diagnostic potential of histology and cytology toward unbelievable limits, e.g. in detecting various infections and possibly also allowing earlier detection of cancer in certain cases.

Increased sophistication and automation of many steps involved in PCR technology will decrease turnaround time and costs. A set of new thermostable DNA polymerases[37,38] with different properties have been isolated from bacteria and might provide amplification reactions when Taq polymerase fails to amplify. A polymerase from *Thermus thermophillus*[39] can reverse transcribe RNA in the presence of $MgCl_2$. The DNA activity of the same enzyme can be stimulated by chelating $MnCl_2$ and adding $MgCl_2$. New methods also allow use of PCR and related techniques for specific *in situ* detection of DNA and RNA sequences within cytological histological specimens (see also the chapter 18 by Zehbe *et al* in this book).[22] All these new developments will result in a potentially increased use and provide a wider diagnostic capability.

References

1. K.B. Mullis, The polymerase chain reaction: Why it works, *in*: Current Communications in Molecular Biology: Polymerase Chain reaction, H.A. Erlich, R. Gibbs, H.H. Kazazian, eds., Cold Spring Habor Laboratory Press, New York, 237 (1989).
2. R.K. Saiki, S. Scharf, F. Faloona, K.B. Mullis, G.T. Horn, H.A. Erlich, and N. Arnheim. Enzymatic amplification of ß-globin genomic sequences and restriction site analysis for diagnosis of sickle cell anemia, *Science* **230**:1350 (1985).
3. K.B. Mullis and F. Faloona, Specific synthesis of DNA *in vitro* via a polymerase catalysed chain reaction, *Methods Enzymol* **155**:335 (1987).
4. R.K. Saiki, D.H. Gelfand, S. Stoffel, S.J. Scharf, R. Higuchi, G.T. Horn, and H.A. Erlich, Primer-directed enzymatic amplification of DNA with a thermostable DNA polymerase, *Science* **239**:487 (1988).
5. K.J. Friedman, W.E. Highsmith, T.W. Prior, T.R. Perry, L.M. Silverman, Cystic fibrosis deletion mutation detected by PCR-mediated mutagenesis, *Clin Chem* **36**:695 (1990).
6. S. Kwok, D.H. Mack, K.B. Mullis, B. Poiesz, G. Erlich, D.Blair, A. Friedman-Kien, and J.J. Sninsky, Identification of human immunodeficiency virus sequences by using in vitro enzymatic amplification and oligomer cleavage detection, *J Virol* **61**:1690 (1987).
7. H. Ehrenfeld, C. Bottner, R. Seelig, M. Renz, Diagnosis of Cystic Fibrosis detection of the gene deletion delta F508, *in*: PCR Topics: Usage of Polymerase Chain Reaction. A. Rolfs, H.C. Schumacher, P. Marx, eds., Springer, New York, 84 (1991).
8. D.A. Nickerson, R. Kaiser, S. Lappin, J. Stewart, L. Hood, and U. Landegren, Automated DNA diagnostics using an ELISA-based oligonucleotide ligation assay, *Proc Natl Acad Sci USA* **87**:8923 (1990).
9. C.Y. Ou, S.H. McDonough, D. Cabanas, T.B. Ryder, M.Harper, J. Moore, and G. Schochetman, Rapid and quantitative detection of enzymatically amplified HIV-1 DNA using chemiluminescent oligonucleotide probes, *AIDS Res Hum Retroviruses* **6**:1323 (1990).
10. F.F. Chehab and Y.W. Kan, Detection of specific DNA sequences by fluorescence amplification: A color complementation assay, *Proc Natl Acad Sci USA* **86**:9178 (1989).
11. U.B. Gyllensten and H.A. Erlich, Generation of single-stranded DNA by the polymerase chain reaction and its application to direct sequencing of the HLA-DQA locus, *Proc Natl Acad Sci USA* **85**:7652 (1988).
12. P. Keohavong, A.G. Kat, N.F. Carello, and W.G. Thilly, DNA amplification *in vitro* using T4 polymerase, *DNA* **7**:63 (1988).
13. C.H. Oste, PCR Automation, *in*: PCR Technology: Principles and Applications, H.A. Erlich, ed., Stockton, New York, 23 (1989)
14. D.R. Engelke, P.A. Hoener, and F.S. Collins, Direct sequencing of enzymatically amplified human genomic DNA, *Proc Natl Acad Sci USA* **85**:544 (1988).
15. E.S. Kawasaki, S.S. Clark, M.Y. Coyne, S.D. Smith, R. Champlin, O.N. Witte, and F.P. McCormick, Diagnosis of chronoc myeloid and acute lymphocytic leucemias by detection of leukemia-specific mRNA sequences amplified *in vitro*, *Proc Natl Acad Sci USA* **85**:5698 (1988).
16. R. Higuchi, C.H. Beroldingen, G.F. Sensabaugh, and H.A. Erlich, DNA typing from single hair, *Nature* **332**:543 (1988).
17. H. Li, U.B. Gyllensten, X. Cui, R.K. Saiki, H.A. Erlich, and N. Arnheim, Amplification and analysis of DNA sequence in single human sperm and diploid cells, *Nature* **335**:414 (1988).

18. C.C. Imprain, R.K. Saiki, H.A. Erlich, and R.L. Teolitz, Analysis of DNA extracted from formalin-fixed, paraffin-embedded tissues by enzymatic amplification and hybridization with sequence-specific oligonucleotides, *Biochem Biophys Res Commun* **142**:710 (1987).

19. E.S. Kawasaki and A.M. Wang, Detection of Gene Expression, *in*: PCR Technology: Principles and Applications, H.A.Erlich, ed., Stockton, New York, 89 (1989).

20. A.M. Wang, M.V. Doyle, and D.F. Mark, Quantitation of mRNA by the polymerase chain reaction, *in*: PCR Topics: Usage of Polymerase Chain Reaction in Genetic and Infectious Diseases. A. Rolfs, H.A. Schumacher, P. Marx eds., Springer, New York, 3 (1991).

21. A.T. Haase, E.F. Retzel, and K.A.Staskus, Amplification and detection of lentiviral DNA inside cells, *Proc Natl Acad Sci USA* **87**:4975 (1990).

22. I. Zehbe, G.W. Hacker, J. Sällström, E. Rylander, and E. Wilander, *In situ* polymerase chain reaction (*in situ* PCR) combined with immunperoxidase and immungold-silver staining (IGSS) techniques. Detection of single copies of HPV in SiHa cells, *Anticancer Res* **12**:2165 (1992).

23. R.K. Saiki, The design and optimization of the PCR, *in*: PCR Technology: Principles and Application, H.A. Erlich, ed., Stockton, New York, 7 (1989).

24. A. Chien, D.B. Edgar, and J.M. Trela, Deoxyribonucleic acid polymerase from the extreme thermophile *Thermus aquaticus, J Bacteriol* **127**:1550 (1976).

25. D.H. Gelfand, *Thermus aquaticus* DNA Polymerase, *in*: Current Communications in Molecular Biology: Polymerase Chain Reaction. H.A. Erlich, R. Gibbs, H.H. Kazazian, eds., Cold Spring Habor Laboratory Press, New York, 11 (1988).

26. D.H. Gelfand, Taq DNA polymerase, *in*: PCR Technology: Principles and Applications. H.A. Erlich, ed., Stockton, New York, 17 (1989).

27. K.R. Tindall and T.A. Kunkel, The fidelity of DNA synthesis by the *Thermus aquaticus* DNA Polymerase, *Biochemistry* **27**:6008 (1988).

28. J.M. Rommens, M.C. Iannuzzi, B.S. Kerem, M.L. Drumm, G.Melmer, M. Dean, R. Rozmahel, J.L. Cole, D. Kennedy, N. Hidaka, M. Zsiga, M. Buchwald, J.R. Riordan, L.C. Tsui, and F.S. Collins, Identification of the cystic fibrosis gene: Chromosome walking and jumping, *Science* **245**:1059 (1989).

29. J.R. Riordan, J.M. Rommens, B.S. Kerem, N. Alon, R. Rozmahel, Z. Grzelczak, J. Zielenski, S. Lok, N. Plavsic, J.L. Chou, M.L. Drumm, M.C. Iannuzzi, F.S. Collins, and L.C. Tsui, Identification of the cystic fibrosis gene: Cloning and characterization of complementary DNA, *Science* **245**:1066 (1989).

30. B.S. Kerem, J.M. Rommens, J.A. Buchanan, D. Markiewicz, T.K. Cox, A. Chakravarti, M. Buchwald, and L.C. Tsui, Identification of the cystic fibrosis gene: Genetic analysis, *Science* **245**:1073 (1989).

31. R. Higuchi, Simple and rapid preparation of samples for PCR, *in*: PCR Technology: Principles and Applications. H.A. Erlich, ed., Stockton, New York, 31 (1989).

32. S.W.M. John, G. Weitzner, R. Rozen, and C.R. Scriver, A rapid procedure for extracting genomic DNA from leukocytes, *Nucl Acids Res* **19**:408 (1990).

33. D.K. Lahiri and J.I. Nurnberger, A rapid non-enzymatic method for the preparation of HMW DNA from blood for RFLP studies, *Nucl Acids Res* **19**:5444 (1991).

34. A.L. Beaudet, and W.E. O'Brien, Advandages of a two-step laboratory approach for cystic fibrosis carrier screening, *Am J Hum Gen* **50**:439 (1992).

35. M. Olson, L. Hood, C.H. Cantor, and D. Botstein, A common language for mapping of the human genome, *Science* **245**:1434 (1989).

36. P. Little, Mapping the way ahead, *Nature* **359**:367 (1992).

37. C. Elie, S. Sahli, J.M. Rossignol, P. Forterre, and A.M. DeRecondo, A DNA polymerase from a thermoacidophilic archaebacterium: evolutionary and technological interests, *Biochem Biophys Acta* **951**:261 (1988).

38. N.F. Cariello, J.A. Swenberg, and T.R. Skopek, Fidelity of Thermococcus litoralis DNA polymerase (Vent) in PCR determined by denaturing gradient gel electrophoresis, *Nucleic Acids Res* **19**:4193 (1991).

39. T.W. Myers and D.H. Gelfand, Reverse transcription and DNA amplification by a thermophilus DNA polymerase, *Biochemistry* **30**:7661 (1991).

CHAPTER 18

POLYMERASE CHAIN REACTION (PCR) *IN SITU*
HYBRIDIZATION: DETECTION OF HUMAN
PAPILLOMAVIRUS (HPV) DNA IN SiHa CELL
MONOLAYERS

Ingeborg Zehbe,[1] Jan Sällström,[1] Gerhard W. Hacker,[2]
Eva Rylander,[3] Anders Strand,[4] Anton-Helmut Graf,[5] and
Erik Wilander[1]

[1] Department of Pathology, University Hospital, S-751 85 Uppsala, Sweden
[2] Institute of Pathological Anatomy, Immunohistochemistry and
Biochemistry Unit, General Hospital, A-5020 Salzburg, Austria
[3] Department of Gynecology and Obstetrics, University Hospital
S-751 85 Uppsala, Sweden
[4] Department of Venereology and Dermatology, University Hospital, S-751
85 Uppsala, Sweden
[5] Department of Gynecology and Obstetrics, General Hospital, A-5020
Salzburg, Austria

SUMMARY

In situ hybridization (ISH) with labelled cDNA probes has become a valuable tool for
the detection of human papillomavirus (HPV) as it allows direct correlation of virus
infection and morphological diagnosis: This method, however, is limited to detect 10-50
DNA copies per cell which was demonstrated on HeLa cells. In the following study, we
present a more sensitive *in situ* method, PCR *in situ* hybridization (PISH), able to detect

[1]Ingeborg Zehbe, Uppsala University Hospital, Institute of Pathology, S-751 85 Uppsala, Sweeden, Tel: +46-
18-663809; Fax: +46-18-552739

a single HPV-DNA copy per cell. We developed a model with SiHa cells containing 1-2 HPV 16 copies per cell. SiHa cells stain negative with *in situ* hybridization. After amplifying the HPV-DNA *in situ* using PCR prior to the hybridization step, we were able to get positive results using either enzymatic detection methods or immunogold silver staining (IGSS).

INTRODUCTION

Historical Review

Solution phase polymerase chain reaction (PCR) was originally the idea of Karry Mullis of Cetus Corporation in 1984, and was significantly developed by Henry Erlich, Norman Arnheim, Randy Saiki, *et al.*, in the Human Genetic Division at Perkin Elmer Cetus, USA.[1-3] This novel molecular biological technique revolutionized the performance of a wide variety of experiments within clinical molecular genetics. Solution phase PCR was initially performed with Klenow polymerase from *Escherichia coli* DNA polymerase I, which had to be added anew for each cycle because it was not a heat-resistant polymerease. At that time, PCR was a tedious manual procedure with increased risk of contamination. Heat resistant *Taq* polymerase is a DNA polymerase isolated from the thermophilic bacterium *Thermus aquaticus* from a hot spring in Yellowstone National Park that made the PCR procedure much easier and even allowed automation of the process.[4] *Taq* polymerase has no 3'-5' exonuclease activity ("proof reading ability") but has a 5'-3' exonuclease activity during polymerization.[5] Misincorporation is less than 10^{-5} nucleotides per cycle with an improved version of *Taq* polymerase that can withstand repeated exposure to high temperature.[6,7] It retains 50% of its polymerase activity after 130 minutes at 92.5°C, after 40 minutes at 95°C, and after 5-6 minutes at 97.5°C and can, therefore, be regarded as being a heat resistant polymerase allowing automation of the PCR process.[8] Thus, the first thermal cycler was developed by scientists at Perkin Elmer Cetus (USA).

Theoretical Consideration

PCR involves the *in vitro* enzymatic synthesis of a specific DNA segment and, at best, results in a million-fold amplification of the original template. The reaction is based on the annealing and extension of two oligonucleotide primers that flank the target region in double helix DNA. After denaturation of the DNA, each primer hybridizes to one of the separated strands and extension is directed from the 3' hydroxyl end of each primer. The annealed primers are extended with DNA polymerase. Thus, one cycle consists of DNA denaturation (melting), primer annealing and primer extension (polymerization). Accordingly, repeated cycles of denaturation, annealing, and extension result in an exponential amplification of the initial target template. Both double stranded and single stranded DNA molecules can be amplified, and even a reverse transcription from mRNA into a cDNA copy is possible. Therefore, mRNA may also serve as a template.

Since 1990, Haase *et al.*, Nuovo *et al.*, and Chiu *et al.*, have developed a new application of the PCR, termed PCR *in situ* hybridization, which allows amplification of a single copy of viral DNA in cytological preparations demonstrating the presence of viral DNA together with cell morphology.[9-16] According to Haase *et al.*,[9] amplification of virus DNA in cells requires a product of 1200 bp to prevent leakage from the nucleus. This may be accomplished by a multiple primer set (MPS) with overlapping primers. A similar theory was recently presented by Chiu *et al.*, using complementary primer tails.[13] This was challenged by Nuovo *et al.*,[11] who emphasized that the size of the amplification product is not the critical factor but rather the suppression of nonspecific amplification. It is possible, therefore, to detect as few as one viral DNA copy per cell, provided that the specificity and sensitivity of the PCR reaction are so high that maximal amplification of the desired PCR product is obtained. To accomplish this, the so called *hot start* technique is employed.[14] Further reports on PCR *in situ* hybridization using one primer pair have recently been published.[15-17]

The amplified DNA is usually detected by complementary nucleic acid probes. However, another method was previously described by Nuovo *et al.* using labelled nucleotides during the amplification step, making the procedure easier.[16,17] The authors have also reported that this technique may be applied with success on SiHa cell monolayers.[16,17] However, when we applied a comparable assay on clinical material, especially sections from biopsies, we sometimes noticed unspecific staining when omitting the primers.[18] Similar results have also been reported by Long *et al.*[19] We, therefore, urgently advise using specific cDNA probes for the detection step to accomplish the highest possible specificity.

Applications

PCR *in situ* hybridization may be employed for the same purposes as *in vitro* PCR. However, this method was particularly developed for diagnosing viral DNA expression *in situ*. Using this technique, it will now be possible to study the etiology of morphological changes and development of viruses and other pathogenic agents in specific cell types, allowing greater understanding of the behavior of microorganisms. Other areas within cancer research may also be studied, e. g. the DNA/RNA expression of oncogenes and suppressor genes. Latent viral infections classifying certain risk groups and retrospective epidemiological studies on archival material could also be fields for investigation applying *in situ* PCR technology.

In the following, we present a PCR *in situ* hybridization (PISH) method using one primer pair and the *hot start* technique on a cultured SiHa cell model. The method described can potentially be adapted for various other applications specifically analyzing DNA or RNA sequences.

RECOMMENDED PROCEDURE

Cell Cultures

SiHa cells with one HPV 16 copy per cell integrated into the genome (HTB 35, American Type Culture Centre, Rockville, MD, USA) and human foreskin fibroblasts (AGO 1523 B, Human Genetic Mutant Cell Repository, Camden, NJ, USA) are grown in a medium consisting of 90% Eagles, 10% fetal calf serum (FCS) and the addition of antibiotics and glutamine. The cell cultures are harvested with trypsin after 3-4 days. The cell pellets are fixed overnight in 10% neutral-buffered formalin (NBF) (250 ml concentrated formalin and 750 ml phosphate buffer consisting of 50 mM Na_2HPO_4 and 30 mM NaH_2PO_4, pH 7.0). Cytospins are prepared by centrifuging the fixed SiHa cells and fibroblasts onto 3-aminopropyl triethoxysilane (APES) coated slides in a Shandon cytospin centrifuge at 450 rpm for 5 min. They are air-dried for 10 min and thereafter stored in 95% ethanol at 4 °C. Immerse the slides twice, 2 min each, in 99% ethanol and air-dry them before processing to protease treatment.

Other fixatives like acetone, acetic acid/ethanol and 4% paraformaldehyde in PBS were tried, but did not result in positive staining in our particular setup. Evidently, the type of fixative may have profound effects on the efficacy of HPV DNA detection. Accordingly,

NBF appears to be the fixative of choice for HPV DNA detection among commonly used fixatives. This is in agreement with Nuovo et al.[20-21] as well as Thompson and Rose.[22] One possible reason for this could be that the biopsies are heated to 60°C during the paraffinization process which may result in denaturation of the double-stranded DNA and subsequent cross-linking of aldehyde groups to amino groups of single-stranded DNA.

CONTROLS

Positive and negative controls must be routinely included.

Positive Controls

1. Using this assay on clinical material, e.g. on cytological material or imprints, a HPV-positive specimen control with very low HPV-copy content such as SiHa cells, must follow the procedure to assure that the PCR-reaction was satisfactory.
2. Hybridization control with known HPV-positive cases to check the hybridization step alone.

Negative Controls to Monitor Unspecific Amplification

1. SiHa cell monolayers treated with the amplification mix but without the Taq polymerase, $MgCl_2$ or primers, should not lead to any staining reaction.
2. Human foreskin fibroblasts, prepared and fixed in the same manner as the SiHa cells and treated with the amplification mix including all ingredients, can also serve as negative controls.

Equipment

In our study, a thermal Cycler *PHC-3* (Techne, Cambridge, UK), specially equipped with a solid, smooth heating-block for up to 5 slides, an Eppendorf Thermomixer 5436, and an Eppendorf centrifuge 5415 C (Eppendorf, Hamburg, FRG) were used. Use of aerosol resistant tips available for pipettors from various sources (e.g. Perkin Elmer Cetus, Eppendorf or Gilson). Round coverslips 19 mm in diameter and rectangular ones 24x50 mm, toothpicks, Dako Pen (code 2002, Dakopatts, Glostrup, DK, and Carpineria, CA, USA) are helpful utensils in the procedure.

Consensus Primers

The 20-mer oligonucleotide consensus primers (GeneAmplimer HPV primers, Perkin Elmer Cetus, at Roche, Branchburg, NJ, USA) are degenerate primers that enable detection of a 450-bp product of many types of HPV: 6, 11, 16, 18, 31, 33, 35, 39, 40, 42, 51-55, 57, 59 and from at least 25 other as yet uncharacterized HPV-types. Dermal HPV-types 1, 5, 8, 26, 27, 41, 47, and 48 are also amplified with these primers. The sequences of these primers derive from the open reading frame L 1 while the genital HPV types share interspersed regions of DNA sequence homology.[23-34] By comparing the DNA sequences of the genital HPV types 6, 11, 16, 18 and 33, regions of homology 20 to 25 bp in length were identified.[25-29] The sequences of these primers are as follows: Primer 1 (MY09), the negative strand primer: 5'-CGTCCMARRGGAWACTGATC-3', and primer 2 (MY11), the positive strand primer: 5'-GCMCAGGGWCATAAYAATGG-3'. (M = A or C, R = A or G, W = A or T and Y = C or T).

Pretreatment

The proteolytic treatment regarding type and time prior to the amplification process is a crucial factor. In our particular application, the use of 2 mg pepsin (code P-7012, Sigma, St. Louis, MO, USA) per ml 0.1 N HCL in 37°C for 15 min, has been found to work satisfactorily. In cases where detection with the direct immuno-peroxidase (IMP) is preferred, we advise that the specimens be treated with aqueous 6% H_2O_2 in 37°C for 10-20 min, to quench endogenous peroxidase activity. This is not necessary for detection with fluorescent, alkaline phosphatase, or immunogold silver staining (IGSS) techniques.

Amplification Mix

The amplification mix consists of 1.0 μM of each primer; 200 μM each of dATP, dCTP, dGTP, and dTTP; 4.5 mM $MgCl_2$, 50 mM KCl, 10 mM tris HCL, pH 8.3, 7.5 U Ampli*taq* per 50 μl amplification mix, and glass distilled water. Reagents are briefly vortexed before preparing the amplification mix. All reagents except the primers, may be purchased as "GeneAmp PCR Core Reagents Kit" (N808-0009, Perkin Elmer Cetus) and should be stored at - 20°C in small aliquots. We advise heating the amplification mix (without the Ampli*taq*) for 5 min at 95°C on the Eppendorf thermomixer to degrade possible endonucleases.

DNA-Amplification with Hot Start

The cell monolayers have to be lightly circumscribed with a circle of the same diameter as the round coverslip using the Dako Pen (code 2002, Dakopatts, Glostrup, Denmark, and Carpineria, CA, USA). With this setup, the use of nail polish described by Nuovo[10-12] can be omitted, making the technique much easier to handle.[17] The slides are placed on the heating-block of the thermal cycler and preheated to 82°C. Meantime, the amplification mix is simultaneously heated on the thermomixer (85°C), the Ampli*taq* polymerase added, briefly vortexed, and thereafter 10 μl of the amplification mix is rapidly pipetted onto the cell monolayer. A round coverslip, 19 mm in diameter, is immediately added, arranged exactly on the thin Dako pen circle with the aid of a toothpick and overlaid with 300 μl mineral oil (code M 3516, Sigma, St. Louis, MO, USA) that was preheated (85°C) on the thermomixer. An additional coverslip, 24x32 mm, spreads the oil evenly over the surface, preventing the samples from drying out.

Thermocycling

Thereafter follows an initial denaturation step for 3 min at 95°C and 25-30 cycles, consisting of annealing and amplifying for 2 min at 58°C and denaturation for 1 min at 95°C. Subsequently, the mineral oil is removed from the slides by immersing them in xylene twice, 2 min each, followed by dehydration in 99% ethanol and air-drying.

In Situ Hybridization

The amplified viral DNA is then hybridized to a genomic digoxigenin-labeled HPV 16 cDNA probe (Kreatech, Amsterdam, NL), preferably overnight at 37°C, followed by two washes, 5 min each in 2xSSC, and detected with one of the described systems (see below). Instead of a digoxigin-labelled probe, systems utilizing biotin as a label may also be amplified.

DETECTION SYSTEMS

All following steps are performed at room temperature. Several systems may be used to detect the digoxigenin-labeled cDNA probes, some of which are described below. If a biotin-labelled cDNA probe has been used, anti-biotin systems must be applied accordingly.

1. Direct Immunoperoxidase (IMP) Detection Method.[30]

After PCR *in situ* hybridization, the samples are to be rinsed in tris buffered saline (TBS, 100 mM TRIS-HCL and 150 mM NaCl, pH 7.5 at 20°C) for 2 min and treated with blocking reagent (code 1093 657, Boehringer Mannheim, Mannheim, FRG), 0.5% w/v for 30 min. They are then rinsed in TBS for 2 min and incubated with the peroxidase-labeled sheep anti-digoxigenin antibody (code 1207 733, Boehringer Mannheim), diluted 1/100 in TBS (without sodium-azide) containing 1% heat-inactivated fetal calf serum (FCS) for 30-60 min. Leaving the antibody overnight is not recommended since after dilution, it is only stable for a few hours. After washing in TBS (3x10 min) the slides are to be developed with diaminobenzidine-tetrahydrochloride (DAB) in buffer pH 7.2 containing H_2O_2. Light counterstaining in Mayer's hematoxylin is recommended. Slides may be dehydrated in graded ethanols, cleared in xylene, and mounted with appropriate mounting medium (e. g. Eukitt, DPX, Cytoseal or Permount).

2. Direct Alkaline Phosphatase.

Slides are treated as above but incubated with the alkaline phosphatase- labeled sheep anti-digoxigenin antibody (code 1207 733, Boehringer Mannheim, FRG), diluted 1:100 in TBS containing 1% FCS for 30-60 min. This is followed by washing 3x10 min in TBS and 2 min in TBS-MgCl$_2$ (100 mM Tris-HCl solution, 100 mM NaCl, 50 mM MgCl$_2$, pH 9.5, at 20°C). The slides are then immersed in water and ethanol insoluble nitroblue tetrazolium (NBT) and 5-brome-4-chlorine-3-indoxylphosphate (BCIP) solution for 1-2 hours and counterstained in 1% eosin or nuclear fast red. For this chromogen combination, aqueous remounting medium should be used (e.g. *Glycergel*, Dakopatts, Glostrup, DK, and Carpinteria, CA, USA).

3. Immunogold-Silver Staining (IGSS) (Figure 1).

The IGSS technique with silver acetate autometallography[31-35] can be applied using mouse monoclonal antibodies to digoxigenin: After *in situ* PCR hybridization, the dried specimens are immersed in 0.05 M TBS pH 7.6 containing 0.1% cold water fish gelatin (Auxion, Wageningen, NL), incubated with normal goat serum at a dilution of 1:20 in TBS-gelatin for 5 min, drained off and incubated with unlabeled mouse monoclonal antibodies against digoxigenin (code 1333 062, Boehringer Mannheim, FRG) diluted 1/10 to 1/50 in PBS (150 mM NaCl, 8 mM Na$_2$HPO$_4$, 2.6 mM KCL and 3.6 mM KH$_2$PO$_4$, pH 7.4) or diluted in TBS-gelatin containing 0.1% BSA, for 1 h at 4°C up to 18 h at room temperature.

This is followed by washing in TBS containing gelatin as described above (3x3 min) and incubation with goat anti-mouse immunogold reagents for 1-2 hours. Gold particle diameters used are 1 and 5 nm, each immunogold-reagent diluted 1/50 and mixed in equal amounts in TBS gelatin also containing 0.1% bovine serum albumin (BSA) (codes RPN 424, GAM-IgG G5 EM-grade and RPN 471, Auroprobe One GAM-IgG, Amersham, U.K.). Instead of the indirect cycle described, 0.8 nm gold-labeled sheep anti-digoxigenin antibodies may also be applied (Aurion, Wageningen, NL; overnight incubation at 4°C) as a direct detection system. After gold-layer incubation, cytospins are washed in TBS gelatin for 2x3 min, followed by PBS pH 7.2 for 3 min and postfixation in 2% glutaraldehyde in PBS for 2 min.

Thorough and repeated washing with deionized or, better, glass-distilled water is required for at least 15 min before silver enhancement of the gold-label using silver acetate autometallography. For this, the specimens are immersed in a freshly prepared solution consisting of 100 mg silver acetate (code 85140, Fluka, FRG) in 50 ml deionized, glass distilled water, and 250 mg hydroquinone in 50 ml citrate buffer (23.5 g trisodium citrate dihydrate and 25.5 g citric acid monohydrate in 850 ml deionized, glass distilled water, pH 3.8).[34,35] Counterstaining may be performed with Mayer's hematoxylin and/or eosin and/or nuclear fast red. Dehydrate in graded alcohols, clear in xylene and mount in DPX (BDH-Biochemicals, UK) or Canada balm or cytoseal (Curtin Matheson Scientific In., Wayne, NJ, USA). In contrast to Eukitt, these suggested mounting media have proven not to affect silver staining by subsequent oxidation or reduction reactions (See also the chapter by Grimelius, in this book). Detergents and Lugol's iodine affect the staining yield in *in situ* PCR and should therefore be avoided for this particular application.

If, in the hybridization step, a biotinylated cDNA probe is used, the same assay can be applied, but using 1 nm gold-labeled anti-biotin antibodies (Auro-Probe One antibiotin (Amersham, UK) or streptividin-gold (1 and/or 5 nm gold diameter, Amersham, UK) with subsequent silver intensification.[36]

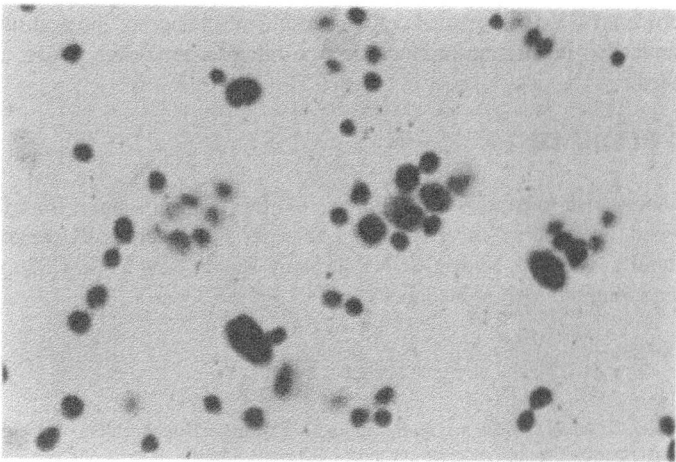

Figure 1. PCR *In Situ* hybridization for human papilloma virus (HPV) in SiHa cells using consensus primers recognizing the genome of most HPV types. Detection with *in situ* hybridization and immunogold-silver staining using silver acetate autometallography and a combination of 0.8, 1 and 5 nm gold-labelled antibodies. Positive nuclei appear in black, a few negative nuclei are grey in this photograph. Counterstaining with nuclear fast red and eosin. Original magnification x 420.

DISCUSSION

A model for PCR *in situ* hybridization (PISH) has been described. The methods described are reproducible and can be finished within one to two working days, depending on the detection system used. After standardization of the assay, it is easy to comprehend and may be performed by skilled technicians. Usually, in contrast to *in vitro* PCR, only one working area is needed, but otherwise the same precautions as with traditional PCR must be used. Gloves should be worn throughout the procedure and sterility of the reagents and glass slides is necessary to avoid inhibition of the amplification process. Reagents should be aliquoted, and repeated thawing and refreezing avoided. It is very important that the $MgCl_2$ solution be homogenous after being thawed. We recommend heating it up to 95°C before adding it to the amplification mix especially in cases where controls show a tendency to stain weaker than usual. No tedious DNA extraction is required and, thus, there is less risk for contamination from one sample to the other.

Our present amplification protocol is adjusted to *in situ* HPV-DNA detection in cultured cells with PISH using consensus primers. In the future, however, it should be possible to apply the assay on a variety of specimens. For clinical evaluation, we are currently performing tests on tissue sections from biopsies or surgical material, cytospins, smears, and imprints. The best results so far were obtained on imprints. Hybridization methods with labelled oligoprobes or the 400 bp generic L1 probe developed by Bauer, Greer and Manos,[37] designed to hybridize to the 450 bp amplification product of the consensus primers are under investigation. This probe is synthesized with nested primers from the L1 PCR fragments of HPV 16, 18 and 31 and detects a broad spectrum of HPV types.

For rapid routine purposes, the authors prefer the enzymatic detection method with immunoperoxidase. IGSS may be somewhat more complicated than the direct IMP technique. However, by using IGSS, more distinct and high contrast results can be achieved. It also has the advantage that no hazardous reagents such as DAB are used. For silver enhancement, silver acetate autometallography was used, allowing the sensitive detection of gold label under daylight conditions. We have also already been able to further process PISH-IGSS-treated cells for resin embedding, for the first time allowing their examination in the transmission electron microscope (Zehbe, Muss, and Hacker, unpublished).

ACKNOWLEDGMENTS

The authors wish to express their appreciation for financial support from the Swedish Cancer Fond, the Lion's Cancer Fond and Selander's Foundation. Furthermore, we are very grateful to Dr. G. J. Nuovo, SUNY at Stony Brook, New York, USA, for his kind assistance during our visit at his laboratory.

REFERENCES

1. R. K. Saiki, S. Scharf, F. Faloona, K. B. Mullis, G. T. Horn, H. Erlich, and N. Arnheim, Enzymatic amplification of globin genomic sequences and restriction site analysis oligonucleotide probes, *Science* **230**:1350 (1985).
2. K. B. Mullis and F. Faloona, Specific synthesis of DNA *in vitro* via a polymerase catalyzed chain reaction, *Methods Enzymol* **155**:335 (1987).
3. R. K. Saiki, D. H. Gelfand, S. Stoffel. S. J. Scharf, R. Higuchi, G. T. Horn, K. B. Mullis, and H. A. Erlich, Primer directed enzymatic amplifiction of DNA with a thermostable DNA polymerase, *Science* **239**:489 (1988).
4. T. D. Brock and H. Freeze, Thermus aquaticus agen. n. and sp. n., a nonsporuling extreme thermophile, *J Bacteriol* **98**:289 (1969).
5. K. R. Tindall and T. A. Kunkel, Fidelity of DNA synthesis of the *Thermus aquaticus* DNA polymerase, *Biochem* **27**:6008 (1988).
6. D. H. Gelfand and T. J. White, Thermostable DNA polymerases, *in:* PCR Protocols: A Guide to Methods and Applications, M. A. Innis *et al.* eds., Academic Press San Diego, CA, USA (1990).
7. K. A. Eckert and T. A. Kunkel, High fidelity DNA synthesis by the *Thermus aquaticus* DNA polymerase, *Nucl Acids Res* **18**:3739 (1990).
8. D. H. Gelfand. *Thermus aquaticus* DNA polymerase, *in:* Current Communications in Molecular

Biology: Polymerase Chain Reaction, H. A. Erlich *et al.* eds., Cold Spring Harbor Laboratory Press, New York, USA (1989).

9. A. T. Haase, E. F. Retzel, and K. A. Staskus, Amplification and detection of lentiviral DNA inside cells, *Proc Natl Acad Sci* **87**:4871 (1990).

10. G. J. Nuovo, Detection of human papillomavirus DNA in formalin-fixed tissues by *in situ* hybridization after amplification by polymerase chain reaction, *Am J Pathol* **139**:847 (1991).

11. G. J. Nuovo, F. Gallery, P. MacConnell, J. Becker, and W. Bloch, An improved technique for the *in situ* detection of DNA after polymerase chain reaction amplification, *Am J Pathol* **139**:1239 (1991).

12. G. J. Nuovo, J. Becker, M. Margiotta, P. MacConnell. S. Comite, and H. Hochmann, Histologic distribution of polymerase chain reaction-amplified papillomavirus 6 and 11 DNA in penile lesions, *Am J Surg Pathol* **16**:269 (1992).

13. K. P. Chiu, S. H. Cohen, D. W. Morris, and G. W. Jordan, Intracellular amplification of proviral DNA in tissue sections using the polymerase chain reaction, *J Histochem Cytochem* **40**:333 (1992).

14. H. A. Ehrlich, D. Gelfand, and J. J. Sninsky, Recent advances in the polymerase chain reaction, *Science* **252**:1643 (1991).

15. O. Bagasra, S. P. Hauptman, D.O.H.W. Lischner, M. Sachs, and R. J. Pomerantz, Detection of human immunodeficiency virus type 1 provirus in mononuclear cells by *in situ* polymerase chain reaction, *N Engl J Med* **326**:1385 (1992).

16. G. J. Nuovo, M. Margiotta, P. MacConnell, and J. Becker, Rapid in situ detection of PCR-amplified HIV-1 DNA, *Diagn Mol Pathol* (in press).

17. I. Zehbe, G. W. Hacker, E. Rylander, J. Sällström, and E. Wilander, Detection of single HPV copies in SiHa cells by *in situ* plymerase chain reaction (*in situ* PCR) combined with immunoperosidase and immunogold silver stainging (IGSS) techniques, *Anticancer Res* **12**:2165 (1992).

18. J. Sällström, I. Zehbe, M. Alemi, and E. Wilander, Pitfalls of *in situ* polymerase chain reaction (PCR) using direct incorporation of labelled nucleotides, *Anticancer Res* (in press).

19. A.A. Long, P. Komminoth, E. Lee, and H. J. Wolfe, Comparison of indirect and direct *in situ* polymerase chain reaciton in cell preparations and tissue sections. *Histochemistry* **99**:151 (1992).

20. G. J. Nuovo and S. J. Silverstein, Comparison of formalin, buffered formalin and Bouin's fixation on the detection of human papillomavirus deoxyribonucleic acid from genital lesions, *Lab Invest* **59**:720 (1988).

21. G. J. Nuovo and R. M. Richart, Buffered formalin is the superior fixative for the detection of HPV DNA by *in situ* hybridization analysis, *Am J Pathol* **134**:837 (1989).

22. C. H. Thompson and B. R. Rose, Deleterious effects of formalin/acetic acid/alcohol (FAA) fixation on the detection of HPV DNA by *in situ* hybridization and the polymerase chain reaction, *Pathology* **23**:327 (1991).

23. H. M. Bauer, Y. Ting, C. E. Greer, J. C. Chambers, C. G. Tashiro, G. J. Chimera, J. Reingold, and M. M. Manos, Genital papillomavirus infection in female university students as determined by a PCR-based method, *J Am Med Assoc* **265**:472-477 (1991).

24. Y. Ting and M. M. Manos, Detection and typing of genital human papillomaviruses, *in*: PCR Protocols. A Guide to Methods and Applications, M. A. Innis *et al* eds., Academic Press San Diago, CA, USA (1990).

25. E. Schwarz, M. Duerst, C. Demankowski, O. Lattermann, R. Zech, E. Wolfersberger, S. Suhai, and H. zur Hausen, DNA sequence and genome organization of genital human papillomavirus type 6b, *EMBO* **2**:2341 (1983).

26. K. Seedorf, G. Krammer, M. Duerst, S. Suhai, and W. G. Rowekamp, Human papillomavirus type 16 DNA sequence, *Virology* **145**:181 (1985).

27. S. T. Cole and R. T. Streeck, Genome organization and nucleotide sequence of human papillomavirus type 33, which is associated with cervical cancer, *J Virol* **58**:991 (1986).

28. S. T. Cole and O. Danos, Nucleotide sequence and comparative analysis of the human papillomavirus type 18 genome. Phylogeny of papillomaviruses and repeated structure of the E6 and E7 gene products, *J Mol Biol* **193**:599 (1987).

29. K. Dartmann, E. Schwarz, L. Gissmann, and H. Zur Hausen, The nucleotide sequence and genome organization of human papillomavirus type 11, *Virology* **151**:124 (1986).

30. P. Nakane, Simultaneous localization of multiple tussue antigens using the peroxidase-labeled antbody method: A study in pituitary glands of the rat, *J Histochem Cytochem* **16**:557 (1968).

31. G. Danscher, Localization of gold in biological tissue. A photochemical method for light and eletron microscopy, *Histochemistry* **71**:81 (1981).

32. C. S. Holgate. P. Jackson, P. N. Cowen, and C. C. Bird, Immunogold-silver staining: new method of immunostaining with enhanced sensitivity, *J Histochem Cytochem* **31**:938 (1983).

33. D. R. Springall, G. W. Hacker, L. Grimelius, and J. M. Polak, The potential of the immunogold-silver staining method for paraffin sections, *Histochemistry* **81**:603 (1984).

34. G. W. Hacker, L. Grimelius, G. Danscher, G. Bernatzky, W. Muss, H. Adam, and J. Thurner, Silver acetate autometallography: an alternative enhancement technique for immunogold-silver staining (IGSS) and silver amplification of gold, silver, mercury and zinc in tissues, *J Histotechnol* **11**:213 (1988).

35. G. W. Hacker, Silver-enhanced colloidal gold for light microscopy, *in*: Colloidal Gold: Principles, Methods, and Applications, Vol 1, M. A. Hayat ed., Academic Press Orlando, USA (1989).

36. G. W. Hacker, A. H. Graf, C. Hauser-Kronberger, G. Wirnsberger, A. Schiechl, G. Bernatzky, U. Sonnleitner-Wittauer, H. Su, H. Adam, J. Thurner, G. Danscher, and L. Grimelius, Application of silver acetate autometallography and gold-silver staining methods for *in situ* DNA hybridization, *Chinese Med J* **106**:83-92 (1993).

37. H. M. Bauer, C. E. Greer, and M. M. Manos, Determination of genital human papillomavirus infection by consensus PCR amplification, *in*: Diagnostic molecular pathology, Herrington and McGee ed., vol. 1-2, Oxford Press (1992).

CHAPTER 19

AXONAL TRANSPORT TRACING COMBINED WITH NEUROTRANSMITTER AND PEPTIDE IMMUNOCYTOCHEMISTRY

Huici Su,[1] **Jiang Gu,**[2] **Gerhard W. Hacker**[3]

Department of Histology and Embryology
Fourth Military Medical University
Xián, Shoanxi, P.R. China

SUMMARY

To trace neurotransmitter or neuropeptide specific pathways, especially peptide-containing neuropathways, various combinations of axonal transport tracing and transmitter immunocytochemistry may be applied. Combining fluorescent retrograde tracing and immunofluorescence is a widely used technique. True blue, fast blue, rhodamine beads, and fluorogold are fluorescent tracers often used. Neurons labelled with a fluorescent tracer and stained by immunofluorescence are visualized simultaneously by using appropriate excitation wave-lengths for each marker. Retrograde tracing with horseradish peroxidase (HRP) or wheat germ agglutinin (WGA) conjugated to HRP can be successfully combined with HRP-based immunostaining. However, the products of enzyme reactions must have different appearances, e.g. granular and homogeneous, or show different colors. When a colloidal gold conjugate of WGA is coupled to enzymatically inactive (apo) HRP, WGA(apo)HRP-gold, is introduced as a retrograde tracer, the labelled neurons can be demonstrated by silver enhancement. Since the tracer contains an inactive form of HRP, this tracing method can be combined with immuno-cytochemistry using enzymatically active HRP as a marker. PHA-L is an efficient anterograde tracer. Coexistence of transported PHA-L and neurotransmitters may be demonstrated by double immuno-staining using two primary antisera raised in different

---

[1] Address for Correspondence: Huici C. Su, M.D., Ph.D., Professor, Department of Histology and Embryology, Fourth Military Medical University, Xian, P.R. China. Fax: 86-29-323-4516.

species. Retrograde or anterograde tracing combined with immunocytochemistry can also been employed at electron microscopic level to illustrate the organization of neurons. Each of the combined methods has its own advantages and limitations. The degree of specificity of retrograde or anterograde labelling and evaluation of immunostaining require careful consideration when results are interpreted.

INTRODUCTION

In the late sixties and early seventies, a nerve tracing method based on *in vivo* intra-axonal transport of tracers was developed.[1-3] When certain substances, e.g. fluorescent dyes or horseradish peroxidase (HRP), are injected into a terminal field of neurons, they can be taken up by nerve terminals through nonspecific endocytosis and transported from the terminals to the parent cell bodies. The tracer-labelled cell bodies can then be visualized. This technique is known as retrograde tracing (*Fig. 1*). Other compounds, e.g. radioactively labelled amino acids[4] or phaseolus vulgaris-leukoagglutinin (PHA-L),[5] when delivered into the nervous system, can be taken up by neurons and subsequently transported anterogradely along nerve axons towards the terminals and this is called anterograde tracing (*Fig. 2*). The axonal transport tracing, as an ingenious and powerful technique, has now become a convenient method for mapping nerve pathways and tracing connections of central neurons, peripheral ganglia and the nerves supplying peripheral organs.

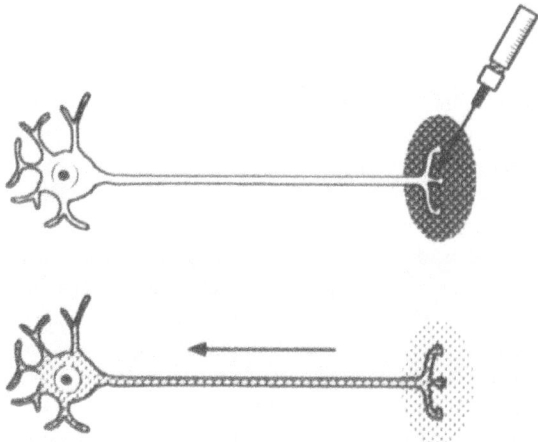

Figure 1. A diagram illustrating the principle of retrograde nerve tracing.

Methods that have been used to detect numerous chemical compounds at the cellular and subcellular level of neurons are direct and indirect immunofluorescence,[6,7] immunoenzymatic staining,[8] PAP method,[9] ABC technique,[10] immunogold labelling, and immunogold-silver staining technique[11-13]. As a result, in addition to classical neurotransmitters, neuropeptides in growing numbers can be localized in neurons, some of which are now strong candidates for neurotransmitter status.[14-16]

In order to trace neurotransmitter-specific pathways especially the peptide-containing neuropathways, axonal transport tracing combined with neurotransmitter immunocyto-chemistry can be applied. In principle, almost all the combination methods are two-step procedures.[17] First, the tracers are taken up and transported by neurons regardless of their neurotransmitter content. In the second step, neurotransmitters or neurotransmitter candidates in the tracer-labelled neurons are demonstrated by immunocytochemistry. The interpretation of the results is based on the assumption that the neurotransmitter present in the cell body is the same as the one stored and released at its terminals. These combined techniques have been extensively used to study the projections of neurons on the bases of their neurotransmitters in both the central and peripheral nervous system. Numerous valuable findings have been added to our understanding of neuroanatomy.

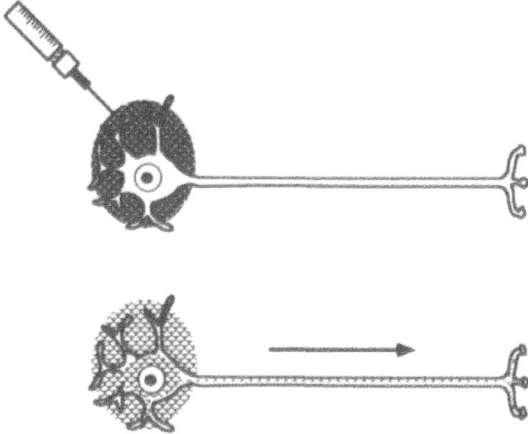

Figure 2. A diagram illustrating the principle of anterograde nerve tracing.

In this chapter, we will focus on the methodology of the combinations and present four protocols. The choice of methods and some general problems when analyzing the results are discussed.

THEORETICAL AND PRACTICAL CONSIDERATIONS

Retrograde Tracing with Fluorescence Tracers Combined with Immunofluorescence

Among various combinations used for tracing neuropeptide specific pathways, fluorescent retrograde tracing combined with immunofluorescence is the most widely used technique. This is due to the simplicity, sensitivity, and easy reproducibility of the technique. Various fluorescent substances can be used for retrograde tracing study.[18-21] However, when combined with immunocytochemistry, the tracer must resist diffusion during the immunostaining procedure and must be reliable for visualization concomitant with positive immunofluorescence staining. Fluorescent tracers, e.g. true blue, fast blue, rhodamine beads, and fluorogold are suitable for this purpose.

Using the above mentioned method for the visualization of enkephalin, Hökfelt *et al.* demonstrated immunoreactive neurons in the medulla oblongata projecting into the spinal cord.[22] A simple method for simultaneous localization of an antigen and a retrogradely transported fluorescent dye (true blue) in single neurons was described in more detail by Sawchenko and Swanson.[23] The method is based on 1) the efficiency of retrograde neuronal labeling with the fluorescent marker true blue, 2) the near quantitative persistence of retrogradely transported true blue localization after subsequent processing of the tissue for immunohistochemistry and 3) the possibility of distinguishing clearly between true blue and immunocytochemically stained cells simply by using appropriate excitation wave-lengths for each. This method has been extensively used to study the projections of neuropeptide specific neurons in both the central and peripheral nervous system.[24-42] However, the method has some disadvantages. It labels fibers poorly, if at all, and their fluorescence fades with prolonged viewing making repeated inspection and photography difficult. The material deteriorates over time, usually after several months. In addition, the tracer is taken up by damaged and intact fibers of passage, making considerable caution necessary in the planning and interpretation of experiments.

Fast blue, when compared with true blue, has similar properties regarding retrograde transport and visualization. It has also been combined with immunofluorescence to demonstrate directly antigen and retrograde marker within the same cell.[43-48] However, fast blue washes out of the cells to a greater extent than true blue during the immunostaining procedure. It has been reported that up to 33% of fast blue-labelled cells can be lost during immunostaining. Hence, it is necessary to perform photography prior to immunostaining.[49] This process is both tedious and time consuming, since it requires photographing many cells that may not be subsequently stained for the particular antigen of interest.

Rhodamine beads (rhodamine-labelled fluorescent latex microspheres, 0.02-0.2 μm diameter) were first used as retrograde neuronal markers for *in vivo* and *in vitro* studies of visual cortex by Katz *et al.*[50] They have some distinct advantages. When injected into brain tissue, these microspheres show little diffusion and consequently produce small, sharply defined injection sites. The tracer is apparently not taken up by undamaged axons of passage. The label persists *in vivo* for at least 10 weeks when transported back to neuronal somata, and is still visible up to one year after fixation. Microspheres have no obvious cytotoxicity or phototoxicity as assessed by intracellular recordings and staining of retrogradely labelled cells in a cortical brain slice preparation. In addition, there is virtually no fading during the examination or photography provided that the sections are well hydrated. The primary disadvantage of rhodamine beads is the incomplete morphological profile of the labelled neurons. Another disadvantage of the method relates to the red fluorescence against a dark background which reduces the chance of spotting a labelled cell, particularly when scanning at low magnification. In combination with immunofluorescence, rhodamine beads should be used with FITC labelled secondary antibodies.[49]

Fluorogold is a fluorescent dye that has been successfully demonstrated to undergo retrograde axonal transport by Schmued and Fallon.[51] It has distinct advantages over the above mentioned tracers. Fluorogold gives an intense fluorescence. The appearance of the retrogradely transported label is characterized by fluorescent gold-colored granules in the neuronal cytoplasm and neuronal processes at neutral and basic pH. A shorter survival duration (1-2 days) typically shows distinct vesicles within the cytoplasm and axon. Longer survival periods (4 days to 4 weeks) lead to accumulations of vesicles in the somata and an extensive filling of dendritic processes. Even prolonged exposure of labelled tissue to ultraviolet light causes relatively little fading, despite months of intermittent examination. Cut or damaged axons appear to take up the dye and transport it retrogradely, while intact fibers do not show such a phenomenon. Survival periods can range widely from 2 days to 2 months. Fluorogold is fully compatible with immunocytochemical techniques,[52-57] since it does not diffuse from the labelled cells during immunostaining. A shorter wavelength excitation is required for it than for FITC and rhodamine. Since the above fluorescent tracers are not electron dense, they are generally not useful in ultrastructural analysis.

Retrograde tracing with Horseradish Peroxidase (HRP) or Conjugated HRP, e.g. HRP Conjugated to Wheat Germ Agglutinin (WGA), Combined with Immunoperoxidase Methods

Horseradish peroxidase (HRP) as an axonal tracer was first reported by Kristensson and

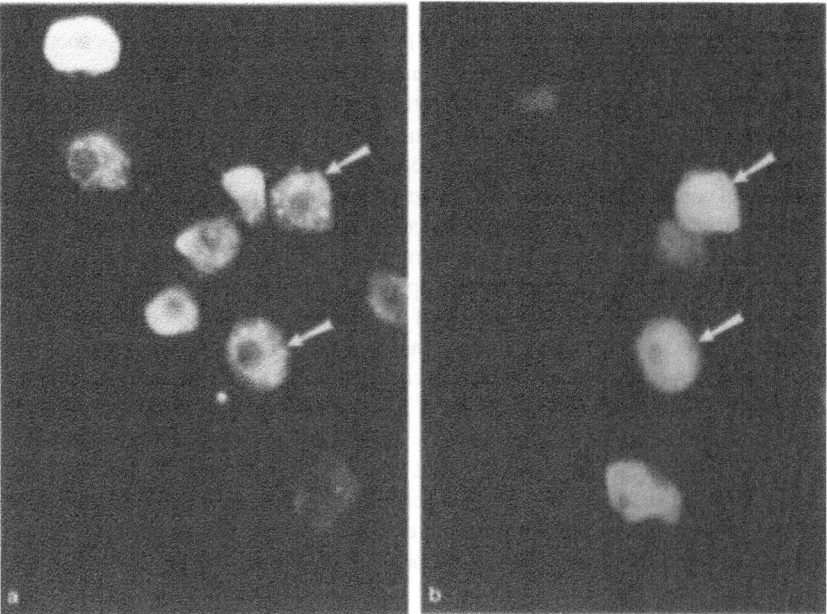

Figure 3. Section from a rat L6 dorsal root ganglion immunostained for CGRP (Fig. 3a). The same field viewed for True Blue fluorescence after injection of the dye into the bladder wall (Fig. 3b). Arrows indicate neuronal cell bodies retrogradely labelled with the True Blue and displaying CGRP-immunoreactivity. Original magnification X420.

Olsson[2] and La Vail.[3] When injected in a terminal field, HRP can be taken up by nerve terminals and retrogradely transported to the parent cell bodies. The rate of retrograde transport of HRP has been estimated at 48-120 mm per day in euthermic animals.[56] After having reached the perikaryon, the labelled endocytic organelles mainly concentrate near the Golgi complex in a perinuclear position and fuse with primary and secondary lysosomes. The labelled neurons can then be detected by HRP histochemical reactions. The reaction medium contains hydrogen peroxide and a chromogen, usually a benzidine derivative. The most commonly used chromogens are diaminobenzidine tetrahydrochloride (DAB), tetramethylbenzidine (TMB)[58] and benzidine dihydrochloride (BDHC).[59,60] The chromogen is oxidized to a colored product by the hydrogen peroxide-peroxidase reaction. Since HRP in the labelled neurons is degraded by lysosome enzymes, it is no longer detectable after four to eight days. HRP can also be taken up by perikarya and dendrites of neurons for subsequent anterograde transport, although it is not efficient.

Many other macromolecules, such as the lectin wheat germ agglutinin (WGA),[61,62] cholera toxin,[63] tetanus toxin,[64] and others have been used for retrograde tracing since late 1970s, often conjugated to HRP. Lectin and bacterial toxin conjugated HRP remain at the injection site for a considerably longer period than HRP alone, which is rapidly removed and usually completely eliminated from the injection site after seven days. WGA-HRP and cholera toxin-HRP, in contrast, can still be identified in the injection site after 13 days of survival. Therefore, these tracers may be available for uptake and transport for longer time periods than free HRP. Lectins or bacterial toxins show little extracellular spread, thereby producing reliable small injection sites. Lectin and bacterial toxin conjugated HRP show greater efficiency of endocytosis and slower degradation or elimination of the substance than free HRP. Therefore, in nerve tracing experiments, lectins, bacterial toxins, or their conjugates used as tracers are consistently more sensitive than free HRP. They usually result in more complete dendritic filling of retrogradely labelled neurones. They are also more suitable for visualizing the anterograde and transganglionic transport and can also be combined with immunocytochemistry.[56,65-69]

Successful combination of HRP or conjugated HRP retrograde tracing with HRP-based immunocytochemistry, e.g. peroxidase anti-peroxidase (PAP) has been reported.[70-80] In this combination, both steps involve development procedures based on the presence of HRP. However, tracer-labelled and neuropeptide positive stainings can be identified by either different appearance or different colors of the enzyme-reaction products. In general, retrogradely transported HRP appears as punctate granules, whereas the HRP demonstrating immunoreactivity often appears as diffuse homogeneous staining in the cytoplasm.[73,79] Sometimes, immunoreactive neurons also give a granulated rather than diffuse appearance, which is often seen in neurons immunostained with antibodies to peptides. In this case, using different colors of the HRP histochemical reaction product is necessary in order to identify double labelled cells. Using cobalt chloride to intensify the DAB product of the first peroxidase reaction, it is possible to increase the color difference between the retrogradely transported HRP (black) and the antibody-coupled HRP (brown).[70] Alternatively, the two HRP products can be easily distinguished by the use of two different chromogens, i.e. DAB and BDHC.[60] Between pH 6 and 7, the BDHC reaction product is bluish-green and crystalline which can easily be differentiated from the brown DAB reaction product. Even at the electron microscopic level, the crystalline BDHC reaction product can be differentiated from the flocculent DAB product. The solubility and instability of BDHC reaction products make it imperative to react the tissue first with DAB to identify the transported HRP, and then with BDHC for immunocytochemistry.[56] Although the TMB method is more sensitive than DAB for demonstrating HRP retrogradely labelled neurons, the low pH required to maintain stability of the chromogen and of the reaction product is not always compatible with immunocytochemistry.

WGA(apo)HRP-gold is a colloidal gold conjugated WGA coupled to enzymatically inactive HRP. The tracer has many features that make it ideal for retrograde transport studies.[81,82] There is very limited diffusion from the site of injection. The tracer can be detected in tissue fixed with a variety of fixatives including paraformaldehyde or glutaraldehyde. The tracer persists in cells for long periods of time (up to 9 months after injection without diminution in retrograde labeling). Anterograde transport, which often makes interpretation of retrograde labelling difficult, is not present by using this complex. Most importantly, because the complex contains an inactive form of the HRP molecule, it can be combined in various double-labelling studies that use enzymatically active HRP as a histochemical marker. The WGA(apo)HRP-gold retrogradely labelled neurons can be demonstrated by silver enhancement, and the antigen in the labelled neurons can be visualized with HRP-labelled secondary antisera, or by PAP or ABC methods.[81,82] The colloidal gold is electron dense and can be detected with or without silver intensification at the electron microscopic level. When the density of label is low, retrogradely labelled cells can be easily detected when silver-enhancement is examined at the electron microscopic level. The appearance of silver particles is distinct from the positive HRP reaction products, the technique is also suitable for neuropeptide identification of retrogradely labelled neurons at the electron microscopic level with HRP-based immunocytochemistry.

Biotin and colloidal gold can also be used to label WGA. Combined biotin-WGA[83] and gold-WGA[84] retrograde with immunocytochemistry has been reported to demonstrated specific neuropathways.

Anterograde Tracing with PHA-L Combined with Immunocytochemistry

PHA-L obtained from the red kidney bean is an efficient anterograde tracer.[5] When PHA-L is delivered into the central nervous system by iontophoreses, it appears to label only those neurons that incorporate the tracer at the injection site. It is subsequently transported by a slow axonal transport mechanism (4-6 mm per day) in the anterograde direction. The uptake of the tracer by fibers of passage and the retrograde axonal transport of the tracer are negligible. Labelled neurons and their axonal projections are demonstrated by an immunocytochemical procedure using a specific antibody against PHA-L, which provides an accurate visualization of cellular and axonal morphology. In the labelled neurons, PHA-L does not enter the lysosomal system and hence is not easily degraded. As a result, the number of labelled neurons does not appear to vary during the survival period, even when it is very long. PHA-L can be localized with standard immunocytochemical techniques, double immunocytochemical labelling at light and electron microscopic levels is feasible. It can not only reveal neuroanatomical and chemical characterization of neuronal targets of PHA-L labelled afferents, but also identify neurochemical properties of PHA-L labelled efferents.[85-93]

There are a few limitations to this method. First, pressure injections or direct applications of the tracer do not always result in effective anterograde labeling. Pressure injections often result in some degree of cellular destruction presumably caused by the toxic effect of PHA-L at high concentrations. The ability to specifically determine the neurons that incorporate PHA-L at the injection site, one of the best attributes of this method, is considerably compromised by such injections. Secondly, PHA-L has only a limited role in the developmental study as in early postnatal rats the tracer is invariably taken up by astroglia as well as by neurons.

Recommended Protocols

Combination of Fluorescent Retrograde Tracers and Immunofluorescence (Based on Sawchenko and Swanson[22] and Skirbolo et al.[49])

Preparation of Tracer Solution or Suspension. Fluorescent tracers are used in solution or suspension in concentrations varying from 1 to 5% (W/V) in water for the dyes and 1:1 to 1:4 in phosphate-buffered saline (PBS) for the rhodamine beads. The suspensions are sonicated for 10 min in order to reduce the particle size as much as possible. Although the solution can be stored in the refrigerator for several weeks, a fresh solution is preferred.

Tracer Injection. Inject the tracer with a Hamilton microsyringe into the terminal field under investigation in anesthetized animals. For small volume injection (0.05-0.20 µl), the syringe can be added to a glass-capillary. Injection must be given slowly (10 nm/min). Leave the needle in place after every injection for 10-20 min to minimize leakage of the tracer along the needle-track.

Survival Period. The animals should be allowed to survive for an optimal time which depends on the fluorescent tracer used and the fiber system studied. It is advisable to empirically determine the optimal time for each system under investigation, but it usually ranges from 24 hours to 10 days.

Perfusion and Histology. Although all dyes are retained to some degree in unfixed tissue, perfusion is necessary for adequate immunocytochemical staining. During anesthesia, the animal is transcardially perfused with PBS followed by 4% ice cold paraformaldehyde. Other fixatives may also be used. Relevant tissues are collected, immersed in the same fixative for 1 hour and then washed in 0.1 M PBS (pH 7.4) containing 15% sucrose for at least 24 hours. Frozen sections, 10-30 μm thick, are cut on a cryostat and thawed onto poly-L-lysine (PLL)-coated slides.[94]

Immunocytochemistry. Sections are processed for indirect immunofluorescence procedures. Briefly, the sections are incubated with the primary antiserum at 4°C for 24-48 hours in a humid atmosphere. After rinsing in PBS (5 min x 3), the sections are incubated with FITC conjugated antibody for 30 min at room temperature, rinsed in PBS (5 min x 3) and mounted with PBS:glycerol (1:3).

Inspection and Evaluation. The sections are examined under a fluorescent microscope with different filter combinations for tracers and FITC immunofluorescence according to their excitation wavelengths (*Table 1*). Photomicrographs are taken from the same field by only switching the filter systems without altering even very slightly the plane of focus and the location of the slide. Only those cells which exhibit a similar location and morphology with both stains when viewed in a single plane of focus are considered double-labelled.

Table 1.

	Excitation Filter	Barrier filter
Fast blue	330-380 nm	460 nm
True blue	340-360 nm	420-430 nm
FITC	460-485 nm	510-545 nm
TRITC	530-550 nm	580 nm
Fluorogold	323 nm	408 nm

Combination of HRP Retrograde Tracing and PAP Immunocytochemistry Method (Based on Bowker et al 1982)[70]

Tracer injection. HRP or WGA conjugated to HRP (WGA-HRP) is injected in anesthetized animals via a glass micropipette (tip 50-100 μm) attached to a Hamilton syringe. The quantity of HRP injected may vary depending upon the system studied. For example, small quantities (0.1-0.2 μl of a 25-50% solution of HRP or 0.1% solution of WGA-HRP are deposited into the spinal cord at each penetration (6-15 penetrations may be made bilaterally over 2-3 spinal segments).

Perfusion. The animals are deeply anesthetized with pentobarbital, and the limbs affixed to a wire grate with tape. After exposure of the heart, the right ventricle (or right atrium) is cut open. The left ventricle is punctured by a blunt 14-18 gauge needle connected to the perfusion apparatus containing saline solution. The needle is held in the ascending aorta. For rats, 300-500 ml of warm physiological saline is infused until the effluent remains clear for 30-60 seconds. This phase of the procedure usually takes 2.5-5 min. For cats and monkeys, 800-1,000 ml of warm saline is infused. The perfusion solution is then changed from warm saline to 3.0-3.8% cold paraformaldehyde. The connecting tube is placed in an ice bath to maintain the temperature of the fixative between 5-12°C. The rate of perfusion of the fixative is adjusted to ensure that 1,100-1,200 ml of infusion is completed in 40-60 min. One liter of cold (5-12°C) 30% sucrose solution is then infused to prepare the tissues for sectioning. Perfusion with the sucrose solution is usually completed within 60-90 min. For larger animals, e.g. cats and monkeys, 2.5-3.0 liters of both fixative and sucrose solution are employed.

Sectioning. After perfusion, the brain and the spinal cord are quickly removed and placed in a small amount of the sucrose solution. Tissue sectioning (20-25 μm) on a freezing microtome can begin immediately or can be delayed until the next day.

HRP Histochemistry. The sectioned tissues are collected in cold phosphate buffer and rinsed two to three times in Tris/HCl buffer prior to incubating them in 0.5% CoCl$_2$ for 5-10 min. The sections are then rinsed two to three times in Tris/HCl buffer and two to three times in phosphate buffer before HRP histochemistry. At this phase of the procedure, the sections have a very light gray or bluish cast to them. The sections are then incubated in 0.02% DAB for 10-15 min without H$_2$O$_2$ and then 15 min in 0.02% DAB with H$_2$O$_2$. To ensure complete exposure of tissues to the DAB solution, the tissue sections are agitated frequently during the reaction. The sections are rinsed in 0.1 M phosphate buffer two to three times at the end of the histochemical reaction. To abbreviate this procedure, the sections can be incubated in DAB with H$_2$O$_2$ for 15 min. At this time, retrograde labelling can be evaluated by examining a few of the wet sections under the microscope (low magnification, bright-field optics). Usually, in the wet tissue sections, the black HRP granules are visible only in large neurons. The tissue sections are next processed for immunocytochemistry.

PAP Immunocytochemistry

1. Tissue sections are placed in hollow polyethylene stoppers and are incubated in a primary antiserum (e.g. from rabbit) for 24-28 hrs at room temperature with continuous or frequent agitation.
2. After several rinses, sections are transferred to a small compartmentalized (12x12x12 mm) tray having a fine nylon mesh attached to its base. This tray with its fine mesh permits thorough washing of the tissue sections and then allows the incubation of the sections in the immunocytochemical reagents during the PAP staining procedure. In addition, by having the tray compartmentalized, tissue sections from one or more animals can be processed simultaneously.
3. After washing 6-10 times in 1% normal goat serum in 0.1 M PBS, the tissue sections are incubated in 3% normal goat serum, 0.75% gelatin in PBS for 30 min. This is to "block" nonspecific staining. (This step may be omitted if nonspecific staining is not a problem.)
4. Incubate the sections with goat anti-rabbit IgG diluted in 0.1 M PBS containing 1% normal goat serum for 30 min.
5. Wash 6-10 times with 0.1 M phosphate buffer containing 1% normal goat serum.
6. Incubate sections in 3% normal goat serum, 0.75% gelatin in 0.1 M PBS for 30 min (see above).
7. Incubate tissue sections in rabbit PAP complex diluted 1:80 in 1% normal goat serum in 0.1 M PBS for 30 min.
8. Wash 6-10 times in 0.1 M phosphate buffer.
9. Incubate sections in 0.02% DAB and 0.01% H$_2$O$_2$ in 0.1 M phosphate buffer for 6 min.
10. Wash thoroughly in 0.1 M phosphate buffer. Note: Instead of DAB, 3-amino-9-ethylcarbazole (AEC) may be used as the chromogen. Dissolve 20 mg of AEC in 5 ml of dimethylformamide and mix with 95 ml of 0.05 M acetate

315

buffer (pH 5.0) containing 30-50 µl of 30% H_2O_2. Allow the reaction to proceed for 3 min, check in microscope, and rinse in distilled water. Reaction product is bright red instead of brown which is seen with DAB.

11. Mount and cover slip in alcohol, xylene, and Permount. The red reaction product of 3-amino-9-ethylcarbazole reaction must be mounted in glycerin and distilled water (4:1).

Combination of WGA(apo)HRP-Gold Retrograde Tracing and Immunocyto chemistry (Based on Basbaum and Menitrey, 1987)[81]

Preparation of Gold Particles

Gold particles (approximately 10-12 nm) are prepared according to Muhlpfordi's procedure.[95] Briefly, 100 ml of chloroauric acid (Sigma, FRG; 0.01% w/v in distilled water) are boiled while being vigorously stirred. Reducing agents (2 ml of sodium citrate and 100 µl of tannic acid, both 1% w/v in distilled water) are mixed in a separate beaker and rapidly added to the boiling solution. The reaction is complete (30-50 seconds) when the solution quickly turns dark violet and then wine red. Boiling and stirring are continued for 5 min and then the solutions are cooled under running water. Solutions can be kept at 4°C for several weeks if supplemented with 2% sodium azide (0.2% v/v).

Preparation of Gold-Labelled WGA(apo)HRP

Lyophilized WGA(apo)HRP (Sigma, FRG. Code No. L0390) is coupled to colloidal Au according to a standard protocol.[96] The pH of the Au solution is raised to 8.4 by adding small quantities (approximately 4 µl/ml solution) of potassium carbonate (0.2 M) and checked on a 1 ml aliquot in the presence of 45 µl of 1% (w/v) polyethylene glycol (PEG; MW 20,000; Sigma, FRG). The Au solution is added to diluted WGA(apo)HRP (1 mg protein per ml of distilled water) and coupled while being vigorously stirred. After 5 min, filtered PEG is added in a proportion of 1% (v/v) of the complete volume of the solution to prevent aggregation of the complex. The complex is centrifuged for 120 min at 18,000 rpm. The supernatant is aspirated and discarded and the soft pellet containing WGA(apo)HRP-Au complex is gently resuspended in distilled water, to be microcentrifuged later.

Tracer Injection

Animals are deeply anesthetized with pentobarbital (60 mg/kg, i.p.). Tracer injections (0.2-1.0 µl) are made by pressure through glass micropipettes with tip diameters of 15-30 µm. The WGA(apo)HRP-Au is injected without further diluting the pellet that was obtained after centrifuging the complex. Based on the concentration of protein originally used, and given that a pellet volume of about 10 µl is typically obtained, it is possible to estimate the protein concentration that is injected. This is generally about 0.5-1.0%, which is approximately one-fifth of the concentration obtained with WGA-HRP.

Perfusion

After an optimal survival time, the animals (e.g. rats) are perfused intracardially with 200 ml of phosphate-buffered saline (pH 7.6) followed by 500 ml of a fixative solution. If LM analysis is planned, the fixative is a 0.1 M phosphate-buffered 4% paraformaldehyde solution. Tissues from these animals are postfixed in the same fixative for several hours and then cryoprotected in a 30% buffered sucrose solution. For EM immunocytochemical analysis, the animals are perfused with 4% paraformaldehyde and 0.2% glutaraldehyde. The tissues are postfixed in the same solution and then cut either on a freezing microtome (for LM) or on a vibratome. Sections are collected in 0.1 M phosphate buffer, without saline. Other fixatives such as formalin (10%) or acrolein (5%) can also be used. Generally, the choice of fixative does not interfere with the quality or extent of the retrograde labelling.

Silver Enhancement (Autometallography)

The sectioned tissue undergoes a silver-intensification procedure in the dark, at 18-22°C. After a rinse (5 min) in sodium citrate buffer (0.1 M, pH 3.8), the sections are dipped in autometallographic solution for 60 min. Each 100 ml of physical developer consists of 60 ml gum arabic (50% solution in distilled water) as a reducing agent, and 15 ml silver lactate (Fluka; 0.7% in distilled water) solution to supply silver ions. The silver lactate solution is protected against light and added just before the developer is used. After development, sections are washed once (5 min) in 0.1 M phosphate buffer (pH 7.4), transferred into sodium thiosulfate (2.5% solution in 0.1 M phosphate buffer, pH 7.4) for 5 min, and then rinsed for 5 min in 0.1 M phosphate buffer pH 7.4. This silver-amplification procedure has been adopted from Danscher[96-98] and replaced water rinses with buffer to improve tissue preservation. For silver enhancement, various other methods based on Danscher's autometallography[97,98] can be used (see the chapters by Hacker *et al.* in this volume). Self-made developer based on silver lactate or silver acetate are preferable to commercially available silver enhancement kits. They are more sensitive and more specific. Silver acetate autometallography allows the visual control of the staining intensity in the bright field light microscope.

Immunocytochemistry

For immunocytochemical analysis, the silver-enhanced sections are first washed extensively and then processed either with PAP or ABC methods.

For combined EM immunocytochemical analysis, the sections are immunostained after the gold-toning procedure.[99] The gold toning may overcome the problem of a possible loss of silver precipitate through oxidation by osmium. The standard protocols of the gold-toning are as follows: The sections are first incubated for 10 min in the dark in 0.05% chloroauric acid (Sigma) and washed three times in distilled water. Then the gold is reduced with 0.05% oxalic acid for 2 min. After three additional distilled water washes, the residual, unreduced metal (silver and/or gold) is eliminated by washing the sections in 1% sodium thiosulfate (3x20 min). The gold-toning steps are all performed at 4°C with continuous agitation. After the gold-toning procedure, the sections are immunostained and then osmicated for 1 hour (1% osmium in 0.1 M phosphate or 0.1 M s-collidine). Sections are then dehydrated in graded alcohols, and flat embedded in plastic.

Gold toning has a mild deleterious effect on immunostaining presumably by reducing antigenicity.

Combination PHA-L Anterograde Tracing and Immunocytochemistry (Based on Gerfen *et al.*)[89]

Deliver PHA-L into the area under investigation by iontophoresis. Solution of PHA-L (2.5% Victor Labs) dissolved in 0.01 M sodium phosphate buffer (pH 7.4-8.0) is loaded into a glass micropipette with a tip diameter of less than 15 μm. With the animal under anesthesia, the tip is stereotaxically positioned in an area of choice in the brain, and a 5 to 10 μA positive current is applied for every other 7 seconds for 15-20 min using a constant-current source that is capable of generating up to 2,000 V (CS-3 current source, Transkinetics systems, Inc., Canton, MA). To obtain larger injections which may be necessary for larger animals such as primates, currents up to 10 μA are used. It is also possible to make multiple injections to increase the size of the injection area.

Survival period. The estimated axonal transport rate for PHA-L is 4-6 mm per day. The survival period is thus dependent on the length of the pathway to be traced. Typical survival periods are between 7 and 21 days, and can be up to 3-4 weeks for primates. There appears to be negligible degradation of neuronally incorporated PHA-L with survival periods of up to 5 weeks.

Perfusion. The stability of the PHA-L allows the use of a wide range of fixatives. Virtually any paraformaldehyde perfusion fixation method can provide excellent results. However, it is advisable to adjust the fixative for optimal labelling of the second neurochemical antigen. The standard perfusion protocol involves deeply anesthetizing the animal before transcardial perfusion with normal physiological saline (0.9% NaCl) to rinse out blood, followed by the infusion of fixative (approximately 500 ml fixative per 300 g rat over 15-30 min).

Tissue preparation. The organ of interest is then removed and postfixed in the fixative plus 20-30% sucrose for 6-72 hours. Because of the stability of the PHA-L, long postfixation periods may be used. Cryostat sections are prepared. Organs that are to be cut with a vibratome do not need to be postfixed in sucrose.

Immunohistochemistry. In order to demonstrate PHA-L and the neurotransmitters, tissue sections are incubated in primary antisera directed against PHA-L and against the neurotransmitter for 24-48 hours at 4°C. It is essential that the two primary antisera are raised in different species. For example, if the primary antiserum directed against the neurochemical antigen is raised in rabbits, the primary antiserum directed against PHA-L should be raised in a different species, such as guinea pig.

Following incubation in the primary antisera, sections are rinsed and incubated in a mixture of fluorescently labelled affinity-purified secondary antisera. Again, it is essential that the secondary antisera do not crossreact with the inappropriate primary antiserum. For example, rhodamine-labelled goat antiserum directed against guinea pig IgG (GaGP-TRITC, Cappel Labs, diluted 1:200) is mixed with fluorescein-labelled goat antiserum directed against rabbit IgG(GaR-FITC, Sigma Chemical, diluted 1:200) and the sections are incubated for 45-60 min at room temperature. Sections are then rinsed and coverslipped with glycerol buffer (pH 8.5).

To demonstrate the neurotransmitter characteristics of neuronal targets of PHA-L-labelled afferents, the sections may be incubated in a mixture of the primary antisera or sequentially in separate dilutions of the antisera. However, to identify neurotransmitters in PHA-L-labelled efferents, it is an absolute requirement to mix the primary antisera during incubations and similarly to mix the secondary antisera. This is essential since it appears that if the antisera are used sequentially, the labeling is affected by steric hindrance within the same structure. When these precautions are taken, the secondary antisera provide reliable labelling of the appropriated antigens without cross activity.

Microscopy. Sections are examined under a fluorescent microscope with filters to view FITC (excitation filter 460-485 nm, barrier filter 510-545 nm), TRITC (excitation filter 535-550 nm, barrier filter 580 nm).

DISCUSSION

Choice of Methods

Many different combinations of axonal transport tracing and immunohistochemistry have been developed. Each of the combination methods has its own advantages and limitations. It is difficult to select one particular approach as the method of choice for a particular problem. The choice of the methods depends on the aim of the study, the previous experience of the investigators and the equipment available in the laboratory. It is advisable to use fluorescent retrograde tracing combined with immunofluorescence to trace neurotransmitter or neuropeptide specific pathways at light microscopic level, since the technique is simple, sensitive and reproducible. In the literature, True Blue[24-42] and fluorogold retrograde[52-56] tracing combined with immunohistochemistry has been widely used to identify neurotransmitter specific pathways in both central and peripheral nervous system. However, in light and electron microscopic study of nerve tracing combined with transmitter localization, HRP, conjugated HRP, or WGA(apo)HRP-gold tracing combined HRP-based immunocytochemistry can be used. The analysis of transmitter content in the postsynaptic target of afferent terminals requires PHA-L anterograde tracing combined with immunohistochemistry at the electron microscopic level.

When analyzing the results of combined axonal tracing and immunocytochemistry, the degree of specificity of tracer labelling and evaluation of immunostaining require careful consideration.

Specificity of Tracer Labelling

False-positive labelling may result from leakage of tracers alongside the pipette track creating difficulties in interpretation. The problem is especially pronounced when using pressure injections or tracers with a high diffusion rate. Various efforts have been made to reduce the diffusion of tracer along the pipette track. It seems that a slow withdrawal of the pipette reduces the risk of contamination. To use a pipette with a long thin taper and tip is also an advantage. Unintended labelling through the pipette track can be avoided by injection of HRP through a cannula implanted 1-10 days previously.[100]

Many tracers, e.g. True Blue, HRP, are avidly taken up not only by nerve terminals, but also by damaged and undamaged fibers of passage. It causes obvious problems associated with interpreting the origins of terminal fields. However, this feature can be used to advantage in some instances, such as when labelling a peripheral nerve or a fiber tract in the central nervous system. It is important to avoid tracer being injected into the blood vessels. The tracer being transported to other regions by the vascular system may cause false positive labelling. When investigating the origins of nerves in peripheral organs, the tracer must not spread to contaminate adjacent organs. A barrier formed from a plastic wound spray (pyroxylin solution, New Skin) is recommended for application to the surface of injected organs.[101]

It has been reported that some tracers, e.g. HRP, WGA, and tetanus toxin, undergo transganglionic or transcellular transport. The occurrence and degree of transneuronal labelling are dependent on many variables including survival time and amount of the tracer injected and possibly, on the type of synapse involved.[56] Although this phenomenon can be used in a more systematic fashion for functional-anatomic studies, it is important to keep in mind when analyzing the specific neuropathways.

False negative results may result from too short or too long survival periods. When working with nerve tracing, it is of primary importance to determine the postinjection survival duration that provides optimal neuronal labelling. The optimum time period is that in which the compound may accumulate in the cell in the absence of discernible intracellular breakdown. Different factors determine the optimal survival time for different tracers, and there are no rules that can be applied to all tracers. In practice, the optimal duration will vary widely depending on the tracer, the system, and the species being studied as well as the methods of tissue processing. The optimal survival period also depends on the rate of tracer transport, the length and diameter of the fiber system being studied, and the age of the animal. The survival period should also be directed by the goal of the investigation, as the rate of anterograde and retrograde transport are different. Degradation of the tracer differs depending on whether it takes place in the cell bodies or in the terminals. When the transport rate of the tracer is slow, too short a survival period may result in an insufficient labelling. When the tracer, e.g. HRP, is relatively easily degraded in the labelled cells, too long a survival period may cause false negative results.

Evaluation of Immunocytochemistry

Fluorescent tracers usually do not mask positive immunostaining. However, retrogradely transported HRP has a tendency to obscure the presence of immunostaining. It is possible that some apparently nonimmunoreactive cell bodies may contain very low concentrations of neuropeptides which are immunocytochemically undetectable. Immunocytochemical visualization of some peptides requires prior treatment of the animals with colchicine to inhibit axonal transport and induce accumulation of peptides in the cell bodies.[102,103] Since colchicine treatment also arrests the axoplasmic transport, it is advisable to administer the colchicine 1-3 days after tracer injection. After tracer injection, depending on the tracer used and the length of the pathway projection, the animal should be perfused after an additional survival time of about 24 hours. A significant limitation of immuno-cytochemistry in the study of neurotransmitters is the fact that only material stored in the cell at the moment of death can be visualized. The results do not reflect the biosynthetic activity of the cell. *In situ* hybridization has been employed to demonstrate various types of mRNA in neurons.[104] Compared to immuno-cytochemistry, *in situ* hybridization offers the advantage of localizing the anatomic site for protein synthesis, not merely detecting the presence of peptide or protein. *In situ* hybridization may also have a higher degree of specificity. The combination of axonal tracing and *in situ* hybridization has been applied to study mRNA levels of different neuronal populations with known projections.[105-107]

ACKNOWLEDGEMENTS

This work was supported by the grant from the Swedish Medical Research Council (No. 102). The authors are grateful to Professor Lars Grimelius (Uppsala, Sweden) for his valuable comments and criticism.

References

1. A.E. Hendrickson, Electron microscopic autoradiography: Identification of origin of synaptic terminals in normal nervous tissue, *Science* **165**:194 (1969).
2. K. Kristesson and Y. Olsson, Retrograde axonal transport of protein, *Brain Res* **27**:363 (1971).
3. L.H. La Vail and M.M. La Vail, Retrograde axonal transport in the central nervous system, *Science* **176**:1416 (1972).
4. S.B. Edwards and A. Hendrickson, The autoradiographic tracing of axonal connections in the central nervous system, *in*: "Neuroanatomical Tracing Methods" p 171, L. Heimer and M.J. RoBards eds., Plenum Press, New York (1983).
5. C.R. Gerfen and P. Sawchenko, An anterograde neuroanatomical tracing method that shows the

detailed morphology of neurons, their axons and terminals: Immunohistochemical localization of an axonally transported plant lectin, Phaseolus vulgaris leucoagglutinin (PHA-L), *Brain Res* **290**:219 (1984).

6. A.H. Coons, E.H. Ledue, and J.M. Connolly, Studies on antibody production I. A method for the histochmical demonstration of specific antibody and its application to a study of the hyperimmune rabbit, *J Exp Med* **102**:42 (1955).

7. A.H. Coons, Fluorescent antibody methods, *in*: "General Cytochemical Methods" P 399, J.F. Danielli ed., Academic Press, New York (1958).

8. P.K. Nakane and G.B.Jr. Pierce, Enzyme labelled antibodies for the light and electron microscopic localization of tissur antigens, *J Cell Biol* **33**:307 (1967).

9. L.A. Sternberger, P.H.Jr. Hardy, J.J. Cuenis, and H.G. Meyer, The unlabelled antibody enzyme method of immunohistochemistry. Preparations and properties of soluble antigen-antibody complex (horseradish peroxidase-antihorseradish peroxidase) and its use in identfication of spirochetes, *J Histochem Cytochem* **18**:315 (1970)

10. S.M. Hsu, L. Raine, and H. Fanger, The use of antiavidin antibody and avidin-biotin-peroxidase complex in immunoperoxidase technics, *Am J Clin Pathol* **75**:816 (1981).

11. J. DeMey, M. Moeremansm G. Geuens, R. Nuydens, and M. De Brabander, High resolution light and elactron microscopic localization of tubulin with the IGS (Immuno-Gold Staining) method, *Cell Biol Int Rep* **5**:889(1981).

12. G.W. Hacker, D.R. Springall, S. Van Noorden, A.E. Bishop, L. Grimelius, and J.M. Polak, The immunogold-silver staining method: a powerful tool in histopathology, *Virchows Arch Path Anat* **406**:449 (1985).

13. C. Holgate, P. Jackson, P. Cowen, and C. Birs, Immunogold-silver staining: new method of immunostaining with enhanced sensitivity, *J Histochem Cytochem* **31**:938 (1983).

14. T. Hökfelt, O. Johansson, A. Ljungdahl, J.M. Lundberg, and M. Schultzberg, Peptidergic neurons, *Nature* **284**:515 (1980)

15. J.M. Polak and S.R. Bloom, Immunocytochemistry of the diffuse neuroendocrine system, *in*: "Immunocytchemistry" 2nd edn. P 328, J.M. Polak and S.Van Noorden eds., Wright, Bristol (1986).

16. I.J. Llewellyn-Smith, Neuropeptides and the microcircuitry of the enteric nervous system, in: "Regulatory Peptides" p 247, J.M.Polak ed., Birkhäuser Verlag, Basel, Boston, Berlin (1989).

17. T. Hökfelt, G. Skagerberg, L. Skirboll, and A. Björklund, Combination of retrograde tracing and neurotransmitter histochemistry, *in*: "Handbook of Chemical Neuroanatomy, Vol. 1: Methods in Chemical Neuroanatomy, p228, A. Bjorklund and T. Hökfelt, eds., Elsvier Science Publisher, B.V., (1983).

18. M. Bentivoglia, H.G.J.M. Kuypers, C. Catwman-Berrevoets, and O. Dann, Flourescent retrograde neuronal labelling in rat by means of substances binding specifically to adeninethymine rich DNA, *Neurosci Lett* **12**:235 (1979).

19. M. Bentivoglio, H.G.J.M. Kuypers, and C. Catsman-Berrevoets, Retrograde neuronal labelling by means of hisbenzimide and nuclear yellow (Hoechst, S 76121): measures to prevent diffusion of the tracers out of retrogradely labelled cells, *Neurosci Lett* **18**:19 (1980).

20. H.G.J.M. Kuypers, M. Bentivoglio, D. Van der Kooy, and C.E. Catsman-Berrevotes, Retrograde transport of bisbenzimide and propidium iodide through axons to their parent cell bodies, *Neurosci Lett* **12**:1 (1977).

21. H.G.J.M. Kuypers, M. Bentivoglio, D. Van der Kooy and C.E. Catsman-Berrevoets, Retrograde axonal transport of fluorescnt substances in the rat's forebrain, *Neurosci. Lett* **6**:127 (1979).

22. T. Hökfelt, L Terenius, H.G.J.M. Kuypers, and O. Dann, Evidence for enkephalin immunoreactive neurons in the medulla oblongata projecting to the spinal cord, *Neurosci Lett* **14**:55 (1979).

23. P.E. Sawchenko and L.W. Swanson, A method for tracing biochemically defined pathways in the central nervous system using combined fluorescence retrograde transport and immunohistochemical techniques, *Brain Res* **210**:31 (1981).

24. P. Alm and L.M. Lundberg, Co-existence and origin of peptidergic and adrenergic nerves in the guinea pig uterus. Retrograde tracing and immunocytochemistry, effects of chemical sympathectomy, capsaicin treatment and pregnancy, *Cell Tissue Res* **254**:517 (1988).

25. C. Bennett-Clarke, R.D. Mooney, N.L. Ciaia, and R.W. Rhoades, A substance P projection from the superior colliculus to the parabigeminal nucleus in the rat and hamster, *Brain Res* **500**:1 (1989).

26. L. Edvinsson, H. Hara, and R. Uddman, Retrograde tracing of nerve fibers to the rat middle cerebral artery with true blue: colocalization with different peptides, *J Cereb Blood Flow Metab* **9**:212 (1989).

27. H.J. Enfiejian, N.L. Chiaia, G.J. Macdonal, R.W. Rhoades, Neonatal transection alters the percentage of substance-P-positive trigeminal ganglion cells that contribute axons to the regenerate infraorbiral nerve, Somatosens, *Mot Res* **6**:537 (1989).

28. T. Grunditz, R. Hakanson, F. Sundler, and R. Uddman, Neuronal pathways to the rat thyroid revealed by retrograde tracing and immunocytochemisty, *Neurosci* **24**:321 (1988).

29. C.M. Klein and Burden H.W., Substance P- and vasoactive intestinal polypeptide (VIP)-immunoreactive nerve fibers in relation to ovarian postganglionic perikarya in para- and prevertebral ganglia: Evidence from combined retrograde tracing and immunocytochemistry, *Cell Tissue Res* **252**:403 (1988).

30. G.C. Kwiat and A.I. Basbaum, Organization of tyrosine hydroxylase-and serotonin-immunoreactive brainstem neurons with axon collaterals to the periaqueductal gray and the spinal cord in the rat, *Brain Res* **548**:83 (1990).

31. A. Luts, R. Uddman, T. Grundits and F. Sundler, Peptide-containing neurons projectiong to the vocal cords of the rat: Retrograde tracing and immunocytochemistry, *J Auton Nerve Syst* **30**:179 (1990).

32. T.P. O'Connor and D. Van der Kooy, Enrichment of a vasoactive neuropeptide (carcitonin gene related peptide) in the trigeminal sensory projection to the intracranial arteries, *J Neurosci* **8**:2468 (1988).

33. R.W. Rhoades, R.D. Mooney, N.L. Chiaia, and C.A. Bennett-Clarke, Development and plasticity of the serotoninergic projection to the hamster's superior colliculus, *J Comp Neurol* **299**:151 (1990).

34. M. Saji and M. Miura, Coexistence of glutamate and choline acetyltransferase in a major subpopulation of laryngeal motoneurons of the rat, *Neurosci Lett* **123**:175 (1991).

35. Su H.C., A.E. Bishop, R.F. Power, Y. Hamada and J.M. Polak, Dual intrinsic and extrinsic origins of CGRP- and NPY-immunoreactive nerves of rat gut and pancreas, *J Neurosci* **7**:2674 (1987).

36. Su H.C., J. Wharton, J.M. Polak, P.K. Mulderry, M.A. Ghatei, S.J. Gibson, G. Terenghi, J.F.B. J.F.B. Morrison, J. Ballesta, and S.R. Bloom, Calcitonin gene-related peptide-immunoreactivity in afferent neurons supplying the urinary tract: Combined retrograde tracing and immunocytochemistry, *Neurosci* **18**:727 (1986).

37. F. Sundler, T. Grunditz, R. Hakanson, and R. Uddman, Innervation of the thyroid. A study of the rat using retrograde tracing and immunocytochemistry, *Acta Histochem Suppl* **37**:191 (1989).

38. E.K. Tayo and R.G. Williams, Catecholaminergic parasympathetic efferents within the dorsal motor nucleus of the vagus in the rat: a quantitative analysis, *Neurosci Lett* **90**:1 (1988).

39. R. Uddman and L. Edvinsson, Neuropeptides in the cerebral circulation, *Cerebrovasc Brain Metab Rev* **1**:230 (1989).

40. R. Uddman, T. Grunditz, A. Larsson, and F. Sundler, Sensory innervation of the ear drum and middle-ear mucosa: Retrograde tracing and immunocytochemistry, *Cell Tissue Res* **252**:141 (1988).

41. R. Uddman, H. Hara, and L. Edvinsson, Neuronal pathways to the rat middle meningeal artery revealed by retrograde tracing and immunocytochemistry, *J Auton Nerv Syst* **26**:69 (1989).

42. F.A. White, C.A. Bennett-Clarke, G.J. Macdonald, H.L. Enfiejian, N.L. Chiaia, and R.W. Rhoades, Neonatal infraorbital nerve transection in the rat: Comparison of effects on substance P immunoreactive primary afferents and those recognized by the lectin Bandierea simplicifolia-I, *J Comp Neurol* **300**:249 (1990).

43. S.Y. Baek, M. Yamano M, Y. Shiotani, and M. Tohyama, Distribution and origin of vasoactive intestinal polypeptide-like immunoreactive fibers in the central amygdaloid nucleus of the rat: An immunocytochemical analysis, *Peptides* **9**:661 (1988).

44. K. Fukami, H. Hiyama, Y. Shiotani, and M. Tohyama, Neurotensin-containing projections from the retrosplenial cortex to the anterior ventral thalamic nucleus in the rat, *Neuroci* **26**:819 (1988).

45. T.R. Stratford and D. Wirstshafter, Ascending dopaminergic projections from the dorsal raphe nucleus in the rat, *Brain Res* **511**:173 (1990).

46. K. Takatsuji and M. Tohyama, Geniculo-geniculate projection of enkephalin and neuropeptide Y containing neurons in the intergeniculate leaflet of the thalamus in the rat, *J Chem Neuroanat* **1**:19 (1989).

47. M. Tanaka, N. Takeda, E. Senba, M. Tohyama, T. Kubo, and T. Matsunaga, Localization and origins of calcitonin gene-related peptide containing fibers in the vestibular end-organs of the rat, *Acta Otolaryngol Suppl* (Stockh) **468**:31 (1989).

48. M. Yamano, C.J. hillyard, S. Girgis, P.C. Emson, I. MacIntyre, and M. Tohyama, Projection of neurotensin-like immunoreactive neurons from the lateral parabrachial area to the central amygdaloid nucleus of the rat with reference to the coexistence with calcitonin gene-related peptide, *Exp Brain Res* **71**:603 (1988).

49. L.R. Skirboll, K. Thor, Helke C., T. Hökfelt, B. Robertson, and R. Long, Use of retrograde

fluorescent tracers in combination with immunohistochemical methods, *in*: "Neuroanatomical Tract-Tracing methods 2 - Recent progress", p 5, L. Heimer and L. Zaborszky eds., Plenum Press, New York (1989).

50. L.C. Katz, A. Burkhalter, and W.J. Dreyer, Fluorescent latex microspheres as a retrograde neuronal marker for the *in vivo* and *in vitro* study of visual cortex, *Nature* **310**:498 (1984).

51. L.C, Schmued and J.H. Fallon, Fluoro-gold: A new fluorescent retrograde axonal tracer with numerous unique properties, *Brain Res* **377**:147 (1986).

52. C.A. Bennett-Clarke, R.D. Mooney, N.L. Chiaia, and R.W. Rhoades, Serotonin immunoreactive neurons are present in the superficial layers of the hamster's, but not the rat's, superior colliculus, *Exp Brain Res* **85**:587 (1991).

53. T.J. Mahalik and G.H. Clayton, Specific outgrowth from neurons of ventral mesencephalic grafts to the catecholamine-depleted striatum of adult hosts, *Exp Neurol* **113**:18 (1991).

54. I. Merchenthaler, Neurons with access to the general circulation in the central nervous system of the rat: a retrograde tracing study with fluoro-gold, *Neuroci* **44**:655 (1991).

55. R.D. Mooney, C.A. Bennett-Clarke, T.D. King, and R.W. Rhoades, Tectospinal neurons in hamster contain glutamate-like immunoreactivity, *Brain Res* **537**:375 (1990).

56. L. Zaborszky and L. Heimer, Combinations of tracer techniques, especially HRP and PHA-L, with transmitter identification for correlated light and electron microscopic studies, *in*: "Neuroanatomical Tract-Tracing methods 2 - Recent progress", p 49, L. Heimer and L. Zaborszky eds., Plenum Press, New York (1989).

57. D.C. Yeomans and H.K. Proudfit, Projection of substance P-immunoreactive neurons located in the ventromedial medulla to the A7 noradrenergic nucleus of the rat demonstrated using retrograde tracing combined with immunocytochemistry, *Brain Res* **532**:329 (1990).

58. M.M. Mesulam, Tetramethyl benzidine for horseradish peroxidase neurohistochemistry: A non-carcinogenic blue reaction-product with superior sensitivity for visualizing neural afferents and efferents, *J Histochem Cytochem* **26**:106 (1978).

59. M. Peschanski and H. J. Ralston III, Light and electron microscopic evidence of transneuronal labelling with WGA-HRP to trace somatosensory pathways to the thalamus, *J Comp Neurol* **236**:29 (1985).

60. S. Lakos and A.I. Basbaum, Benzidine dihydrochloride as a chromogen for single- and double-label light and electron microscopic immunocytochemical studies, *J Histochem Cytochem* **34**:1047 (1986).

61. D.M. Lechan, J. Nestler, and S.J. Jaxobsen, Immunohistochemical localization of retrogradely and anterogradely transported wheat germ agglutinin (WGA) within the central nervous system of the rat: Application to immunostaining of a second antigen within the same neuron, *J Histochem Cytochem* **29**:255 (1981).

62. M.E. Schab, F. Javoy-Agid, and Y. Agid, Labelled wheat germ agglutinin (WGA) as a new, highly sensitive retrograde tracer in the rat brain hippocampal system, *Brain Res* **152**:145 (1978).

63. X.S.T. Wan, J.Q. Trojanowski, and K.O. Vonatas, Cholera toxin and wheat germ agglutinin conjugates as neuroanatomical probes: Their uptake and clearance, transganglionic and retrograde transport and sensitivity. *Brain Res* **243**:215 (1982).

64. K. Stockel, M.E. Schwab, and H. Thoenen, Role of gangliosides in the uptake and retrograde axonal transport of cholera and tetanus toxin as compared to nerve growth factor and wheat germ agglutimin, *Brain Res* **132**:173 (1977).

65. B.N. Cardozo, R. Huijs, and J. Van der Want, Glutamate-like immunoreactivity in retinal terminals in the nucleus of the optic tract in rabbits, *J Comp Neurol* **309**:261 (1991).

66. S. Shioda, Y. Nakai, M. Iwase, and I. Homma, Electron microscopic studies of medullary synaptic inputs to vasopressin-containing neurons in the hypothalamic paraventricular nucleus. *J Electron Microsc (Tokyo)* **39**:501 (1990).

67. J.H. McLean, M.T. Shipley, W.T. Nickell, G. Aston-Jones, and C.K. Reyher, Chemoanatomical organization of the noradrenergic input from locua coeruleus to the olfactory bulb of the adult rat, *J Comp Neurol* **285**:339 (1989).

68. C.I. de Zeeuw, J.C. Holstege, F. Calkoen, T.J. Ruigrok, and J. Voogd, A new combination of WGA-HRP anterograde tracing and GABA immunocytochemistry applied to afferents of the cat infereior olive at the ultrastructural level, *Brain Res* **447**:369 (1988).

69. C.I. de Zeeuw, J.C. Holstege, T.J. Ruigrok, and J. Voogd, Ultrastructural study of the GABAergic, cerebellar, and mesodiencephalic innervation of the cat medial accessory olive: Anterograde tracing combined with immunocytochemistry, *J Com Neurol* **284**:12 (1989).

70. R.M. Bowker, K.N. Westlund, M.C. Sullivan, and J.D. Coulter, A combined retrograde transport and immunocytochemical staining method for demonstrating the origin of serotonergic projections, *J Histochem Cytochem* **30**:805 (1982).

71. R.M. Bowker, H. Stteinbush, and J.D. Coulter, Serotonin and peptidergic projections to the spinal cord demonstrated by a combined retrograde HRP histochemical and immunocytochemical staining method, *Brain Res* **211**:412 (1981).

72. R.M. Bowker, K.N. Westlund, and J.D. Coulter, Serotonergic projections to the spinal cord from midbrain of the rat: An immunocytochemical and retrograde transport study, *Neurosci Lett* **24**:221 (1981).

73. R.M. Bowker, K.N. Westlund, and J.D. Coulter, Origins of serotonergic projections to the spinal cord in rat: An immunocytochemical-retrograde transport study, *Brain Res* **226**:187 (1981).

74. P. Brodal, G. Mihailoff, B. Border, O.P. Ottersen, and J. Satorm-Mathisen, GABA-containig neurons in the pontine nuclei of rat, cat and monkey. An immunocytochemical study, *Neurosci* **25**:27 (1988).

75. Y. Kawai, H. Takagi, K. Yanai, and M. Tohyama, Adrenergic projection from the caudal part of the nucleus of the tractus solitarius to the parabrachial nucleus in the rat: Immunocytochemical study combined with a retrograde tracing method, *Brain Res* **459**:369 (1988).

76. Y.Q. Li, Z.R. Raom, and J.W. Shi, Serotoninergic projections from the midbrain periaqueductal gray to the nucleus accumbens in the rat, *Neurosci Lett* **98**:276 (1989).

77. J.V. Priestley, P. Somogyi, and C. Cuello, Neurotransmitter specific projection neurons revealed by combined immunohistochemistry with retrograde transport of HRP, *Brain Res* **320**:231 (1981).

78. M.V. Sofroniew and U. Schrell, Evidence for a direct projection from oxytocin and vasopressin neurons in the hypothalamic paraventricular nucleus to the medulla oblongata: immunohisto-chemical visualization of both the horseradish peroxidase transported and the peptide produced by the same neurons, *Neurosci Lett* **22**:211 (1981).

79. B.H. Wainer and D.B. Rye, Retrograde horseradish peroxidase tracing combined with localization of choline acetyltrasferase immunoreactivity, *J Histochem Cytochem* **32**:439 (1984).

80. B.R. Wang, E. Senba, and M. Tohyama, Ultrastructural investigation of substance P-, leucine-enkephalin- and 5-hydroxytryptamine-like immunoreactive terminals in the area of cremaster motoneurons of the male rat, *Neurosci* **28**:711 (1989).

81. A.I. Basbaum and D. Menetrey, Wheat germ agglutinin-apoHRP gold: A new retrograde tracer for light- and electron microscopic single- and double-label studies, *J Com Neurol* **261**:306 (1987).

82. A.I. Basbaum, A rapid and simple silver enhancement procedure for untrastructural localization of the retrograde tracer WGA(apo)Hrp-Au and its use in double-label studies with post-embedding immunocytochemistry, *J Histochem Cytochem* **37**:1811 (1989).

83. S. Shiosaka, S. Shimada, and M. Tohyama, Sensitive double-labelling technique of retrorade biotinized tracer (biotin-WGA) and immunocytochemistry: Light and electron microscopic analysis, *J Neurosci Meth* **16**:9 (1986).

84. D.A. Morilak, P. Somogyi, R.A. McIlhinney, and J. Chalmers, An enkephalin-containing pathway from nucleus tractus solitarius to the pressor area of the rostral ventrolateral medulla of the rabbit, *Neurosci* **31**:187 (1989).

85. F.M. Clark and H.K. Proudfit, The projection of noradrenergic neurons in the A7 catecholamine cell group to the spinal cord in the rat demonstrated by anterograde tracing combined with immunocytochemistry, *Brain Res* **547**:279 (1991).

86. C. Gerfen and P.E. Sawchenko, A method for anterograde axonal tracing of chemically specified circuits in the central nervous system: Combined Phaswolus vulgaris-leucoagglutinin (PHA-L) tract tracing and immunohistochemistry, *Brain Res* **343**:144 (1985).

87. F.M. Clark and H.K. Proudfit, The projection of locus coeruleus neurons to the spinal cord in the rat determined by anterograde tracing combined with immunocytochemistry, *Brain Res* **538**:231 (1991).

88. P.G. Luiten, F.G. Wouterlood, T. Matsuyama, A.D. Strosberg, B. Buwalda, and R.P. Gaykema, Immunocytochemical applications in neuroanatomy, *Histochem* **90**:85 (1988).

89. C.R. Gerfern, P.E. Sawchenko, and J. Carlsen, The PHA-L anterograde axonal tracing method, *in*: "Neuroanatomical Tract-Tracing methods 2 - Recent progress", p 19, L. Heimer and L. Zaborszky eds., Plenum Press, New York (1989).

90. F.G. Wouterlood and R.P. Gaykema, Innervation of histaminergic neurons in the posterior hypothalamic region by medial preoptic neurons. Anterograde tracing with phaseolus vulgaris leucoagglutinin combined with immunocytochemistry of histidine decarboxylase in the rat. *Brain Res* **455**:107 (1988).

91. D.M. Wallace, D.J. Magnuson, and T.S. Gray, The amygdalo-brainstem pathway: Selective innervation of dopaminergic, noradrenergic and adrenergic cells in the rat, *Neurosci Lett* **97**:252 (1989).

92. F.G. Wouterlood, R.P. Gaykema, H.W. Steinbusch, T. Watanabe, and H. Wada, The connections between the septum-diagonal band complex and histaminergic neurons in the poeterior hypothalamus of the rat. Anterograde tracing with Phaseolus vulgaris-leucoagglutinin combined with immunocytochemistry of histidine decarboxylase, *Neurosci* **26**:827 (1988).

93. L. Zaborszky and W.E. Cullinan, Hypothalamic axons terminate on forebrain cholinergic neurons: an ultrastructural double-labeling study using PHA-L tracing and ChAT immunocytochemistry, *Brain Res* **479**:177 (1989).

94. W.M. Huang, S.J. Gibson, P. Focer, J. Gu, and L.M. Polak, Improved section adhesion for imuoctochemisry using high molecular weight polymers of L-lysine as a slide coating, *Histochem* **77**:275 (1983).

95. H. Muhlpfordi, The preparation of colloidal gold particles using tannic acid as an additional reducing agent, *Experientia* **38**:1127 (1982).

96. D. Menetrey and C.L. Lee, Retrograde tracing of neural pathways with a protein-gold complex II: Electron microscopic demonstration of projections and collaterals, *Histochem* **83**:525 (1985).

97. G. Danscher, Localization of gold in biological tisse. A photochemical method for light and electon microscopy, *Histochem* **71**:81 (1981).

98. G. Danscher, Histochemical demonstration of heavy metals, *Histochem* **71**:1 (1981).

99. A. Fairen, A. Peters, and J. Saldanga, A new procedure for examining Golgi impregnated neurons by light and electon microscopy, *J Neurocytol* **6**:311 (1977).

100. C. Wajefuekd and N. Shonnard, Observations of HRP labelling following injection through a chronically implanted cannula - a method to avoid diffusion of HRP into injured fibers, *Brain Res* **168**:221 (1979).

101. E.A. Fox and T.L. Powley, Tracer diffusing gas exaggerated CNA maps of direct preganglionic innervation of pancreas, *J Auton Nerve Syst* **15**:55 (1986).

102. A. Dahlström, Effect of vinblastine and colchicine on monoamine-containing neurons of the rat, with special regard to axoplasmic transport of amine granules, *Acta Neuropath* **5**:226 (1971).

103. D. Dube and G. Pelletier, Effect of colchicine on the immunohistochemical localization of somatostatin in the rat brain: Light and electron microscopic studies, *J Histochem Cytochem* **27**:1557 (1979).

104. J.T. McCabe, J.I. Morrell, R. Ivell, H. Schmale, D. Richter, and D.W. Pfaff, *In situ* hybridization technique to localize rRNA and mRNA in mammalian neurons, *J Histochem Cytochem* **34**:45 (1986).

105. A.G. Watts and L.W. Swanson, The combination of *in situ* hybridization with immunohistochemistry and retrograde-tracing, in: "Methods in Neurosciences Vol 1: Genetic Probes" p 127, P.M. Conn ed., Academic Press, New York (1989).

106. E.T.Jr. Cunningham, D.M. Simmons, L.W. Swanson, and P.E. Sawchenko, Enkephalin immunoreactivity and messenger RNA in a discrete projection from the nucleus of the solitary tract to the nucleus ambiguous in the rat, *J Comp Neurol* **307**:1 (1991).

107. B.M Chronwall, M.E. Lewis, J.S. Schwaber, and T.L. O'Donohue, *In situ* hybridization combined with retrograde fluorescent tract tracing, in: "Neuroanatomical Tract-Tracing Methods 2 - Recent Progress, p 265, L. Heimer and L. Zaborszky, eds., Plenum Press, New York (1989).

CHAPTER 20

AUTOMETALLOGRAPHIC (AMG) NERVE TRACING: DEMONSTRATION OF RETROGRADE AXONAL TRANSPORT OF ZINC SELENIDE IN ZINC-ENRICHED (ZEN) NEURONS

Gorm Danscher[1]

Department of Neurobiology
The Steno Institute
University of Aarhus
DK-8000 Aarhus C, Denmark

SUMMARY

Autometallography (AMG) is the technique whereby minute (>0.2 nm) crystal lattices of gold or the selenides and sulphides of silver, mercury and zinc are enlarged by silver amplification to dimensions that can be visualized in the light microscope. The main advantages of the method are its exquisite sensitivity and demonstration of the precise location of the metal lattices which initiate the amplification. Several different AMG protocols are available for both light and electron microscopical use. The basic principles of AMG are demonstrated in *Figure 1*. These apply to all methods of silver amplification. In short, silver ions adhere to the initiating crystal lattices and are subsequently reduced to silver atoms. The developing process continues as long as there is an adequate supply of silver ions and reducing molecules in the vicinity of the expanding silver grains.

As a practical example of the use of AMG, silver amplification of zinc selenide crystal lattices which have been retrogradely transported and located in lysosomes of zinc enriched neurons (ZEN) will be described. These lattices are created in synaptic vesicles of ZEN neurons by application of exogenous selenium, either as intra-cerebral application of sodium selenide or intraperitoneal (IP) injections of sodium selenite. One to two hours after an IP injection, the chelatable zinc in the synaptic vesicles of the ZEN terminals will

---

[1]Address for Correspondence: Gorm Danscher, Professor, Department of Neurobiology, The Steno Institute, University of Aarhus, DK-8000 Aarhus C, Denmark. Telephone: 45-8612-8066, Fax: 45-8618-4093.

have been transformed to zinc selenide crystal lattices. AMG development then provides a detailed map of the ZEN terminal fields throughout the brain. If the animal is allowed to survive for 24 hours, some of the ZEN vesicles (or their content of zinc selenide) have been retrogradely transported to the perinuclear lysosomes, where crystal lattices can be silver amplified.

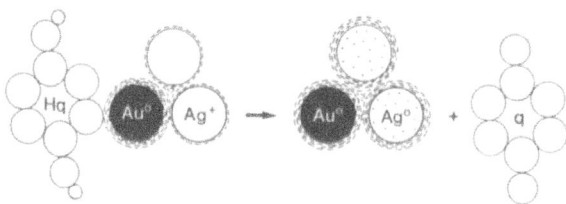

Figure 1. Catalytic crystal lattices, e.g. made up of gold, can catalyze a reduction of adhering silver ions if reducing molecules are available. The autometallographic developer consists of silver ions (Ag$^+$) and hydroquinone (Hq) in an aqueous solution. An adhering hydroquinone molecule releases two electrons into the valens cloud of the crystal lattice and then is disconnected. The two surplus electrons eventually will be captured by adhering silver ions that, in turn, will become atomic silver. As metallic silver does not disturb the catalytic efficacy of the original particle the process will go on as long as the capacity of the AMG developer allows.

INTRODUCTION

Historical Review

Zeiger[1] demonstrated that silver sulphide molecules are able to provoke the reduction of silver on their surfaces in the presence of a developer containing silver ions. This paved the way for Timm's practical application as a histochemical technique -- the sulphide silver method. In the original method,[2] sulphide ions were added to the fixative (*in casu* alcohol) which resulted in the formation of metal sulphides in tissue blocks. Sections from the blocks were then exposed to Liesegang's "physical developer."[3,4]

Liesegang knew how a photographic plate, where the silver bromide crystals had been removed by a thiosulfate solution, could be developed by a developer containing both reducing molecules and silver ions.

In a photographic plate rinsed with thiosulfate, the image is present as 10^{-10}m silver particles which result from the photon-induced reduction of silver ions in the silver bromide crystals. If such a plate is developed with a normal chemical developer, it is completely blank. However, if the developer is enriched with silver ions, the pattern of atomic silver becomes visible. This is because the small silver particles trigger a process whereby silver ions in the "physical" developer are reduced to metallic silver and deposited on the surface of the particles. The electrons responsible for this conversion originate from the reducing molecules. In Liesegang's developer and most of its modifications, the reducing molecule is hydroquinone.

Liesegang wanted to develop a method whereby Ramón y Cajal's tissue block silver staining method could be applied to histological sections. Although he accomplished this, for reasons not known to the present author, the technique never became popular.[5] However, Liesegang had another goal. He wanted to demonstrate that silver present in tissues from animals and humans could be silver-amplified by physical development.[4] He did not, however, succeed in this aim, possibly because he did not wait long enough after the silver exposure.[6] Nevertheless, the mere concept of demonstrating a metal in tissue sections by a photographic development established his pioneering role within the field.

Roberts[7] was the first to follow this lead. He attempted to demonstrate gold in sections from animals which had been exposed to gold thio compounds. Roberts placed the sections in a Liesegang developer which he exposed to the light from a 1000 W bulb. Because of heavy autocatalytic activity, the developer turned dark within seconds and therefore had to be changed several times.[7] Nevertheless, he finally succeeded in demonstrating differences between the sections that were supposed to contain gold and the control sections.

Due to the inherent ability of certain metallic ions to encyst themselves with metallic silver when exposed to a photographic developer containing silver ions, the demonstration of metal in tissue sections had finally been accomplished. Querido was the first to realize that exposure to light was important.[8] He thought that gold was present in the tissues as gold sulphides and that these accumulations were activated by light. Although incorrect, this straightforward hypothesis reflects the subtle development of thoughts and understanding that took place over the years and resulted in the present rather widespread use of silver amplification of gold, and in particular colloidal gold.[9-14]

Timm knew about Roberts' and Querido's work and this was instrumental in his development of the sulphide silver method. Another factor was the ancient knowledge that most metal sulphides are insoluble and colored. If the tissues were treated with sulphide the metals, if present in sufficient amounts, could be observed directly (Timm, personal communication 1981).

Interestingly, the metal that Timm demonstrated with his new method was zinc, which makes an invisible white precipitate with sulphide. Four years later he elaborated the AMG method for the demonstration of mercury in tissues from animals exposed *in vivo* to mercury.[15]

It is now possible to demonstrate four different metals histochemically by AMG at light and electron microscopical levels. These are metallic gold and the sulphides and selenides of silver, mercury and zinc. Furthermore, techniques to differentiate between these metals have been developed.[16-19]

The possibility of detecting mercury, silver and zinc selenides by AMG was introduced in 1982.[13,14,20,21] This extension in the application of AMG as a tool for neurochemical and neurotoxicological studies was created on notable observations made by Shirabe. He used electron microscopic X-ray microanalysis to demonstrate the presence of mercury, sulphide and *selenium* in the lysosomes of cerebellar neurons from a Minamata diseased victim.[22] Aaseth et al. used crystallography to demonstrate the presence of silver selenide crystals in the basal lamina of the glomerular capillaries of the kidney from an argyrotic patient.[23] These data demonstrated that mercury and silver bind to both selenium and sulphide. It was logical therefore to develop the AMG selenium method.[20]

THEORETICAL CONSIDERATIONS

Two major problems concerning the reproducibility of Timm's sulphide silver method and its many modifications have been encountered: 1) that the amount of sulphide ions supplied to the tissues is not matched to the quantity of free or loosely bound zinc ions available to be transformed to zinc sulphide molecules; 2) that the Liesegang developers use silver nitrate, which dissociates rapidly, as a silver ion donor.

Concerning the first problem, sulphide which is surplus to that required to bind zinc, will become loosely associated with other molecules within the tissues. These "non-metal bound" sulphide ions will bind to silver ions in the AMG developer and create silver sulphide crystal lattices. In accordance with Zeigers' finding,[1] such silver sulphide lattices will be silver amplified. Possibly because the excess sulphide ions bind to particular molecules, these "false" AMG grains are arranged in a very orderly fashion and in some peculiar way reflect a kind of specificity.

This "surplus sulphide problem" is most noticeable in the Timm-Haug method, and has caused a lot of misinterpretations, some of which have been made by the present author. It also has led to the false assumption that the Timm method demonstrates several different metals in normal non-contaminated tissue. However, Timm's original belief that his method specifically demonstrates zinc is undoubtedly true.[17] In the Neo-Timm method,[24] the correct level of sulphide ions is achieved by prior chelation *in vivo* of the tissue metal by the water soluble, low-toxic chelator, sodium diethyldithio-carbamate. The tissue is then exposed by transcardial perfusion to the highest level of sulphide ions which does not result in AMG grains following autometallographical development, i.e. unstained sections.

Regarding the second problem, the disadvantage of using silver nitrate as a silver donor is most evident in the Timm-Haug method. The almost immediate and total dissociation of silver nitrate results in a drastic increase in the unspecific staining following AMG development. This problem can be resolved by using silver lactate [13,14,24] or silver acetate[25,26] as the silver ion donors. Neither of these silver salts are completely dissociated in the developer and, at any given moment, there is only a limited amount of silver ions available for reduction to metallic silver on the surfaces of the initiating crystal lattices. As the silver ions are progressively reduced to metallic silver, more ions will be supplied from the bound pool.

Timm's original methods, the sulphide silver method[2] and the mercury method[15] have since been further developed to a level whereby gold and the sulphides and selenides of silver, mercury and zinc, can be specifically located in tissues by AMG at both the light and the electron microscopical levels.[13,14]

Two methods are presently used to demonstrate the zinc content of ZEN neurons. The classical method of Timm[2] has been modified and improved so that it specifically demonstrates zinc in the terminals of ZEN neurons at both light and electron microscope levels.[13,24] The *in vivo* selenium method demonstrates both terminals and somata of ZEN neurons.[20,27-30]

PRACTICAL CONSIDERATIONS

Cleanliness of Glassware

It is of great importance that the apparatus used to perform the AMG technique is extremely clean. The glassware and other tools are therefore carefully washed in Farmer's solution, as described under Optimal Procedure. It is also imperative that high purity chemicals (pro analysi) are used in all procedures.

Although cleanliness is the most important element in preparing high quality AMG preparations, two additional steps can be taken, both of which are described in *Optimal Procedure*: 1) The sections can be covered with gelatine before AMG development. Auto development grains in the developer which adhere to the surfaces of the sections can then be removed by warm water (45°C) after development. 2) The glass slides can be placed in Farmer's solution to remove silver after development. The treatment with Farmer's solution removes AMG silver grains in the most superficial parts of the section, which results in a most satisfactory AMG staining.

Significance of Light during AMG Development

The acid silver lactate developer described here is not inordinately light sensitive. Development for up to 30 min will not cause too much blackening of the developer. Nevertheless, it is recommended that the developing vials be wrapped in aluminum foil throughout development, or that the water bath be covered by a light-tight box. Since there is plenty of time, the slides can be placed in the cups and covered by the AMG developer in full daylight and then placed in the water bath through a door in the light-tight box.

It should be stressed that the silver acetate developer introduced by Skutelsky et al.[31] and Hacker et al.[25] is an alternative to the silver lactate developer. The latter has been found to be less light sensitive in AMG developers that do not contain gum arabic. However, we always include gum arabic in the Hacker AMG developer since it is our experience that this provides the best results for our AMG tasks.

AMG and Osmium Fixation

A particularly important technical consideration might be the possibility of being able to fix the AMG sections in osmium tetroxide. If a high quality of tissue preservation and contrast in EM pictures is important, the tissue blocks can be stained with uranyl acetate and osmium tetroxide after AMG development.

As described in Optimal Procedure, 200 μm Vibratome sections are AMG developed and rinsed. The blocks to be studied in the electron microscope are cut using a stereo microscope, stained with osmium tetroxide and/or uranyl acetate, embedded and cut in the conventional way. The ultrathin sections are stained lege artis.

Use of Nickel Grids for Ultrastructural AMG

When AMG development is performed directly on the grids, these must be made of nickel, since nickel is not affected by the acid silver lactate or silver acetate developers. Also, they do not influence the AMG silver amplification process, and they are cheap and easy to handle.[32]

Use of AMG in Combination with Other Histological Techniques

Sections from tissues that have been exposed to sulphide by perfusion (Neo-Timm), selenium in vivo (selenium method) or gold, silver and mercury in vivo can be stained with most other histological methods, e.g. monoamine oxidase, acetylcholinesterase, horseradish peroxidase, Fink-Heimer, Nauta, and several counterstaining methods. If the particular method allows fixation with a fixative containing paraformaldehyde (e.g. methods for AMG detection of gold, silver, mercury and vesicular zinc selenide), AMG treated sections can be subjected to immunocytochemical techniques.

OPTIMAL PROCEDURE

AMG Demonstration of All ZEN Neuronal Somata in the Brain[20,29]

1. Selenite solution for IP injection: Anhydrous sodium selenite (Na_2SeO_3) (Sigma; St. Louis, MO) is dissolved in deionized water at a concentration of 1 mg.ml^{-1}.
2. The animals are injected with sufficient sodium pentobarbital to keep them anesthetized for 2 hrs.

3. The animals are injected IP with 8 mg.kg^{-1} body weight sodium selenite and allowed to survive for 12-48 hrs.

4. The animals are given an overdose of pentobarbital and either decapitated (for cryostat sectioning) or fixed by transcardial perfusion with an 0.1 M phosphate buffer containing 2% glutaraldehyde and 1% para-formaldehyde. Alternatively, 3% glutaraldehyde in a 0.1 M phosphate buffer is used. Tissue blocks from the fixed brains are embedded in methacrylate or resin.

5. The brain and/or spinal cord are removed and subjected to one of the following procedures: 1. frozen with gaseous carbon dioxide for subsequent sectioning on a cryostat; 2. placed in the fixative for at least one hour and then embedded in paraffin, methacrylate or in resin; 3. sectioned by a vibratome in 200 µm thick slices, which are then subjected to AMG as described below. After AMG development, the structures of interest are dissected from the slices and stained with uranyl acetate and fixed in osmium tetroxide before being embedded.

6. a) 30 µm cryostat sections are placed on glass slides, dried, fixed in alcohol and dehydrated. Before AMG development, the sections are dipped in a 0.5% gelatine solution and allowed to dry in the air.
   b) Epon embedded sections are cut to a thickness of approximately 3 µm and placed on glass slides, where they are coated with 0.5% gelatine and dried.

AMG Demonstration of Separate ZEN Pathways[20,28,30]

1. 0.2% sodium selenide (Na$_2$Se) (Ventron; Karlsruhe, FRG) is dissolved in 0.1 M phosphate buffer. Before adding the selenide, the buffer is deoxygenated by continuous bubbling with nitrogen for 20 min. The selenide solution has to be kept in a sealed air-tight glass flask. If the solution is not kept free of oxygen, it will turn red indicating that selenide ions have been oxidized to selenium atoms.

2. The selenide solution can either be injected by pressure or, preferably, applied intracerebrally by iontophoresis.

Iontophoresis

Glass micropipettes with tip diameters of 40-50 µm are filled with 0.2% sodium selenide solution. An ejecting current of 4 µA delivered intermittently for a total of 25 s over a 5 min period has been successfully used in our laboratory.

The technical procedures for processing the brains after axonal transport are described above.

AMG DEVELOPMENT[13,14,20,24,33]

Preparation of Developer

The AMG solution contains the following substances:
A. **Protective colloid (gum arabic).** One kg of non-refined acacia resin is dissolved in 2 l deionized water at room temperature. The viscous, pulp-like solution is stirred intermittently throughout the next five days. The solution is then filtered through six layers of gauze and the colloid is divided into suitable portions and frozen in jars at -17°C, where it can be stored for at least one year.
B. **Citrate buffer.** 25.5 citric acid 1 H$_2$O and 23.5 g trisodium citrate 2 H$_2$O are dissolved in 100 ml deionized and distilled water.
C. **Reducing agent (hydroquinone).** 0.85 g hydroquinone is dissolved in 15 ml deionized and glass-distilled water immediately before use.
D. **Silver ion donor (silver lactate).** In a flask wrapped in aluminum foil to exclude the light, 0.11 g silver lactate is added to 15 ml deionized and glass-distilled water.

Mixing the Ingredients of the AMG Developer

Solutions A (60 ml), B (10 ml), and C (15 ml) are carefully mixed in a 100 ml beaker.

The glass slides are placed in staining cups and developer poured over. The cups are placed in a water bath adjusted to 26°C and covered by e.g. aluminum foil.

The optimum developing times are dependent on the thickness of sections, size of the initiating particles and the magnification required. In our laboratory, developing periods from 30-120 min are commonly used.

Post-developmental Treatment of Cryostat Sections

The staining cups are placed under gently running tap water at 45°C for 20 min in order to remove the gum arabic and the thin layer of gelatine. The sections are then carefully rinsed twice in distilled water for 10 min.

If some unspecific AMG grains are present on the surface of the sections, a short (10 s) dip in a 2% Farmer's solution (nine parts 2% sodium thiosulfate, one part 2% potassium ferricyanide) followed by a rinse in distilled water is recommended.

Counterstaining can be performed with any suitable stain. We usually use neutral red or toluidine blue. After dehydration, the sections can be mounted in Canada balsam, Dammar resin, DPX (BDH Chemicals, Poole, Dorset, England) or any other non-acid resin.

Post Developmental Treatment of Epon Sections

The glass slides in their developing cups are placed under gently running tap water at 45°C for 20 min and then immersed in 10% Farmer's solution (10% sodium thiosulfate and 10% potassium ferricyanide) for 10 s followed by careful rinsing in distilled water.

Sections for light microscopical studies can be counterstained with toluidine blue. In our laboratory we usually have three sections on each glass slide, one of which is counterstained.

After analyzing (and photographing) the sections in the light microscope, one of the sections is covered with a drop of Epon and a blank Epon block is placed on top of it. After polymerization, the block is removed from the slide by quickly heating on a hot plate (100°C). The ultrathin sections are made conventionally and stained with uranyl acetate and lead.

AMG Development of Ultrathin Sections Placed on EM Grids[14,27,35]

Ultrathin sections are placed on nickel grids which are coated with formvar and carbon. Such sections can be AMG developed in three different ways:

1. Under a safe light and using the "loop technique," the grids are covered by a monolayer of silver bromide crystals. Ilford L4 autoradiographic emulsion is suitable for this purpose. After the emulsion has been applied, the grids are placed upside down on a drop of chemical developer. We have used Kodak D 19 with developing times of 10-25 min. After development, the grids are rinsed in 20% sodium thio-sulphate to remove surplus silver ions from the emulsion. The autoradiographic emulsion can be removed either by dipping in distilled water at 45°C or by enzymes. We have used Subtilisin A 0.5% (NOVO Nordic, Denmark) for 20 min followed by careful rinsing with distilled water. After AMG the sections are stained conventionally with uranyl acetate and lead.

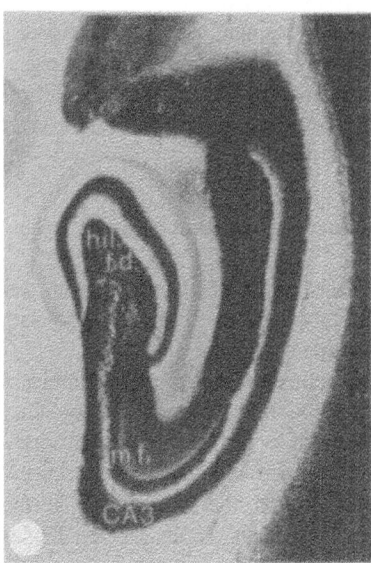

Figure 2. Light micrograph of 30 μm cryostat section from hippocampus from rat perfused transcardially with a 0.1% sodium sulphide solution for 7 min at a pressure of 120 mm Hg. The section was exposed to the silver lactate AMG developer for 60 min at 26°C. Note the highly laminated pattern of ZEN terminals. hil.f.d., hilus fasciae dentatae; m.f., mossy fibers. Original magnification x27.

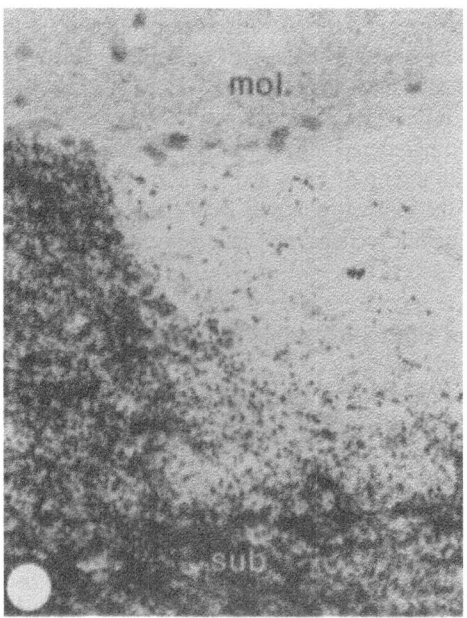

Figure 3. 30 μm thick cryostat section from a rat IP treated with 18 mg sodium selenite per kg body weight and allowed to survive for 1 h. The picture demonstrates the difference in size of the ZEN neuronal terminals. mol., stratum molecular of fasciae dentatae; sub., subiculum. Original magnification x470.

334

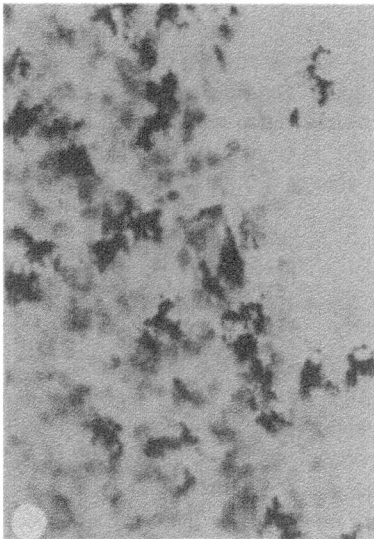

Figure 4. Frontal section of the cerebellum of selenium treated rat. The bizarre AMG silver formations in the granular layer have been found to be mossy fibre boutons (Danscher, unpublished). Original magnification x691.

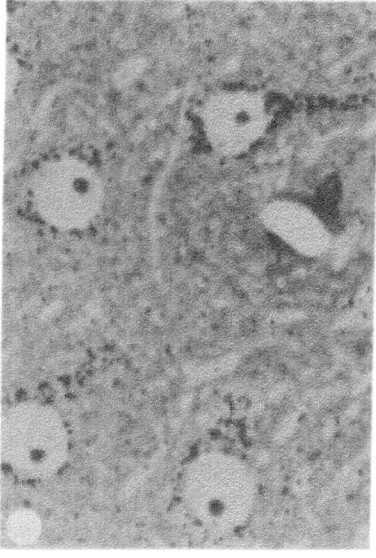

Figure 5. Light micrograph of 3 μm thick Epon section from neocortex of a rat that was treated IP with 8 mg sodium selenite per kg body weight 24 h before being sacrificed. Retrogradely loaded neuronal somata from layer III demonstrate AMG silver amplified zinc selenide accumulations. Original magnification x1,500.

2. In normal daylight, the grids can be placed upside down on a drop of commercial AMG developer (silver amplification kit). After careful rinsing in distilled water, the sections are stained as in 1. above.

3. The grids are briefly dipped in 0.5% gelatine and allowed to dry for 15 min. Thereafter, the grids are placed on a drop of gum arabic containing the AMG silver lactate developer described above. After developing for 10-40 min, depending on the magnification required, the grids are rinsed very carefully in distilled water at 45°C to remove the gelatine and gum arabic. The sections are treated as 1. and 2. above.

DISCUSSION

ZEN Neurons and Their Axonal Terminals

In the mammalian brain, the densest populations of ZEN terminals are found in telencephalic structures (*Figs. 2,3*). However, zinc-containing terminals also invade cerebellum (*Fig. 4*), hypothalamus, thalamus, mesencephalon, putamen, medulla oblongata, and the spinal cord. Until the beginning of the 1980s, most studies on the differentiated and highly laminated patterns of ZEN terminals in telencephalic structures were conducted with Timm modifications that were not specific for zinc. Because the brains were exposed to excess sulphide, the sections contained false AMG grains. As mentioned in "Theoretical Considerations," these false silver grains are organized in an orderly fashion and cannot be distinguished from the silver-amplified zinc sulphide crystal lattices. Thus, in addition to a description of the fascinating patterns of ZEN terminals, results based on the Timm-Haug method contain a description of the superimposed pattern of the "surplus sulphide ion staining."

Species investigated using the zinc-specific Timm method include man, rat, hedgehog, rabbit, mouse, birds, reptiles, and fish[20,29,36-47] (Montagnese *et al.*, in preparation; Molowny *et al.*, in preparation). All these species have been found to contain ZEN neurons and to demonstrate a complicated pattern of zinc-containing terminals.

As described initially, *in vivo* injections of sodium selenite or sodium selenide will result in accumulations in the lysosomes of ZEN neurons 24 hrs later.[27-30] The rate at which zinc selenide crystal lattices are created in the synaptic vesicles is different for individual ZEN pathways. Alternatively, it may be dependent upon the density of ZEN synaptic vesicles per volume brain tissue.

By using the "AMG retrograde selenium" technique it is now possible to detect all ZEN pathways and this work is presently being conducted in several different laboratories. As an example, we have found large populations of ZEN neurons in layers II, III (**Fig. 5**) and VI of all parts of neocortex.[29,49] The cerebellum is completely devoid of ZEN neuronal somata, while ZEN neurons are present in the medulla oblongata and spinal cord of the rat (Danscher, in preparation).

The significance of the synaptic vesicular zinc (ZEN zinc) is not presently known. It could serve as a stabilizer of macromolecules, as in the proinsulin containing vesicles of the β-cells of the islets of Langerhans in the pancreas. The rather widespread presence of ZEN vesicles in different secretory cells (e.g. acinar excretory cells in the pancreas, granular convoluted tubule cells of murine salivary glands, somatotroph cells of the pituitary, prostate, mast cells and Paneth cells) might reflect a general function of zinc ions in ZEN vesicles similar to that in the β-cells, i.e. reversible aggregation of molecules in order to make them osmotically inert for transport and/or a store for later exocytosis.

In the brain and spinal cord, therefore, the function of ZEN vesicles could be to transport one or more important molecules to the synapses of ZEN boutons (e.g. growth or maintenance of synapses). Approximately 10% of the synaptic vesicles of ZEN neurones contain zinc and release it by exocytosis. In locations like pancreas and pituitary, they are released together with the hormone and they might be functionally inert until being taken up by the cells again. In the CNS, however, the release of zinc ions into the synaptic clefts might have a modulatory effect on postsynaptic receptors as well. The demonstration of a modulatory binding site for zinc on the $GABA_A$ receptor complex in cultured rat neurons[50] and the demonstration of an antagonizing effect of zinc on NMDA and GABA responses on hippocampal neurons[51] support this hypothesis.

Autometallography

Because of the high technical quality that can be obtained when AMG is applied to cryostat, paraffin, methacrylate or Epon sections, it is possible to quantify the density of AMG silver grains in the sections. There are, however, some facts that should be borne in mind: 1) The size of the initiating crystal lattices has no influence on whether the AMG process of silver amplification actually commences;[52] 2) The speed of amplification is dependent on temperature, concentration of silver ions and concentration of reducing molecules (*in casu* hydroquinone); 3) The size and amount of catalytic crystal lattices per mm^3 tissue influences not only the colour but also the density of the AMG grains.

In 1911, Liesegang demonstrated that shades observed in the light microscope ranged from yellow for the smallest grains, to orange, red, brown, olive and black with successively larger grains. It should be stressed therefore that the AMG technique does not demonstrate a quantitative distribution of the specific crystal lattice per se. The possibility that the yellow stained areas contains the highest concentration of the catalyst cannot be excluded in advance. However, if one is aware of these short- comings, a semi-quantification can be obtained. For instance, the number of AMG silver grains in individual synaptic vesicles is very constant--most often one, occasionally two and seldom three grains. This information can be used to estimate the regional concentrations of ZEN zinc in a given area.[54] Since the human eye is relatively insensitive to the shades of AMG, electronic imaging can be used with advantage to transfer densities to artificial colors.

The prospect of the simultaneous use of AMG, not only with most other histochemical and immunohistochemical methods, but also in combination with new quantitative methods such as electron energy loss spectroscopy (EELS), laser microprobe mass analyses (LAMMA) and electron probe X-ray microanalysis (EPMA) makes the autometallographic technique most promising.

REFERENCES

1. Zeiger, Physikochemische Grundlagen der histologischen Methodik, *Wiss Forschungsber* **48**:55 (1938).
2. F. Timm, Zur Histochemie der Schwermetalle, das Sulfid-Silber-Verfahren, *Dtsch Z Gesamte Gerichtl Med* **46**:706 (1958).
3. R.E. Liesegang, Die Kolloidchemie der histologischen Silber-Färbungen, *Kolloid Beihefte* **3**:1 (1911).
4. R.E. Liesegang, Histologische Versilberungen, *Z Wiss Mikrosk* **45**:273 (1928).
5. G. Danscher, A silver method for counterstaining plastic embedded tissue. *Stain Technol* **58**:365 (1983).
6. G. Danscher, Light and electron microscopic localization of silver in biological tissue, *Histochemistry* **71**:177 (1981).
7. W.J. Roberts, A new procedure for the detection of gold in animal tissue, *Proc R Acad Amsterdam* **38**:540 (1935).
8. A. Querido, Gold intoxication of nervous elements. On the permeability of the blood-brain-barrier, *Acta Psychiatr* **12**:151 (1947).
9. E. Gilg, A photochemical method for microdetection of gold in tissue sections, *Acta Psychiatr Scand* **27**:43 (1952).

10. J.L. Doré, The demonstration and distribution of gold in tissue sections. Thesis (London) (1974).

11. J.L. Doré and B. Vernon-Roberts, A method for the selective demonstration of gold in tissue sections, *Med Lab Sci* **33**:209 (1976).

12. G. Danscher, Localization of gold in biological tissue. A photochemical method for light and electron microscopy, *Histochemistry* **71**:81 (1981).

13. G. Danscher, Detection of metals in tissues. Histochemical tracing of zinc, mercury, silver and gold, *Prog Histochem Cytochem* **23**:273 (1991).

14. G. Danscher, Applications of autometallography to heavy metal toxicology, *Pharmacology and Toxicology* **69**:414 (1991).

15. F. Timm, Histochemische Lokalisation und Nachweis der Schwermetalle, *Acta Histochem (Jena)* Suppl **3**:142 (1962).

16. F. Timm, Der histochemische Nachweis des Kupfers im Gehirn, *Histochemie* **2**:332 (1961).

17. G. Danscher, G. Howell, J. Pérez-Clausell, and N. Hertel, The dithizone, Timm's sulphide silver and the selenium methods demonstrate a chelatable pool of zinc in CNS, *Histochemistry* **83**:419 (1985).

18. G. Danscher and J. Rungby, Differentiation of histochemically visualized mercury and silver, *Histochem J* **18**:109 (1986).

19. J.D. Schiønning, G. Danscher, M.M. Christensen, E. Ernst, and B. Møller-Madsen, Differentiation of silver-enhanced mercury and gold in tissue sections, *Histochemical J* (in press 1992).

20. G. Danscher, Exogenous selenium in the brain. A histochemical technique for light and electron microscopical localization of catalytic selenium bonds, *Histochemistry* **76**:281 (1982).

21. G. Danscher and B. Møller-Madsen, Silver amplification of mercury sulfide and selenide. A histochemical method for light and electron microscopic localization of mercury in tissue, *J Histochem Cytochem* **33**:219 (1985).

22. T. Shirabe, Identification of mercury in the brain of Minamata disease victims by electron microscopic X-ray microanalysis, *Neurotoxicology* **1**:349 (1979).

23. J. Aaseth, A. Olsen, J. Halse, and T. Hovig, Argyria-tissue deposition of silver as selenide. *Scand. J Clin Lab Invest* **41**:247 (1981).

24. G. Danscher, Histochemical demonstration of heavy metals. A revised version of the sulphide silver method suitable for both light and electron microscopy. *Histochemistry* **71**:1 (1981).

25. G.W. Hacker, L. Grimelius, G. Danscher, G. Bernatzky, W. Muss, H. Adam, and J. Thurner, Silver acetate autometallography: An alternative enhancement technique for immunogold-silver staining (IGSS) and silver amplification of gold, silver, mercury and zinc in tissues. *J Histotechnol* **11**:213 (1988).

26. G.W. Hacker, G. Danscher, A.H. Graf, G. Bernatzky, A. Schiechl, and L. Grimelius, The use of silver acetate autometallography in the detection of catalytic tissue metals and colloidal gold particles bound to macromolecules, *Prog Histochem Cytochem* **23**:286 (1991).

27. G. Danscher, Dynamic changes in the stainability of rat hippocampal mossy fiber boutons after local injection of sodium sulphide, sodium selenite, and sodium diethyldithiocarbamate, *in:* "The Neurobiology of Zinc. Part B: Deficiency, Toxicity, and Pathology," Alan R. Liss, Inc., New York (1984).

28. G.A. Howell and C.J. Frederickson, A retrograde transport method for mapping zinc-containing fiber systems in the brain, *Brain Res* **515**:277 (1989).

29. L. Slomianka, G. Danscher, and C.J. Frederickson, Labeling of the neurons of origin of zinc-containing pathways by intraperitoneal injections of sodium selenite, *Neuroscience* **38**:843 (1990).

30. M.K. Christensen, C.J. Frederickson, and G. Danscher, Retrograde tracing of zinc-containing neurons by selenide ions: A survey of seven selenium compounds, *J Histochem Cytochem* **40**:575 (1992).

31. E. Skutelsky, V. Goyal, and J. Alroy, The use of avidin-gold complex for light microscopic localization of lectin receptors, *Histochemistry* **86**:291 (1987).

32. Y.-D. Stierhof, B.M. Humbel, R. Hermann, M.T. Otte, and H. Schwarz, Direct visualization and silver enhancement of ultra-small antibody-bound gold particles on immunolabeled ultrathin resin sections, *Scanning Microscopy* **6**:1009 (1992).

33. G. Danscher, G.W. Hacker, L. Grimelius, and J.O.R. Nørgaard, Autometallographic silver amplification of colloidal gold, *J Histotechnol*, in press (1993).

34. G. Danscher, Autometallography. A new technique for light and electron microscopic visualization of metals in biological tissues (gold, silver, metal sulphides and metal selenides), *Histochemistry* **81**:331 (1984).

35. G. Danscher and J.O.R. Nørgaard, Ultrastructural autometallography: A method for silver amplification of catalytic metals, *J Histochem Cytochem* **33**:706 (1985).

36. F.M.S. Haug, Heavy metals in the brain. A light microscope study of the rat with Timm's sulphide silver method. Methodological considerations and cytological and regional staining patterns, *Adv Anat Embryol Cell Biol* **47**:1 (1973).

37. F.A. Geneser-Jensen, F.-M.S. Haug, and G. Danscher, Distribution of heavy metals in the hippocampal region of the guinea-pig. A light microscope study with Timm's sulphide silver method, *Z Zellforsch* **147**:441 (1974).

38. M.D. Cassell and M.W. Brown, The distribution of Timm's stain in the nonsulphide-perfused human hippocampal formation, *J Comp Neurol* **22**:461 (1984).

39. B. Friedman and J.L. Price, Fiber systems in the olfactory bulb and cortex: A study in adult and developing rats, using the Timm method with the light and electron microscope, *J Comp Neurol* **223**:88 (1984).

40. M.J. West, F.B. Gaarskjaer, and G. Danscher, The Timm-stained hippocampus of the European hedgehog: A basal mammalian form, *J Comp Neurol* **226**:447 (1984).

41. W.K. Schwerdtfeger, G. Danscher, and H. Geiger, Entorhinal and prepiriform cortices of the European hedgehog. A histochemical and densitometric study based on a comparison between Timm's sulphide silver method and the selenium method, *Brain Res* **348**:69 (1985).

42. A. Molowny and C. Lopez-Garcia, Estudio citoarquitectonico de la corteza cerebral de reptiles. III. Localización histoquimica de metales pesados y definición de subregiones Timm-positivas de la corteza de Lacerta, Chalcides, Tarentola y Malpolon, *Trab Inst Cajal Invest Biol* **70**:55 (1978).

43. C. Lopez-Garcia, E. Soriano, A. Molowny, J.M. Garcia Verdugo, P. Berbel, and J. Regidor, The Timm positive system of axonic terminals of the cerebral cortex of Lacerta, *in:* "Ramón y Cajal's Contribution to the Neurosciences," S. Grisolia, C. Guerri, F. Samson, S. Norton and F. Reinoso-Suarez, eds., Elsevier Science Publishers, B.V., Amsterdam (1983).

44. J. Pérez-Clausell, Organization of zinc-containing terminal fields in the brain of the lizard *Podarcis hispanica*: A histochemical study, *J Comp Neurol* **167**:153 (1988).

45. H. Faber, K. Braun, W. Zuschratter, and H. Scheich, System-specific distribution of zinc in the chick brain. A light and electron microscopic study using Timm's method, *Cell and Tissue Res* **258**:247 (1989).

46. W.J.A.J. Smeets, J. Pérez-Clausell, and F.A. Geneser, The distribution of zinc in the forebrain and midbrain of the lizard *Gekko gekko*. A histochemical study, *Anat Embryol* **180**:45 (1989).

47. C. Piñuela, E. Baatrup, and F.A. Geneser, Histochemical distribution of zinc in the brain of the rainbow trout, *Oncorhyncus myciss*: I. The telencephalon, *Anat Embryol* **186**:275 (1992).

48. L. Slomianka, Neurons of origin of zinc-containing pathways and the distribution of zinc-containing boutons in the hippocampal region of the rat, *Neuroscience* **48**:325 (1992).

49. C.J. Frederickson and G. Danscher, Zinc-containing neurons in hippocampus and related CNS structures, *Prog Brain Res* **83**:71 (1990).

50. T.G. Smart, Uncultured lobster muscle, cultured neurons and brain slices: the neurophysiology of zinc, *J Pharm Pharmacol* **42**:377 (1990).

51. G.L. Westbrook and M.L. Mayer, Micromolar concentrations of Zn^{2+} antagonize NMDA and GABA responses of hippocampal neurons, *Nature* **328**:640 (1987).

52. J.F. Hamilton and P.G. Logel, The minimum size of silver and gold nuclei for silver physical development, *Photogr Sci Eng* **18**:507, (1974).

53. I.E. Holm, A. Andreasen, G. Danscher, J. Pérez-Clausell, and H. Nielsen, Quantification of vesicular zinc in the rat brain, *Histochemistry* **89**:289 (1988).

CHAPTER 21

RECEPTOR DETECTION

Wolfgang Kummer,[1] Axel Fischer,[2]
Sebastian Bachmann[2]

[1]Institute for Anatomy and Cell Biology
Philipps-University, Marburg, FRG (WK)
[2]Institute for Anatomy and Cell Biology Ruprecht-Karls-University
Heidelberg, FRG (AF, SB)

SUMMARY

Receptors for hormones and neurotransmitters can be histochemically detected along their biosynthetic pathway. This may extend from intranuclear transcription and splicing of mRNA to the cytoplasmatic translation into the receptor protein 7, which when referring to the receptors of hydrophilic ligands, is inserted into the cell membrane. The receptor mRNA can be detected by *in situ* hybridization. The receptor protein can be labeled with a ligand or by receptor antibodies. The following chapter summarizes the various techniques of non-radioactive labeling of ligands and describes the different methods of how to obtain receptor antibodies. The primary techniques for non-radioactive receptor detection are a) *in situ* hybridization, and 2) immunohistochemistry. Detailed protocols for these methods are provided. In addition to these direct techniques of histochemical demonstration of receptors, they can be indirectly traced by visualization of the intracellular events triggered by stimulation via the ligand. These techniques provide a suitable supplement to the direct techniques. Their basic principles are reviewed.

---

[1]Address for Correspondence: Professor Wolfgang Kummer, MD, Institut für Anatomie und Zellbiologie, Philipps-Universität, Robert-Koch Str. 6, D-35033 Marburg, F.R.G. Phone: 07421/284035; Fax: 06421/285783.

Modern Methods in Analytical Morphology, Edited by
J. Gu and G.W. Hacker, Plenum Press, New York, 1994

INTRODUCTION

Historical Review

For a long time, receptors for hormones and neuronal messengers have been operationally defined by their binding characteristics to natural ligands, synthetic agonists and antagonists. In contrast to the chemical structure of many of the naturally occurring ligands, the primary structure of receptor proteins has been successfully analyzed only during the last decade.

Thus, over a long period, the only available approach to receptors were the ligands themselves. Accordingly, detection of receptors at the histological level started with monitoring of the binding of labeled ligands to tissue sections. Highly sensitive methods had to be developed to visualize bound ligands since the numbers of receptor molecules per cell, and consequently, also that of bound ligands are usually very low. It has been calculated that human T lymphocytes carry about 30,000 receptor molecules for substance P.[1] As for other applications in molecular biology, radiolabeling of the ligands proved to be most sensitive and versatile. Receptor autoradiography using radiolabeled ligands was the first technique to be used, and dominated the field of histological receptor detection for a long time (see also the chapter by Skofitsch, *et al.* in this book). Later, various types of non-radioactive labeling of ligands were developed, the advantages and disadvantages of which will be discussed below.

Binding of a hormone or neuromediator to its receptor causes - by definition - some kind of intracellular event. Bioassays and biochemical assays had used this "target response" for quantitation of receptors long before its potential value for histochemistry was recognized. Since analyses of this kind still do not require a detailed knowledge of the primary structure of the receptor, histochemical techniques based upon this approach have been used to localize structurally poor and non-defined receptors. In practice, tissues were stimulated with appropriate ligands. The activity of receptor-coupled enzymes (e.g. adenylate cyclase[2]) or the intracellular content of second messengers (e.g. cAMP[3]) were monitored by histochemistry and immunohistochemistry, respectively.

Finally, direct approaches to the receptor protein were made by using receptor antisera. In the beginning, antisera were obtained without precise knowledge of the receptor structure by immunizing with 1) more or less purified protein fractions,[4-6] 2) anti-ligand antibodies [production of anti-idiotypic antibodies[7,8]], and 3) peptides being complementary to a ligand of the peptide class.[9,10] Antibodies obtained by either of these ways have also been used to purify the receptor protein, e.g. by affinity chromatography.[10,11]

This technical approach to receptor purification has been almost totally replaced by molecular cloning techniques. Currently, the primary structures of many receptors are available and their number is still growing rapidly. Based upon these data, synthetic peptides corresponding to selected fragments of the receptor protein have been used to raise site-specific antisera being suitable for routine immunohistochemistry.[12,13] The most recent developments in this field include the construction of nucleic acid probes for application in *in situ* hybridization.[14,15]

THEORETICAL CONSIDERATIONS

The receptors addressed in this chapter are proteins which are inserted into the cell membrane with at least one membrane spanning region. Histochemical techniques allow detection of the receptor all along protein synthesis and degradation (*Fig. 1*).

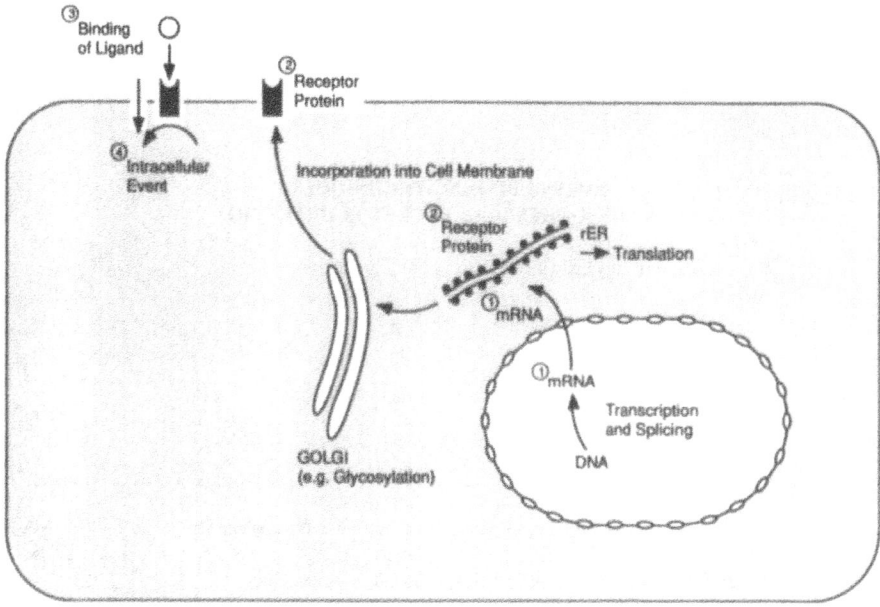

Figure 1. Pathway of intracellular synthesis and trafficking of a receptor protein for hydrophilic ligands. Four major histochemical techniques allow to follow this pathway morphologically: the receptor mRNA can be detected by in situ hybridization (1), the receptor protein can be detected by immunohistochemistry (2) and by labeled ligands (3), and, finally, the intracellular event triggered by binding of an agonist to its receptor can also be visualized by various approaches (4).

Receptor mRNA generated by splicing from the primary transcript in the nucleus can be detected by *in situ* hybridization. The few reports on this technique focus upon radioactive *in situ* hybridization.[14,15] It should be kept in mind, however, that the mRNA might not be translated, so that successful demonstration of receptor mRNA by *in situ* hybridization does not necessarily imply the presence of the receptor protein within the same cell. Furthermore, the intracellular distribution of mRNA often differs greatly from that of its corresponding protein, which is particularly evident in the nervous system. Although *in situ* hybridization techniques in a few cases have revealed labeling of neuronal processes,[16] a signal will usually be obtained only in the perikaryon where the mRNA is translated. The receptor protein then enters some intracellular traffic, e.g. axonal transport, and may be destined for a site located several centimeters distant from the cell body where it has been synthesized.

The receptor protein itself can be addressed either by immunohistochemistry using antisera directed against the receptor protein or by a labelled natural or synthetic ligand. The classical technique of radiolabeling of ligands followed by autoradiography having been the topic of several recent reviews will not be discussed further here (e. g. see the chapter by Skofitsch, *et al.* in this volume). The number of non-radioactive markers which have been introduced to label ligands is rather high (*Table 1*), but the number of successful applications for each of them is low, clearly indicating the numerous technical difficulties connected with these methods. A major problem of these techniques is that there are at least four critical steps: 1) labeling of the ligand must not affect significantly its binding to the receptor; 2) tissue preparation should not destroy the binding properties of the receptor protein; 3) following the initial binding of the ligand to its receptor, there should be no dissociation of this complex during further processing of the tissue or section, and 4) visualization of the marker must be highly sensitive due to the small number of receptor molecules and, hence, bound ligands per cell. It should be noted that there are a few reports in which an unlabelled ligand of the peptide class bound to its receptor was detected by immunohistochemistry using antibodies directed against the non-receptor binding terminal of the peptide.[17,18] In our hands, attempts to apply this approach for the detection of substance P-receptors (= NK-1 receptors) have been unsuccessful.

TABLE 1

**LIGAND BINDING TECHNIQUES
FOR HISTOCHEMICAL RECEPTOR DETECTION**

Labelling of ligands with

- Radioactive Isotopes

- Fluorochromes

- Biotin

- Colloidal Gold

- Carrier Proteins, to be Detected by Immunohistochemistry

IHC Detection of Unlabelled Ligand Bound to its Receptor

Antisera against the receptor protein itself may be obtained by several ways. Sometimes they occur naturally, e.g. in the case of myasthenia gravis, an autoimmune disease characterized by the presence of autoantibodies against the nicotinic acetylcholine receptor. Two approaches have been developed which need neither purification of nor knowledge about the primary structure of the receptor protein: 1) the anti-idiotypic approach, and 2) the anti-complementary peptide approach. Both techniques and their application to the histochemical detection of substance P-receptors were described in detail in an earlier review,[19] so only the basic principles will be briefly described here.

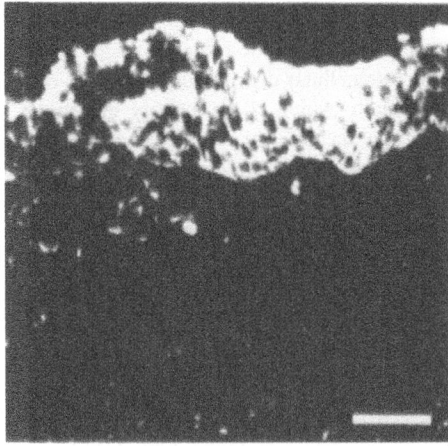

Figure 2. Demonstration of substance P-receptor-immunoreactivity at the human tracheal epithelium by immunofluorescence utilizing an anti-idiotypic antibody. Unpublished micrograph from Fischer et al. (1992). Methodological details are provided in "Optimal Procedures". Bar = 50 μm

The anti-idiotypic approach is based upon the consideration that an antibody against the receptor-binding region of a ligand has an identical, or at least very similar structural configuration, at its Fab-region as the ligand-binding site of the receptor protein. Thus, such antibodies may be used instead of the receptor protein as immunogens for raising anti-anti-ligand antibodies (anti-idiotypic antibodies), which then should recognize the receptor protein as well. An example of the use of such an anti-idiotypic antiserum for immunohistochemistry is provided in *Fig. 2*. Raising anti-idiotypic antibodies is more difficult than raising antibodies against the ligands: titers are usually low and immunizations often end without success. In the case of successful immunization, the antibodies are expected to be directed against the ligand-binding site of the receptors.[19] Thus, their specificity can be tested, both pharmacologically (acting as agonists and/or antagonists; displacement assays[7]) and histochemically (e.g. preincubation of tissue sections by ligands[20]). Discrimination between subtypes of a receptor family is difficult or even impossible,[19,20] probably because the ligand-binding site of the receptor is rather conserved among the different subtypes.

The anti-complementary peptide approach can be applied - as its name implies - only to peptide receptors. Starting with the known nucleotide sequence encoding the peptide ligand, the complementary sequence is deduced and the peptide encoded is synthesized. This complementary peptide is then used for raising antibodies by routine immunization protocols. This approach originated from the assumption that complementary nucleic acid sequences encode interacting peptides.[10] Since many peptide receptors have been cloned and their primary structure is known, it is now possible to compare the sequences of the complementary peptide with that of the naturally occurring receptor proteins. As for the complementary peptide of substance P, however, we found only 36% and 18% homology with the reported sequences of the rat and murine substance P-receptor (NK-1 receptor) and neurokinin A-receptor (NK-2 receptor), respectively. Nevertheless, antisera raised against this complementary peptide have been shown to bind radiolabelled substance P,[19,21] and have been used for immunohistochemistry.[19] Similar reports have been given for antisera against the complementary peptides of ACTH, endorphin, and luteinizing hormone-releasing hormone.[10]

Antisera against the receptor protein can be most easily obtained if the purified protein is available (*Fig. 3*). Even if the receptor has not been isolated to purity, receptor protein-enriched fractions can be used for raising of monoclonal antibodies, provided that an appropriate screening system for identification of receptor antibodies has been

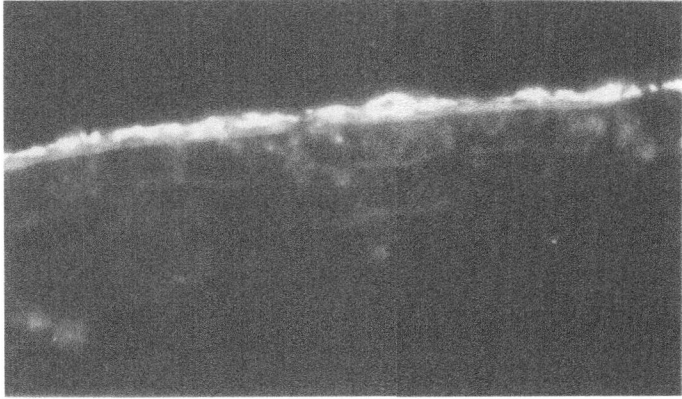

Figure 3. Demonstration of immunoreactivity to muscarinic cholinergic receptors at endothelial cells of the guinea-pig pulmonary artery by immunofluorescence utilizing a mouse monoclonal IgM antibody (M35, Chemunex, Maisons-Alfort, F) obtained by immunization against affinity purified calf forebrain receptor. Bar = 10 μm

developed.[4,22] Alternatively, synthetic peptides - usually 10 to 20 amino acid residues long - can be constructed, corresponding to fragments of the receptor protein, and used for immunization (*Fig. 4*). This latter technique offers the advantage of raising site-specific antisera for predicted which receptor subtypes will be recognized.[23] In contrast, due to the presence of highly conserved regions among different receptor subtypes, immunization with full receptor proteins may result in more or less cross-reacting antisera.[24] However, immunization with the full receptor protein offers another advantage: antibodies will be directed against exposed epitopes, since the receptor protein is presented in naturally occurring tertiary structure. On the other hand, amino acid sequences corresponding to synthetic peptides may be masked in the full protein, e.g. by overlaying parts of the protein or by posttranslational modification.

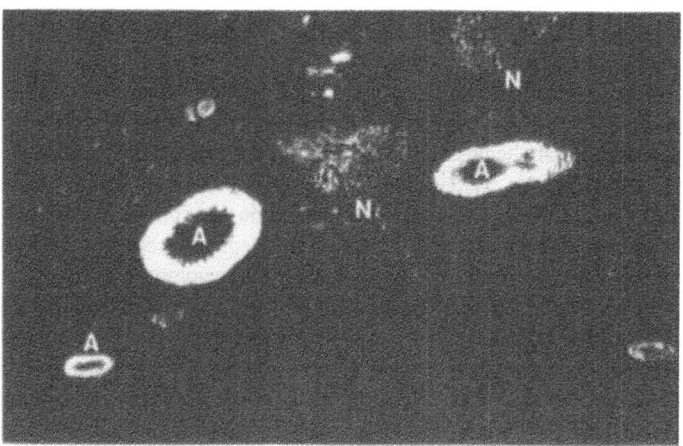

Figure 4. Immunoreactivity to ß$_2$-adrenoreceptors at vascular smooth muscle cells and Schwann cells in the adventitial layer of the guinea pig cervical trachea demonstrated by immunofluorescence utilizing an antiserum raised against a synthetic peptide. This antiserum was kindly provided by Dr. C. D. Strader, Merck Sharp and Dohme, Rahway, NJ, USA. The detailed protocol is provided under "Optimal Procedures". A = artery, N = nerve fiber bundle. Bar = 50 μm

Immunohistochemistry using antisera against the receptor protein is a very powerful tool that allows us to gain high spatial resolution. Immunohistochemistry may demonstrate the receptor all along its biosynthetic pathway, from the ribosomes at the perikaryon to its insertion into the cell membrane and possibly also along its intracellular degradation. Respectively, electron microscopical studies have revealed receptor immunoreactivity, not only at the cell membrane but also at various intracellular locations[19,25,26] (*Fig. 5*). It should be noted, however, that posttranslational processing such as glycosylation[27] may affect epitopes recognized by the antibody, so that the receptor protein might be detected only within a short "window" of this pathway.

Since the optimal conditions of fixation and tissue processing for immunohistochemical receptor labelling are often compatible with the conditions of fixation and tissue processing of the corresponding ligand (especially in case of peptides), simultaneous demonstration of the receptor and its ligand can be achieved by routine double-labeling protocols.[28]

Finally, receptors can be indirectly demonstrated by tracing the intracellular events triggered by them in response to activation via their ligand. In living cells, changes in intracellular calcium concentrations can be monitored by calcium-sensitive dyes and fluorescence microscopy. In conjunction with patch clamp studies, this technique has, for example, led to the characterization of neuropeptide Y-receptors on primary afferent

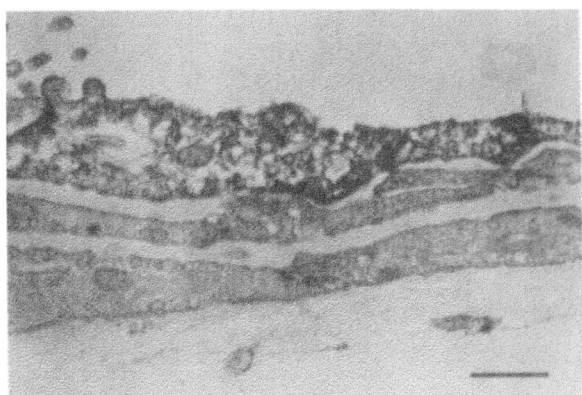

Figure 5. Ultrastructural demonstration of ß₂-adrenoreceptor-immunoreactivity using an indirect peroxidase method as described in detail in the corresponding chapter of this volume ("Pre-embedding Immunohistochemistry"). Immunoreaction product is almost evenly distributed in the cytoplasm of an endothelial cell within a rat dorsal root ganglion. Bar = 0.5 μm

neurons.[29] Intravital stimulation is also necessary, if the intracellular accumulation of a second messenger shall be monitored.

Both cAMP and cGMP have been demonstrated by means of immunohistochemistry: an intracellular increase of cAMP-immunoreactivity was taken as an indicator of the presence of vasoactive intestinal peptide-receptors,[3] and atrial natriuretic peptide-receptors were indicated by enhanced cGMP-immunoreactivity.[30,31] Similarly, soluble guanylyl cyclases, which act as receptor for nitric oxide, can be traced by cGMP-immunohistochemistry.[30] In addition to immunohistochemistry, adenylyl and guanylyl cyclases can be histochemically demonstrated on the basis of their catalytic activity. Since adenylyl cyclase is coupled via G-proteins to a variety of receptors, specific stimulation by ligands and comparison of non-stimulated and stimulated tissues is necessary to utilize adenylyl cyclase histochemistry for receptor localization. This approach has been successfully applied to vasoactive intestinal peptide-receptors in human sweat glands.[2] Guanylyl cyclases, however, are not coupled via G-proteins to the receptor protein, but the receptor protein itself is enzymatically active. Membrane bound guanylyl cyclases represent the receptors of the members of the natriuretic peptide family and of the newly discovered peptide, guanylin, soluble guanylyl cyclases function as receptors for nitric oxide.[32,33] Therefore, the histochemical localization of guanylyl cyclases might allow some conclusion as to the localization of specific receptors, even without stimulation experiments.

PRACTICAL CONSIDERATIONS

In Situ Hybridization for Detection of Receptor mRNA

Receptor mRNAs are usually present in low amounts and their detection by *in situ* hybridization is more difficult than that of their ligands (prepro-peptides) and ligand synthesizing enzymes. Therefore, highly sensitive detection systems have to be utilized. ^{35}S-labeled probes have been used in the first attempts.[14,15] Among non-radioactive *in situ* hybridization techniques, labeling of nucleotides with digoxigenin and subsequent immunohistochemical detection of hybridized probes by digoxigenin antibodies have reached a similar sensitivity. Using a digoxigenin labeled cRNA, we have been able to

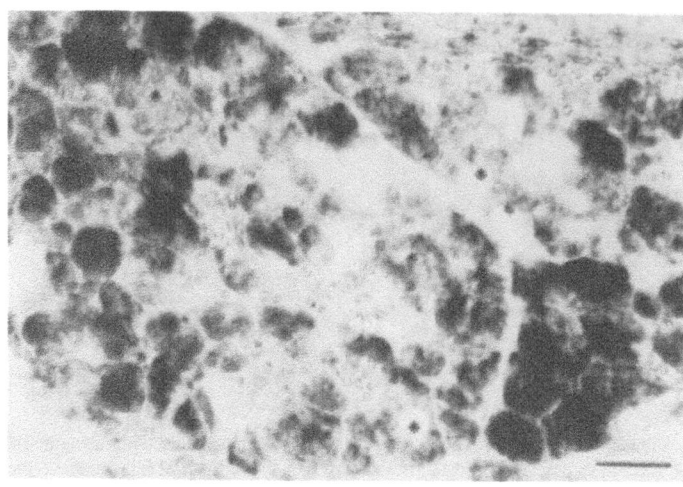

Figure 6. Labeling of small and medium sized rat dorsal root ganglion cells by in situ hybridization using a digoxigenin-labeled ß₂-adrenoreceptor cRNA (kindly provided by Dr. R. Strasser, Heidelberg, FRG). Some of the largest neurons (asterisk) are unlabeled. Technical details are provided under "Optimal Procedures". Bar = 50 μm

localize ß₂-adrenoreceptor mRNA (Fig. 6, vide infra). However, attempts to apply this technique for the A- and B-types of atrial natriuretic peptide receptors have failed so far, despite the successful use of the same cRNAs with radiolabeled nucleotides.

Receptor Detection by Labeled Ligands (non-radioactive)

In contrast to the broad range of commercially available radioactive ligands, only a few non-radioactive labeled ligands can be purchased directly. Consequently, users of this technique, in most cases, have to label the ligands by themselves. Several labeling procedures, e.g. biotinylation, can be easily performed even in those laboratories which do not routinely utilize biochemical techniques, since ready-for-use reagents are commercially available.

Any procedure by which a reporter molecule is introduced by covalent binding to the ligand might affect receptor binding. Thus, the binding properties of the labeled ligand have to be checked before application of the conjugate to tissue sections. Of course, it is possible to try receptor detection without this test; if labeling will be obtained, its specificity can be verified by appropriate controls, e.g. displacement by unlabelled antagonists and agonists. In fact, this is the usual way that radioactive ligands are employed. Severe problems, however, will occur if no specific labeling is obtained on the first try. As outlined under *Theoretical Considerations*, there are at least four critical steps associated with this technique, binding properties of the labeled ligand being only one of them. Thus, trouble-shooting in case of negative results is very laborious, even hopeless, if started with inactive ligands! Therefore, we strongly recommend the conjugate be tested for its receptor binding ability. In laboratories primarily equipped for histochemical techniques, this might cause more technical difficulties than the labeling of the ligand. Appropriate methods consist of e.g. displacement of a radiolabeled ligand by the new conjugate in binding assays and any kind of bioassay. As demonstrated below for colloidal gold conjugates, it is also for covalently labeled ligands of utmost importance that free, i.e. unlabeled ligands, have been fully separated from the conjugates. Otherwise, the unlabeled ligands will mimic biological activity in the assay, but will compete with the labeled ligand in histochemistry, leading to negative histochemical results.

Instead of covalent binding of a marker to a ligand, the ligand can be attached non-covalently to colloidal gold. Maintenance of receptor binding is very likely under these conditions, but several other problems do occur: the smaller the ligand, the smaller the chance of successful conjugation to colloidal gold! Therefore, this technique might be largely restricted to peptides, preferentially to those consisting of more than 10 amino acid residues. Successful studies have been performed using insulin[34] and somatostatin.[35] Furthermore, colloidal gold is negatively charged at its surface and peptides must carry positive charges to stick to it. Hence, the pH has to be adjusted slightly higher than the isoelectric point of the peptide. Since many neuroactive peptides are rather basic, this implies working at basic pH, e.g. in case of calcitonin gene-related peptide (CGRP) we were able to obtain stable conjugates only at pH 10.8. We tried to overcome this problem by using cationic colloidal gold particles (colloidal gold coated with polycationes such as poly-L-lysine) and working at physiological pH, but stable conjugates with CGRP could not be obtained this way.

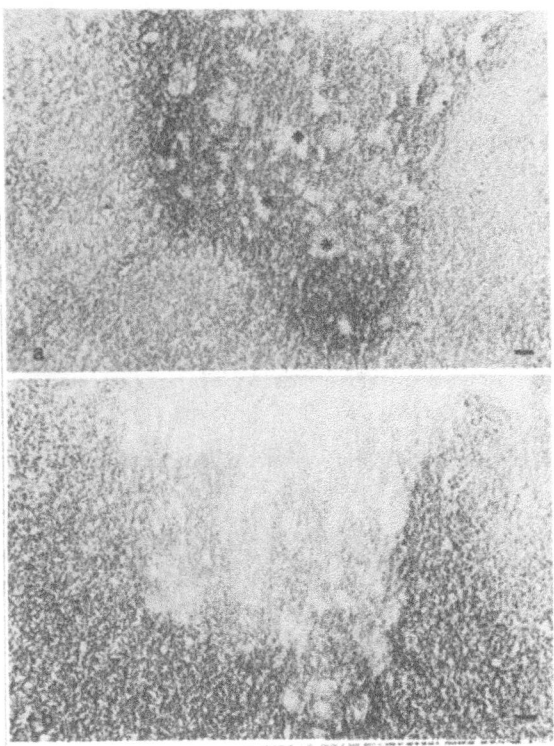

Figure 7. a) Binding of CGRP-colloidal gold to the neuropil of the ventral horn of the guinea-pig spinal cord is demonstrated by silver enhancement. Large neuronal cell bodies (asterisks) are spared. b) Adjacent section processed identically but in the presence of free CGRP (10^{-6} M). Labeling of the neuropil is inhibited, non-specific silver precipitation is seen over the surrounding white matter. Bar = 50 μm

As a net result of the attempts in our laboratory, successful conjugation to colloidal gold was achieved for large proteins, small peptides which were covalently bound to carrier proteins (e.g. substance P bound to bovine serum albumin, but this conjugate was biologically inactive), CGRP (37 amino acids), and salmon calcitonin (32 amino acids]; attempts to conjugate smaller peptides such as substance P (11 amino acids) and [leu]⁵-enkephalin (5 amino acids) directly to colloidal gold were unsuccessful. It should be noted that even in the case of CGRP rather variable results were obtained. When a successful conjugation was achieved, however, the CGRP-colloidal gold conjugates proved to be biologically active (measured by their positive inotropic effect on isolated rat heart atria) and resulted in histochemical labeling of the respiratory epithelium and smooth muscle of the guinea-pig trachea, and parts of the central nervous system, (e. g. the ventral horn of the spinal cord) (*Fig. 7*). Gold conjugates were visualized by commercially available enhancers (Amerham, UK), as found in immunogold silver staining (see chapter by Danscher in this book). We were unable to fully separate unbound ligands from the gold conjugates by ultracentrifugation (*Fig. 8*). This is most likely due to spontaneous dissociation of peptides from the colloidal gold in media without free ligand.

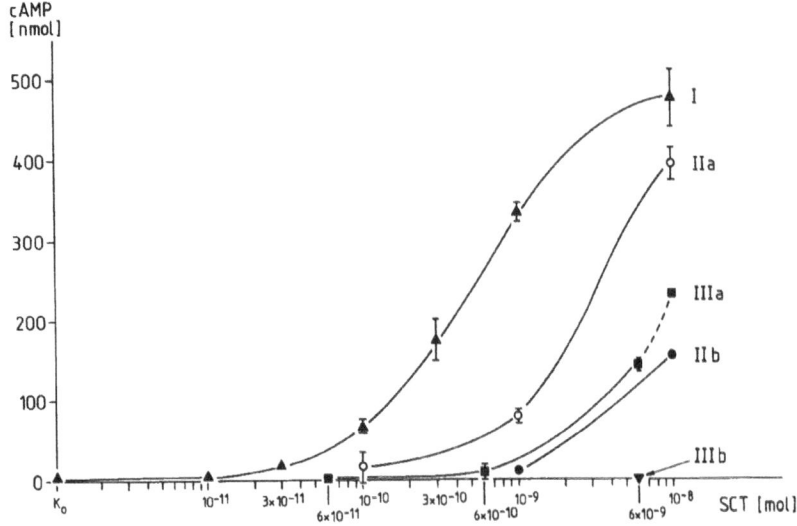

Figure 8. Biological activity of colloidal gold conjugates of salmon calcitonin (SCT) measured by cAMP stimulation in T47D cells. I) free SCT, IIa) SCT-colloidal gold after repeated ultracentrifugation, IIb) supernatant of IIa, which, theoretically, should be devoid of activity, IIIa) SCT-colloidal gold after ultracentrifugation plus dialysis, IIIb) buffer against which dialysis in IIIa was performed. This figure demonstrates the presence of still free ligand during the preparation of ligand-colloidal gold complexes.

The problems associated with the labeling and binding capabilities of the ligand are complicated by additional problems. Both the labeled ligand and the receptor must retain their binding properties during tissue processing and histochemical incubation. As viewed from the receptor's side, this implies that a very mild if any fixatives can be applied. As viewed from the ligand's side, it is important to note that an active ligand may loose its activity during this incubation: some ligands spontaneously oxidize (e.g. substance P); they may also be inactivated by enzymes contained within the tissue section. Therefore, the appropriate peptidase inhibitors appear to be mandatory components of the incubation medium. In the case of substance P, blockade of the neutral endopeptidase 24.11 by thiorphan or phosphoramidon is adequate.[20] The interaction of a ligand with its receptor is not a covalent binding as it tends to dissociate during the washing steps, as known from quantitative studies using radiolabeled ligands.[36] Chemical cross-linking of the ligand to its receptor may be helpful. We obtained positive results with CGRP-gold conjugates only after cross-linking with disuccinimidyl-suberate (DSS; 1 mM DSS in 2% dimethyl-sulfoxide in phosphate buffered saline). (Note: DSS must be dissolved in pure dimethylsulfoxide). Systematic studies on the effects of various cross-linkers, however, are lacking. General recommendations cannot be provided at present.

Immunohistochemistry of Receptor Proteins

From a practical point of view, immunohistochemical demonstration of receptors does not differ much from that of other cellular proteins. Dependent upon the antibody and receptor of interest, some applications will require the use of snap-frozen, unfixed tissue sections,[20,37] whereas in other cases, paraffin sections may be used.[27,37] For the most part, any routine protocol for immunohistochemistry can be applied. It should be noted that detergents may extract membrane bound receptors from unfixed, frozen tissue sections, and should not be used under these conditions.[20,37] The specificity of immunolabelling should be checked, preferentially by preabsorption of the antiserum with its corresponding antigen. Often, this is not available, especially in those cases in which the receptor protein still has not been purified. As discussed above, most of the receptor antisera raised against structurally unknown receptor proteins are directed against the ligand binding site of the receptor. Therefore, a full[20] or partial[37] prevention of immunolabelling may be obtained by preincubating the tissue section with free ligand before application of the receptor antiserum. Moreover, free ligand should be added also to the incubation medium containing the primary antiserum. Using unfixed tissue sections and an anti-idiotypic antiserum against the substance P-receptor, we found it necessary to add inhibitors of substance P-degrading enzymes to all media containing the free ligand.[20] On the other hand, peptidase inhibitors could be omitted when paraffin sections were used for the demonstration of VIP receptors.[37] In many instances, however, immunolabelling cannot be prevented by preincubation with free ligands. Such cases include, of course, antibodies which are directed against epitopes distant from the ligand binding site of the receptor, but also some antisera recognizing the binding site itself. This might be explained by different affinities of the receptor antiserum and the ligand to the receptor.

Visualization of the Intracellular Event Triggered by the Ligand

As discussed above, studies of this kind mostly require intravital stimulation - either *in vivo* or *in vitro* - of the tissue by the appropriate ligand. The optimal experimental conditions for this stimulation will vary considerably from one application to another. General guidelines cannot be outlined. A major difficulty in this kind of study is to discriminate the primary from the secondary effects caused by the release of endogenous mediators from the tissue in response to stimulation by the exogenously applied ligand. The use of various inhibitors of substances likely to be involved reduces the likelihood of the occurrence of secondary effects. For example, Lazarus et al.,[3] working on freshly excised tracheal segments, added indomethacin into the bathing solution to prevent the injury-induced formation of prostaglandins by cyclooxygenase. Then, stimulation was performed by exogenous application of VIP, and the intracellular cAMP content was documented by immunohistochemistry. Controls included the addition of propanolol which provided evidence that the observed rise in cAMP was not due to secondary effects of ß-adrenoreceptor agonists.[3] Similar protocols may be developed for other tissues and receptor types.

The full range of methodologies which have to be summarized in this category cannot be dealt with here. Two points concerning the rather common techniques of immunohisto-chemical demonstration of cAMP and cGMP may be of general importance: 1) despite their small molecular size, these cyclic nucleotides can be sufficiently retained within the tissue by glutaraldehyde-free fixatives (e.g. 4% paraformaldehyde plus 0.2% picric acid[30]); cAMP is less stabile with this respect than cGMP;[38] 2) sodium azide, which is often added as a preservative to buffers and other solutions used in histochemistry, stimulates the formation of endogenous cGMP[39] and, therefore, should be omitted when detecting cGMP.

OPTIMAL PROCEDURES

Demonstration of Receptor mRNA by *In Situ* Hybridization with Digoxigenin-Labelled ß₂-adrenoreceptor cRNA Probes

The cRNA probe was constructed by Dr. R. Strasser (Heidelberg, FRG). *Fig. 6* provides an example of an application of this method.

Tissue Preparation

1. Fixed tissue should be obtained by perfusion of animals with 3% paraformaldehyde in PBS and subsequent rinsing *in situ* with 800 mOsm sucrose in PBS as cryoprotectant.
2. Snap-freeze unfixed and fixed tissues in isopentane cooled by liquid nitrogen.
3. Cut tissues at -25 to -30°C at 5-10 μm in a cryostat and transfer sections onto washed, gelatin-coated slides.
4. Postfix sections by immersion of the slides in 4% paraformaldehyde for 10 min followed by 3 washes in PBS.
5. Slides may now be stored in 70% ethanol at 4°C for a maximum of 4 days.
6. Allow slides to warm up to room temperature before beginning protocol at room temperature (r.t.).
7. Wash 2 times for 10 min each in distilled water.
8. Deproteinate the slides by immersion in 0.1 M HCl for 10 min.
8. Wash 2 times 10 min in PBS.
9. Block unspecific RNA binding-sites by immersion in 200 ml of 0.1 M triethanolamine (pH 8.0) for 20 min with 0.5 ml of acetic acid anhydride added.
10. Wash 2 times in PBS for 5 min.
11. Dehydrate with graded alcohols (70%, 80%, 95%) for 5 min each step.
12. Air dry slides for 30 min under dust protection.

Hybridization

The following hybridization steps should be performed in a moist chamber containing filter paper soaked with a 1:1 mixture of formamide and PBS:

1. Cover each section with 150 μl prehybridization mixture and incubate for 60 min at 37°C.
2. Take sections out of the moist chamber, discard prehybridization mixture by dropping onto filter paper.
3. Cover sections with 35 μl of hybridization mixture, then coverslip sections with siliconized coverslips.
4. Close moist chamber using adhesive tape and allow sections to incubate overnight at 37°C.

The following washing steps should be performed at temperatures of 48-58°C (if problems of background staining occur, choose higher temperatures within this range):

- 30 min sodium-chloride/sodium-citrate (SSC; 2x concentrated) without formamide.
- 60 min SSC (1x conc.) containing 50% formamide.

352

- 90 min SSC (0.5x conc.) containing 50% formamide.
- 60 min SSC (0.1x conc.) containing 50% formamide.

At room temperature:
- 10 min SSC (0.2x conc.), 2 times.
- 10 min SSC (0.1x conc.).

Visualization of the Hybridized Digoxigenin-labeled Riboprobe

1. Equilibrate sections in buffer A (see appendix) for 10 min at r.t.
2. Block nonspecific protein binding sites with modified blocking-solution for 30 min at r.t.
3. Discard exceeding blocking solution onto filter paper.
4. Cover each section with 40 µl of alkaline phosphatase (AP)-conjugated anti-digoxigenin antibody (Boehringer, Mannheim, FRG) diluted 1:250 and 1:500 in modified blocking solution, coverslip sections with siliconized coverslips and incubate overnight at 4°C in a moist chamber.
5. Wash slides with buffer A for 15 min, 2 times.
6. Wash slides with buffer B for 3 min.
7. Discard exceeding buffer onto filter paper and cover each section with 100-150 µl of AP developing solution containing levamisole as blocker of endogenous phosphatases, incubate overnight at 4°C in a moist chamber.
8. Stop reaction with buffer C for 15 min, 2 times.
9. Wash in distilled water for 10 min.
10. Discard excessive water.
11. Coverslip with carbonate buffered glycerol (0.5 M carbonate buffer, pH 8.6 : glycerol = 1:2).

Appendix

Preparation of glass slides and coverslips for *in situ* hybridization:
1) Siliconization of coverslips:
- Immerse coverslips in Sigmacote (Sigma, St. Louis, MO, USA) for 30 sec.
- Wash in distilled water, 10 min.
- Rinse briefly in 95% ethanol.
- Dry coverslips for 2-3 h at 80°C.
2) Cleaning glass slides:
- Immerse glass-slides for 30 min in acetone or 1 M HCl.
- Rinse in distilled water.
- Immerse glass-slides for 30 min in 95% ethanol and airdry.
3) Gelatin-coating - as described for immunohistochemistry
4) Siliconization of glass slides:
- Immerse overnight in distilled water containing 1% of aminopropyltriethoxysilan (70°C; adjust pH to 3.45).
- Rinse in distilled water and dry overnight at 100°C.
- Immerse in 10% glutaraldehyde for 30 min at r.t.
- Rinse in distilled water containing 0.1 M sodium-metaperiodate.
- Rinse in distilled water and airdry.

Solutions for *In Situ* Hybridization:

1) Paraformaldehyde fixation solution:
   as described for immunohistochemistry

2) Phosphate buffered saline (PBS):
   40 g NaCl, 1 g KCl, 1 g KH_2PO_4 x 2 H_2O in 5000 ml distilled water, adjust pH to 7.4

3) Sodium-chloride-/sodium-citrate-solution (SSC), 2x concentrated:
   87.65 g NaCl, 44.1 g Na-citrate in 5000 ml distilled water, adjust pH to 7.4 and autoclave

4) Denhardt's solution (50x concentrated):
   5 g Ficoll, 5 g polyvinylpyrolidine, 5 g bovine serum albumin in 500 ml of distilled water, filter through a membrane, store in aliquots of 25 ml at -20°C

5) Basic hybridization mixture (10 ml):
   - 2.0 ml TRIS-HCl (1 M), pH 7.5
   - 200 µl EDTA (0.5 M)
   - 20 ml Denhardt's (50x)
   - 2.0 ml tRNA (25 mg/ml)
   - 1.0 ml poly-A-RNA (10 mg/ml)
   - 2.8 ml DEPC-H_2O (0.1% diethylpyrocarbonate, stir for 30 min at r.t., autoclave)

6) Prehybridization mixture (100 ml):
   50 ml formamide, 30.5 ml distilled water, 5 ml Denhardt's (50x), 5 ml EDTA(0.5 M), 5 ml TRIS-HCl (1 M, pH 7.6), 2.5 ml tRNA (10 mg/ml), 2 ml NaCl (1 M)

7) Hybridization mixture:

Stock solution	Dilution	Final Concentration
Formamide	1:1	50%
Basic hybrid. mixture (10x)	1:10	1x
RNA probe		0.25-1 ng/µl
lDDT (2 M)	1:20	0.1 M
NaCl (5 M)	1:15	0.33 M
Dextran sulfate (50%)	1:5	10%

(RNA probe first, then heat for 3 min, finally add dextran sulfate, mix, and cool on ice)

Solutions for Alkaline Phosphatase Reaction
1) Buffer A:
   100 mM TRIS-HCl, 150 mM NaCl, pH 7.5
2) Modified blocking solution:
   1% blocking reagent (Boehringer, Mannheim, FRG), 0.5% bovine serum albumin (fraction V), dissolved in buffer A
3) Buffer B:
   100 mM TRIS-HCl, 100 mM NaCl, 50 mM $MgCl_2$, pH 9.5
4) Alkaline phosphatase developing solution:
   - 45 µl nitro blue tetrazolium (NBT) solution (75 mg/ml NBT in 70% dimethylformamide).
   - 35 µl X-phosphate solution (50 mg/ml 5-bromo-4-chloro-3-indolyl-phosphate in dimethylformamide).
   - 2.5 mg levamisole.
   - 10 ml buffer C.

5) Buffer C:
100 mM TRIS-HCl, 1 mM EDTA, pH 8.0

IMMUNOHISTOCHEMICAL DEMONSTRATION OF RECEPTOR PROTEINS USING ANTI-IDIOTYPIC ANTIBODIES

The anti-idiotypic antiserum recognizing the substance P-receptor was prepared and provided by Dr. J.-Y. Couraud, Gif-sur-Yvette, France. An example for an application of the method is provided in Fig. 2.

Tissue Preparation

- Tissues obtained should be as fresh as possible, e.g. fresh specimen from the operation theater or tissues removed immediately after sacrifice of laboratory animals.
- Transfer tissues into a drop of OCT-compound (Miles, Elkhart, IN, USA) on a strip of aluminum foil. Lumina or holes in the tissues should be filled with OCT-compound.
- Snap freeze tissues in isopentane (= n-methyl-butane; Merck, Darmstadt, FRG) cooled by liquid nitrogen.
- Replace the aluminum foil by a marked filter paper using OCT-compound. Warming tissues to more than -20°C should be avoided.
- Until further processing, tissues should be stored at -85°C.
- Cut tissues on a cryostat at 10 - 16 μm and mount sections on chromalum/gelatin coated glass-slides (0.5% gelatin powder and 0.05% potassium chromium sulfate-12-hydrate in distilled water, heat to 40°C, immerse glass-slides for 10 min, dry overnight at 40°C).
- Postfix the sections for 10 min in cold acetone.
- Allow sections to airdry for 30 min before starting incubation.
- Sections may be stored at -30°C for a few days, although an immediate proceeding is preferable.

Immunohistochemistry

- Block nonspecific protein binding with phosphate buffered saline (PBS; 0.1 M NaH_2PO_4, 0.1 M Na_2HPO_4, 0.15 M NaCl in distilled water, pH = 7,4) containing 10% normal swine serum and 1% bovine serum albumin for 30 min at 4°C. Avoid addition of any kind of detergent, e.g. Tween 20 or Triton-X-100, to the blocking solution.
- Remove the blocking solution from the glass slides and cover the sections with the anti-SP anti-idiotypic antiserum from rabbit (dilution 1:50 in PBS, 2 hours at room temperature).
- Wash sections in PBS 3 times for 10 min.
- Remove exceeding PBS without allowing the sections to dry and apply a fluorescein isothiocyanate (FITC)-conjugated anti-rabbit immunoglobulin G (IgG) for one hour at room temperature; most commercially available antisera are appropriately diluted 1:50 to 1:200 under these conditions.
- Wash sections in PBS 3 times for 10 min.
- Remove extra PBS and coverslip sections with a small volume (30-50 μl) of carbonate buffered glycerol (0.5 M carbonate buffer, pH 8.6 : glycerol = 1:2).
- Sections should be examined with a epifluorescence microscope equipped with a filter suitable for demonstration of FITC, e.g. a Leica B2 module (excitation filter BP 450-492 nm; barrier filter BP 520-560 nm).

Controls

All controls should be performed as described above, except that the primary antiserum should be replaced by one of the following solutions:

a)   Serum of the same rabbit obtained before immunization (=pre-immune serum)

b)   pretreatment of the sections with the inhibitor of neutral endopeptidase (E.C. 3.4.24.11), phosphoramidon (Sigma, St. Louis, MO, USA; concentration 10^{-6} M), incubation with 1 μM Sp or 1 μM NKA for 20 min in presence of phosphoramidon (as above), and subsequent application of the anti-SP anti-idiotypic antibody, still in presence of Phosphoramidon and either tachykinin

c)   Preabsorption of the anti-SP anti-idiotypic antibody with either a monoclonal C-terminal specific anti-SP-antibody (clone NC1/34HL, Dunn, Asbach, FRG; dilution 1:200) or a polyclonal N-terminal specific antiserum (own laboratory; dilution 1:200) overnight at 4°C.

IMMUNOHISTOCHEMICAL DEMONSTRATION OF RECEPTORS USING ANTISERA RAISED AGAINST PURIFIED PROTEIN OR SYNTHETIC PEPTIDES CORRESPONDING TO FRAGMENTS OF THE RECEPTOR PROTEIN

The antiserum against a synthetic peptide of the ß$_2$-adrenoreceptor was provided by Dr. C. D. Strader, (Rahway, NJ, USA) (Application, see *Figs. 4, 5*).

Tissue Preparation

Tissues from laboratory animals can be fixed either by perfusion with 4% paraformaldehyde in 0.1 M phosphate buffer, pH 7.4, (total time of fixation: 1 - 4 h) preceded by appropriate rinsing solution (see appendix) or by immersion for 18 h at 4°C in 2% formaldehyde/15% saturated picric acid in 0.1 M phosphate buffer (pH 7.4). Specimens are then washed at least three times for 1 - 2 h in 0.1 M phosphate buffer (pH 7.4, 4°C), immersed overnight (4°C) in the same buffer containing 18% sucrose as cryoprotectant, mounted with OCT-compound (Miles, Elkhart, IN, USA) on filter paper, and frozen with liquid nitrogen. They can be stored at -75°C for several years, provided that they are tightly wrapped with Parafilm^R or Nescofilm^R to prevent drying. Sections are cut with a cryostat at 10 - 16 μm and collected on slides as described above, and air-dried for 30 - 60 min.

Immunohistochemical Incubation

The incubation protocol is exactly the same as described for anti-idiotypic antibodies, except that 0.5% Tween 20 is added to the blocking solution preceding the incubation with the primary antiserum. Using these incubation times and temperatures, appropriate dilutions of polyclonal antisera are mostly in the range of 1:200 - 1:1000.

Controls

The most appropriate control is preincubation of the primary antiserum with its corresponding antigen overnight at 4°C at a concentration of 20 - 50 μg antigen per ml of antiserum diluted to working concentration. Subsequent use of this serum for immunohistochemistry should not result in any immunolabelling. It is mandatory to run a positive control (not preabsorbed) in the same series of incubation!

Appendix

Composition of rinsing solution[40]

9.0 g NaCl, 25.0 g polyvinylpyrrolidone (PVP, molecular weight 40,000; Sigma, PVP-40), 20,000 IU heparin, 5 g procaine hydrochloride; add a.d. to a final volume of 1 liter.

DISCUSSION

The methodologies described here can be subdivided into direct techniques targeting either the receptor mRNA or the receptor protein and indirect techniques, tracing the intracellular event evoked by binding of a ligand to its corresponding receptor. Obviously, a direct technique is preferable if the main goal of the study is to localize the receptor. Demonstration of the receptor mRNA by *in situ* hybridization enables identification of the cell type expressing receptor mRNA, but this does not necessarily imply that the mRNA will be translated into the receptor protein. Therefore, studies at mRNA level, optimally, should be supplemented by investigations at protein level. Moreover, the intracellular localizations of mRNA and translated protein differ considerably. In most cases, that of the protein is more relevant for understanding of the function of a receptor, e.g. it has important implications if a receptor is located in the apical or basolateral membrane of an epithelial cell. There are, however, other features making *in situ* hybridization increasingly attractive for receptor research. Currently, molecular cloning techniques are most commonly employed for biochemical characterization of receptors. Consequently, synthetic oligonucleotides, cRNA and cDNA probes are available earlier than receptor antibodies. The use of non-radioactive *in situ* hybridization procedures offers a slightly higher resolution than autoradiography, although in the nervous system, the latter technique also suffices to identify single cellular elements. The rather rough structural resolution with concomitant intense signal of autoradiography using X-ray films can be utilized in analyses of gross structures such as the brain, where region specific distributions of receptor mRNAs can be mapped by this technique.[41] Finally, quantitative *in situ* hybridization may provide to be a useful tool in the analysis of receptor dynamics under various conditions. Currently, the stoichiometric properties of autoradiographic techniques make them favorable for such quantitative studies, but it should be noted that quantitative *in situ* hybridization has also been performed using digoxigenin labeled probes.[42]

The direct demonstration of the receptor protein should be attempted in any kind of study which primarily addresses the localization and intracellular pathway (biosynthetic and metabolic) of the receptor. The classical technique of autoradiography using radiolabeled ligands has provided numerous valuable data, and is still the method of choice if 1) the primary structure of the receptor is unknown, and 2) the demands on structural resolution are not too high. A typical example is the evaluation of regional distribution patterns of receptors in the brain.[43,44] Since most ligands are hydrophilic and are susceptible to diffusion from their receptor in aqueous solutions, receptor autoradiography is mostly performed using X-ray films instead of emulsions, thus further limiting spatial resolution. Non-radioactive labeling techniques of ligands are advantageous in that they have a much higher potential spatial resolution, are less time consuming, and avoid the safety hazards that occur when working with radioactive isotopes. However, as described in detail above, a lot of technical difficulties are encountered using these methodologies. Despite several successful studies having been reported, it is impossible to provide a standardized dependable protocol. Therefore, non-radioactive labeling of ligands is not recommended for the histochemical detection of receptor proteins.

Each method addressing the ligand binding site of a receptor protein including labeled ligands as well as antibodies directed against this site, e.g. generated by the anti-idiotypic approach, is likely to reveal more than one receptor subtype of a family. Further specificity may be obtained by displacing experiments utilizing unlabeled receptor specific ligands. This problem can be circumvented by the use of site specific antibodies, which are preferentially obtained by immunization against synthetic peptides. The latter correspond to selected amino acid sequences of the receptor protein. Currently, such antibodies are the most versatile and a specific tool for the localization of receptor proteins. If available, we consider them as the first choice for this kind of study.

Techniques that allow the visualization of intracellular events in response to application of a ligand are important tools for the investigation of the cell biology of target cells. Their value for the histochemical localization of receptors is limited because of their indirect nature, and the many variables involved. However, in case the primary structure of the receptor is unknown, such techniques (e.g. cAMP-immunohistochemistry) are still applicable, provided that it is known to which intracellular event the receptor type in question is coupled. The spatial resolution obtained by these methodologies is usually higher than that of receptor autoradiography. Furthermore, it should be stressed that neither histochemical technique alone is able to prove the existence of a molecule. Thus, the use of several independent techniques to demonstrate a receptor greatly enhances the strength of the conclusions drawn from these results. Proteins encoded by highly homologous mRNAs will also have highly homologous amino acid sequences. Thus, the possibility of cross-reactivities might be given at both nucleic acid and protein level. The analysis of intracellular events, however, is largely independent from such possible similarities. this provides a good supplement to receptor localization at either mRNA or protein level.

ACKNOWLEDGEMENT

The author's studies were supported by the Deutsche Forschungsgemeinschaft. We thank the colleagues listed in the text for generously providing antisera, cRNA probes, and technical support.

References

1. D.G. Payan, D.R. Brewster, A. Missirian-Bastian, and E.J. Goetzl, Substance P recognition by a subset of human T lymphocytes, *J Clin Invest* **74**:1532 (1984).
2. H. Tainio, Cytochemical localization of VIP-stimulated adenylate cyclase activity in human sweat glands, *Br J Dermatol* **116**:323 (1987).
3. S.C. Lazarus, C.B. Basbaum, P.J. Barnes, and W.M. Gold, cAMP immunocytochemistry provides evidence for functional VIP receptors in trachea, *Am J Physiol* **251**:C115 (1986).
4. J. Pichon, M. Hirn, J.M. Muller, P. Mangeat, and J. Marvaldi, Anti-cell surface monoclonal antibodies which antagonize the action of VIP in a human adenocarcinoma cell line (HT29), *EMBO J* **2**:1017 (1983).
5. A. Pfeiffer, R. Simeler, G. Grenningloh, and H. Betz, Monoclonal antibodies and peptide mapping reveal structural similarities between the subunits of the glycine receptor of rat spinal cord, *Proc Natl Acad Sci USA* **81**:7224 (1984).
6. M.L. Organist, J. Harvey, J.P. McGillis, M. Mitsuhashi, P. Melera, and D.G. Payan, Characterization of a monoclonal antibody against the lymphoblast substance P receptor, *J Immunol* **139**:3050 (1987).

7. J.Y. Couraud, E. Escher, D. Regoli, V. Imhoff, B. Rossignol, and P. Pradelles, Anti-substance P anti-idiotypic antibodies, *J Biol Chem* **260**:9461 (1985).

8. R. Schulz, and C. Gramsch, Polyclonal anti-idiotypic opioid receptor antibodies generated by the monoclonal ß-endorphin antibody 3-E7, *Biochem Biophys Res Commun* **132**:658 (1985).

9. L.R. Smith, K.L. Bost, and J.E. Blalock, Generation of idiotypic and anti-idiotypic antibodies by immunization with peptides encoded by complementary RNA: a possible molecular basis for the network theory, *J Immunol* **138**:7 (1987).

10. K.L. Bost, L.R. Smith, and J.E. Blalock, A molecular recognition code: its use for purification of ACTH, endorphin and LHRH receptors, *in*: "Anti-idiotypes, receptors and molecular mimicry", S. Linthicum and N. Farid, eds., Springer-Verlag, New York, Berlin, Heidelberg, pp 35-44 (1987).

11. J.P. McGillis, M.L. Organist, and D.G. Payan, Immunoaffinity purification of membrane protein constituents of the IM-9 lymphoblast receptor for substance P, *Analyt Biochem* **164**:502 (1987).

12. C. Aoki, ß-adrenergic receptors: astrocytic localization in the adult visual cortex and their relation to catecholamine axon terminals as revealed by electron microscopic immunocytochemistry, *J Neurosci* **12**:781 (1992).

13. C. Aoki, B.A. Zemcik, C.D. Strader, and V.M. Pickel, Cytoplasmic loop of ß-adrenergic receptors: synaptic and intracellular localization and relation to catecholaminergic neurons in the nuclei of the solitary tracts, *Brain Res* **493**:331 (1989).

14. R. Elde, M. Schalling, S. Ceccatelli, S. Nakanishi, and T. Hökfelt, Localization of neuropeptide receptor mRNA in rat brain: Initial observations using probes for neurotensin and substance P receptors, *Neurosci Lett* **120**:134 (1990).

15. A.P. Nicholas, V.A. Pieribone, R. Elde, and T. Hökfelt, Initial observations on the localization of mRNA for α and ß adrenergic receptors in brain and peripheral tissues of rat using in situ hybridization, *Mol Cell Neurosci* **2**:344 (1991).

16. O. Steward and G. Banker, Getting the message from the gene to the synapse: sorting and intracellular transport of RNA in neurons, *Trends Neurosci* **15**:180 (1992).

17. R. Ravid, D.F. Swaab, and C.W. Pool, Light microscopic immunocytochemical localization of vasopressin binding sites in the rat kidney, *J Endocrinol* **105**:133 (1985).

18. R. Ravid, D.F. Swaab, T.P. Van der Woude, and G.J. Boer, Immunocytochemically stained binding sites for oxytocin and alpha-MSH in rat brain following ventricular administration, *Brain Res* **379**:404 (1986).

19. J.Y. Couraud, S. Maillet, D. Zouaoui, W. Kummer, J. Grassi, P. Pradelles, and M. Conrath, Immunocytochemistry of neuropeptide receptors using anti-idiotypic antibodies and antibodies to peptides encoded by complementary RNA, *in*:" NATO ASI Series, Vol. H 58 Neurocytochemical Methods", A. Calas and D. Eugène, eds., Springer-Verlag, Berlin Heidelberg, pp 225-240 (1991).

20. W. Kummer, A. Fischer, U. Preissler, J.Y. Couraud, and Ch. Heym, Immunohistochemistry of the guinea-pig trachea using an anti-idiotypic antibody recognizing substance P receptors, *Histochemistry* **93**:541 (1990).

21. D.W. Pascual, J.E. Blalock, and K.L. Bost, Antipeptide antibodies that recognize a lymphocyte substance P receptor, *J Immunol* **143**:3697 (1989).

22. K. Maderspach, K. Németh, J. Simon, S. Benyhe, M. Szücs, and M. Wollemann, A monoclonal antibody recognizing κ- but not μ- and δ-opioid receptors, *J Neurochem* **56**:1897 (1991).

23. C.D. Strader, I.S. Sigal, A.D. Blake, A.H. Cheung, B.S. Register, E. Rands, B.A. Zencik, M.R. Candelore, and R.A.F. Dixon, The carboxyl terminus of the hamster ß-adrenergic receptor expressed in mouse L cells is not required for receptor sequestration, *Cell* **49**:855 (1987).

24. C.P. Moxham, S.T. George, M.P. Graziano, H.J. Brandwein, and C.C. Malbon, Mammalian beta$_1$ and beta$_2$-adrenergic receptors. Immunological and structural comparisons, *J Biol Chem* **261**:14562 (1986).

25. A. Wanaka, C.C. Malbon, M. Matsumoto, and M. Tohyama, Presence of catecholaminergic axon terminals containing beta-adrenergic receptor in the periventricular zone of the rat hypothalamus, *Brain Res* **479**:190 (1989).

26. L.J. Martin, C.D. Blackstone, R.L. Huganir, and D.L. Price, Cellular localization of a metabotropic glutamate receptor in rat brain, *Neuron* **9**:259 (1992).

27. R. Sprengel, T. Braun, K. Nikolics, D.L. Segaloff, and P. Seeburg, The testicular receptor for follicle stimulating hormone: structure and functional expression of cloned cDNA, *Mol Endocrinol* **4**:525 (1990).

28. W. Kummer, Simultaneous immunohistochemical demonstration of vasoactive intestinal polypeptide and its receptor in human colon, *Histochem J* **22**:249 (1990).

29. D. Bleakman, W.F. Colmers, A. Fournier, and R.J. Miller, Neuropeptide Y inhibits Ca^{2+} influx into cultured dorsal root ganglion neurones of the rat via a Y_2 receptor, *Br J Pharmacol* **103**:1781 (1991).

30. Z.Z. Wang, L.J. Stensaas, J. de Vente, B. Dinger, and S.J. Fidone, Immunocytochemical localization of cAMP and cGMP in cells of the rat carotid body following natural and pharmacological stimulation, *Histochemistry* **96**:523 (1991).

31. Z.Z. Wang, L.J. Stensaas, W.J. Wang, B. Dinger, J. de Vente, and S.J. Fidone, Atrial natriuretic peptide increases cyclic guanosine monophosphate immunoreactivity in the carotid body, *Neuroscience* **49**:479 (1992).

32. D. Koesling, E. Böhme, and G. Schultz, Guanylyl cyclases, a growing family of signal-transducing enzymes, *FASEB J* **5**:2785 (1991).

33. M.G. Currie, K.F. Fok, J. Kato, R.J. Moore, F.K. Hamra, K.L. Duffin, and C.E. Smith, Guanylin: an endogenous activator of intestinal guanylate cyclase, *Proc Natl Acad Sci USA* **89**:947 (1992).

34. G.A. Ackerman, and K.W. Wolken, Histochemical evidence for the differential surface labeling, uptake and intracellular transport of a colloidal gold-labeled insulin complex by normal human blood cells, *J Histochem Cytochem* **29**:1137 (1981).

35. R. Mentlein, C. Buchholz, and B. Krisch, Binding and internalization of gold-conjugated somatostatin and growth hormone-releasing hormone in cultured rat somatotropes, *Cell Tissue Res* **258**:309 (1989).

36. J.M. Palacios, W.S. Young, and M.J. Kuhar, Autoradiographic localization of GABA receptors in rat cerebellum, *Proc Natl Acad Sci USA* **77**:670 (1980).

37. A. Fischer, W. Kummer, J.Y. Couraud, D. Adler, D. Branscheid, and C. Heym, Immunohistochemical localization of receptors for vasoactive intestinal peptide and substance P in human trachea, *Lab Invest* **67**:387 (1992).

38. J. de Vente, H.W.M. Steinbusch, and J. Schipper, A new approach to immunocytochemistry of 3'5'-cyclic guanosine monophosphate, *Neuroscience* **22**:361 (1987).

39. C.A. Briggs, Potentiation of nicotinic transmission in the rat superior cervical sympathetic ganglion: effects of cyclic GMP and nitric oxide generators, *Brain Res* **573**:139 (1992).

40. W.G. Forssmann, S. Ito, E. Weihe, A. Aoki, M. Dym, and D.W. Fawcett, An improved perfusion fixation method for the testis, *Anat Rec* **188**:307 (1977).

41. M. Pompeiano, J.M. Palacios, G. Mengod, Distribution and cellular localization of mRNA coding for $5-HT_{1A}$ receptor in the rat brain: correlation with receptor binding, *J Neurosci* **12**:440 (1992).

42. E. Robbins, F. Baldino, J.M. Roberts-Lewis, S.L. Meyer, D. Grega, and M. E. Lewis, Quantitative non-radioactive in situ hybridization of preproenkephalin mRNA with digoxigenin-labeled cRNA probes, *Anat Rec* **231**:559 (1991).

43. R. Quirion, C.W. Shults, T.W. Moody, C.B. Pert, T.N. Chase, and T.L. O'Donohue, Autoradiographic distribution of substance P receptors in rat central nervous system, *Nature* **303**:714 (1983).

44. J.A. Danks, R.B. Rothman, M.A. Cascieri, G.G. Chicchi, T. Liang, and M. Herkenham, A comparative autoradiographic study of the distributions of substance P and eledoisin binding sites in rat brain, *Brain Res* **385**:273 (1986).

Chapter 22

LIGHT MICROSCOPIC *IN VITRO* RECEPTOR
AUTORADIOGRAPHY: X-RAY FILM VISUALIZATION
OF NEUROPEPTIDE RECEPTOR BINDING SITES
IN RAT BRAIN

Gerhard Skofitsch[1]

Histopharmacology Unit
Department of Zoology
University of Graz
Universitätsplatz 2
A-8010 Graz, Austria

SUMMARY

A general and simple autoradiographic procedure to visualize neuropeptide receptor binding sites in rat brain tissue sections is described in detail. The method utilizes unfixed slide-mounted cryostat tissue sections of the rat brain which are incubated with ^{125}I labeled neuropeptides. Visualization of binding sites is achieved by direct exposure of radiolabeled tissue sections to X-ray film in autoradiography cassettes. For example, binding of ^{125}I-porcine galanin to coronal sections of the rat brain is shown. Details of the autoradiographic procedure are discussed.

INTRODUCTION

Using immunohistochemical methods, the discrete anatomical localization of neurotransmitters, synthesizing enzymes, neuropeptides, neurohormones or neuro-modulators could be assessed, if the desired substances were able to serve as antigens and antibodies could be raised against them. As further attempts to elucidate and understand the role and function of a neurochemical were made, it became more and more evident that information about the localization of a neurochemical and also its sites of action and release, the so called receptor binding sites, is important. The main advantage of

---

[1]Address for Correspondence: Gerhard Skofitsch, Ph.D., Histopharmacology Unit, Department of Zoology, University of Graz, Universitätsplatz 2, A-8010 Graz, Austria.
Telephone:(0316) 380-5608 or 380-5595; Telefax:(0316) 381-255

autoradiographic methods over conventional membrane binding studies is their ability to quantitatively locate receptor binding sites on defined structures of tissue section even at cellular level.

Historical Review

Early autoradiographic approaches to study cholinergic muscarinic receptors,[1-4] opiate receptors,[5,6] or dopamine receptors[7-9] were based on the systemic *in vivo* administration of radioactive steroid hormones to animals.[10-13] To minimize diffusion of reversibly bound ligands, cryostat sections were freeze-dried and then mounted in the dark onto slides previously coated with photographic emulsion (dry-mount), or cryostat sections were directly thaw-mounted on emulsion coated slides (thaw-mount), followed by rapid drying with a stream of cold air. The latter method is more convenient, although chemographic artifacts produced by the melting process on the photographic emulsion are possible.[7,13] Following postfixation and photographic processing of the emulsion, tissue sections could also be used for immunohistochemistry[14] or formaldehyde induced catecholamine fluorescence.[15,16]

Along with many advantages, *in vivo* autoradiography has many disadvantages. The primary disadvantage is the need to administer extremely high concentrations of ligands to overcome vasculature-tissue (especially blood-brain) barriers and diffusion problems. The short half life of ligands in circulation, due to enzymatic breakdown followed by non-specific or unwanted binding of radiolabeled fragments, is also a problem. Certain binding parameters are beyond experimental control, resulting in considerable background. *In vitro* autoradiography hardly allows biochemical characterization of binding sites or quantitation.

An alternative to *in vivo* autoradiography is the method of receptor binding to slide-mounted tissue[17] which is further discussed as *in vitro* autoradiography. Several advantages of the *in vitro* autoradiographic method make it superior to the *in vivo* autoradiographic method, especially its suitability for receptor studies with human post mortem material as emphasized by Young and Kuhar.[18] With this method, concentration of radiolabeled ligands can be lowered considerably as circulation-tissue barriers and diffusion problems will no longer exist and binding sites are almost directly exposed to the ligand. The problem of enzymatic breakdown of the ligand can be easily controlled by enzyme inhibitors which in a living animal would cause life threatening problems. Binding parameters can be controlled in the association and dissociation phase, resulting in a decrease of nonspecific binding. Biochemical characterization of the binding parameters and the binding site is possible. Adjacent sections of the same tissue specimen can be used for comparative studies and for incubation in different ligands. The *in vitro* receptor autoradiographic method is a precise and highly reproducible method and thus quantification is possible.

Methodological Considerations

The selection of the proper autoradiographic method for visualization of the *in vitro* radiolabeled receptor binding sites is of great importance as it determines the resolution and the reliability of quantitation.

The coverslip method is an early and reliable technique which was invented by Roth and coworkers[19] and further adapted and extensively described by Kuhar[20,21] and Kuhar and Unnerstall.[22] This method utilizes direct exposure of labeled slide-mounted tissue sections to coverslips which were previously coated with photographic emulsion. Coverslips are glued onto one end of the slides to ensure proper realignment of exposed coverslip in register with tissue section. The coverslip has to be bent away from the slide by a spacer during photographic processing of the emulsion and staining of the tissue section. This method is laborious and time consuming. One of its main disadvantages is that the tissue section can be exposed only once to a coverslip. If the chosen exposure time was not correct, the experiment has to be repeated. Another disadvantage is a longer exposure time than with other methods, e.g. X-ray film. The main advantages of this method are high resolution even at cellular level, high reproducibility, and the easy and reliable quantification by simply counting silver grains within a defined area, either by direct microscopic evaluation or by computer assisted image analysis.

Direct dipping of slide-mounted and radiolabeled tissue sections into photographic emulsion[23] is commonly used in *in situ* hybridization histochemistry, as RNA/RNA or DNA hybrids are extremely stable and survive rough chemical treatment of sections. This method is less laborious and time consuming than the coverslip method; it also allows resolution at cellular level. This method, however, is not recommended for reversibly bound ligands, since dipping into molten emulsion at 45°C may result in loss or at least diffusion, of ligand from the binding site. Under normal conditions, this may be a critical issue. It was extensively discussed in the pioneering work of Stumpf and Roth[13] and Roth and Stumpf.[11] This method may be suitable for irreversibly bound ligands or neuropeptides which are irreversibly cross-linked to the tissue section by paraformaldehyde-vapor fixation following the labeling procedure.[24] Limitations of this method are obvious. It is difficult to obtain an evenly distributed emulsion layer on the rough surface of a cryostat section. Hydrophilic and hydrophobic constituents of the tissue section influence adhesion of the emulsion. Thus, authors recommend defatting tissue sections (demyelinate brain sections) by subsequent immersion in ascending concentrations of alcohol and xylene, which also removes unfixed diffusible radioligand.[24-26] Because of the possibility of uneven distribution of the emulsion layer and the rough handling of the tissue sections following radiolabeling of binding sites, the coverslip method, despite laborious handling, appears to be more reliable than the direct emulsion dipping, when it comes to quantitation.

Direct exposure of radiolabeled tissue sections to X-ray film is now the most commonly used method in receptor autoradiography.[27,28,29] This excellent, fast and easy method will be further described in detail in this issue. As this is a method which utilizes exposure of air dried tissue sections (with or without postfixation following radiolabeling), the ligand is immobilized and diffusion from the binding site is prohibited. Commercially available high sensitive (³H-sensitive) X-ray film assures equal distribution, and even thickness, of the emulsion layer of the film. The major advantage of the X-ray film method is its quantification by microdensitometry[28-33] rather than by silver grain counting. This is made possible because of the high density of very fine silver particles in the high resolution X-ray films. Densitometry by computerized image analysis allows color coding of optical density and gross-quantitation by eye. The advantages of the film autoradiography are limited by resolution. Spatial, but not cellular resolution, can be achieved which, however, is sufficient.

At present, several excellent reviews and book articles have been published on the various autoradiographic methods.[11,14,15,20-22,24] A short autoradiographic procedure for X-ray film receptor autoradiography based on these articles, is given below.

OPTIMAL PROCEDURE

Most of our experience of *in vitro* receptor autoradiography originates from demonstration of neuropeptide receptor binding sites using ¹²⁵I-labelled neuropeptides. We used in this method unfixed, cryostat sectioned, slide-mounted vertebrate (e.g. rat) brain tissues, visualizing the binding sites by direct exposure of tissue sections to X-ray film.

A simple *in vitro* autoradiographic procedure, that is easy to perform and has been used extensively in our laboratory in the past will be described.[34-41] It works best in our hands for the visualization of neuropeptide receptor binding sites in rat brain. This method can be easily adapted for other species and tissues. An example of the binding of [^{125}I]-porcine galanin (^{125}I-pGAL) to coronal rat brain sections is demonstrated.

Tissue Preparation

Animals were decapitated and the brain and spinal cord removed immediately. Some remaining serum and blood cells can attract radiolabeled compounds non-specifically and may give false readings. Therefore, it is desirable to remove the blood to a great extent from the vasculature by perfusion which improves tissue preservation during freezing and cryosectioning procedures.

Before perfusion, animals are anesthetized with sodium pentobarbital (Nembutal, 50 mg kg^{-1}, intraperitoneally). The chest is opened and a blunt cannula introduced via the left ventricle into the ascending aorta where it is secured with a clamp. The right atrium is opened by an incision and the animal is perfused with precooled (4°C) phosphate buffered saline (PBS) containing 0.05% sodium nitrite and 10% sucrose until the liver becomes uniformly light yellowish in color. Then the animals are decapitated and the organs of interest are removed immediately.

The brain is placed into a brain block made of plexiglass; using a razor blade, 1 to 1.5 cm pieces are cut in the coronal planes. The spinal cord is cut into 0.5 cm pieces if coronal sections are to be made. For sagittal (longitudinal) sections, we prefer to cut sagittally through one hemisphere to obtain even and maximum support on the specimen holder. For horizontal sectioning, brains are either mounted with the dorsal or ventral surface onto the specimen holder.

Tissue blocks are mounted at room temperature onto cryostat specimen holders. First, a small drop of OCT-mounting medium is placed on the specimen holder, then the tissue block is positioned over the drop and slightly pressed onto the specimen holder to allow the tissue to form good contact with the rugged surface of the specimen holder. Then the specimen holder with the tissue block is transferred onto dry ice and covered with powdered dry ice for approximately 3 minutes. Thereafter, the blocks are transferred into the cryostat microtome and cut at an average temperature setting of -18 to -20°C into 20 μm sections. For a general overview of receptor distribution in the brain, 20 μm sections were found to be optimal, as it is easy to cut serial sections without loosing a section in between.

Tissue sections are thaw-mounted onto chromalum gelatin coated slides with a frosted end. This eases identification by pencil markings. After cutting a section, it is directly taken up from the blade of the knife with a slide kept at room temperature, positioned parallel and lowered towards the knife blade until the section flips from the blade directly onto the slide where it melts. The slide, with the melted section, is allowed to air-dry for at least 10 to 60 min at room temperature before it is placed into a slide box. This is stored either at room temperature or 4°C (for use within a few days) or within the cryostat at -20°C. For further storage, some desiccant is added to the slide boxes which are sealed and kept at -20 to -60°C for months.

Autoradiographic Procedure

Frozen slide boxes are placed into a cryostat at -10°C and the temperature was allowed to equilibrate. Slide boxes are opened within the cryostat compartment to avoid condensation of humidity on slides stored within the boxes; slides were pulled out for one experiment, placed onto a laboratory bench, marked with pencil and air dried for several minutes before they are placed into a staining jar (Hellendahl jar). Endogenous ligands are removed by preincubation of slides in plain 50 mM Tris buffer (pH 7.7) containing 5 mM $MgCl_2$ and 2 mM EGTA at room temperature for two times 15 min.

Preincubation is followed by a 20 to 45 min incubation at room temperature with the same buffer but the addition of 0.5% gelatin (dissolved by heating to 60°C). After cooling

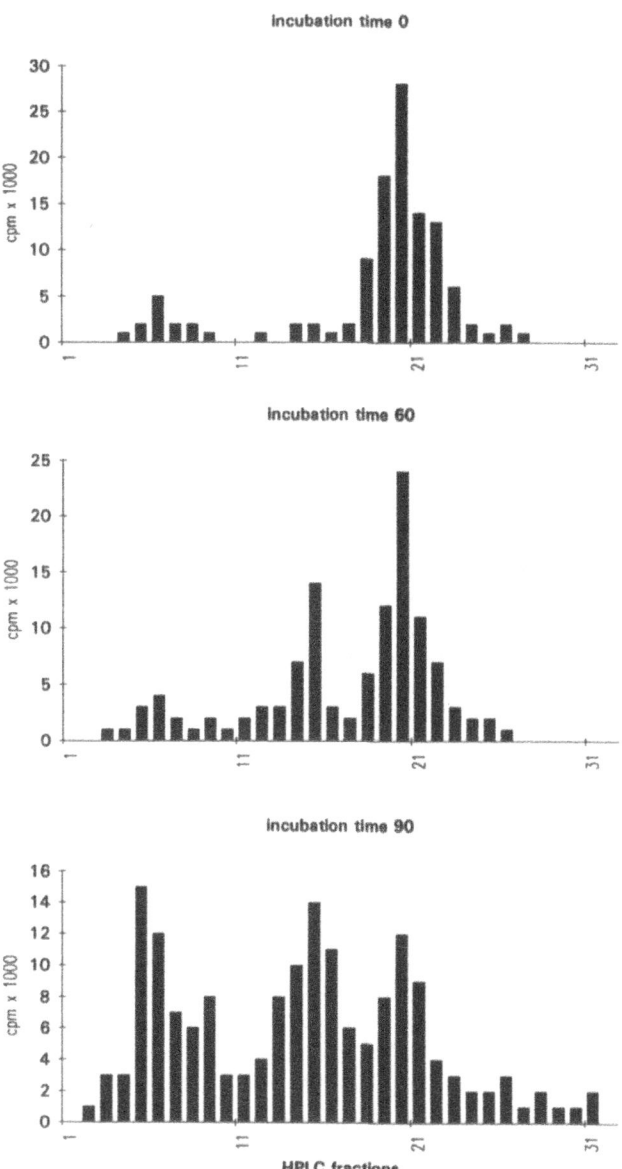

Figure 1. HPLC analysis of a 100 µl sample of incubation buffer containing approximately 0.25 nM[125]I-pGAL at various incubation times with rat hindbrain sections showing degradation of the ligand during incubation.

to room temperature, 0.5% bovine serum albumen (BSA), 2000 KIU ml^{-1} aprotinin (Tasylol, Bayer Leverkusen) and the appropriate amount of radioactive labeled neuropeptide are added. Nonspecific binding is assessed in a parallel experiment by incubation of alternate sections in the same incubation buffer. Further addition of unlabeled peptide in a concentration of 1 to 2 μM (approximately 1,000 x KD) should be sufficient to displace the radioactive ligand from the high affinity binding site.

Gelatin and BSA are added to prevent adhesion of the tracer to glass or plastic surfaces and to cover non-specific adhesion sites on the tissue section as well. The Trasylol is used as a protease inhibitor. Commercially available ^{125}I-labelled peptides usually have a specific activity of 2,000 Ci mmol^{-1} (approximately 80 TBq mmol^{-1}). For initial studies of high affinity neuropeptide binding sites (in the lower nanomolar to femtomolar range), we recommend using 30,000 to 50,000 counts per minute (cpm) of labeled peptide per 100 μl of incubation buffer. This equals a concentration of approximately 0.5 nM. Incubation time varies depending on binding characteristics and stability of the radioactive ligand during incubation with tissue sections. Checking the stability of the ligand by high performance liquid chromatography (HPLC) at several time points during incubation is strongly recommended because the ligand may undergo a decay catastropy when it is dissolved. In addition, exposure to enzymes which are not completely blocked by inhibitors may break it down to small particles exhibiting other binding characteristics than does the full peptide.

High affinity binding sites for ^{125}I-pGAL in rat brain have been reported with dissociation constant values, ranging from 0.2 to 1.9 nM, and a maximum number of binding sites of 50 to 110 fmol mg^{-1} protein.[37,39,40,42,43] For autoradiographic demonstration of binding sites, we used ^{125}I-pGAL (New England Nuclear, Du Pont de Nemours, Germany) with a specific activity of 2,200 Ci mmole^{-1} (81.4 TBq mmole^{-1}). The radioligand was added to the incubation buffer to result in 39,000 cpm in 100 μl of incubation buffer (if counted in a gamma counter). This is equal to a concentration of approximately 0.25 nM. For assessment of unspecific binding in a parallel experiment together with the radioactive ligand, unlabeled synthetic porcine galanin was added to a final concentration of 2 μM. This results in an almost complete dissociation of the labeled ligand from the binding sites. Incubation was carried out at room temperature for 45 min. HPLC analysis of incubation buffer has shown that the radioactive ligand is stable for at least 60 min if incubated with rat posterior medulla sections (*Fig. 1*). In this experiment, we found that after 90 min of incubation with tissue sections (i.e. rat medulla), considerable loss of ligand and increase in radiolabelled break down products occurs. This is most likely responsible for the biphasic time association curve as shown in *Fig. 2*.

After incubation with the radioactive ligand, slides with tissue sections are rinsed with 50 mM Tris buffer, followed by two 10 min washes in the same buffer or PBS and two rinses in distilled water with the pH adjusted to 7.4 at room temperature. Thereafter, slides are placed onto single side plastic coated absorbent paper towels and dried with a stream of air from a fan.

Air-dried slides are placed into autoradiography cassettes (Ilford Gevamatic Standard, 24 x 30 cm). Binding sites are visualized by direct exposure of tissue sections to X-ray film. Usually for ^{125}I-labeled peptides, exposure times of 4 to 16 days at room temperature and an average relative humidity of less than 50% are sufficient for most purposes and prevent diffusion of the ligand. Even if quantitation of autoradiograms is not desired, it is strongly recommended to include a slide-mounted, ^{125}I-micro scale (Amersham micro scales) to each autoradiography cassette to determine the degree of film saturation. The thickness of the microscale should be the same as the tissue sections.

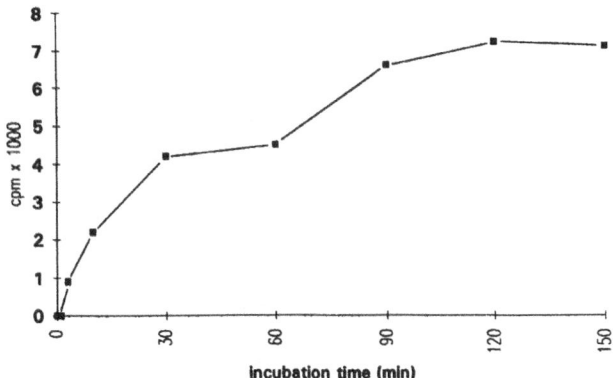

Figure 2. Time course of association of approximately 0.25 nM ^{125}I-pGAL to rat hindbrain sections showing a biphasic curve. The first plateau is possibly due to the intact ligand, the second plateau obviously is due to binding of radiolabeled break down products to nonspecific sites.

For visualization of ^{125}I-pGAL-binding sites, we have chosen direct exposure of tissue sections to 3H sensitive single coated film (Hyperfilm-3H, Amersham) which, in our hands, gives high resolution and sharp images. Incubation time was 6 days at room temperature.

X-ray films are developed by standard X-ray film developers such as Kodak D19, Ilford D19, Dupont 30D or GAF30 for about 4 min and can be fixed, after a rinse in tap water, in commercial fixatives for at least 10 min. This is followed by a 30 min wash in running tap water. Thereafter, films are air-dried and subjected to examination.

For examination, X-ray films are placed onto a light box and inspected by eye or with a binocular microscope. It is recommended that after optimal exposure to X-ray film has been achieved, exposed tissue sections be subjected to standard histological staining procedures. This allows for the parallel observation of stained sections, the identification of morphological structures, and comparison with the X-ray film image. We routinely use immersion of air dried sections into a thionine or Nissel stain for 1 to 10 min, followed by several rinses in running tap water. Thereafter, slides are air dried, immersed into xylene and mounted in permount (Entellan, Merck).

For documentation, either the X-ray film is used directly as a negative [and positives are generated which show binding sites as white dots on dark background (*Fig. 3*)] or pictures are taken from the X-ray film image [and positives show binding sites in black on white background (*Fig. 4*)].

However, even for semiquantitative evaluation, a computer assisted image analyzing system is helpful. We use a Zeiss-Kontron "Vidas" computer assisted image analyzer. This enables us to digitize a black and white video signal taken from the X-ray film image with a high resolution video camera, and to correct for uneven illumination by the light box and background fog (*Fig. 5*). For documentation, complete subtraction of non specific binding from X-ray film images is not recommended, as background structures that are helpful in morphological orientation are lost. Furthermore, it is possible to transform grey values according to optical density into false colors which eases semiquantitation or even quantitation if compared to the microscales.

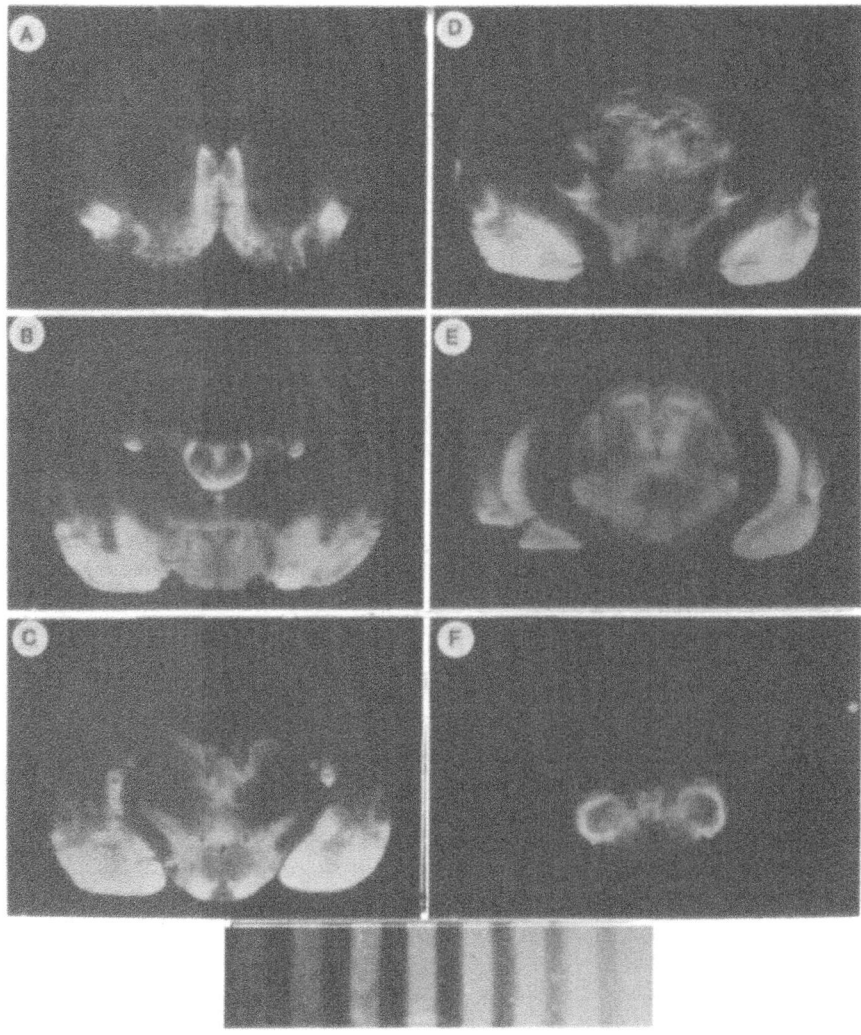

Figures 3-6. GAL binding sites on the same coronal 20 μm sections of the rat brain at different levels[37,39,40] visualized by direct X-ray film autoradiography.

Figure 3.   The X-ray film is used as a negative for generating pictures.

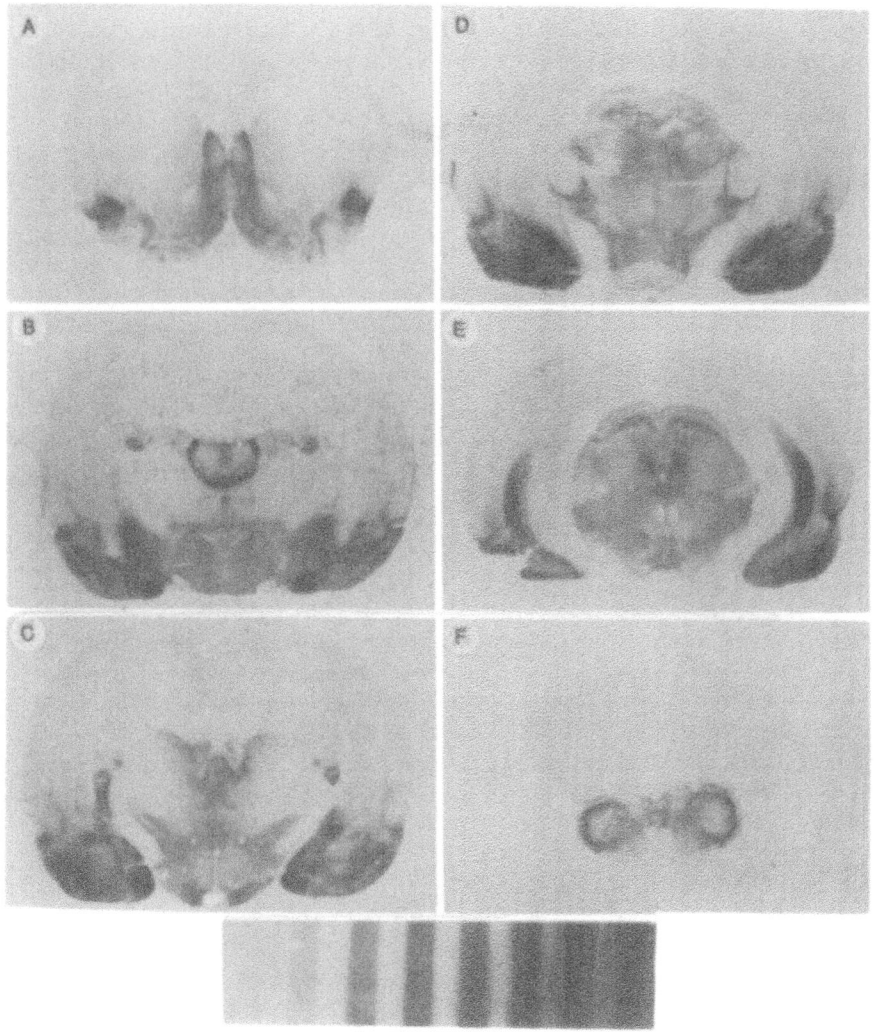

Figure 4.   Photographs are taken from the X-ray film to generate pictures.

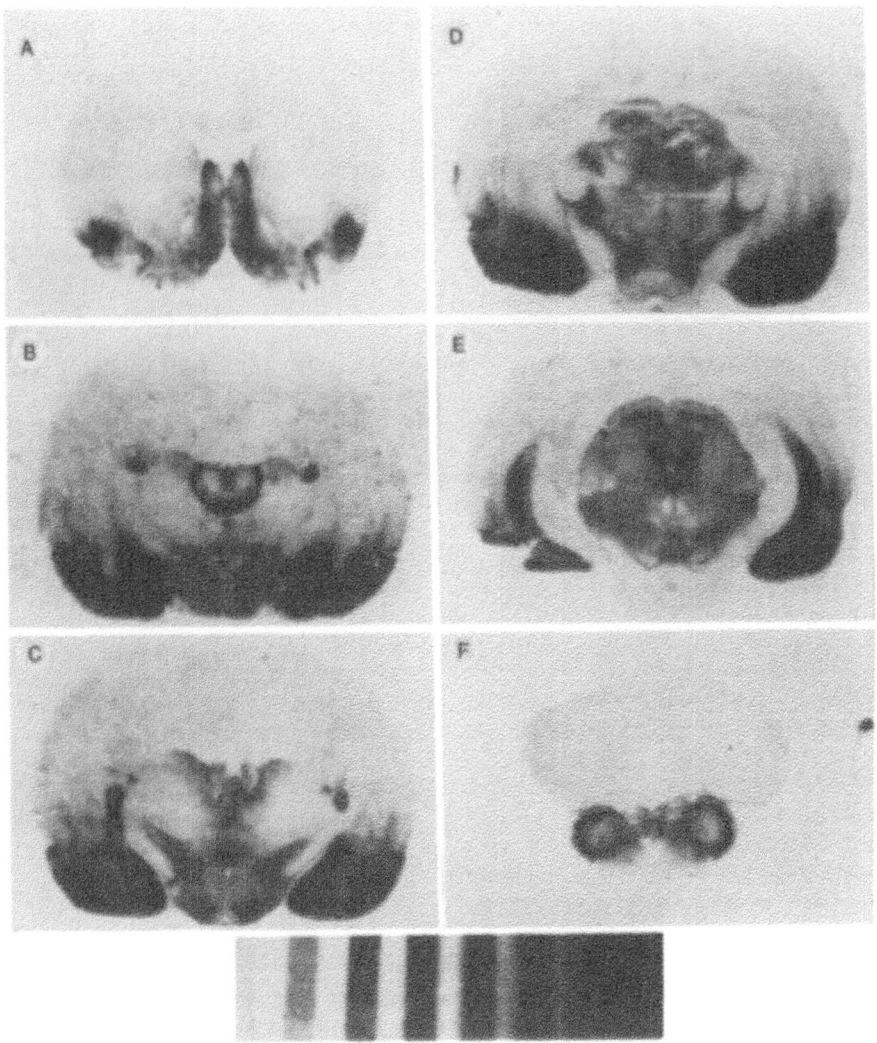

Figure 5. Pictures are generated by the image analysis system after correction of background and uneven illumination.

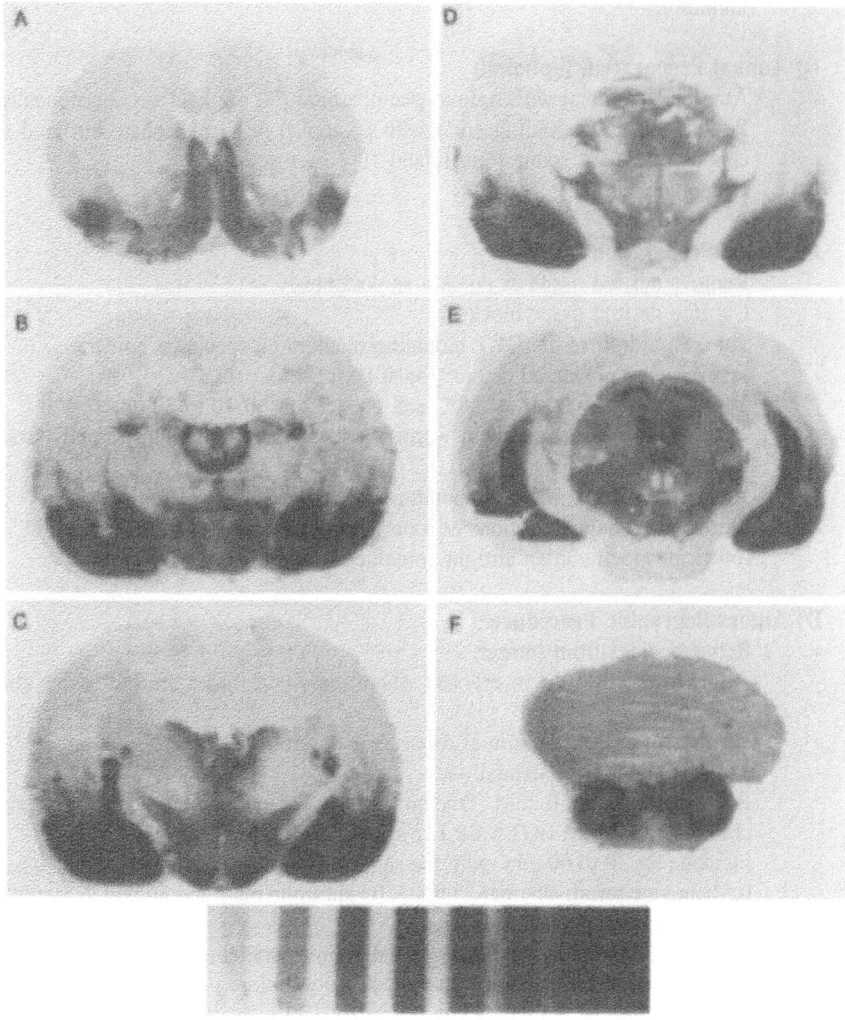

Figure 6. Eighteen steps of optical density are identified and grey-level or color coded for easy gross quantitation by the image analysis system.

STEP BY STEP PROCEDURE FOR X-RAY FILM RECEPTOR AUTORADIOGRAPHY

A) Preparation of Slides
- Clean glass slides by dipping into 0.3% nitric acid.
- Wash slides by several dips in distilled water.
- 5.0 g of powdered gelatin are dissolved in 1 liter of distilled water by heating to 70°C on a magnetic stirrer; 0.5 g chromium potassium III sulfate are added.
- Slides are then dipped into this solution for 10 sec, transferred to slide holders and air-dried under a dust protective coating overnight.

- For best results, the dipping procedure is repeated once.
- Air-dried slides can be stored in slide boxes at room temperature and low humidity.

B) Animal Preparation (optional)
- Anesthetize animal with sodium pentobarbital (50 mg kg-1 intraperitoneally).
- Perfuse via the ascending aorta with precooled (4°C) phosphate buffered saline containing 0.05% sodium nitrite and 10% sucrose.
- Decapitate animal.

C) Tissue Preparation
- Remove desired tissue as quickly as possible.
- Cut into desired tissue blocks.
- Put onto a drop of the OCT mounting medium on specimen holders.
- Freeze under powdered dry ice for at least 3 min.
- Transfer to cryostat set at -20°C and cut sections at 10 to 20 µm.
- Take up frozen sections with warm (room temperature) slides directly from the blade of the knife.
- Air-dry sections at room temperature for at least 10 (to 60) min.
- Store slide mounted sections at room temperature or 4°C for a week or at -20 to -60°C for months in an airtight container with desiccant.

D) Autoradiographic Procedure
- Remove slides from freezer.
- Allow to reach -10°C in cryostat where individual slides are pulled out for one experiment.
- Air-dry for at least 10 min at room temperature and label.
- Transfer slides to Hellendall jars.
- Preincubate with 50 mM Tris buffer, pH 7.7, containing 5 mM magnesium chloride and 2 mM EGTA for two times 15 min.
- Incubate for 30 to 60 min with the same buffer plus the addition of 0.5% gelatin, 0.5% bovine serum albumen, 2,000 KIU aprotinin ml-1 and 30,000 to 50,000 cpm of labeled neuropeptide
- (For determination of non specific binding in a parallel experiment, add 1 to 2 µM synthetic unlabeled peptide to the incubation mixture.)
- (Draw samples of the incubation buffer for HPLC analysis to determine stability of the ligand at 15 min intervals.)
- Rinse with Tris-preincubation buffer.
- Wash with Tris-preincubation buffer twice for 2 to 10 min.
- Rinse twice with distilled water.
- Dry with a stream of air.

E) Visualization of Binding Sites
- Place air dried slides into autoradiography cassettes.
- Add appropriate slide mounted microscale.
- Apply emulsion layer of the X-ray film directly onto tissue sections under safe light conditions.
- Expose for 4 to 16 days at room temperature.
- Develop film with standard X-ray film developer for 4 min.
- Rinse in tap water.
- Fix with standard fixatives for 10 min.

- Rinse with running tap water for 30 min.
- Air-dry film.

F) Evaluation
- View X-ray film on light box or under stereo microscope.
- Compare X-ray film image to corresponding tissue section stained by standard staining procedures (0.1% thionin in water for 10 to 30 sec; wash out excessive stain under running tap water).
- Apply image analysis.

DISCUSSION

Although qualitative autoradiography is a simple standard procedure and can be easily performed in most histochemical laboratories, some special suggestions follow.

Whole brain blocks were found to be difficult to handle especially in coronal sectioning. Primarily, this is because the cryostat microtome often does not allow a tissue block of 4 cm in height to be mounted between knife and specimen holder; second, it would need extensive support by tissue paste or mounting medium to freeze-mount a whole brain onto the specimen holder, with the olfactory bulb or the medulla and cerebellum attached. In some cases, we found the mounting media to interfere with the binding; third, high but relatively small diameter tissue blocks are subjected to extreme forces during cryosectioning which can result in a tilted, or even broken, specimen. Thus, we prefer tissue blocks which are not higher than the average diameter of the tissue block.

In general, the thicker and larger the average diameter of a tissue block, the warmer the temperature setting should be, the thinner and smaller the section, the colder the cryostat temperature setting. Use of motor driven cryostats is not recommend. Best results are obtained by manually controlling the cutting force and velocity, in response to tissue resistance which can be felt at the turning wheel. If the temperature is too cold or the cutting velocity too high, the cutting resistance increases forming cracks in the tissue sections. This can be perceived with the hand on the turning wheel of the cryostat. Sections often appear stretched or extended in the direction of knife or tissue block movement; this can be prevented by decreasing cutting velocity. If temperature setting is too warm or cutting velocity is too low, a moderate cutting resistance and adhesion of the knife to the tissue can be felt and tissue sections look depressed in the direction of the knife or tissue block movement. This can be prevented by increasing the cutting velocity. Manual control of cutting resistance and control of cutting velocity and tissue block temperature are essential for optimal sectioning and good preservation of tissue structure. We use a Leitz cryostat (model 1740 digital) in our laboratory. It has the advantage of a large interior, adequate for depositing precooled slide boxes, etc. and the ability to control the temperature of the specimen holders independent from the room temperature of the cryostat.

Some authors stress that for optimal tissue preservation, unfixed cryostat sections should be air-dried at subzero temperatures without freezing in an environment which is free of humidity.[24] They use drying chambers that are filled with desiccant and immersed in an ice/salt bath to provide a temperature of about -15°C; often a mild vacuum is applied. Optimal air-dried sections prepared in the above mentioned manner are described as having a "glassy, transparent appearance". Sections dried at warmer (room) temperatures are described as appearing translucent. They may disintegrate during subsequent incubation and wash-cycles. However, we and others[20-22] have not experienced disintegration of slides dried at room temperature during incubation and wash steps, provided that air-drying is complete before sections are refrozen and stored. We agree, however, that freeze-dried sections which appear frosty white are not suitable for *in vitro* receptor autoradiographic purposes. We also allow sections to reach room temperature and again air-dry before preincubation in Tris buffer.

Several preincubation and incubation buffers are recommended in the literature. In our hands, the simplest one, a hypotonic 50 mM Tris-HCl buffer containing 5 mM magnesium chloride and 2 mM EGTA, works best and serves as a preincubation and wash buffer. This is the basis for the incubation buffer. The addition of salts does not improve results in most cases and has to be verified for each ligand.

Preincubation with calcium rich buffer to deplete endogenous stores of the ligand as recommended for classical transmitters, is not recommended for neuropeptides or hormones. As we know from physiological and binding experiments, neuropeptides and some hormones bind for a long time (sometimes for hours) at the receptor site without being released. Thus, released endogenous ligands would compete with the radioactive ligand at the binding site during experiments. Therefore, we strongly recommend working in all incubation steps with calcium free medium to avoid release of endogenous neuropeptides.

In order to prevent breakdown of radiolabeled neuropeptides during incubation with unfixed tissue sections, we previously used a cocktail of commonly used proteinase inhibitors in receptor autoradiography: leupeptin (0.05%), pepstatin A (0.001%), bacitracin (0.1 mM), aprotinin (100 KIU ml^{-1}). Most of the named proteinase inhibitors are expensive and can hardly be dissolved in aqueous media at neutral pH. Thus, we recently switched to aprotinin, supplied in aqueous alcoholic solution (Trasylol, Bayer Leverkusen, 100,000 KIU in 5 ml ampullae). It was found to be sufficient for most neuropeptides. We monitor radioligand stability by HPLC during the incubation period independently from the addition of protease inhibitors.

For HPLC analysis, we use a gradient forming system consisting of two computer controlled HPLC-pumps, a manual injector with a variable sample loop, and a two channel UV-detector set at 280 and 210 nm respectively. An autosampler allows us to collect the column effluent at 1 min time intervals in small test tubes which can be counted in a gamma counter. The solid phase consists of a Waters μBondapak C18 reverse phase column. The mobile phase is a linear gradient from 20 to 60% acetonitrile in distilled water and 0.1% trifluoroacetic acid. It is applied to the solid phase at a flow rate of 1 to 1.5 ml min^{-1}. There are few peptides which need steeper gradients and less polarity in the mobile phase (20-80% acetonitrile or methanol) to wash from the solid phase, forming an acceptable peak. Peak detection of the unlabeled peptide was achieved either by radioimmunoassay of the freeze-dried column effluent or by UV-detection, if the peptide sample was in the μg range. Radioactivity of the labeled compound was measured in a gamma counter.

For preliminary experiments, we recommend following the step by step procedure given above. For more elaborate studies and especially for quantitative studies and receptor characterization, determining appropriate time of incubation and washing periods as well as the optimal radioligand concentration and pharmacologically characterizing the receptor binding site is recommended.[22,44]

First, a region of the tissue (brain) has to be selected which exhibits a fair amount of moderate to dense receptor accumulations. Serial sections are prepared on a variety of slides. The optimal concentration of radiolabeled ligand should be close to its KD-value. This can often be obtained from the literature or, in the case of preliminary experiments, has to be assumed to be in the lower nanomolar range.

Determination of optimal incubation times (time course of association) is essential. Following preincubation, slides are incubated at the same radioligand concentration for different time periods and processed for a short, but consistent, wash period. After the last rinse with distilled water, sections were wiped off the slides with filter paper strips and counted for radioactivity. (Instead of wiping off the slide, sections were air-dried and exposed to film which is subjected to image analysis.) Each time point should be verified in triplicate (at least) and the stability of the ligand should be monitored for the whole duration of the experiment. Association of ligand to tissue sections increases with time and reaches a plateau phase at a time point which is selected as optimal incubation time. In *figure 2*, the time course of association of ^{125}I-pGAL shows two subsequent plateau phases. The first plateau phase reflects the time course of association of the intact ligand. The second plateau is most likely due to the association of radiolabelled breakdown products of the ligand to the tissue section, as the progressive breakdown was monitored by HPLC analysis (*Fig. 1*).

After preincubation, slides are incubated at the same radioligand concentration: (1) with (determination of nonspecific binding), or (2) without (determination of total binding) the addition of unlabeled synthetic peptide. After having reached the optimal incubation time, slides of both experimental setups are further processed for various washing periods. Specific binding is calculated by subtraction of non-specific binding from total binding. A washing time is selected where the washout reaches a plateau and dissociates well between specific and nonspecific binding (usually a 5 to 10 min wash is optimal).

Furthermore, a saturation study should prove that, at optimal incubation and wash conditions, the binding of a radioligand depends on the ligand concentration only to a certain point at which binding sites are saturated; increase in ligand concentration does not increase binding to the tissue any more. A Scatchard plot of saturation data should then reveal different classes of receptors, the KD-value and the maximum number of binding sites. Dissociation studies should demonstrate relative displacing potencies of closely related compounds or drugs. If reliable binding data can be obtained from the literature, there is no need to repeat these studies.

In general, after the incubation and wash procedure, the slides should be air-dried as quickly as possible to prevent diffusion of the ligand. In most cases, it is not as critical for neuropeptides as it is for other ligands. We use an ordinary hair-dryer and a stream of warm air applied at a distance of about 50 to 100 cm to the slides, lying on a single side plastic-coated absorbent bench paper towels in a hood.

For ligands of low affinity or ligands which require a short wash time, air-drying is a critical step. Several methods have been developed to overcome the problem of ligand diffusion by passing a stream of air through Drierite and liquid nitrogen or dry ice-acetone traps to produce a stream of cold dry air. Thereafter, cross-linking of ligands with formaldehyde vapor is described.[20-22,24]

There are several methods available for the visualization of radiolabeled receptor binding sites on tissue sections. They range from macroscopic to electron microscopic resolution. We will focus on the method of direct exposure of radiolabeled tissue sections to X-ray film. This is the easiest way to visualize radiolabelled receptor binding sites and gives only spatial but no cellular resolution. On the other hand, this method provides an excellent overview about the anatomical distribution of binding sites in brain tissue sections. Further advantages include the ease in producing photographic images for documentation and the ability to subject autoradiograms to densitometry or image analysis for quantitation. Other techniques, e.g. dry or wet emulsion methods, allow resolution of binding sites at cellular or subcellular level. The wet emulsion technique, however, can only be applied to covalently linked ligands or following cross-linking of the ligands by formaldehyde vapor fixation. Otherwise, the ligand will diffuse from the binding site during the procedure of defatting and dipping of the section into warm photographic emulsion.[24] In the dry emulsion technique, coverslips are coated with photographic emulsion and positioned over the air-dried tissue section. This reduces the problems of ligand diffusion. Coverslips are glued on one side to the slides to ensure that the autoradiogram is kept in register with the tissue section.[20-22]

In any case, adding a standard to each experimental setup even if quantitation is not desired, is recommended (for recent advances in quantitative analysis of radioligand binding data evaluation[45] and for advances in image processing for autoradiography [46]). Standards consist of tissue paste containing certain amounts of radioactivity; these are frozen, cut at the same thickness and mounted to slides the same way as are tissue sections.

Several companies distribute standardized microscales for different radioactive isotopes. Microscales consist of 10 layers of polymer with different amounts of radioactivity. They are separated from each other by non-radioactive layers. Microscales can be either purchased as precut strips, 20 μm thick or as blocks from which strips of the desired thickness can be cut with a microtome. Optical density, as a result of silver grain formation in a X-ray film, is directly related to the amount of radioactivity within a certain area at a given exposure time. The receptor density (B max) can be calculated by comparing optical density or silver grain formation over a certain area of labelled tissue section to the standardized microscale. The microscale shows the limits of the X-ray film. At a certain incubation time, optical density form a plateau at a certain concentration of radioactivity. Optimal exposure time is achieved when the darkest spot on the X-ray film does not reach the plateau phase of optical density of the film. This ensures that a wide range of optical densities (receptor densities) can be visualized.

Film selection is also an important matter. A variety of single and double coated X-ray films are available, with or without anti scratch protection layer. Selection of film has to be based on the energy of the emitter. The most commonly used radioisotopes in receptor autoradiography are ^{125}I, ^{3}H, and ^{14}C. A weak gamma- and a weak to moderate beta-emitter hardly penetrates the emulsion layer of the film and the film base. Thus, intensifying screens would not be helpful to either enhance sensitivity or decrease exposure time. (The use of intensifying screens would also reduce sharpness of the image. This is due to the radial emission and long penetration ranges from the emitting source. They pass the film through to the screen and again back to the emulsion layer of the film.) As ^{125}I and ^{14}C easily penetrate the emulsion layer of the film but not the film base, it makes no difference whether a single or double coated film is used, with or without an anti scratch protection layer. Sharp images are generated in the emulsion layer directly exposed to the labeled tissue sections. However, ^{3}H is such a weak beta emitter that its energy is fully adsorbed by the silver halide within the emulsion layer. Penetration of an antiscratch protection layer would again adsorb lots of energy. Thus, special ^{3}H-sensitive film (Hyperfilm, Ultrafilm, Amersham, LKB) were designed to allow direct exposure of the emulsion layer of the film to the labeled tissue section without an anti scratch protection layer in between. These films also produce very sharp images with other emitters. This is due to the minimal loss of energy and less distance between emitting source and emulsion layer.

For gross surveys to determine optimal incubation and wash procedures as well as for preliminary overviews, we use Kodak X-OMAT XAR-5 film (20.3 x 25.4 cm). This is a low cost multipurpose double coated film which can be used for the detection of beta and gamma emitters. Double coating means that both sides of the film base are coated with film-emulsion. The X-OMAT film also has a scratch protective coating which allows the film to be developed in an automatic X-ray film processing machine. For high definition images, we use single coated films without scratch protective coating (Hyperfilm-3H, Amersham). This film has a rounded corner for identification at the top left side which indicates that the emulsion coated side of the film is facing the user. This side is directly exposed to the tissue sections. These hyperfilms, however, have to be developed by hand in tanks, as the automatic film processing machine would scratch the delicate emulsion layer.

Because of its increased anatomical resolution and sensitivity, quantitative receptor autoradiography has substantial advantages over conventional membrane binding studies. When specifically studying the brain and its complex organization, the distribution of receptors gives more information and leads to more functional suggestions than does the localization of the substance itself. Together, substance and receptor localization provide a complete view of the organization and distribution of a neurochemical system in the brain. This information may guide researchers to promising regions for studying e.g. neuropeptide functions in brain. However, growing information about the discrete anatomical localization of neurotransmitter and neuropeptides and their receptors in the central nervous system revealed that the distribution of neurochemical containing nerve endings often does not match the distribution of the corresponding receptors; or a so-called "mismatch problem". This was extensively reviewed and discussed by Herkenham.[47] His observations question the commonly held hypothesis that neurochemicals bind to post- or presynaptic receptors, located at the sites of the termination of pathways containing relevant neurochemicals. "The weight of evidence suggests that (neuro-) peptides and their receptors are not anatomically related and, instead, seem to be independently distributed in brain," Herkenham concluded.[47] It demonstrated that neuropeptide containing nerve terminals and corresponding receptor binding sites often occupy adjacent, but not overlapping or interconnected, brain areas.

Several explanations are given for the mismatch problem. It could be a technical failure to demonstrate binding.[47] Another explanation is that neurochemicals and their receptors are located on different neurons. Receptor expression is most likely to be confined not only to the synaptic cleft but over the entire neuron.[48] Biochemical explanations might consist in that the binding sites being occupied by endogenous ligands and not available for radiolabeled markers; low affinity binding sites might remain unrecognized.[47] Interference of the radioactive isotope coupled to the ligand molecule with receptor binding sites could also account for the failure of revealing specific receptor binding sites as seen with tachykinins.[49] The possibility that high affinity binding sites are not located at synapses and can be reached only by diffusion is discussed.[50-52] An assumption that is favored by the finding that many nerves in the central nervous system show vesicle-containing varicosities and nerve endings that lack specialized synaptic contact.[53] Schultzberg and Hökfelt[54] suggested the apparent lack of correlation of receptors and related neurochemicals might be due to the fact that in many neurons, several neurochemicals are co-stored and co-released and that their receptors may be expressed in coupled form. Both receptors may also exist in areas where only one of the normally coupled neurochemicals occur.

Despite the "mismatch problem", receptor autoradiography is a helpful tool which, in combination with immunocytochemistry, provides essential basic information for further physiological and pharmacological studies to understand neuropeptide function in the brain.

ACKNOWLEDGEMENTS

This work was supported by the Austrian Scientific Research Fund: Grant Number P8378-Med and the Austrian National Bank Fund: Grant Number 3378.

References

1. M.J. Kuhar and H.I. Yamamura, Light microscopic autoradiographic localization of cholinergic muscarinic sites in rat brain, *Proc Soc Neurosci* **4**:29 (1974).
2. M.J. Kuhar and H.I. Yamamura, Light autoradiographic localization of choloinergic muscarinic receptors in rat brain by specific binding of a potent antagonist, *Nature* **253**:560 (1975).
3. M.J. Kuhar and H.I. Yamamura, Localization of cholinergic muscarinic receptors in rat brain by light microscopic autoradiography, *Brain Res* **110**:229 (1976).
4. H.I. Yamamura, M.J. Kuhar, and S.H. Snyder, *In vivo* identification of muscarinic cholinergic receptor binding in rat brain, *Brain Res* **80**:170 (1974).

5. C.B. Pert, M.J. Kuhar, and S.H. Snyder, Autoradiographic localization of the opiate receptor in rat brain, *Life Sci* **16**:1849 (1975).

6. C.B. Pert, M.J. Kuhar, and S.H. Snyder, The opiate receptor: autoradiographic localization in rat brain, *Proc Natl Acad Sci USA* **73**:3729 (1976).

7. M.J. Kuhar, L.C. Murrin, A.T. Maluoff, and N. Klemm, Dopamine receptor binding in vivo: the feasibility of autoradiographic studies, *Life Sci* **22**:203 (1978).

8. P.M. Laduron, P. Janssen, and J.E. Leysen, Spiperone: a ligand of choice for neuroleptic receptors. 2. Regional distribution and *in vivo* displacement of neuroleptic drugs, *Biochem Pharmacol* **27**:317 (1978).

9. V. Hollt and P. Schubert, Demonstration of neuroleptic sites in mouse brain by autoradiography, *Brain Res* **151**:149 (1978).

10. D.W. Pfaff, Autoradiographic localization of radioactivity in rat brain after injection of tritiated sex hormones, *Science* **161**:1355 (1968).

11. L.J. Roth and W.E. Stumpf. "Autoradiography of Diffusible Substances", Academic Press, New York (1969).

12. W.E. Stumpf, Estradiol-concentrating neurons: Topography in the hypothalamus by dry-mount autoradiography, *Science* **162**:1001 (1968).

13. W.E. Strumpf and L.G. Roth, High resolution autoradiography with dry-mounted, freeze-dried frozen sections: comparative study of six methods using two diffusible compounds, 3H-estradiol and 3H-mesobilirubinogen, *J Histochem Cytochem* **14**:274 (1966).

14. M. Sar and W.E. Stumpf, Localization of steroid hormone target neurons by autoradiography. Anatomical relationships to aminergic and peptidergic systems by combined autoradiography and immunohistochemistry, *in*: "Handbookl of Chemical Neuroanatomy, Vol. 1: Methods in Chemical Neuroanatomy", A. Björklund and T. Hökfelt, eds., Elsevier Science Publishers B.V., New York (1983).

15. L.D. Grant and W.E. Stumpf, Combined autoradiography and formaldehyde-induced fluorescence methods for localization of radioactivelv labeled substances in relation to monoamine neurons, *J Histochem Cytochem* **29/1A**:175 (1981).

16. A.S. Heritage, L.D. Grant, and W.E. Stumpf, [3H]-estradiol in catecholamine neurons of rat brain stem: combined localization by autoradiography and formaldehyde-induced fluorescence, *J Comp Neurol* **176**:607 (1977).

17. G.J. Polz-Tejera, J. Schmidt, and H.J. Karten (1975), Autoradiographic localization of α-bungarotoxin-binding sites in the central nervous system, *Nature* **258**:349 (1975).

18. W.S. Young and M.J. Kuhar, A new method for receptor autoradiography: [3H]opioid receptors in rat brain, *Brain Res* **179**:255 (1979).

19. L.J. Roth, I.M. Diab, M. Watanabe, and R.J. Dinerstein, A correlative radioautographic, fluorescent, and histochemical technique for cytopharmacology, *Mol Pharmacol* **10**:986 (1974).

20. M.J. Kuhar, Autoradiographic localization of drug and neurotransmitter receptors, *in*: "Handbook of Chemical Neuroanatomy. Vol. 1: Methods in Chemical Neuroanatomy", A. Björklund and T. Hökfelt, eds., Elsevier Science Publishers B.V., Amsterdam-New York-Oxford (1983).

21. M.J. Kuhar. Receptor localization with the microscope, *in*: "Neurotransmitter Receptor Binding", H.I. Yamamura *et al.*, ed., Raven Press Ltd., New York (1985).

22. M.J. Kuhar and J.R. Unnerstall, Receptor autoradiography, *in*: "Methods in Neurotransmitter Receptor Analysis", H.I. Yamamura *et al.*, eds., Raven Press Ltd., New York (1990).

23. W.M. Cowan, D.I. Gottlieb, A.E. Hendrickson, J.L. Price, and T.A. Woosley, The autoradiographic demonstration of axonal connections in the central nervous system, *Brain Res* **37**:21 (1972).

24. M. Herkenham and C.B. Pert, Light microscopic localization of brain opiate receptors: A general autoradiographic method which preserves tissue quality, *J Neurosci* **2**:1129 (1982).

25. A. Hendrickson and S.B. Edwards, The use of axonal transport for autoradiographic tracing of pathways in the central nervous system, *in*:"Neuroanatomical Research Techniques", R.T. Robertson, ed., Academic Press, New York (1978).

26. J.L. Kent, C.B. Pert, and M. Herkenham, Ontogeny of opiate receptors in rat forebrain: visualization by *in vitro* autoradiography, *Dev Brain Res* **2**:487 (1981).

27. B. Larsson and S. Ullberg, A rapid film for gross autoradiography with tritium, *Acta Pharmacol Toxicol (Copenh)* **41**:48 (1977).

28. J.M. Palacios, D.L. Niehoff, and M.J. Kuhar, Receptor autoradiography with tritium-sensitive film: potential for computerized densitometry, *Neurosci Lett* **25**:101 (1981).

29. R. Quirion, R.P. Hammer, Jr., M. Herkenham, and C.B. Pert, The phencyclidine (angelk dust)/ "opiate" receptor: its visualization by tritium sensitive film, *Proc Natl Acad Sci USA* **78**:5881 (1981).

30. C. Goochee, W. Rasband, and L. Sokoloff, Computerized densitometry and colour coding of [^{14}C] deoxyglucose autoradiographs, *Ann Neurol* **7**:359 (1980).

31. J.B. Penney, Jr., H.S. Pan, A.B. Young, K.A. Frey, and G.W. Dauht, Quantitative autoradiography of [^{3}H] muscimol binding in the rat brain, *Science* **214**:1036 (1981).

32. T.C. Rainbow, W.V. Bleisch, A.Biegon, and B.S. McEwen, Quantitative densitometry of neurotransmitter receptors, *J Neurosci Methods* **5**:127 (1982).

33. J.R. Unnerstall, D.L. Niehoff, M.J. Kuhar, and J.M. Palacios, Quantitative receptor autoradiography using [^{3}H]-Ultrofilm: Application to multiple benzodiacepine receptors, *J Neurosci Methods* **6**:59 (1982).

34. G. Skofitsch, T.R. Insel, and D.M. Jacobowitz, Binding sites for corticitropin releasing factor in sensory areas of the rat hindbrain and spinal cord, *Brain Res Bull* **15**:519 (1985).

35. G. Skofitsch and D.M. Jacobowitz, Autoradiographic distribution of ^{125}I-calcitonin gene-related peptide binding sites in the rat central nervous system, *Peptides* **6**:975 (1985).

36. D.M. Jacobowitz and G. Skofitsch, Calcitonin gene-related peptide in the central nervous system: Neuronal and receptor localization, biochemical characterization and functional studies, *in*: "Neuronal and Endocrine Peptides and Receptors", T. Moody, ed., Plenum Press, New York, pp 247 (1986).

37. G. Skofitsch, M.A. Sills, and D.M. Jacobowitz, Autoradiographic distribution of 125I-galanin binding sites in the rat central nervous system, *Peptides* **7**:1029 (1986).

38. G. Skofitsch and D.M. Jacobowitz, Atrial natriuretic peptide in the central nervous system of the rat, *Cellular and Molecular Neurobiology* **8**:339 1988.

39. G. Skofitsch and D.M. Jacobowitz, Galanin in the central nervous system: A review, *in*: "Current Aspects in the Neurosciences", N.N. Osborne, eds., MacMillan Press, Oxford, (1990).

40. G. Skofitsch and D.M. Jacobowitz, Distribution of galanin binding sites in the central nervous system, *in*: "Galanin: A new Multifunctional Peptide in the Neuro-Endocrine System. Wenner-Gren International Symposium Series, Vol. 58", T. Hökfelt, T. Bartfai, D.M. Jacobowitz, and D. Ottoson, eds., MacMillan Academic and Professional Ltd., London (1992).

41. G. Skofitsch and D.M. Jacobowitz, Calcitonin and calcitonin gene-related peptide: Receptor binding sites in the central nervous system, *in*: "Handbook of Chemical Neuroanatomy, Vol. 11", A. Björklund and T. Hökfelt, eds., Elsevier Science Publishing Co., New York (1991).

42. G. Fisone, C.F. Wu, S. Consolo, Ö. Nordström, N. Brynne, T. Bartfai, T. Melander, and T. Hökfelt, Galanin inhibits acetylcholine acetylcholine release in the ventral hippocampus of the rat: Histochemical, autoradiographic *in vivo*, and *in vitro* studies, *Proc Natl Acad Sci USA* **88**:7339 (1989).

43. A.L. Servin, B. Amiranoff, C. Royerfessard, K. Tatemoto, and M. Laburthe, Identification and molecular characterization of galanin receptor sites in rat brain, *Biochem Biophys Res Commun* **144**:298 (1987).

44. J.M. Palacios, W.S. Young III, and M.J. Kuhar, Autoradiographic localization of GABA receptors in rat cerebellum, *Proc Natl Acad Sci USA* **77**:670 (1980).

45. D.L. McEachron, C.R. Gallistel, and O.J. Tretjak, Issues in quantitative imaging, *in*: "Three Dimensional Neuroimaging", A.W. Toga, ed., Raven Press, New York (1990).

46. J. Nissanov and D.L. McEachron, Advances in image processing for autoradiography, *J Chem Neuroanat* **4**:329 (1991).

47. M. Herkenham, Mismatches between neurotransmitter and receptor localizations in brain: Observations and implications, *Neuroscience* **23**:1 (1988).

48. M.J. Kuhar, The mismatch problem in receptor mapping studies, *Trends Neurosci* **8**:190 (1985).

49. C.J. Helke, J.E. Krause, P.W. Mantyh, R. Couture, and M.J. Bannon, Diversity in mammalian tachykinin peptidergic neurons: multiple peptides, receptors, and regulatory mechanisms, *FASEB* **4**:1606 (1990).

50. A.C. Cuello, Nonclassical neuronal communications, *Fedn Proc Fedn Am Soc Exp Biol* **42**:2912 (1983).

51. F.O. Schmitt, Molecular regulators of brain function: a new view, *Neuroscience* **13**:991 (1984).

52. E.S. Vizi. "Non Synaptic Interactions Between Neurons: Modulation of Neurochemical Transmission", Wiley, New York (1984).

53. R.P. Barber, J.E. Vaughn, J.R. Slemmon, P.M. Salvaterra, E. Roberts, and S.E. Leeman, The origin, distribution and synaptic relationships of substance P axons in rat spinal cord, *J Comp Neurol* **184**:331 (1979).

54. M. Schultzberg and T. Hökfelt, The mismatchproblem in receptor autoradiography and the coexistence of multiple messengers, *TINS* **9**:109 (1986).

CHAPTER 23

IMAGE CYTOMETRY OF DNA-PLOIDY

Doris Mack,[1] and Gerhard W. Hacker[2]

[1] Department of Urology
[2] Institute of Pathological Anatomy
Immunohistochemistry and Biochemistry Unit
Salzburg General Hospital
Salzburg, Austria

SUMMARY

Prognosis in tumors depends on various potentialities of the neoplasm. Morphological characteristics of parenchyma, cells and nuclei have proven to reflect malignant behavior and are used to evaluate neoplasms with several grading systems. However, the descriptive and often subjective nature of such systems sometimes may result in disturbingly low reproducibility. Digital image analysis of parenchyma and cell characteristics in microscopic images may be applied to quantitatively study the histological and cytological changes of the neoplasms in a more objective and reproducible manner. Basically, there are two methods for quantitating the DNA content in cell nuclei: static (image) cytometry, and flow cytometry. In the latter, nuclei in suspension are used to determine the relative DNA content, whereas in image cytometry, Feulgen-stained nuclei on slides can be analyzed from cytological and sometimes even from histological material. Using image cytometry, structural alterations can be recognized better, whereas in flow cytometry minor DNA content alterations in the tumor stem line may be described more exactly. It is important to know that nuclear visual selection in flow cytometry is not possible. Image cytometry offers pathologists the opportunity to preview or review the measured objects. The number of cells needed for analysis is much higher in flow cytometry than in static cytometry. All these aspects makes image cytometry more favorable. The technique can be used in smears, tumor imprints, or

---

[1] Address for Correspondence: Dr. Doris Mack, M.D., Landeskrankenanstalten Salzburg, Salzburg General Hospital, Department of Urology, Muellner Hauptstr. 48, A-5020 Salzburg, Austria. Telephone: +43-662-4482, Beeper 216; Fax: +43-662-4482-881

Modern Methods in Analytical Morphology, Edited by
J. Gu and G.W. Hacker, Plenum Press, New York, 1994

cytospin preparations obtained from paraffin embedded archival material, fresh fixed tissue or liquid samples such as urine. DNA staining with the Feulgen procedure is commonly employed, where the integrated optical density of the nucleus stoichiometrically correlates with the DNA content. To apply DNA ploidy as a routine method, standardized fixation, staining, selection and measuring techniques are essential. It is extremely important to use a high quality image analyzing computer program specialized for DNA ploidy measurements.

INTRODUCTION

Prognosis for a patient with a malignant tumor depends on the various properties of the neoplasm. Morphological characteristics can reflect malignant behavior and are used in several grading systems which occasionally give low reproducibility due to the subjective nature of histological grading. Computer image analysis can yield additional information on the histological and cytological changes of the neoplasm. It has been shown that measurements of neoplastic DNA ploidy may be an important prognostic marker for the malignancy and outcome of the tumor. However, in literature, the value of DNA ploidy has been discussed controversially and the major rationale for this appears to be the non-standardization of the techniques applied.

HISTORY

As early as 1868, a substance called "nuclein" from cell nuclei was isolated.[1] These experiments performed more than a century ago initiated the discovery of DNA as the main constituent of chromosomes and as the carrier of genetic code information. In 1924, Feulgen and Rossenbeck introduced their "nucleal reaction," a histochemical staining which can be applied for DNA detection.[2] Twelve years later, Caspersson at the Stockholm Karolinska Hospital showed that the nucleic acid content of chromosome doubles during the mitotic cell cycle.[3]

The ultimate experiment which established DNA as the carrier of genetic coding information was published in 1944 by Avery et al.[4] Four years later,[1] it was shown that the DNA content of haploid, diploid, and tetraploid cell nuclei corresponds to the chromosome sets in a ratio of 1:2:4. In the fifties, Leuchtenberger and Atkin discovered that malignant tumors may contain increased amounts of nuclear DNA when compared with normal tissue.[5,6] These observations led to great optimism about the possibilities of using cytochemical methods for objective differentiation between benign and malignant lesions. Since the early seventies, deviations from the normal DNA content (aneuploidy) have often been regarded as a parameter for neoplastic cell transformation.

THEORETICAL CONSIDERATIONS

"Cytometry" is a term introduced more than fifty years ago for procedures that are designed to quantify cyto- or histochemically detectable substances in their biological locations. "DNA cytometry" is reserved for assessments of the nuclear DNA distribution pattern of normal, hyperplastic, dysplastic and neoplastic cells. For the interpretation of a DNA ploidy histogram characterized by DNA cytometry, it is necessary to know the mitotic cell cycle. Non-replicating cells are assumed to be in the G0/G1-phase of the cell cycle, whereas replicating cells have to synthesize DNA molecules, moving from the G0/G1-phase to the G2/GM position. The amount of DNA between that of the non-dividing cells (2x23 chromosomes) and that of the dividing cells at the G2/GM phase

(4x23 chromosomes) is refered to being that of synthesis phase (S phase).

Malignant tumors often show highly increased DNA amounts, at least in some of their nuclei. To objectively measure these deviations from the normal DNA content, several techniques may be applied. Because of the minuteness of the cells, microspectrophotometric procedures can be regarded as being ideal for the assessment of an intracellular substance. All such techniques are based on the *Lambert-Beer's law*: the intensity of light passing through a substance diminishes exponentially with the layer thickness and the penetration of a ray through a solution. This is expressed as the extinction (or the negative logarithm of transmission) which is linearly dependent on the layer thickness and the concentration of the substances. If the volume is known, the quantity of a substance absorbing the ray can be determined.

In 1970, Caspersson and Lomakka made a pioneering break-through in cell biology when they developed the so-called high resolution scanning technique: a monochromatic light beam - very thin (0.5 μm) in comparison with the structure to be analyzed — is directed in a scanning mode over the object. The extinction of a substance specifically absorbing this monochromatic light can be measured. According to Lambert-Beer's law the concentration of that particular substance can be assessed in each measuring point during the scanning procedure.[7] Today, with modern microspectrophotometric scanning instruments, cells fixed on slides and specifically stained can be examined either under a microscope or on a video monitor. This allows visual inspection and identification of the cells. Data acquisition, analysis and storage can all be performed with a computer.

SAMPLING TECHNIQUES

DNA ploidy measurements can be made by a number of different procedures. Today's standard techniques are the flow cytometry and the image (static) cytometry. The first method uses cells or nuclei in suspensions. The second uses cells or nuclei adhered to glass slides.

With image cytometry, cytological as well as histological material may be investigated. One can therefore perform image measurements on cytospins of cell suspensions, smears, fine needle aspirations, and imprints and, with some reservation, also on tissue sections. In flow cytometry, the amount of material needed is comparatively high. Clinical tissue specimens with unknown proportions of the normal and the tumor cells, either well-preserved or damaged, can make the interpretation of flow cytometric results rather difficult. According to the method described by Hedley *et al.*; these problems become even more pronounced when disaggregated paraffin embedded material is used.[8] Therefore, evaluation of DNA ploidy by use of image cytometry based on processed nuclear images screened by a video camera mounted on a light microscope and processed by a computer, should possibly be performed on freshly obtained cytological preparations. Imprints, cytospins, and smears prepared in a smooth, standardized way, are optimal.

Cytological preparations made from paraffin-embedded material may exhibit cellular alterations or selective cell destruction related to initial handling and fixation of the surgical biopsies. Varying sensitivities of tumor and non-tumor cells to enzyme treatment during the disintegration procedure are an additional problem. Thus, in order to obtain reliable results using flow- or static cytometric techniques, it is important to carefully control and standardize each step from cell sampling to preparation of cell suspensions or cells adhered to glass slides. Optimally, cytological material of fresh tumor tissue should be cytospun, or rapidly smeared on a PLL-coated glass slide,[9] air-dried for at least one hour, and fixed in 10% neutral buffered formalin before further cytochemical staining procedures are conducted. A valuable way to obtain single whole cells for DNA measurements is also to work with imprints. Fresh unfixed tumor tissue is cut with a knife, and the excess liquid from the cut surface is absorbed by filter paper. Then the sticky cut surface is pressed slightly for several times onto the PLL-coated glass slide and/or smeared on the slide. In order to obtain enough control or reference cells on the slide, it is also advisable to imprint cells from a fresh lymph node onto the same slide, preferably from the same patient (see below).

For retrospective research on archival specimens preserved as paraffin blocks, it is advisable to isolate nuclei from histologic sections. For this purpose, one can use the preparation techniques described in literature.[10,11] The method used routinely in our laboratory is a slighly modified version of a technique worked out by U. Falkmer at the Stockholm Karolinska Hospital (*Protocol 1*).

PROTOCOL 1. Disaggregation of Cells from Paraffin Sections

1. Material: If possible, more than 80% of the nuclei should be derived from tumor cells (check in hematoxylin and eosin stained section). Exclude normal tissue, especially lymphoid tissue, by using a scalpel blade.

2. Deparaffination: 1-4 paraffin sections each approximately 50 μm thick, depending on the whole size of the tumor in the section, are transferred into a centrifuge vial made of a xylene-resistent material, e.g. glass. After a mechanical disintegration into smaller pieces, using syringes, 2 ml of xylene is added and kept for 10 min. Then, most of the xylene is removed with a Pasteur pipet, and the process is repeated.

3. Rehydration: 2 ml of concentrated ethanol are added, mixed well, kept for 10 min and then removed again with a pipette. This process is repeated once with concentrated ethanol, twice with 70% ethanol and once with 50% ethanol. In between these rehydration steps, let the tissue pieces sediment (do not centrifuge yet!), and remove the remaining ethanol with a pipette.

4. Washing: 2 ml phosphate buffered saline (PBS) pH 7.2 are added. The tissue pieces are sedimented again, and PBS is removed with a pipette. Repeat this process once more.

5. Enzyme predigestion: 2 ml 0.05% pronase in PBS are added. Incubate for 15-30 min at 37°C in a thermomixer (e.g., Eppendorf type 5436, Hamburg, FRG). The length of this process depends on the type of tissue and the enzyme batch used and must be tested before routine use. After digestion, very briefly centrifuge at room temperature, remove enzyme solution with a pipette, add 2 ml cold PBS, centrifuge again and remove excess PBS. Repeat wash with PBS once.

6. Filtration: The non-disintegrated tissue pieces have to be filtered away using a mounted filter syringe. Depending on the average nuclear size, filter sizes of 30-90 μm are used. 2 ml carbowax [10g polyethyleneglycol (Merck no. 807489) dissolved in 500 ml of 50% ethanol] are added, the solution is then vortexed and pressed through a millipore filter using a 2 ml volume syringe.

7. Cytocentrifugation: The remaining nuclei are well vortex-mixed and suspended with carbowax (light opalescence), usually with 2 ml. The suspension is then emptied into the cytospin containers (diluted 1:3 and 2:3) of a centrifuge and centrifuged for 15 min at 1.200 x g onto PLL-coated glass slides. After 5 hours of air drying, nuclei are fixed with formalin and Feulgen stained as outlined in *Protocol 2*.

Pronase: Sigma P 8083, type VII. To obtain a 0.05% solution, 50 mg of pronase VII are dissolved in 100 ml PBS. Prepare freshly for each enzyme treatment.

Even histological tissue sections may be used directly for image ploidy analysis. In this case, sections have to be cut in a thickness depending on the diameters of the nuclei contained. If the section is too thin, only parts of the nuclei are available for measurement; if the section is too thick, too many overlapping nuclei are the result. However, it has to be taken into consideration that many of the nuclei in the section are not contained in the whole but only in part because they may have been cut. Therefore, ploidy values obtained from such nuclei may be lower than the DNA amounts obtained from an intact nucleus. Still, this kind of analysis also gives valuable information on a possibe aneuploidy, defining nuclei with increased DNA amounts. It is also a valuable technique if only a very small carcinoma *in situ* is present, and visual control of histo-morphology is necessary.

DNA STAINING METHODS

Several staining methods have been widely used for the histochemical assessment of the nuclear DNA content.[12] The staining method described in 1924 by Feulgen and Rossenbeck still serves as a standard to which the other methods can be compared.[2] This Feulgen staining technique includes acid hydrolysis of DNA, which removes purine bases and thereby unmasks the aldehyde groups of desoxyribo-pentose molecules. Decolorized or leuco Schiff's reagent, is then allowed to react with the aldehyde groups and is thereby converted into its colored form and covalently bound to DNA. RNA and proteins usually show no color reaction with Schiff's reagent, and therefore, the Feulgen staining procedure does not demand removal of these cell components. Feulgen staining has a high degree of specificity, thus allowing important conclusions to be made about DNA.[13] Many investigators regard Feulgen staining as the most reliable stoichiometric method in histo- and cytochemical studies of DNA, although other methods for DNA detection also exist. (Also see the chapter by Nöhammer *et al.*, in this book). With the Feulgen staining method, nuclear DNA is stained in violet/pinkish color; the cytoplasm remains unstained. The technique used in our laboratory is outlined in *Protocol 2*, and has been adapted following guidelines by U. Falkmer (Karolinska Hospital, Stockholm, S).

PROTOCOL 2. Feulgen Staining Method

1. Air dried cytological specimens are fixed overnight in 4% buffered formaldehyde solution.
2. Rinse slides carefully in distilled water (2x).
3. Immerse in 5 M HCl, 1 hour.
4. Rinse slides carefully in distilled water.
5. Immerse in Schiff's reagent, in dark, at room temperature, for 2 hours.
6. Rinse slides carefully in distilled water.
7. Immerse in sodium sulfite solution, 3x10 min.
8. Rinse slides in running tap water, 5 min.
9. Rinse slides in distilled water, dehydrate in ethanol, clear in xylene, and coverslip.

The sodium sulfite solution used in Feulgen's staining has to be freshly prepared from: 180 ml distilled water plus 10 ml 1 M HCl plus 10 ml 10% sodium sulfite. Schiff's reagent is ready made up from Merck (Darmstadt, FRG) code no. I 9033, containing pararosanilin, which is a highly carcinogeneous substance. Therefore, use gloves and avoid contact with this solution! Keep in dark!

FLOW CYTOMETRY AND IMAGE CYTOMETRY: ADVANTAGES AND DISADVANTAGES

Both methods seem to have different specificities. In addition to DNA ploidy measurment, image cytometry recognizes structural alterations, whereas flow cytometry describes minor DNA content alterations in the tumor stem line more exactly.[14,15] Another important difference is the inability of nuclear visual judgement and selection in flow cytometry. Microscopic cytometry offers the investigator the opportunity to preview or review the measured objects. For each nucleus or cell, analytic data and microscopic images can be stored in a computer. This method not only reduces noise in measuring specific objects, but also can assist the investigator during screening the slide. The number of cells or nuclei needed for flow cytometry is much higher than in image cytometry. These facts favor the latter method for using cytological or histological slides. Image cytometry is invaluable when the sampled material contains only a few measuable nuclei, for example from fine needle biopsies of the prostate or the kidney, urine cytology, or gynecological smears. The reproducibility is high in both methods. Costs are high in flow cytometry and moderate in static cytometry. Measuring time per sample is minutes for flow cytometry and minutes to one hour in image analysis, depending on computer system and software used.

NOMENCLATURE AND ANALYSIS OF CYTOMETRIC RESULTS

The commonly used term *DNA PLOIDY* is based on a concept describing the nuclear DNA content assessed by means of cytometric procedures. It has been recommended to describe normal human non-replicating tissue cells as *DNA DIPLOID*. Normal DNA contents are called *DNA-EUPLOID*, and abnormal DNA contents of cells are *DNA ANEUPLOID*. With the exception of germinal cells, all other human cell nuclei have double sets of 23 chromosomes. The nuclear DNA amount corresponding to these 46 chromosomes is called "*2c*." Before a cell gets into the actual dividing process, the G2-phase of the cell cycle, there is a DNA amount corresponding to 2x46 chromosomes, i.e. "*4c*". Normal cells with DNA values between 2c and 4c are assumed to be in the S-phase (synthesis phase). In several tissues, there is an *EUPLOID POLYPLOIDY*, as observed in the hypertrophic cardiac muscle, liver cells, endocrine parenchyma, mesothelial cells, urothelial cells, and sometimes even after cytologically proven viral infections. The same can be observed in some benign tumors (Böcking, personal communication).

Numeric and structural alterations from these chromosome sets, according to their DNA contents, are called *ANEUPLOIDY*. Chromosomal aneuploidy provides an excellent marker for neoplastic cells.[16,17] However, it has be considered that because of irradiation or cytostatic agents, aneuploidy can occur also in benign tissues. If one wants to use aneuploidy as a marker for malignant cells, such influences have to be excluded. It is also true that the whole population of proliferating benign cells would never become aneuploid because of irradiation or cytostatics. So the existence of one aneuploid stemline is often understood as proof for malignancy.

A nuclear DNA content is called aneuploid, when it does not accord with a normal euploid or polyploid chromosome set. Therefore, one could conclude that non-polyploid tissue with a nuclear DNA content over 4c is aneuploid, but we have to consider the whole error of Feulgen DNA cytophotometry, due to possible hardware (camera) problems. The degree of aneuploidy has been expressed as 5c-exceeding rate and various other measures. It was often stated therefore that only nuclei with a DNA content of more than 5c should be valued as aneuploid.[18] In tissues with euploid polyploidy up to 8c, e.g. urothelium or some liver cells, DNA contents of more than 9c were a marker for aneuploidy.[19-21] Today, it appears to be sensible only to call those distributions aneuploid where peaks or values clearly dividing from normal ploidy values are obtained.

DNA HISTOGRAM

A DNA histogram is obtained from image cytometry or flow cytometry, that represents the distribution of the DNA amounts of cells in the tissues in different phases of the cell cycle. In these ploidy histograms, the cell cycle can be divided into three basic compartments: 1) G0G1 or resting phase, 2) S or synthesis phase and 3) the G2M-phase, in which nuclear DNA is doubled and cell division occurs during mitosis. The peaks in the histogram reflect the distribution of cells in the compartments of cell cycle. A normal dividing cell has a diploid 2c peak (G0G1 phase), a tetraploid 4c peak (G2M phase) and some cells betweeen these two peaks (S-phase). It has been proposed that it is possible to automatically calculate the percentage of cells in the S-phase from DNA histograms.[22] Diploidy (*Fig. 1*) implies that the main peak in the DNA histogram is approximately in the 2c region. This DNA pattern resembles the distribution of the cells over the cell cycle phases of a normal cell population. Aneuploidy, or nondiploidy (*Fig. 2*), refers to DNA histograms with main peaks outside the 2c region or the presence of one or more abnormal peaks in the DNA distribution,[23] especially when values are obtained that are higher than 5c (and Böcking, personal communication).

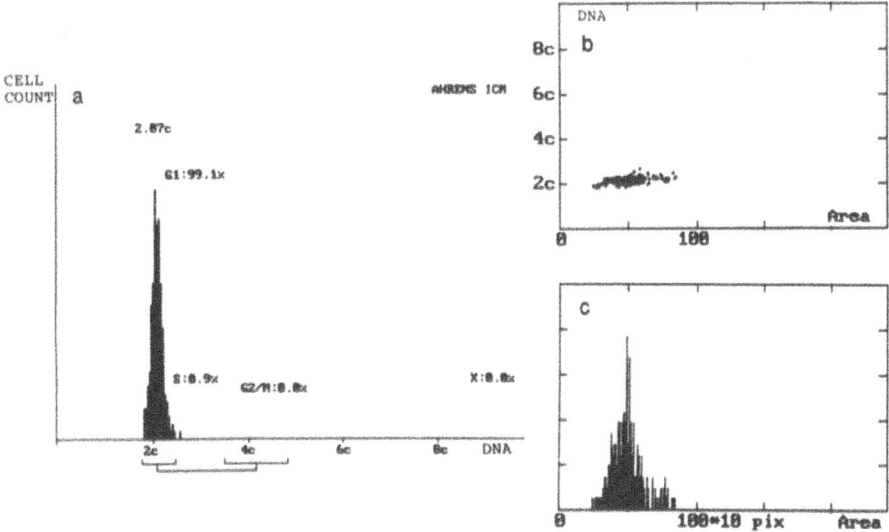

Figure 1 a-c. Highly differentiated carcinoma of the prostate gland. This case was a diploid, euploid tumor with an average ploidy of 2.1. Fig. 1a shows the ploidy histogram with a well-defined peak in the 2c region. Accordin to Auer, this tumor is classified as Type I and is likely to have a good prognosis. Fig. b shows the ploidy values plotted against the relative area of the nuclei, and Fig. 1c is a histogram of the relative area of the tumor cell nuclei.

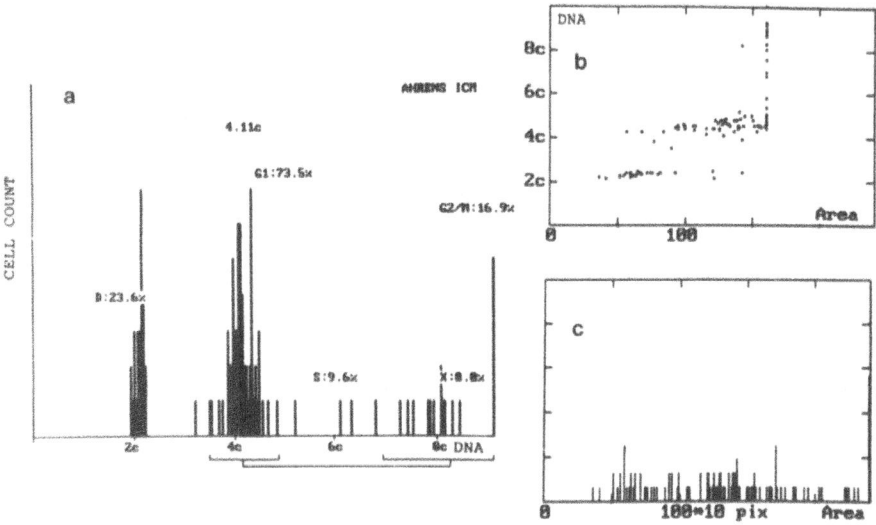

Figures 2 a-c: Highly malignant lobular breast cancer, grade III, showing a tetraploid-aneuploid ploidy histogram (Fig. 2a) which is Type IV according to the Auer classification and is likely to have poor prognosis. In this diagram, even values above 10c were obtained. The main ploidy peak is at 4.1c, the 5c-exceeding rate is at 19.1%, the proliferation index is 27%, and the calculated percentage of tumor cells in S-phase at 9.6%. Fig. 2b shows the area/ploidy plot, and different populations of tumor cells can be distinguished. Fig. 2c shows a highly varying size of the tumor cell nuclei measured.

The definition of diploid and aneuploid is not always clear. Besides the role "outside the main peak" related to the 2c region, the percentages of cells in different peaks also determine ploidy. A normal cell population is characterized by a diploid pattern with approximately ninety percent of the cells in the 2c peak. To improve the reproducibility of sometimes subjective interpretations of the DNA histogram, several additional parameters may be defined:

2cDI:	2c deviation index; deviation from 2c reference value.
5cER:	5c exceeding rate; cells with more than 5c content (minus 2c exponents) 1.757 log (2cDI+1).[18]
DNA-MG:	DNA malignancy grade; logarithmic calculation of 2cDI in a scale from 0 to 3. The lowest variance of 0 is malignancy grade 0, and the highest watched value of a osteosarcoma of 51 is malignancy grade 3.[17,18]
PB:	Ploidy balance; difference between percentage of euploid and aneuploid cells.
PI:	Proliferation index; percentage of cells between the peaks.[24]
DI:	DNA index; place of peak mean relative to 2c.
CV:	Coefficient of variation; peak tightness.[25]

Gerd Auer *et al.* (Karolinska Hospital, Stockholm) perform their visual interpretation with a special grading: Types I and II are euploid (diploid or tetraploid), TYpe III is "diploid-proliferating", and Type IV is distinctly aneuploid. Type I represents a diploid histogram, in which >95% of the nuclei assessed show a diploid DNA content between 1.8c and 2.2c (or with a DNA index between 0.0 and 1.1). Type II represents a DNA tetraploid histogram: 80-85% of the nuclei show a diploid DNA content, and 15-20% show a tetraploid DNA content between 3.8c and 4.2 c (or with a DNA index betwen 1.9

and 2.1). Type III represents a DNA histogram of a diploid "proliferating cell population: 88-95% of the nuclei show a diploid DNA content, and 5-20% of the nuclei show a DNA content equally distributed between 2c and 4c possibly indicating rapid growth of the tumor. Type IV are DNA histograms where the nuclei show DNA values distinctly outside of the diploid and the tetraploid ranges. It has been shown in breast cancer, that some tumors with a higher grade (according to the Auer-system) have a worse prognosis and a reduced survival time.[26] However, interpretation has to be tumor type-related.

Forsslund and Zetterberg classify their histograms into diploid, tetraploid, and aneuploid:[27]

Diploid:	All cells lower than 2.5c or between 3.5 and 4.5c ('D-type').
Tetraploid:	Like D-type, but with a higher tetraploid peak ('T-type').
Aneuploid:	Non D- or T-type ('A-type').

To determine the normal 2c value in the DNA histogram, measurement of *reference cells* is needed. As a reference for diploid, the cells used are: lymphocytes, granulocytes, fibroblasts, and endothelial cells. It is very important that reference material be stained at the same time and possibly on the same glass slide as the test material. For some of these reference cell types, the resulting 2c value has to be corrected in the final histogram, as it is known that dense nuclei may give a defined measuring error. To determine these correcting factors, various types of reference cells have to be compared when a new system is being installed. Liver cells or non-dividing epidermal cells usually give no or very little measurement error.

QUANTITATIVE MICROSCOPY SYSTEMS

Recent developments in computer technology have enabled fast and precise measurements of objects in microscopic images. A variety of computerized image analysis systems are available. For DNA ploidy measurements by image cytometry, microscopic images are recorded by a black and white video camera using a green filter in the microscope and stored. After shading correction,[28] the image is segmented to define objects of interest.[29,30] Shading can be done by manual contour drawing or interactively with manual adjustment, or with segmentation algorithms. In an interactive system or open system, the operator selects the objects to be measured. Automated systems use a decision scheme to identify objects of interest. Today's interactive systems are very fast and yield reproducible measurements. Subjective selection of objects for measurement may, however, be a reason for inconsistency.[31-35] To possibly exclude such errors and to obtain well-defined tumor stemline peaks, it is recommended that the image cytometric DNA ploidy of at least 100-200 definition nuclei (nuclei from cells which appear to be tumor cells) for each patient, be measured (Böcking, and Falkmer, personal communications).

Böcking (Institute of Pathology, Aachen, FRG) has worked out a check list which can be used to find optimum hard and software for DNA ploidy measurements (Böcking, personal communication). It is important to use a high quality green filter, preferably an interference filter for pararosanilin, 570 ± 10 nm, to improve contrast. The objectives of the microscope should be high quality plan or semiplan type objective. The best magnification of the objective is about 40x on whole nuclei, and 100x for sections. It is also very important to use a voltage stabilizer to keep current, and therefore light intensity, constant during measurements. The TV camera used should give a linear signal and should have a high sensitivity for green light. Shading control should be done automatically and quickly. It must be possible to choose or exclude individually the nuclei for measurement (interactive measurement). Background brightness must be determined individually for each single nucleus and be substracted from the DNA value measured. It should be possible to divide individually the measuring masks of overlapping nuclei ("cut" function). Measurement should be possible from the live-image. In addition to ploidy, other factors of the nuclei should be measured in parallel, such as the nuclear area (morphometrical parameters, *Figs. 1 and 2*). Precision of measurement is only given if the diagram obtained can be corrected for the different reference cell types. The problem is that very dense nuclei have too low transmission light, they are "too black". It should be possible to measure reference and definition cells in parallel, and to remove individual incorrect raw data from the list. Ploidy values for each single cell may be displayed on the TV screen immediately during measurement. The measuring time for each nucleus should be shorter than one second, in order to save time. Software updating should be guaranteed from the maker of the image analysis system. In the computer system used, it should be possible to export the raw data and transfer it to other computer programs, e.g. for better statistical analysis and more individually adjusted histograms. Various DNA parameters should be calculated, such as DNA index, CV's, 5c exceeding rate, etc.

DISCUSSION

The Value of DNA Analysis in Clinical Practice

Since benign or low grade malignant tumors most often exhibit euploid DNA distribution patterns *(Fig. 1)*, while highly malignant tumors are frequently aneuploid *(Fig. 2)*, DNA ploidy measurements have been used as a complement to cytologic and histologic diagnosis in cases where additional information is needed in order to distinguish between benign, low grade, or high grade malignant lesions. Morphological parameters complemented by DNA ploidy measurements in most cases proved to be sufficient in order to obtain reliable diagnostic information concerning preneoplastic and neoplastic lesions.[23] There are types of malignant tumors, all or at least the vast majority of which, exhibit clear-cut aneuploid DNA profiles that can easily be identified. In these tumor types, DNA ploidy measurements can contribute significant diagnostic information over and above that obtained by cytomorphologic or histologic parameters. These tumors include: squamous bronchus carcinoma, small cell lung carcinoma, cervical carcinoma, esophageal carcinoma.[23] On the other hand, in a number of malignant tumors, or in subsets of tumors, closely euploid DNA amounts or only minor deviations are observed.[23] Such tumors cannot be distinguished from normal tissues or benign tumors by means of DNA content, thus limiting the usefulness of DNA measurements as a diagnostic tool. Such findings have been reported for breast cancer, thyroid tumors, prostatic cancer, ovarian tumors, endometric carcinoma, chondrosarcoma, and neuroendocrine tumors. One should, on the other hand, be aware that at least some of these inconsistent findings reported in literature may be the results of non-standardization, especially of the staining technique and computer program used. Only a few image analysis systems available, specialized in interactive DNA ploidy measurement, fulfill the criteria of exact and reproducible measurement. Within the last years, we have tried a variety of systems on the market and found only two of them that performed well.

DNA ploidy measurements may be used as an additional important objective diagnostic parameter complementing conventional clinical and morphological indicators for the detection and diagnosis of premalignant and malignant lesions. It is discussed as being a valuable screening parameter in premalignant and malignant squamous bronchial lesions and in the uterine cervix. DNA ploidy appears to be a diagnostically relevant parameter in tumors of the lung, the uterine cervix, esophagus, pancreas, and the skeleton.[23,36,37]

Prognosis

Numerous publications strongly support a correlation between tumor nuclear DNA content and clinical course in patients suffering from malignant tumors. The value of prognostic information obtainable by DNA measurements depends largely on the predictive power of conventional clinical and histopathological parameters in various tumor types.[38] However, a more accurate prognostic evaluation of the expected tumor aggressiveness in each individual case would, undoubtedly, form a new basis for more individualized and improved tumor therapy. It can also be forseen that the number of low grade malignant tumors being overtreated today would be reduced. It would also be possible to select patients for new therapy principles, in cases where conventional therapy regimens are insufficient or ineffective. It is obvious that optimal grading of malignancy is a most important diagnostic problem.

There are numerous studies which prove that DNA analysis can contribute prognostic information over and above that obtainable by conventional clinical and morphological parameters, e.g. in tumors of the breast, prostate, thyroid, ovary, uterus and skeleton. With DNA measurements these tumors can be classified as euploid (diploid and tetraploid) and aneuploid tumors with an obvious abnormal DNA content, indicating a high degree of karyotypic abnormality and genome instability (Fig. 1 and 2). Analysis of the malignant behaviour, including the clinical follow-up of the two types of tumors in long term retrospective studies, showed that DNA aneuploid tumors generally progressed rapidly and frequently, killing patients within a few months or years, whilst DNA euploid tumors progressed more slowly. It is a proven fact that DNA ploidy, e.g. in prostate cancer, can be a useful and worthwhile method to estimate the outcome of those patients.[39,40] However, it has also to be pointed out that patients with DNA aneuploid malignancies, especially those with low tumor stages, may be cured; but the clinical course in those patients does not always really reflect the potential malignancy of the tumor. This has been observed e.g. in breast cancers with small tumor sizes with postoperatively demonstrated lymph nodes involvement.[36,37] It is also important to consider that depression of tumor defence mechanisms, as observed in AIDS patients or patients who were treated unsuccessfully with anticancer drugs, may change the malignant behaviour of DNA euploid tumors.[41]

The use of DNA ploidy in malignancy grading has been discussed controversially. Reasons for reported inconsistencies include: 1) that the heterogeneity of tumors depends results largely on the method of selection of material used for study. Multiple samples, e.g. several imprints from different areas of the tumor, and clearly defined selection criteria are necessary; 2) the preparation method influences results; standardization of these techniques is mandatory; 3) differences in methods of quantitation, consisting of differences in image recording and processing techniques, and nonuniformity of features used, require a clear description of apparatus and software.

Although the genetic background for DNA euploid and DNA aneuploid tumors is not understood at present, nuclear DNA content assessments undoubtedly provide us with a powerful diagnostic method for discriminating between low grade and high grade malignant tumors. In summary, DNA measurements may be used as an important prognostic indicator, complementary to the traditional clinical and morphological parameters in tumors from breast, prostate, lung, thyroid, ovary, uterus, colon-rectum and skeleton. Its value in the other tumor types remains to be clarified.

ACKNOWLEDGEMENTS

Dedicated to Professor Dr. Julian Frick, Salzburg, on the occasion of his 60th anniversary. We gratefully acknowledge the continuous scientific support on the DNA ploidy matters by Dr. Ursula Falkmer (Stockholm) and Dr. A. Böcking (Aachen, FRG).

Glossary. Cytometry = quantitation of cellular features; Densitometry = staining intensity or static cytophotometry (for example DNA content); Flow cytometry = single cell analysis of cells and nuclei in suspension; Static cytophotometry = densitometric measurements, for example for DNA content quantitation; Feulgen staining = stoichiometric DNA staining with Schiff's reagent; Low resolution techniques = quantitation at lower magnification (for example 20x objective) which results in relatively large pixel size as in histometry and cytometry; Mitotic index = expression of mitotic activity as number of mitotic figures per square mm neoplastic field

References

1. Miescher. *cited in*: "100 Years of Histochemistry in Germany." W. Sandritter, *ed.*, Schattauer, Stuttgart: 1 (1964).
2. R. Feulgen and H. Rossenbeck, Mikroskopisch-chemischer Nachweis einer Nukleinsure vom Typus der Thymonukleinsure und die darauf beruhende Frbung von Zellkernen in mikroskopischen Prparaten, *Hoppe Seyler's Z Physiol Chem* **135**:203 (1924).
3. T. Caspersson, Ber den chemischen Aufbau der Strukturen des Zellkernes. *Scand Arch Physiol* **73**, Suppl 8, (1936)
4. O. T. Avery, C. M. MacLoed, and M. McCarty, Studies on chemical nature of substance inducing transformation of pneumococcal types: Induction of transformation by a desoxyribonucleic acid fraction isolated from pneumococcus type III, *J Exp Med* **79**:137 (1944).
5. C. Leuchtenberger, R. Leuchtenberger, and A. M. Davis, A microspectro-photometric study of the deoxyribose nucleic acid (DNA) content of normal and malignant human tissues, *Am J Pathol* **30**:65 (1954).
6. N. B. Atkin and B. M. Richards, Desoxyribonucleic acid in human tumors as measured by microspectrophotometry of Feulgen stain: A comparison of tumors arising at different sites, *Br J Cancer* **10**:769 (1956).
7. T. Caspersson and G. Lomakka, Recent progress in quantitative cytochemistry: Instrumentation and results. *in*: "Introduction to Quantitative Cytochemistry II", G. L. Wied and G. F. Bahr, *eds.*, Academic Press, New York, London: 27 (1970).
8. D. W. Hedley, C. A. Rugg, A. B. Ng, and I. W. Taylor, Influence of cellular DNA content on disease free survival of stage II breast cancer patients, *Cancer Res* **44**:5395 (1984).
9. W. M. Huang, S. J. Gibson, P. Facer, J. Gu, and J. M. Polak, Improved section adhesion for immunocytochemistry using high molecular weight polymers of L-lysine as a slide coating, *Histochemistry* **77**:275 (1983).
10. G. Mikuz, F. Hofstetter, and R. Delgado, Extraction of cells from paraffin embedded tissue sections for single cell DNA cytophotometry, *Anal Quant Cytol Histol* **7**:343 (1985).
11. A. M. J. V. van Driel-Kulker, W. E. Mesker, I. van Velzen, H. J. Tanke, J. Feichtinger, and J. S. Ploem, Preparation of monolayer smears from paraffin embedded tissue for image cytometry, *Cytometry* **6**:268 (1985).
12. U. V. Mikel, W. N. Fishbein, and G. F. Bahr, Some practical considerations in quantitative absorbance microspectrophotometry. Preparation techniques in DNA cytophotometry, *Anal Quant Cytol* **7**:107 (1985).
13. F. H. Kasten, The Feulgen Reaction: An enigma in cytochemistry, *Acta Histochem* **17**:88 (1964).
14. A. M. Uyterlinde, A. W. M. Smeulders, and J. P. A. Baak, Reproducibility and comparison of quantitative DNA histogram features abtained with a scanning microdensitometer and a flowcytometer in breast cancers, *Anal Quant Cytol Histol* **11**:353 (1989).
15. G. U. Auer, U. Askensten, K. Erhardt, A. Fallenius, and A. Zetterberg, Comparison between slide and flowcytometric DNA measurements in breast tumors, *Anal Quant Cytol Histol* **9**:138 (1987).
16. A. Böcking, R. Chatelain, M. Honge, R. Daniel, A. Gillissen, and D. Wohltmann, Representativity and reproducibility of DNA malignancy grading in different tumors, *Anal Quant Cytol Histol* **11**:81 (1989).
17. A. Böcking, R. Chatelain, U. Orthen, G. Gien, G. von Kalckreuth, D. Joehan, and D. Wohltmann, DNA grading of prostatic carcinoma, *Anticancer Res* **8**:129 (1988).
18. A. Böcking, C. P. Adler, H. H. Common, M. Hilgarth, B. Ganzen, and W. Auffermann, Algorithm for DNA cytophotometric diagnosis and grading of malignancy, *Anal Quant Cytol* **6**:1 (1984).

19. R. Chatelain, G. W. Lhr, H. Common, and A. Böcking, Angioimmunoblastic lymphadenopathy - a malignant disease due to DNA aneuploidy, *Anticancer Res* **9**:129 (1989).

20. R. Chatelain, T. Schmuck, E. Schindler, A. Schindler, and A. Böcking, Diagnosis of prospective malignancy in koilocytic dysplasia of the uterine cervix with DNA cytometry, *J Reprod Med* **34**:505 (1989).

21. R. Chatelain, B. Hoffmeister, F. Hrle, A. Böcking, and C. Mittermayer, DNA grading of oral squamous carcinomas, *Int J Oral Maxillofac Surg* **18**:43 (1989).

22. U. Falkmer. Methodological aspects on DNA cytometry. Thesis, Karolinska Hospital, Stockholm, 9 (1989).

23. G. Auer, U. Askensten, and O. Ahrens, Cytophotometry, *Human Pathol* **20**:6 (1989).

24. M. Opfermann, G. Brugal, and P. Vassilakos, Cytometry of breast carcinoma: Significance of ploidy balance and proliferation index, *Cytometry* **8**:217 (1987).

25. L. G. Koss, B. Czerniak, F. Herz, and R. P. Wersto, Flow cytometric measurements of DNA and other cell compartments in human tumors: a critical appraisal, *Human Pathol* **20**:528 (1989).

26. T. O. Caspersson, G. Auer, A. Fallenius, and J. Kudynowski, Cytochemical changes in the nucleus during tumor development, *Histochem J* **15**:337 (1983).

27. G. Forsslund and A. Zetterberg, A quantitative evaluation of cytophotometric DNA analysis in retrospective studies using archival tumor specimens, *Anal Quant Cytol Histol* **9**:190 (1987).

28. T. K. ten Kate. TV microscopical image analysis for accurate DNA quantification in pathology. Thesis: Amsterdam, (1990)

29. W. Abmayr, E. Mannweiler, D. Sterle, and E. Daml, Segmentation of scenes in tissue sections, *Anal Quant Cytol Histol* **9**:190 (1987).

30. C. E. Liedtke, T. Gahm, F. Kappei, and B. Aeikans, Segmentation of microscopic cell scenes, *Anal Quant Cytol Histol* **9**:197 (1987).

31. E. K. Wong, E. H. Liang, E. K. Lin, D. A. Simmons, and L. G. Koss, A selective mapping algorithm for computeranalysis of voided urine cell images, *Anal Quant Cytol Histol* **11**:203 (1989).

32. C. E. M. Blomjous, W. Voss, N. W. Schipper, A. M. Uyterlinde, J. P. A. Baak, H. J. de Voogt, and C. J. L. M. Meijec, The prognostic significance of selective nuclear morphometry in urinary bladder carcinoma, *Human Pathol* **21**:409 (1990)

33. E. C. M. Ooms, P. H. J. Kurver, R. W. Veldhizen, C. L. Alous, and M. E. Dron, Morphometric grading of bladder tumors in comparison with histologic grading by pathologists, *Human Pathol* **14**:144 (1983).

34. H. G. van der Poel, M. E. Boon, E. A. van der Meulen, and A. Wijsman-Grootenhorst, The reproducibility of cytomorphometrical grading of bladder tumors, Virchows Arch A **416**:521 (1990).

35. Y. Collan, T. Torkkali, V. M. Kosma, E. Pesonen, O. Kosunen, E. Jantunen, G. M. Marinzzi, R. Montioni, F. Marinelli, and G. Colliuma, Sampling in diagnostic morphometry: The influence of variation sources, *Pathol Res Pract* **182**:401 (1987).

36. G. Auer, E. Erikkson, and E. Azavedo, Prognostic significance of nuclear DNA content in mammary adenocarcinomas in humans, *Cancer Res* **44**:393 (1984).

37. A. Fallenius. DNA content and prognosis in breast cancer. Thesis: Karolinska Hospital, Stockholm (1986)

38. O. W. Hedley, M. L. Friedlander, and W. Taylor, Methods for analysis of cellular DNA content of paraffin embedded pathological material using flow cytometry, *J Histochem Cytochem* **31**:1333 (1983).

39. A. Zetterberg, and P. L. Esposti, Prognostic significance of nuclear DNA content levels in prostatic carcinoma, *Scand J Urol Nephrol* **55**:53 (1980).

40. E. C. Jones, J. McNeal, and J. Bruchovsky, and G. de Jong, DNA content in prostatic carcinoma, *Cancer* **66**:752 (1990).

41. M. Sanchez, E. Ames, and K. Erhardt, Analysis of DNA distribution in Kaposi's sarcoma in patients with and without acquired immunodeficiency syndrome, *Anal Quant Cytol Histol* **10**:16 (1988).

CHAPTER 24

QUANTIFICATION OF CELLULAR PROLIFERATION
AND DIFFERENTIATION BY MICROPHOTOMETRY
OF DNA AND PROTEIN

Gerhard Nöhammer,[1] Peter M. Eckl,[2] Frank Girardi,[3]
Verena Hermes,[3] Aldo Paolicchi,[4] Hellmuth Pickel,[3]
Wannee Rojanapo,[5] Roberto Tongiani,[4] and Gerhard
Wirnsberger[6]

[1]Institute of Biochemistry, Karl-Franzens-University Graz, Austria
[2]Institute of Genetics and General Biology, University of Salzburg
Austria
[3]University Clinic of Obstetrics and Gynecology, Landeskrankenhaus
Graz, Austria
[4]Institute of General Pathology, University of Pisa, Italy
[5]National Cancer Institute, Bangkok, Thailand
[6]Medical University Clinic, Department of Nephrology
Landeskrankenhaus Graz, Austria

SUMMARY

Both histochemical DNA staining with 3-hydroxy-2-naphthoic acid hydrazide and
Fast Blue B, and the histochemical protein staining with 2-hydroxy-1-naphtaldehyde were
used for DNA-protein double staining of different cells and tissues after being optimized
and quantified. Integrated nuclear extinctions were determined simultaneously at the

[1]Address for Correspondence: UD Dr. Gehard Nöhammer, Institute of Biochemistry, Karl-Franzens-
University Graz, Halbärthgasse 5/I, A-8010 Graz, Austria. Tel: 0043 316 380-5490

absorption maximum of DNA staining (550 nm) and the absorption maximum of protein staining (420 nm) by scanning microphotometry. The DNA protein scattergrams obtained were used for the characterization and quantification of cellular proliferation and differentiation. Differentiation was investigated with squamous epithelium of sections from human uterine cervix and with rat liver parenchymal cells isolated at different times after birth. Differentiation was shown to be accompanied by a characteristic increase of nuclear proteins relative to DNA. Dedifferentiation was investigated with rat liver cells grown under proliferation promoting conditions. In contrast to differentiating cells cellular dedifferentiation is accompanied by a characteristic decrease of nuclear proteins. Analysis of DNA protein scattergrams obtained from double stained sections of human uterine cervix, breast and liver revealed the existence of three subgroups of malignant tumors. Analysis of DNA-protein scattergrams of basal cells of the normal squamous epithelium of human uterine cervix also showed a "field effect" of adjacent tumors. A characteristic change of the DNA protein scattergram of basal cells in the vicinity of malignant tumors enabled the discrimination of corresponding cells of healthy volunteers.

INTRODUCTION

The idea of simultaneous microphotometric determination of both DNA and protein originated from microphotographs published by Köhler.[1] It was Caspersson[2,3] who first presented definitive evidence of the contribution of both proteins and nucleic acids to the ultraviolet absorption of tissues. He demonstrated that the absorption curves obtained with cytological material could be resolved into two major components, the overlapping absorption of nucleic acids and proteins. A general theory of the role of nuclear nucleic acids and proteins in cytoplasmic protein synthesis was summarized by Caspersson.[4] The applicability of UV-microphotometry, however, proved to be limited due to the superposition of absorption from nucleic acids, proteins, and the nonspecific loss from scattered light. Therefore, UV-microphotometry has been supplanted by simpler methods of staining and photometry with visible light. Direct measurements of cell components have been replaced by indirect microphotometry, accompanied by new complications mainly caused by the histochemical methods used. Different methods have been developed and used for microphotometric quantification of both DNA and protein. Most of the authors used the Feulgen method for DNA staining,[5] sometimes in combination with UV-microphotometry for the determination of both DNA and RNA, and combined it with different methods for protein determination.

Early studies combined the histochemical method for arginine with Feulgen staining.[6,7] This period was followed by DNA protein cytophotometry using Feulgen staining in combination with microinterferometric determinations of dry mass.[8-10] Finally, histochemical methods were used for the microphotometric quantification of proteins in combination with Feulgen staining. Thus, dinitrofluorobenzene was introduced by Mitchell[11] for protein staining. Gaub et al.[12] published a method combining Feulgen staining and Naphtol Yellow S, the histochemical protein staining of Deitch,[13] which was used also by Welsch and Resch[14] and by Mitchell et al.[15] for combined DNA and protein analysis. A method for the combined staining of protein bound sulfhydryl groups and DNA was published by Mitchell.[16] Ninhydrin-Schiff together with Feulgen techniques were used by Fukuda et al.[17] Oud et al.[18] recommended the use of both Light Green or Orange II for quantitative protein staining in combination with the Feulgen method. From the early beginnings, microphotometry of both DNA and protein has been directed to two different fields of application. The first was the processes of proliferation and differentiation including the cell cycle[10,19-23] and the second an attempt to characterize tumor cells.[9,24] Finally, the development of the dual wavelength method (Bioscan) by

Van der Ploeg et al.[25] improved and facilitated microphotometry with double stained samples. Microphotometry is a method for the investigation of single cells or tissues prepared and stained on slides enabling critical observation and selection of distinct cells or sub-cellular compartments. Flow cytometry on the other hand proved to be useful for the rapid investigation and sorting of large numbers of cells.

A number of double staining methods for DNA and protein have been developed using fluorescent dyes for flow cytometry,[26,27] a development which has not yet reached its cumulative point. Monoclonal antibodies raised against 5-bromodeoxyuridine,[28] Ki-67[29] and other proliferation associated nuclear proteins[30] like cyclins, p105 and p34 are used in combination with DNA stains for flow cytometric investigation of the cell cycle. Cellular differentiation, cell type analysis and cell sorting are performed again also in combination with DNA staining, using monoclonal antibodies directed against cytoskeletal proteins, surface proteins, and other cellular components characterizing special states of differentiation.

In any case, quantitative static microphotometry remains the method for single cell studies and especially for the investigation of tissue sections although the classical method is substituted increasingly by image analyzers enabling rapid measurements of larger numbers of cells or nuclei. For static microphotometry, Feulgen DNA staining remained the backbone. The drawbacks of classical Feulgen staining are its still unknown reaction mechanism and the variety of its reaction products.[31] The extent of Feulgen staining depends on several factors, e.g. conditions of hydrolysis,[32,33] method of fixation,[34] mode of staining procedure used,[35] type of cells stained,[35,37,38] and so on. The composition of the dyes generated during Feulgen staining depends on the conditions of the reaction and the dyes that differ in respect to their stability and spectral properties.

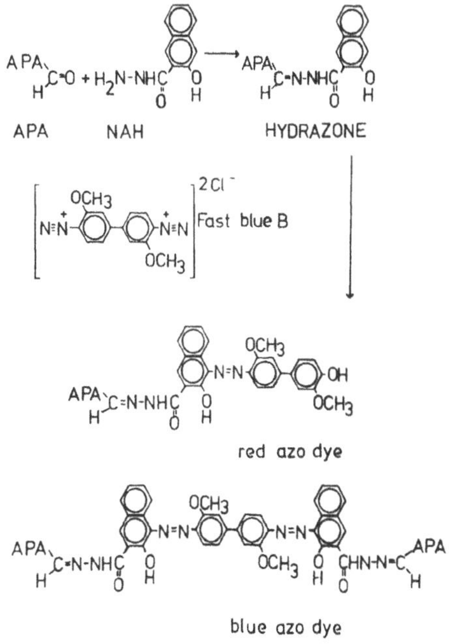

Figure 1. Reaction scheme of the NAH-FB-DNA staining. APA - apurinic acid; NAH = 3-hydroxy-2-naphthoic acid hydrazide

397

The NAH-FB-DNA+HNA-Protein Double Staining: Theoretical Considerations

We aim to establish a reproducible DNA staining that could be combined with a reproducible protein staining. Therefore, we chose the histochemical demonstration of aldehydic groups generated during Feulgen hydrolysis using 3-hydroxy-2-naphtoic acid hydrazide (NAH) and Fast Blue B (FB) mainly due to the simple and clear chemical reaction, the hydrazone formation, followed by azocoupling of β-naphthol bound to the apurinic acid (*Fig. 1*). The second reason for using the NAH-FB-DNA staining has been the formation of a stable insoluble dye that withstands subsequent staining for proteins. Furthermore, this dye has spectral properties enabling separate simultaneous measurements of both DNA and proteins of double stained objects. Unfortunately, Feulgen hydrolysis which is necessary for the generation of aldehydic groups, could not avoided. Moreover, formaldehyde fixation could not be used. Feulgen hydrolysis per se makes all DNA staining methods based upon it open to discussion. Theoretically, Feulgen hydrolysis should generate the so-called apurinic acid state of DNA, where all purine bases are split off and an equivalent of reactive aldehyde groups is available. The kinetics of all the processes occurring during Feulgen hydrolysis are unknown. A huge body of literature exists on Feulgen hydrolysis.[32-48] From this, it is evident that three main processes are occurring in parallel but with different velocities: the rapid hydrolysis and loss of RNA, the moderate hydrolysis of purine bases, and the slow decomposition and loss of DNA macromolecules. The result is the typical shape of the Feulgen hydrolysis curve.[49] The velocities of the different hydrolytic processes depend on several factors, e.g. conditions of hydrolysis, the ratio eu- to heterochromatin, nuclear proteins and especially on the stability of the nuclear matrix. Different kinds of nuclear matrices, supposed to cause the so-called proportionality error,[36-38,45,50-52] are stabilized before Feulgen hydrolysis by crosslinking with formaldehyde.[34] Formaldehyde fixation is recommended for optimum Feulgen staining. However, it is not clear if it reduces the proportionality error or if it yields the proper three-dimensional reaction conditions necessary for optimal Feulgen staining as suggested by Hörmann et al.[53] In any case, formaldehyde fixation or any other aldehydic fixative, has to be avoided for NAH-FB-DNA staining. NAH also reacts with aldehydes bound to proteins[54,55] and thus yields a considerable unspecific background staining. All fixatives other than aldehydes may be used for NAH-FB-DNA staining. We preferentially use either methanol or ethanol-ether (1:1) for fixation of cell smears or freshly frozen tissue sections. It should be emphasized that the most important restriction of the NAH-FB-DNA method is caused by the impossibility to apply it for formalin fixed and paraffin embedded material. The results obtained after NAH-FB-DNA staining may be criticized due to the need of Feulgen hydrolysis and the fixation method required. On the other hand, however, Feulgen hydrolysis curves obtained with different cells and tissues, fixed with ether-ethanol and stained with the NAH-FB-DNA method, indicated that the kinetics of hydrolysis are independent of the extent of chromatin condensation, the state of differentiated and proliferating cells, and the ratio between eu- and heterochromatin[56] (G. Nöhammer, unpublished data). Our data suggest that Feulgen hydrolysis, performed with alcohol fixed cells or freshly frozen tissues, might lead to a comparable generation of aldehydes as well as to a similar loss of decomposed DNA; this would amount to approximately 37% of total DNA at optimum hydrolysis time.[54] In other words, for the present, we rely on the results obtained after optimized NAH-FB-DNA staining[56] but keep in mind all information about Feulgen hydrolysis and the proportionality error in order to avoid misinterpretation and overestimation.

The investigation of changes of DNA during the cell cycle, cancer progression, during differentiation, dedifferentiation, cell growth and cell death, usually implicate measurements on many samples prepared and stained at different times. A DNA staining that is not absolutely reproducible needs a DNA standard, e.g. Feulgen staining, DNA methods used for flow cytometry. Haploid cells like sperm cells or diploid cells, lymphocytes, chicken erythrocytes, squamous epithelial cells, are frequently used as DNA standards. An ideal DNA standard should have a close distribution of its DNA-values, should indicate clearly the 2C-value and, furthermore, should eliminate the proportionality error. Who is able to standardize such a standard? Standardization is based upon the assumption of distinct constant DNA values for distinct kinds of differentiated, nondividing cells. This is a dogma, a hypothesis, no more. One way out of this dilemma is the use of an absolutely reproducible DNA staining method. Then, all the results obtained with this method could be compared directly without the need of any standard. This consideration has been the real reason for the optimization of the histochemical DNA staining using NAH and FB, a method published originally by Ashbel and Seligman[57] and used for DNA staining by Pearse.[58,59] The mechanism of the histochemical reaction with NAH and FB was simple and the reaction products generated had the expected special properties, e.g. known formula, insolubility and stability against a subsequent protein staining and no interference -- either chemically or spectroscopically -- with the subsequent protein staining. The mixture of two azodyes generated after coupling of the 3-hydroxy-2-naphtoic acid hydrazone with Fast blue B exhibited these properties (*Fig. 1*).

The optimization of the original method of Pearse[58,59] for DNA staining affected mainly three steps of the histochemical reaction, i.e. the acid catalysis of the hydrazone formation, the removal of excess NAH, and the conditions of specific azocoupling. The original method suggested acetic acid as a catalyst for hydrazone formation. We found p-toluene sulfonic acid (p-TSA) to be the best catalyst. It is a strong organic acid and improves also the solubility of NAH in the reaction medium used, e.g. 50% ethanol. The equilibrium of the hydrazone formation is adjusted after one hour reaction time at room temperature. Then excess NAH has to be removed from samples. This can be accomplished by two changes of acetone, supplemented with 20 to 30 mg p-TSA/100 ml acetone, followed by three changes of pure acetone. p-TSA competes with acid groups located intracellularly and retaining excess hydrazide by electrostatic forces. Pure acetone finally removes both NAH and p-TSA prior to azocoupling. As suggested originally by Pearse[58,59] we also used Fast blue B for azocoupling. All other coupling reagents tested -- mono- and dicoupling reagents -- either generated lower staining intensities or higher background absorption. One of the most important things to be considered with azocoupling is the pH of the buffer used as a coupling medium. Fast blue B, like other coupling reagents, reacts preferentially and quickly with β-naphthols but also with some amino acids of proteins.[60-65] The unspecific side reaction of Fast blue B with proteins yields a yellowish background staining the intensity of which depends on the pH of the buffer used for azocoupling.[56] At pH 6.5, using 0.1 M phosphate buffer background staining due to azocoupling of proteins reaches its minimum. Fast blue B as a dicoupling reagent yields a mixture of two dyes, a red mono- and a blue dicoupling product (*Fig. 1*). The result is a common and broad absorption maximum that ranges from 590 to 520 nm with the definite maximum at 550 nm. The generation of two dyes might be obstructive to the quantification of the NAH-FB-DNA staining.[54] Our results were obtained with different concentrations of aldehydes that were located intracellularly. They revealed the independence of the composition of reaction products generated under optimum histochemical conditions on the concentration of aldehydic groups to be demonstrated.[54] Thus, we have been able to define the molar absorptivity of the mixture of dyes generated after NAH-FB staining ($\epsilon 550=11094$).[54] It should be stressed that this ϵ-value is no true extinction coefficient in the sense of Beer's law. It is a factor, as is the ϵ-value, which applies to the histochemical conditions of the optimal NAH-FB-DNA staining. Using this factor, the knowledge that 37% of DNA becomes lost during Feulgen hydrolysis at the conditions used (alcohol fixation, 5 M HCl 35 min at 20°C), and the fact that according to Britten and Davidson[66] 9.13×10^8 nucleotide pairs are equivalent to 10^{-12}g DNA, which in turn is equivalent to 1.526 x 10^{-15} moles of aldehydic groups, generated theoretically by Feulgen hydrolysis (apurinic acid state), we could demonstrate a close correlation between the mean integrated extinctions obtained microphotometrically after NAH-FB-DNA staining and published DNA contents of different cells, determined mainly by methods of analytical biochemistry.[54] This correlation, together with experiments performed to check the reproducibility of the optimized NAH-FB-DNA staining of different cells, convinced us to perform serial investigations on cell preparations and tissue sections without using any DNA standard and to compare our data directly.

All the work performed and published in the past, using DNA-protein double staining, independent of the methods of measurement used, e.g. microphotometry or flow cytometry, indicated the necessity to correlate both parameters for an improved interpretation of DNA values. From incipiency, the simultaneous quantification of both DNA and proteins of distinct objects proved to be indicative of the state and kind of both proliferation and differentiation of cells under investigation. In addition, all important findings made on this subject, in histochemistry, molecular biology and cell biology during the last five decades point in that direction. Therefore, it has also been our aim to develop a DNA-protein double staining method to enable a two-dimensional analysis of cellular events. The difference from other comparable methods is the reproducibility of both methods used in DNA-protein double staining. The final goal has been the direct comparison of data obtained from serial measurements, independent of place and time of preparation and measurement.

Theoretically, the NAH-FB-DNA staining could be used directly for DNA-protein double measurements. As discussed above, Fast blue B not only couples with β-naphthols bound as hydrazones to apurinic acid, but, to a reproducible extent, it also couples with some amino acids of proteins. The yellowish unspecific background staining generated could, therefore, be used principally for quantification of proteins (Nöhammer, unpublished results). Unspecific azocoupling of proteins yields no absorption maximum and, furthermore, with respect to DNA staining, it should be reduced to a minimum. Strong background staining with Fast blue B, obtained at higher pH of buffers used for azocoupling, reduces the contrast of DNA staining. It proved to be unstable, tending to generate nitrogen bubbles caused by slow decomposition of excess diazonium groups. In any case, the yellowish background staining after the NAH-FB-DNA reaction increasing continually with decreasing wavelength, does not interfere with the specific maximum absorption of the DNA staining.[56] It is a weak staining with an absorption around 400 nm. We, therefore, looked for a protein staining that would not interfere, either chemically or spectroscopically, with the NAH-FB-DNA staining. This protein staining should work under mild reaction conditions and should have an absorption maximum in the range of minimum of the DNA staining without any absorption at 550 nm -- the absorption maximum of the DNA staining.

A method for histochemical demonstration of protein-bound amino groups has been published by Weiss et al.[66] The authors used 3-hydroxy-2-naphthaldehyde because of the possibility of azocoupling and the generation of a visible reaction product. Instead of the expensive 3-hydroxy-2-naphthaldehyde, we decided on the more economic 2-hydroxy-1-naphthaldehyde (HNA), which cannot be azocoupled. HNA yields a strong, steep double absorption maximum at 420 and 400 nm, respectively.[67] Under the conditions used for optimal HNA reaction,[67] specific primary amino groups of proteins are left for histochemical reaction. Only these primary amino groups can form Schiff-bases which yield the double maximum at 420 and 400 nm. Principally, aldehydes can react with other nucleophilic groups too, e.g. secondary amino groups and thiols.

Since HNA specifically demonstrates primary amino groups of proteins, this method can be used for protein staining. This assumption is confirmed by the correlation between micro-photometrically determined extinction values with HNA stained sections of different tissues and the corresponding protein values of these sections as determined by Lowry's method.[68,69] The interpretation of data obtained after HNA staining, however, should consider the fact that it is a staining of primary amino groups of proteins.

Two different staining procedures using HNA have been optimized, one with HNA dissolved in absolute ethanol; the other with HNA, dissolved in a mixture of 60 ml 95% ethanol and 40 ml 0.1 M acetate buffer at pH 4.0 (HNA-pH4-1d staining). The reaction time at pH 4.0 is 1 day at room temperature.[66] Both HNA methods proved to be reproducible, but only the HNA-pH4-1d method has been used for DNA-protein double staining. The reasons have been the favorable kinetics of the reaction at pH 4, as well as the possibility of using the HNA-pH4-staining solution repeatedly.

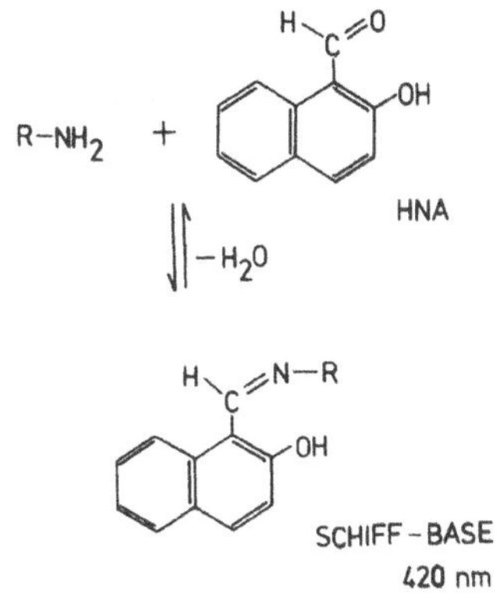

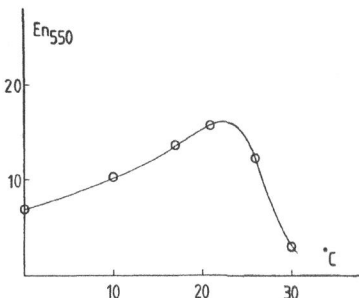

Figure 2. Reaction scheme of the HNA-protein staining. HNA = 2-hydroxy-1-naphthaldehyde

The NAH-FB-DNA+HNA-Protein Double Staining: Practical Considerations

DNA-protein double staining is performed by transferring the fixed preparations from the fixative directly to 5 M HCl for 35 min at 18 - 20°C. Care should be taken that Feulgen hydrolysis is performed at optimum temperature. Especially higher temperatures of non air-conditioned labs during summer impaired the results of NAH-FB-DNA staining (*Fig. 3*).

Figure 3. Dependence of the NAH-FB-DNA staining on temperature of preceding Feulgen hydrolysis (5 M HCl; 35 min). abscissa: Temperature (°C); ordinate: mean integrated nuclear extinctions of NAH-FB-DNA stained Ehrlich ascites tumor cells (En 550)

The use of glass slides precoated with 3-(triethoxysilyl)-propylamine to prevent loss of cells or tissues during Feulgen hydrolysis is recommended. After removal of excess HCl, the slides are put into NAH-solution for 1 h at room temperature. Excess NAH is removed then with acetone + p-TSA and pure acetone. The medium has to be changed prior to azocoupling, performed with Fast blue B freshly dissolved in 0.1 M phosphate buffer at pH 6.5. 10 min of azocoupling. It is followed by rinsing with tap water and an ascending series of 30%, 50% and 70% ethanol 1 min each to adjust the DNA stained cells for the subsequent HNA-pH4-1d reaction. For this purpose, the slides are transferred to the HNA-pH4 staining solution. After 1 day of HNA reaction, excess HNA has to be removed by 100% ethanol and the remaining ethanol by xylene. The double stained preparations, still wet from xylene, are embedded using Merckoglass.

Microphotometric measurement of nuclear DNA and protein is performed by successive scanning of nuclear areas at 550 nm and 420 nm, the absorption maxima of DNA and protein staining. Prior to every measurement at a distinct maximum wavelength, the blank value for 100% transmission had to be determined. A SMPO3-microphotometer (C. Zeiss, Oberkochen, FRG) was used for meandric scanning using a circular measuring diaphragm of 1 μm diameter and a step length and a line distance of 1 μm, respectively. The scanned nuclear areas were rectangular fields. The dimensions of these measuring fields were determined by the nuclear diameters in x- and y-direction. Care was taken that the measured spot did not exceed nuclear diameters in order to reduce unspecific cytoplasmic absorption. The integrated nuclear extinctions obtained microphotometrically at 550 and 420 nm, were used without any corrections as DNA and protein values. Usually, these uncorrected integrated nuclear extinctions (En) were plotted to yield two-dimensional DNA-protein scattergrams with DNA values (En550) on the ordinate and protein values (En420) on the abscissa. According to Van der Ploeg et al,[25] it would be possible, at least theoretically, to calculate the correct DNA values and protein values. Towards this end, the spectral properties of both individual stainings -- the NAH-FB-DNA staining and the HNA-pH4-1d staining -- have to be determined for every kind of cells or tissue measured. The true total integrated nuclear extinctions could be calculated from the integrated nuclear extinctions, measured at 550 nm (En550) and at 420 nm (En420), after DNA-protein double staining by the following formulas (Van der Ploeg et al.[25]):

$$En550 = En550, true + b.En420, true$$
$$En420 = En420, true + a.En550, true$$

En550 and En420 = total integrated nuclear extinctions measured at 550 nm and 420 nm, respectively.

En550, true and En420, true = true (corrected) integrated nuclear extinctions at 550 nm and 420 nm, respectively.

a = the percentage of En550, true measurable at 420 nm.
b = the percentage of En420, true measurable at 550 nm.

a and b have to be determined from dual-wavelength measurements with both NAH-FB-DNA single stained and HNA-pH4-1d single stained preparations. With these values of a and b, the true integrated extinctions could be calculated:

$$En550, true = (En550 - b.En420)/(1-ab)$$
$$En420, true = (En420 - a.En550)/(1-ab)$$

We did not use the method of Van der Ploeg et al.[25] for several reasons: 1) Our dual-wavelength measurements performed on different cells, after single staining, revealed some inconstancy in the extinction increments (a and b) within distinct cell types and between different cell types; 2) for the calculation of the true integrated nuclear extinctions, unspecific light loss that was caused by the object (e.g. light scattering, refraction, defraction, etc.) had to be considered for both wavelengths, as well as the percentage of unavoidable and inconstant cytoplasmic extinctions; 3) in the case of the NAH-FB-DNA+HNA-protein double staining, we had to consider two different protein stainings measured at 420 nm. Microphotometry is performed at the absorption maximum of HNA staining but, as discussed above, azocoupling with Fast blue B also yields some (unspecific) protein staining contributing to the integrated nuclear extinction at 420 nm. Summarizing all the obstacles that were provided by the special methods used for double staining and microphotometry prevented the exact calculation of nuclear DNA and protein values. Our experience, obtained from numerous measurements performed with double stained nuclei of different cells and tissues, taught us that direct information, the uncorrected integrated nuclear extinctions, is indicative enough to reveal changes in proliferation, growth and differentiation.

Proliferation and Differentiation in Terms of Nuclear DNA and Protein Values

To investigate the differences that characterize and discriminate proliferating and differentiated cells in terms of integrated nuclear DNA extinctions and nuclear protein extinctions, we used different models:

Normal squamous epithelium of freshly frozen, fixed and DNA-protein double stained 10 μm sections of the human uterine cervix. Normal squamous epithelium (nEp) is built up by four different layers of cells. The proliferating basal cells (BC) divide to generate parabasal cells (PbC). Parabasal cells are already differentiated cells which usually no longer divide but grow and differentiate further to become intermediate cells (IC). Superficial cells (SC) form the upper layers of nEp. Cell death is the fate of SC. Thus, within normal squamous epithelium both proliferating cells as well as cells at different stages of differentiation can be discriminated and investigated separately. Fig.4a shows a typical DNA-protein scattergram obtained from a DNA-protein double stained normal squamous epithelium of a section from human uterine cervix. Plotted are the integrated nuclear extinctions of basal cells, parabasal cells and intermediate cells measured at 550 nm (En550; DNA; y-axis) and 420 nm (En420; protein; x-axis). The DNA-protein scattergram of proliferating basal cells (symbolized by crosses) is shaped conically. Such conically shaped scattergrams have been found with all proliferating cells investigated until now. The origin of the cone is located at an En550 (DNA)-value of three. Data obtained from different DNA-protein double stained cells indicate that diploid cells seem to start DNA synthesis from a En550 (DNA)-value of six, tetraploid ones from 12, etc. (Nöhammer, unpublished results). Thus, a value of En550 = 3 should indicate a haploid stem cell. The numerous DNA-protein scattergrams obtained from proliferating cells, -- normal as well as tumor cells --, all shaped conically, suggest defining the stem cell of a cell population by using the integrated nuclear DNA and protein extinction values of the origin of the corresponding cone. The nuclei of these stem cells are small and inconspicuous. Proliferating cells can be discriminated from comparable differentiated cells by their higher ratio of DNA to nuclear proteins. Differentiation, on the other hand, seems to be characterized by an increase of nuclear proteins. Increasing differentiation is accompanied by a decreasing DNA:protein-ratio. Cell death (superficial cells not shown in *Fig. 4a*) is accompanied by a loss of both DNA and nuclear proteins. Nuclear pyknosis frequently leads to a loss of nuclear proteins only.

Figure 4b illustrates the distribution of the integrated nuclear extinctions measured at 550 nm (DNA value). The same values are used for the DNA-protein scattergram (*Fig. 4a*). The DNA histogram demonstrates that a mere DNA analysis might lead to some misinterpretation. Thus, a measured DNA value, e.g. En550=13, can belong to a proliferating basal cell and also to differentiated parabasal and intermediate cells. The analysis of DNA protein scattergrams provides better discrimination between proliferating and differentiated cells. In this context, it should be stressed that intermediate cell nuclei are used frequently as DNA standards. Figs. 4a and 4b illustrate the broad distribution of the DNA values of IC-nuclei ranging from haploid to hypertetraploid. The nEp analyzed was taken from a healthy donor. Until now, using the NAH-FB-DNA method, we have been unable to find any cell type which could be used as a reliable DNA standard.

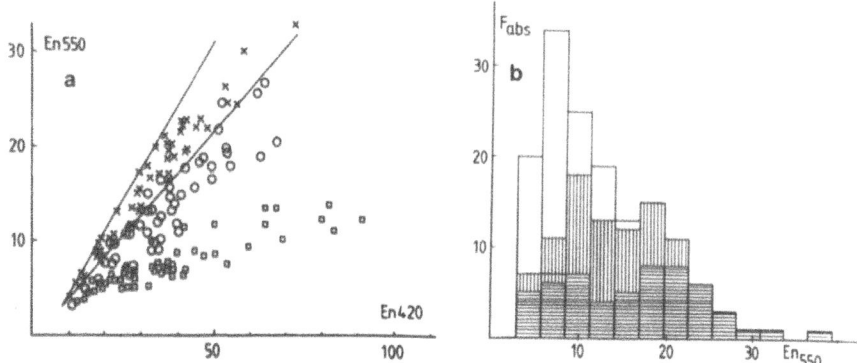

Figure 4a. DNA protein scattergram of NAH-FB-DNA + HNA-protein double stained normal squamous epithelium of human uterine cervix. abscissa: nuclear integrated extinctions measured at 420 nm (En 420; protein) ordinate: nuclear integrated extinctions measured at 550 nm (En 550; DNA) symbols: crosses - basal cell nuclei; circles - parabasal cell nuclei; squares - intermediate cell nuclei
Figure 4b. Frequency distribution of En550-values of DNA protein double stained nuclei of normal squamous epithelium (Figure 4a). Fabs: absolute frequency symbols: horizontal lines - basal cell nuclei; vertical lines - parabasal cell nuclei; open bars - intermediate cell nuclei.

Rat Liver Parenchymal Cells. Fig. 5a shows the DNA-protein scattergram of nuclei of liver parenchymal cells of an adult rat. The cells were isolated from rat liver using the method of Berry[70] and Friend modified by Seglen,[71] fixed with ether-ethanol and double stained for DNA and proteins. Within the stained cells, nuclear integrated extinctions were measured microphotometrically at 550 nm and 420 nm, respectively, and plotted. Liver parenchymal cells of adult rats are highly differentiated. Liver cells partly exhibit extremely low DNA:protein ratios. As shown above with squamous epithelium, cellular differentiation is accompanied by an increase of nuclear proteins relative to DNA. Within the population measured, no proliferating cell could be detected. The DNA-protein scattergram clearly allows discrimination between di-, tetra- and octoploid cells.

On the contrary, the corresponding DNA histogram (*Fig. 5b*) could lead to some misinterpretation. Thus, a DNA-value of En550 = 12 could belong to a diploid cell with high nuclear protein content or a tetraploid cell with low nuclear proteins. Only DNA-protein analysis enables a discrimination. Again, a fairly broad distribution of the DNA values of liver cells of a distinct ploidy is revealed by the DNA histogram (*Fig.5b*).

Growing Rat Liver.   Growing rat   liver might   be used   as a   model to   study differentiation. It is known that liver parenchymal cells,   starting with birth,   not only increase their number and volume but also the state of differentiation   and   polyploidy. We isolated rat liver parenchymal cells at different times after   birth,   fixed the   cells and stained them for DNA and proteins. Figs. 6a, 6b and 6c   illustrate the integrated nuclear extinctions measured successively, at 550 nm and 420 nm within the double stained cells; they were isolated at 1 day, 28 days, and 120 days after birth, respectively. One day after birth, the population of cells consisted mainly of diploid resting cells. These cells have low nuclear protein values. More than 10% of the cells still were cycling. Twenty-eight days after birth, no cycling cells could be found. Diploid and tetraploid cells increased nuclear proteins. Binucleation (symbolized by squares) of diploid cells could be observed. One hundred and twenty days after birth, when liver cells are known to be fully differentiated, all cells -- di-, tetra- and octoploids -- exhibited high nuclear protein values. Binucleation of all diploids and of many tetraploids could be observed.

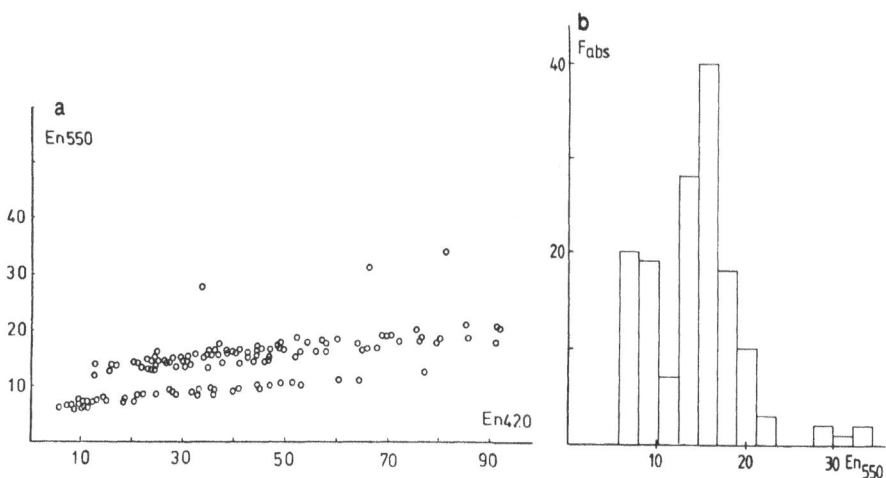

Figure 5a. DNA protein scattergram of NAH-FB-DNA + HNA-protein double stained liver parenchymal cells of an adult rat.
Figure 5b. Frequency distribution of EN550-values of DNA protein double stained rat liver parenchymal cells (*Fig. 5a*)

Fig. 6d illustrates the changes of both mean nuclear DNA and protein values with increasing time after birth. Increasing differentiation of rat liver cells with increasing age is accompanied by an increase of both DNA and nuclear proteins. Nuclear DNA increases due to polyploidization. Fig. 6e illustrates the growth of rat liver cells after birth. The period between day 14 and day 77 is characterized by an exponential increase of total cellular proteins (Ec420). Logically, growth is coupled with differentiation.

Dedifferentiation in Terms of DNA and Nuclear Proteins

In most cases, an unwelcome side effect of cell culture is the dedifferentiation of cells. To follow the process of cellular dedifferentiation, rat liver parenchymal cells were isolated from female Fischer 344 rats by the *in situ* two-step collagenase perfusion technique as described by Michalopoulos *et al.*[72] The isolated hepatocytes were plated at a density of 20,0000 viable cell/cm^2 on collagen-coated glass microscope slides in 90 mm diameter plastic culture dishes as described by Eckl *et al.*[73]

A 15 ml serum free MEM supplemented with non-essential amino acids and gentamicin (50 µg/ml) and 0.4 mM Ca^{++} was used. The plates were then incubated at 37°C, 5% CO$_2$ and 95% relative humidity. After 3 hours incubation, the medium was exchanged for freshly prepared medium, supplemented as described above.

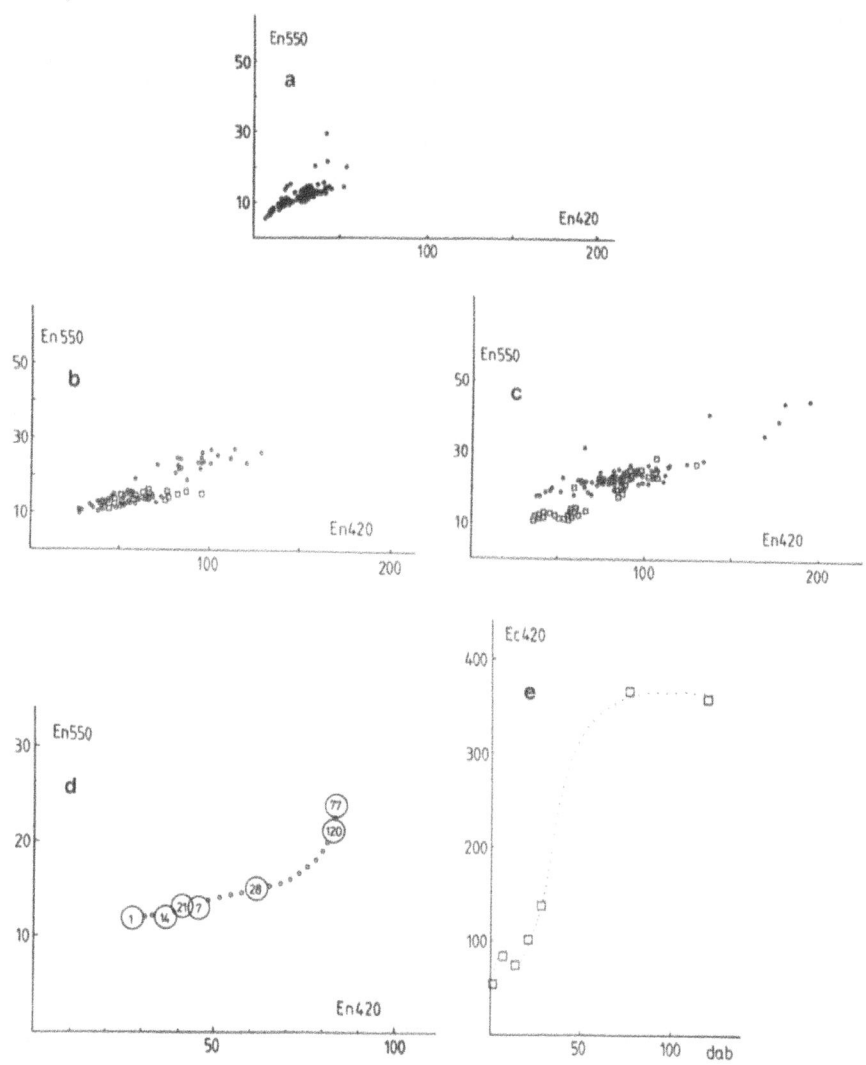

Figure 6a,b,c. DNA protein scattergrams of NAH-FB-DNA + HNA-protein double stained rat liver parenchymal cells isolated 1 day (6a), 28 days (6b), and 120 days (6c) after birth, respectively. Symbols: circles - mononucleated cells; squares - binucleated cells
Figure 6d. Time course of mean nuclear extinctions of DNA protein double stained rat liver parenchymal cells isolated 1, 7, 14, 21, 28, 77, and 120 days after birth respectively. Ordinate: mean integrated nuclear extinctions at 550 nm abscissa: mean integrated nuclear extinctions at 420 nm
Figure 6e. Time course of the mean integrated cellular extinctions measured at 420 nm (Ec420; mean cellular protein content) of rat liver parenchymal cells isolated 1, 7, 14, 21, 28, 77, and 120 days after birth (dab), respectively.

Cultured cells grown on glass slides were taken at distinct times in culture, fixed in methanol-glacial acetic acid (3:1) for 5 min and double stained for DNA and protein. Fig. 7a illustrates changes of both nuclear DNA- and protein values of rat liver parenchymal cells cultured for 3, 5, and 28 days, respectively.

Under the conditions used for cell culture, a small decrease of DNA and a marked decrease of nuclear proteins could be observed, the time course being illustrated in Fig.7b.

Dedifferentiation of cells is accompanied by a marked decrease of nuclear proteins. Finally, the cells gain a special state of dedifferentiation from which they can enter quickly into the S-phase, if stimulated by epidermal growth factor (EGF) (Figs.7c and 7d).

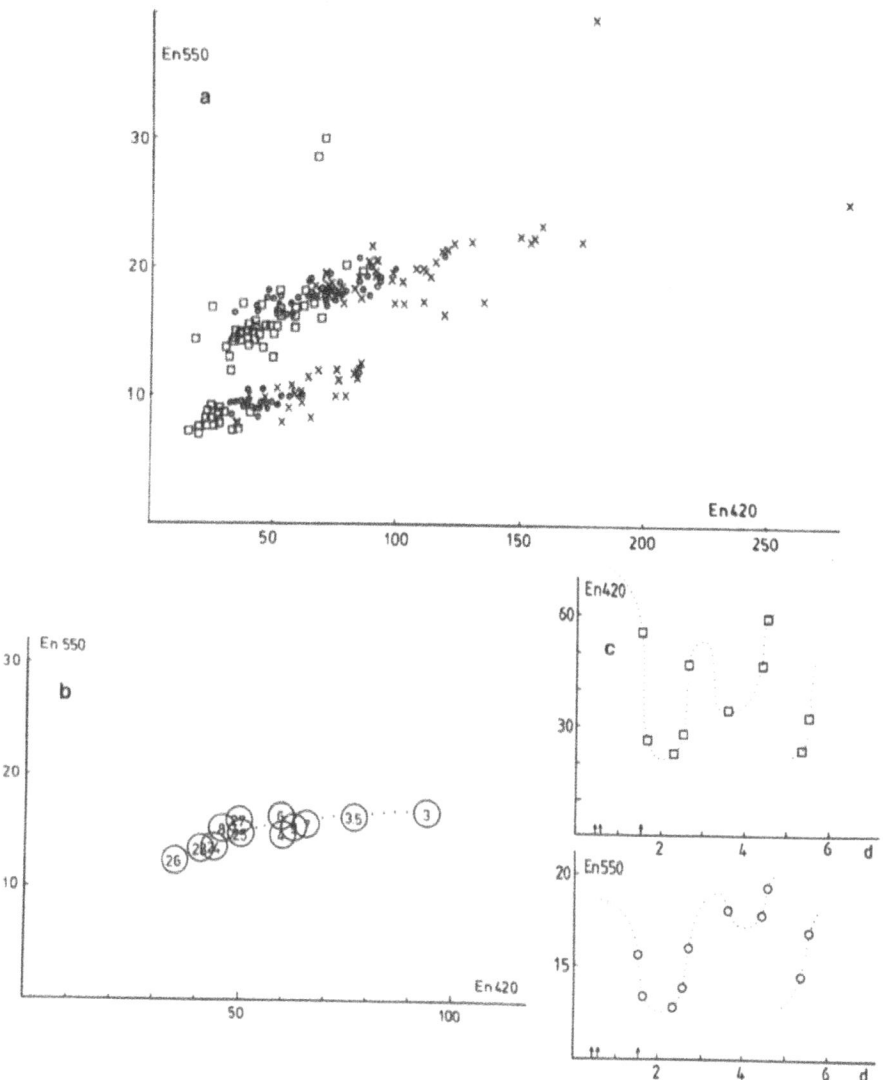

Figure 7a. DNA-protein scattergram of NAH-FB-DNA + HNA-protein double stained rat liver parenchymal cells after 3 hours (crosses), 5 hours (circles), and 28 hours (squares) of primary culture.

Figure 7b. Time course of mean nuclear extinctions of DNA-protein double stained rat liver cells during the period between 3 and 28 hours of primary culture. abscissa: mean integrated nuclear extinctions at 420 nm ordinate: mean integrated nuclear extinctions at 550 nm

Figure 7c. Time course of mean nuclear extinctions measured at 550 nm and 420 nm, respectively, of DNA-protein double stained rat liver cells in primary culture supplemented with EGF 1 day after perfusion. arrows on abscissa (from left to right): time of perfusion, of changing the medium, and of supplementation with EGF, respectively.

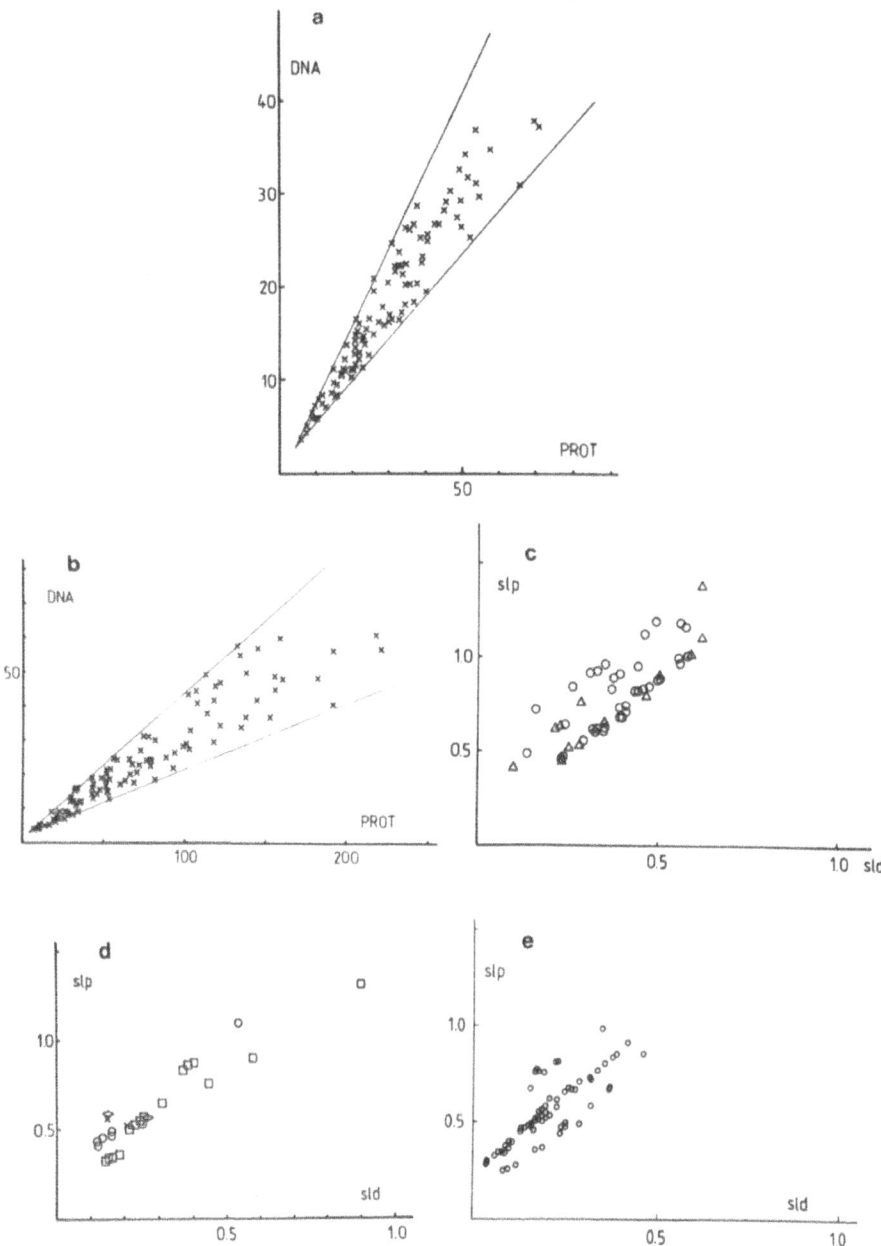

Figure 8. a) DNA-protein scattergram of a CIN III (carcinoma *in situ*). abscissa: PROT (integrated nuclear extinctions at 420 nm) ordinate: DNA (integrated nuclear extinctions at 550 nm); b) DNA-protein scattergram of a squamous cell carcinoma of the uterine cervix. c) Linear correlations between slp and sld of malignant tumors of human uterine cervix. symbols: triangles-moderate and severe dysplasias; circles-carcinoma *in situ*; d). Linear correlations between slp and sld of malignant tumors of human liver. symbols: circles-adenocarcinomas; squares-hepatocellular carcinomas; rhombs-cholangiocarcinomas; crosses-metastases; e) Linear correlations between slp and sld of malignant tumors of human breast.

The first mitoses were observed 1.5 days after the addition of EGF but after 3 days, many mitosis were seen. Fig. 7c demonstrates that proliferation induced by EGF is accompanied by an increase of both DNA and nuclear proteins. It is interesting to note that following mitosis, the cells start proliferation from the same undifferentiated state as gained during the course of dedifferentiation during primary culture without EGF-supplementation. This state should be the G_1-phase and is characterized by special low values of both DNA and nuclear proteins.

Classification of Tumors Using DNA-protein Scattergrams

As demonstrated above, proliferating normal basal cells of human squamous epithelium yielded conically shaped DNA-protein scattergrams (Fig. 4a). In addition to samples from normal tissue of the human uterine cervix, freshly frozen, fixed and DNA-protein-double stained, sections from tissue of patients suffering from different stages of cancer of the human uterine cervix have also been prepared and investigated microphotometrically.

Figs. 8a and 8b illustrate typical DNA-protein scattergrams obtained from CIN III and from an invasive carcinoma. We investigated 49 invasive carcinomas, 37 carcinomas *in situ* (CIN III) and 13 moderate to severe dysplasias. The DNA-protein scattergrams obtained from all these malignant tumors were conically shaped. Looking for a parameter to provide a general characterization of all these different scattergrams, we calculated the slopes of the two lines bordering the cones. The slope of the left line bordered the values with the highest DNA: protein ratios and has been termed "slope of proliferation" (slp). The slope of the right line, bordering the values with the lowest DNA: protein ratios, has been termed "slope of differentiation" (sld). This designation reflects our findings on normal squamous epithelium. The values calculated for both slp and sld do not reflect the proliferation potential or state of differentiation of cells. Fig. 8c illustrates the plot of the paired values of slp and sld as calculated from conically shaped DNA-protein scattergrams obtained from dysplasias (symbol-triangles) and carcinomas *in situ* (symbols-circles). The tumors investigated obviously split into three subgroups characterized by their special linear correlation between slp and sld. Practically, the same three subgroups were obtained with the invasive carcinomas investigated (data not shown). The subdivision of tumors into three groups does not seem to be restricted to tumors of the uterine cervix. Preliminary results obtained from DNA-protein double stained sections of both human liver and human breast yielded a similar (liver) and even the same (breast) subdivision of malignant tumors as observed with tumors from the uterine cervix (Figs. 8d and 8e).

The "Field Effect" of Malignant Tumors

A malignant tumor can be accompanied by characteristic tumor-associated changes of adjacent, apparently normal, tissue. Tumor-associated changes of apparently normal tissue in the neighborhood of a malignant tumor have been termed "field effect" of a tumor.[72-77] Tumor-associated changes observed on normal skin and distant from their primary site tumors have been termed "extended field effect".[75,76] We investigated normal squamous epithelium (nEp) of human uterine cervix from normal patients without tumors. We have also studied the apparently normal squamous epithelium (anEp) in patients suffering from neighboring tumors of the human uterine cervix. As mentioned above, the goals have been to investigate both proliferation and differentiation of epithelial cells, and also to search for tumor-associated changes of both proliferation and differentiation of anEp. Tumors may induce such changes on adjacent normal tissue by paracrine mechanisms. DNA-protein scattergrams from both nEp and anEp were analyzed. The DNA-protein values of proliferating basal cells (*Fig. 4a*), yielded conically shaped scattergrams. Again, the slopes of both the left and the right line bordering the DNA-protein scattergram of basal cells were determined.

Fig. 9 illustrates the linear correlations of both paired values of slp and sld of basal cells in healthy donors and in patients suffering from a malignant tumor of the uterine cervix, located in the neighborhood of the anE$_p$ investigated. Obviously, an adjacent tumor is accompanied by a characteristic change of the linear correlation between slp and sld of DNA-protein scattergrams of basal cells. It is a change of proliferation of apparently normal cells. Interestingly enough, the same correlation as found between slp and sld adjacent to tumors characterizes the first subgroup of tumors (*Fig. 8c*).

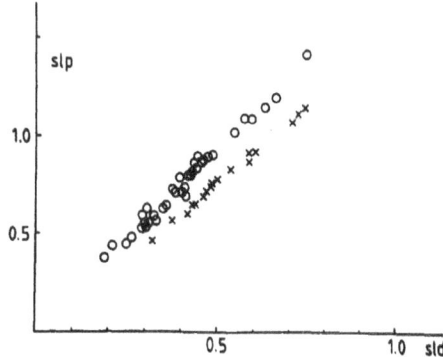

Figure 9. "Field effect" of malignant tumors; linear correlations between slp and sld of basal cells of normal squamous epithelium of healthy donors (crosses) and of apparently normal squamous epithelium adjacent to malignant tumors of human uterine cervix (circles).

OPTIMAL PROCEDURE

The Histochemical DNA-Protein Staining (NAH-FB-DNA+HNA-Protein Method)

Fixation: Cell smears of freshly frozen tissue sections are immediately fixed with either methanol or ethanol-diethylether (1:1) for at least 15 min. Prolonged fixation performed at 4°C (refrigerator) does not impair the results of DNA-protein double staining.

Feulgen-Hydrolysis

Transfer the preparations from fixative to 5M HCl. Hydrolysis is performed at 18-20°C for 35 min. Excess HCl is removed by 4x1 min washes in a mixture of 50 ml 95% ethanol and 50 ml distilled water, supplemented with a trace amount of EDTA (this mixture is stirred prior to use with a glass rod which removes dissolved oxygen).

Hydrazone Formation

After removal of excess HCl, the hydrolyzed preparations are immediately transferred to the NAH-solution [202 mg 3-hydroxy-2-naphthoic acid hydrazide (NAH; Sigma No: H-3627)] and, together with 190 mg toluene-4-sulfonic acid monohydrate (p-TSA; Merck No: 9613), are dissolved by stirring in 50 ml 95% ethanol and then diluted with 50 ml of distilled water. The reaction time in NAH-solution is 1 hour at room temperature.

Afterwards, excess NAH is removed by a 5 min wash in acetone + p-TSA (100 ml acetone, supplemented with 10 to 30 mg p-TSA), followed by 2x10 min acetone + p-TSA and finally by 2x5 min washes with acetone to remove the p-TSA. The slides are then put into distilled water for 3 min to adjust the medium for subsequent azocoupling.

Azocoupling

The slides are transferred to a freshly prepared solution of 100 mg Fast blue B (FB; Merck No: 3191) in 100 ml 0.1 M phosphate buffer pH 6.5 for 10 min. Azocoupling is performed at room temperature. Excess FB is removed by rinsing for 5 min with tap water. To adjust the DNA-stained preparations for subsequent protein staining, the slides are put into 30%, 50% and 70% ethanol for 1 min.

Protein Staining

The NAH-FB-DNA stained slides are transferred from 70% ethanol into HNA-pH4 solution [344 mg 2-hydroxy-1-naphthaldehyde (HNA; Merck-Schuchardt No: 820680) is dissolved by stirring in 60 ml of 95% ethanol and then diluted with 40 ml of 0.1M acetate buffer at pH 4.0]. The reaction time in HNA-pH 4 solution at room temperature is 1 day. Excess HNA is removed afterwards by 3x5 min washes in 95% ethanol.

Finish

Dehydration is performed by 2x1 min in 100% ethanol, followed by 1 min in xylene + ethanol (100%) and 3x1 min in xylene. The double stained preparations are embedded with Merckoglass (Merck No: 3973).

Special Comments on NAH-FB-DNA+HNA-Protein Double Staining Method

5M HCl used for Feulgen hydrolysis is prepared by mixing 100 ml fuming HCl with 144 ml of distilled water. Prior to use, the temperature has to be adjusted to 18-20°C. 5M HCl can be used repeatedly for Feulgen hydrolysis. The p-TSA used for dissolution of NAH in ethanol and for catalysis of the hydrazone formation frequently becomes humid. In this case, an additional 10-30 mg of p-TSA/100 ml NAH-solution helps to improve the dissolution of NAH in ethanol and does not impair hydrazone formation.

Merckoglas was used as a liquid coverglass. If used for embedding, evaporation of part of the solvent (toluene) by keeping the bottle open for a few hours is recommended.

Microphotometry

Microphotometric measurements were performed using a scanning microphotometer SMP 03 (C. Zeiss, Oberkochen, FRG). The SMP 03: Circular measuring diaphragm -- 1 µm diameter; circular iris diaphragm --to 26 µm diameter; step width and line distance of meandric scanning -- to 1 µm, respectively; the objective - to UF 100/1.25; the slit of the monochromator -- to 0.2 mm; computer program used - APAMOS 3T (C. Zeiss, Oberkochen, FRG).

Rectangular areas circumscribing the DNA-protein double stained cell nuclei are scanned to yield total integrated nuclear extinctions. Scanning is performed first at the extinction maximum of the NAH-FB-DNA staining at 550 nm, followed by scanning of the same nuclear area at 420 nm -- the absorption maximum of the HNA-protein staining. Prior to scanning at a distinct wavelength, the value for 100% transmission is adjusted at a point of the preparation free of any absorbing material in the vicinity of the measured nuclei. The size of the scanned rectangular nuclear areas was defined as the diameters of the nuclei measured in x- and y-direction. Care was taken that the measuring spot did not pass over these nuclear borders, in order to reduce unavoidable cytoplasmic absorption.

The microphotometrically obtained integrated nuclear extinction values at 550 and 420 nm were used directly and without any correction as the DNA- and protein values. Plotting of the integrated DNA-extinction values on the y-axis and the corresponding nuclear protein extinctions on the x-axis yielded DNA-protein scattergrams useful for further analyses.

DISCUSSION

The combination of two optimized and quantified methods -- the NAH-FB-DNA staining[54,56] and the HNA-protein staining,[67,69] -- was used for DNA-protein double staining to enable serial investigations on different samples, prepared and measured at different times. Only reproducible methods enable direct comparison of results. In addition, a reproducible DNA-staining does not require standardization. Experiments performed with reproducible methods enable the investigation of the time course of cellular processes, e.g. cell cycle traverse, differentiation, dedifferentiation, tumorigenesis and tumor progression. Small, subtle changes of nuclear DNA and proteins can be recognized exclusively by using reproducible methods. The simultaneous measurements of both DNA and proteins, mainly performed at the present time by flow cytometry, proved to be superior to DNA measurements. It is evident that cellular processes with specific changes -- of both DNA and nuclear or cytoplasmic proteins -- can be characterized by quantification of these parameters. For this purpose, a reproducible DNA-protein double staining method is recommended; both parameters should be determined simultaneously on one and the same object, to have access to processes occurring in single individual cells or subcellular structures. Our attempts, presented partly in this chapter, are based on the following working hypothesis: "Distinct ratios exist between DNA and nuclear proteins as characterizing distinct states of proliferating cells within the cell cycle, as well as distinct states of differentiation in G_o." If this hypothesis is correct, then the microphotometric quantification of both nuclear DNA and protein should enable cell-typization, the determination of the state of a cell within the cell cycle and the state of cellular differentiation.

Because both staining methods used for double staining have been quantified,[25,54,69] we never used the microphotometric integrated nuclear extinction values for the calculation of either the mass of DNA, or of proteins located within distinct nuclei. Too many obstacles would prevent the exact calculation and, furthermore, it would not be necessary. If an integrated nuclear extinction can be used for computation, this value must be proportional to the mass of the absorbing substance located within a nuclear area. Thus, for the investigation of changes of DNA and nuclear proteins caused by the above mentioned cellular processes, it proved to be sufficient to operate with uncorrected integrated nuclear extinction values.

The NAH-FB-staining requires a preceding Feulgen hydrolysis. Much more, hydrolysis has to be performed with cells or tissue sections fixed without aldehydes, e.g. formaldehyde or glutaraldehyde. The product generated by Feulgen hydrolysis is not well defined (although termed apurinic acid). The number of aldehydic groups liberated for histochemical reaction is said to be dependent on the kind of DNA and the nuclear matrix proteins. Feulgen hydrolysis remains a necessary evil and all the DNA staining methods based upon it are open to discussion. A typical example in this discussion is that of the so-called proportionality error. Another example is that of the usual interpretation of the fairly broad distribution of DNA values as determined microphotometrically with cells which theoretically should give a constant value characteristic for diploids. Such distributions are believed to be due mainly to differences in the nuclear matrix impairing Feulgen hydrolysis and, partly, to be caused by the Feulgen reaction itself. On the other hand, it should be kept in mind that Feulgen hydrolysis determines the specificity of the DNA staining methods requiring it.

Our results obtained with DNA-protein double stained cells and tissue sections may be used as arguments in favor of both the Feulgen hydrolysis and the DNA staining used. The first argument is the reproducibility of the method, on which all our serial investigations are based. If a DNA staining by Feulgen hydrolysis and simple chemical reaction for the aldehydes generated, proved to be reproducible with different kinds of cells and tissues (including all stages of proliferation and differentiation), the extent of hydrolysis should not depend on the kind of cells investigated. A second argument in favor of the method used consists of the results obtained with differentiated cells, e.g. intermediate cells of squamous epithelium (*Fig.4a*) and rat liver parenchymal cells (*Fig. 5a*). The plot of nuclear DNA and protein values reveals that the observed distribution of DNA-values of this kind of cells, frequently used as DNA-standards, is functionally correlated with that of corresponding nuclear protein values. Thus, we tend to interpret our data obtained with differentiated cells in such a way that differentiation is the escape of cells from the cell cycle, accompanied by a functional sequence of increases and decreases of both DNA and nuclear proteins which determine the state and kind of differentiation. A third argument in favor of Feulgen hydrolysis and the method used consists of the following:

1) Dedifferentiation stops when a cell reaches a state characterized by distinct low nuclear DNA and protein values. This could be the point at which cells enter the S-phase. Within the DNA-protein scattergram of double stained cells, this point is defined by a special value of the integrated nuclear DNA extinction, e.g. 6 for diploid cells and 12 for tetraploid cells. The corresponding nuclear protein values may depend on the special type of cell. Surprisingly, we could not find any measurable difference between these basic DNA-values of man, mouse and rat.

2) Malignant tumors of the human uterine cervix, breast and liver split into three distinct subgroups, if the paired values of slp and sld obtained from DNA-protein scattergrams of these tumors are plotted (*Figs. 8c, 8d, 8e*). Also, other tumors investigated, e.g. Ehrlich ascites tumor, Yoshida ascites tumor, and sarcoma 180 fit into these subgroups (Nöhammer, unpublished results).

3) The paired values of slp and sld obtained from DNA-protein scattergrams of basal cells of normal squamous epithelium of healthy donors are correlated linearly. Another linear correlation has been found between the paired values of slp and sld, obtained from DNA-protein scattergrams of basal cells of apparently normal squamous epithelium adjacent to malignant tumors of human uterine cervix (*Fig.9*)

CONCLUSION

The method used for microphotometric quantification of both DNA and protein -- the DNA-protein double staining described in this chapter together with scanning microphotometry -- might be used for cell typing, the determination of the state of a cell within the cycle as well as of the state of differentiation. The method might also be used to improve the interpretation of microphotometric DNA-values, to discriminate between proliferating and differentiated cells. A new kind of tumor classification might provide some contribution to tumor diagnosis and prognosis, although the meaning of the subgroups found is still unclear. The difference found between basal cells of normal squamous epithelium of healthy donors and of apparently normal squamous epithelium adjacent to malignant tumors of human uterine cervix could be used for tumor diagnosis. Tumors exert a "field effect", a tumor-associated change of morphologically inconspicuous tissue. The method used proved capable of revealing such subtle changes. In addition, the method can be used to follow changes associated with differentiation and dedifferentiation of cells in culture. In addition to the time necessary for DNA-protein double staining, another drawback of the method used is the low staining intensity, especially that of the NAH-FB-DNA-staining. 6 pgm DNA yielding an integrated nuclear extinction of 6. Therefore, a diploid nucleus with 100 μm^2 nuclear area has a mean extinction per unit area (1 μm^2) of 0.06, a faint staining. In this respect, the method claims an improvement. Furthermore, the interpretation of the results obtained with this method need to be compared with results obtained with other comparable double staining methods, especially with those that do not need Feulgen hydrolysis.

Acknowledgements

The authors (G.N. and W.R.) are grateful to the Association for International Cancer Research (AICR) for having supported this international collaboration.

References

1. A. Köhler, Mikrophotographische Untersuchungen mit ultraviolettem Licht, *Z wiss Mikroskp* **21**:273 (1904).
2. T. Caspersson, Über den chemischen Aufbau der Strukturen des Zellkernes, *Skand Arch Physiol* **73(Suppl 8)**:1 (1936).
3. T. Caspersson and J. Schultz, Nucleic acid metabolism of the chromosomes in relation to gene reproduction, *Nature* **142**:294 (1938).
4. T. Caspersson, "Cell Growth and Cell Function", Norton, New York (1950).
5. R. Feulgen and H. Rossenbeck, Mikroskopisch-chemischer Nachweis einer Nucleinsäure vom Typus der Thymonucleinsäure und die darauf beruhende elektive Färbung von Zellkernen in mikroskopischen Präparaten, *Z Physiol Chem* **135**:203 (1924).
6. C.R. Hoover and L.E. Thomas, Microspectrophotometric studies of the Feulgen and arginine histochemical reactions, *Proc Histochem Soc J Natl Cancer Inst* **12**:219 (1951).
7. R. Vendrely and C. Vendrely, Arginine and deoxiribonucleic acid content of erythrocyte nuclei and sperms of some species of fishes, *Nature* **172**:30 (1953).
8. W. Sandritter, D. Müller, and H. G. Schreiner, Über den Nucleinsäuregehalt und das Trockengewicht von haploiden und diploiden Zellen, *Verhandl Anat Ges (Jena)* **55**:146 (1958).
9. W. Sandritter and D. Kleinhans, Über das Trockengewicht, den DNS- und Histonproteingehalt von menschlichen Tumoren, *Z Krebsforsch* **66**:333 (1964).
10. A. Zetterberg, Synthesis and accumulation of nuclear and cytoplasmic proteins during interphase in mouse fibroblasts *in vitro*, *Exp Cell Res* **42**:500 (1966).
11. J. P. Mitchell, Combined protein and DNA measurements in plant cells using the dinitrofluorobenzene and Feulgen techniques, *J Microsc Soc* **87**:375 (1967).
12. J. Gaub, G. Auer, and A. Zetterberg, Quantitative cytochemical aspects of a combined Feulgen Naphthol Yellow S staining procedure for the simultaneous determination of nuclear and cytoplasmic protein and DNA in mammalian cells, *Exp Cell Res* **92**:323 (1975).
13. A.D. Deitch, Microspectrophotometric study of the binding of the anionic dye, Naphthol Yellow S, by tissue sections and by purified proteins, *Lab Invest* **4**:324 (1955).
14. R.M. Welsch and K. Resch, Application of Naphthol Yellow S cytophotometry in deoxyribonucleic acid and protein determinations on larval material of D. melanogaster, *J Histochem Cytochem* **11**:675 (1963).
15. J.P. Mitchell, M. Van der Ploeg, and P. van Duijn, Combined staining procedures for cytophotometry of protein and DNA Feulgen-Naphthol Yellow S and dinitrofluoro-benzene-Feulgen, *Histochemistry* **73**:211 (1981).
16. J. P. Mitchell, Combined staining of protein bound sulfhydryl groups and DNA in polyacrylamide model systems, *Histochem J* **7**:283 (1975).

17. M. Fukuda, K. Nakanishi, N. Böhm, J. Kimura, K. Harada, and S. Fujita, Combined proteins and DNA measurements by the ninhydrin-Schiff and Feulgen techniques, *Histochemistry* **63**:35 (1979).

18. P.S. Oud, J.B.J. Henderik, A.C.L.M. Huysmans, M.M.M. Pahlplatz, H.G. Hermkens, J. Tas, J. James, and G.P. Vooijs, The use of Light Green and Orange II as quantitative protein stains, and their combination with the Feulgen method for the simultaneous determination of protein and DNA, *Histochemistry* **80**:49 (1984).

19. D.P. Bloch and G.C. Godman, A microphotometric study of the synthesis of desoxyribonucleic acid and nuclear histone, *J Biophys Biochem Cytol* **1**:17 (1955).

20. H.S. Di Stephano, H.F. Diermeier, and J. Tepperman, Effects of growth hormone on nucleic acid and protein content of rat liver cells, *Endocrinology* **57**:158 (1955).

21. J. Woodward, Intracellular amounts of nucleic acids and protein during pollen grain growth in Tradescantia, *J Biophys Biochem Cytol* **4**:383 (1958).

22. J. Woodward, B. Gerber, and H. Swift, Nucleoprotein changes during the mitotic cycle in Paramecium aurelia, *Exp Cell Res* **23**:258 (1961).

23. E.S. Meek, A quantitative cytochemical study of chromosomal basic protein in static and proliferative cell population, *Exp Cell Res* **33**:355 (1964).

24. D.P. Bloch and G.C. Godman, Evidence of differences in the desoxyribonucleoprotein complex of rapidly proliferating and non-dividing cells, *J Physiol Biochem Cytol* **1**:531 (1955).

25. M. Van der Ploeg, K. Van der Broek, and J.P. Mitchell, Dual wavelength scanning cytophotometry (Bicoscan), *Histochemistry* **62**:29 (1979).

26. W. Göhde and W. Dittrich, Simultane Impulsfluorimetrie des DNS und Proteingehaltes von Tumorzellen, *Z Anal Chem* **252**:328 (1970).

27. H.A. Crissman and J.A. Steinkamp, Rapid, one step staining procedures for analysis of cellular DNA and protein by single and dual laser flow cytometry, *Cytometry* **3**:84 (1982).

28. H.G. Gratzner, Monoclonal antibody to 5-bromo- and 5-iododeoxyuridine: A new reagent for detection of DNA replication, *Science* **218**:474 (1982).

29. J. Gerdes, U. Schwab, H. Lemke, and H. Stein, Production of mouse monoclonal antibody reactive with a human nuclear antigen associated with cell proliferation, *Int J Cancer* **31**:13 (1983).

30. S. Bruno, H.A. Crissman, K.D. Bauer, and Z. Darzynkiewicz, Changes in cell nuclei during S-phase: Progressive chromatin condensation and altered expression of the proliferation-associated nuclear proteins Ki-67, Cyclin (PCNA), p105 and p34, *Exp Cell Res* **196**:99 (1991).

31. W.A.L. Duijndam and P. van Duijn, The interaction of apurinic acid aldehyde groups with pararosaniline in the Feulgen-Schiff and related staining procedures, *Histochemistry* **44**:67 (1975).

32. G.K.A. Andersson and P.T.T. Kjellstrand, Exposure and removal of stainable groups during Feulgen acid hydrolysis of fixed chromatin at different temperatures, *Histochemie* **27**:165 (1971).

33. G.K.A. Andersson and P.T.T. Kjellstrand, Influence of acid concentration and temperature on fixed chromatin during Feulgen hydrolysis, *Histochemie* **30**:108 (1972).

34. N. Böhm, Einfluß der Fixierung und der Säurekonzentration auf die Feulgen-Hydrolyse bei 28°C, *Histochemie* **14**:201 (1968).

35. W. Graumann, Zur Standardisierung des Schiffschen Reagens, *Z Wiss Mikrosk* **61**:225 (1953).

36. N. Böhm, E. Sprenger, G. Schlüter, and W. Sandritter, Proportionalitätsfehler bei der Feulgen-Hydrolyse, *Histochemie* **15**:194 (1968).

37. K. Noeske, Diskrepanzen von Feulgen-Wert und DNS-Gehalt, *Histochemie* **20**:322 (1969).

38. W.A.L. Duijndam and P. van Duijn, The influence of chromatin compactness on the stoichometry of the Feulgen-Schiff procedure studied in model films. II. Investigations on films containing condensed or swollen chicken erythrocyte nuclei, *J Histochem Cytochem* **23**:891 (1975).

39. T.B. Osbone and F.W. Heyl, The pyrimidine derivatives of nucleide acid, *Amer J Physiol* **21**:157 (1908).

40. C. Tamm, M.E. Hodes, and E. Chargaff, The formation of apurinic acid from the deoxyribonucleic acid of calf thymus, *J Biol Chem* **195**:49 (1952).

41. C. Tamm and E. Chargaff, Physical and chemical properties of the apurinic acid of calf thymus, *J Biol Chem* **203**:689 (1953).

42. P.S. Woods, A chromatographic study of hydrolysis in the Feulgen nucleal reaction, *J Biophys Biochem Cytol* **3**:71 (1957).

43. W. Sandritter, K. Jobst, L. Rakow, and K. Bosselmann, Zur Kinetik der Feulgenreaktion bei verlängerter Hydrolysezeit. Cytophotometrische Messungen im sichtbaren und ultravioletten Licht, *Histochemie* **4**:420 (1965).

44. J.J. Decosse and N. Aiello, Feulgen hydrolysis: Effect of acid and temperature, *J Histochem Cytochem* **14**:601 (1966).

45. N. Böhm and W. Sandritter, Feulgen hydrolysis of normal cells and mouse ascites tumor cells, *J Cell Biol* **28**:1 (1966).

46. G.K.A. Andersson, Histochemical properties of spermatozoa and somatic cells. III. Depolimerization and extraction of DNA during Feulgen acid hydrolysis, *Histochemistry* **48**:315 (1976).

47. J. Erenpreisa and S. Tsvetkova, Feulgen hydrolysis at refrigerator temperature, *Acta histochem* **63**:38 (1978).

48. J.A. Millet, O.A.N. Husain, L. Bitensky, and J. Chayen, Feulgen-hydrolysis profiles in cells exfoliated from the cervix uteri: a potential aid in the diagnosis of malignancy, *J Clin Pathol* **35**:345 (1982).

49. N. Böhm and H. U. Seibert, Zur Bestimmung der Parameter der Bateman-Funktion bei der Auswertung von Feulgen-Hydrolysenkurven, *Histochemie* **6**:260 (1966).

50. B.H. Mayall, Deoxyribonucleic acid cytophotometry of stained human leukocytes, I. Differences among cell types, *J Histochem Cytochem* **17**:249 (1969).

51. K. Noeske, Discrepancies between cytophotometric Feulgen values and deoxyribonucleic acid content, *J Histochem Cytochem* **19**:169 (1971).

52. H.J. Stutte, L. Schroeder, and M. Kikuchi, Cytophotometrische Bestimmung des Feulgen-DNS-Gehaltes von Zellen der roten Milzpulpa, I. Untersuchungen an normalen menschlichen Milzen, *Virchows Arch Abt B Zellpath* **6**:198 (1970).

53. H. Hörmann, W. Grassmann, and G. Fries, Über den Mechanismus der Schiffschen Reaktion, Justus Liebigs, *Ann Chem* **616**:125 (1958).

54. G. Nöhammer, Quantification of the histochemical staining for carbonyles and DNA using 3-hydroxy-2-naphthoic acid hydrazide and Fast blue B, *Histochemistry* **94**:485 (1990).

55. A. Pompella and M. Comporti, The use of 3-hydroxy-2-naphthoic acid hydrazide and Fast blue B for the histochemical detection of lipid peroxidation in animal tissues - a microphotometric study, *Histochemistry* **95**:255 (1991).

56. G. Nöhammer, Optimization of the histochemical demonstration of DNA using 3-hydroxy-2-naphthoic acid hydrazide and Fast blue B, *Histochemistry* **90**:465 (1989).

57. R. Ashbel and A.M. Seligman, A new reagent for the histochemical demonstration of active carbonyl groups. A new method for staining ketonic steroids, *Endocrinology* **44**:565 (1949).

58. A.G.E. Pearse, A review of modern methods in histochemistry, *J Clin Pathol* **4**:1 (1951).

59. A.G.E. Pearse, " Histochemistry. Theoretical and applied", 3rd. edn, Vol 1, Churchill, London (1968).

60. M. Clara and F. Canal, Histochemische Untersuchungen an den Körnchen in den basalgekörnten Zellen des Darmepithels, *Z Zellforsch* **15**:801 (1932).

61. J.F. Danielli, A study of techniques for the cytochemical demonstration of nucleic acids and some components of proteins, *Symp Soc Exp Biol* **1**:101 (1947).

62. M.S. Burstone, An evaluation of histochemical methods for protein groups, *J Histochem Cytochem* **3**:32 (1955).

63. E.A. Barnard, Quantitative cytochemical observations on a specific nucleoprotein reaction in cell nuclei, *Nature* **186**:447 (1960).

64. G. Nöhammer, Die cytospektrometrische Proteinbestimmung mit der Tetrazonium-Kupplungsreaktion. I. Optimierung der Färbemethode, *Acta Histochem* **61**:317 (1978).

65. G. Nöhammer and G. Desoye, The cytospectrophotometrical determination of proteins by the tetrazonium coupling reaction. II. Calibration of the staining method, *Acta histochem* **68**:103 (1981).

66. L.P. Weiss, K.C. Tsou, and A.M. Seligman, Histochemical demonstration of protein-bound amino groups, *J Histochem Cytochem* **2**:29 (1954).

67. G. Nöhammer, Histochemical demonstration of primary amino groups with 2-hydroxy-1-naphthaldehyde (HNA): Optimization of the method, *Acta Histochem* **86**:167 (1989).

68. O.H. Lowry, N.J. Rosebrough, A.L. Farr, and R.L. Randall, Protein measurement with Folin phenol reagent, *J Biol Chem* **193**:265 (1951).

69. G. Nöhammer, The quantification of the histochemical protein staining with 2-hydroxy-1-naphthaldehyde (HNA) demonstrating primary amino groups of proteins, *Acta histochem* **90**:5 (1991).

70. M.N. Berry and D.S. Friend, High-yield preparation of isolated rat liver parenchymal cells, *J Cell Biol* **43**:506 (1969).

71. P.O. Seglen, Preparation of rat liver cells. III. Enzymatic requirement for tissue dispersion. *Exp Cell Res* **82**:391 (1973).

72. G. Nöhammer, F. Bajardi, C. Benedetto, E. Schauenstein, and T.F. Slater, Studies on the relationship between epithlium and stroma in sections of human uterine cervix in different pathological conditions, *in*: "Cancer of the uterine cervix", D.C.H. Mc Brien, and T.F. Slater, eds., Academic Press, London (1984).

73. G. Nöhammer, F. Bajardi, C. Benedetto, E. Schauenstein, and T.F. Slater, Quantitative cytospectrophotometric studies on protein thiols and reactive protein disulphides in samples of normal human uterine cervix and on samples obtained from patients with dysplasia or carcinoma-*in-situ*, *Br J Cancer* **53**:217 (1986).

74. G. Nöhammer, F. Bajardi, C. Benedetto, H. Kresbach, W. Rojanapo, E. Schauenstein, and T.F. Slater, Microphotometric determination of protein thiols and disulphides in tissue samples from human uterine cervix and the skin reveal a "field effect" in the surroundings of benign and malignant tumors, *in*: "Eicosanoids, Lipid Peroxidation and Cancer", S.K. Nigam *et al.*, eds., Springer, Berlin (1988).

75. G. Nöhammer, F. Bajardi, C. Benedetto, H. Kresbach, W. Rojanapo, E. Schauenstein, and T.F. Slater, Histophotometric quantification of the field effect and the extended field effect of tumors, *Free Rad Res Comms* **7**:129 (1989).

76. G. Nöhammer, F. Bajardi, C. Benedetto, H. Kresbach, W. Rojanapo, E. Schauenstein, and T.F. Slater, Histophotometry of protein thiols and disulphides in tissue samples from the human uterine cervix and the skin reveals a "field effect" as well as an "extended field effect" of malignant tumors, *Acta histochem*, **Suppl Band XXXVIII**:247 (1990).

77. C. Benedetto, F. Bajardi, B. Ghiringhello, L. Marozio, G. Nöhammer, P. Phitakpraiwan, W. Rojanapo, E. Schauenstein, and T.F. Slater, Quantitative measurements of the changes of protein thiols in cervical intraepithelial neoplasia and in carcinoma of the human uterine cervix provide evidence for the existence of a biochemical field effect, *Cancer Res* **50**:6663 (1990).

CHAPTER 25

COMPUTER - BASED IMAGE ANALYSIS FOR HISTOCHEMISTRY

Gustav Bernroider[1]

Institute for Zoology
University of Salzburg
A-5020 Salzburg, Austria

SUMMARY

A large variety of imaging methods has been developed to address the location and quantification of signals emerging from a place coded distribution of biological activity. The key question addressed by all such techniques is the extraction of cellular or molecular attributes from image coded information. This requires multiple calibration steps that gradually associate biological attributes with images. In the following text, major requirements or calibration steps that must be met to infer physico-chemical aspects from the space-coded intensity distribution provided by images are outlined. Also included is a comparison between traditional transmission images from radiographic receptor binding and novel approaches using intrinsic emission tomographs from enzyme-labeled ligands. Another major point is that computer-assisted manipulations of images allow the construction of "coincident images" that can combine such diverse signals as low light emissions with high-level light transmission images. Demonstrative examples from neuro-imaging show that coincident images can provide multiple histochemical information in a global, yet spatially selective way.

[1]Address for correspondence: Gustav Bernroider, Ph.D., Institute of Zoology, University of Salzburg, Hellbrunnerstr. 34, A-5020 Salzburg, Tel: ++43-662-8044-5604, Fax: ++43-662-8044-5698

Modern Methods in Analytical Morphology, Edited by
J. Gu and G.W. Hacker, Plenum Press, New York, 1994

INTRODUCTION

Many forms of histochemical research involve the quantification of objects within organic scenes. The "objects" are derived from signals representing a place-coded distribution of cellular and sub-cellular activity. A technical revolution combined light microscopes with electronic imaging devices and computerized imaging techniques to provide the researcher with powerful tools to analyze these signals. Contemporary progress in the use of radioactive, fluorescent or luminescent substances that selectively bind to target molecules in proportion to functional activity has further facilitated the potential to extract functional information from tissue samples. A key question in computer-assisted image analysis seems to be the extraction of molecular attributes from specific, image- coded information. To accomplish this extraction, chemical strategies and imaging methods have to work in unison. In other words, the chemical nature of those processes that underlie the formation of images is inseparable from all features encoded by the image. For example, in transmission imaging the resulting signal represents a measure of concentration, whereas in images derived from emission processes, the signal usually codes the rate of degradation of a substance. Although the resulting images may be indistinguishable, pixel coded information from both images estimates different biological attributes.

The present text begins with a brief review of the major requirements that must be met to allow quantification of images with inferences to underlying physiological processes. This includes an attempt to frame a general theory of biological imaging that accounts for most relevant physico-chemical aspects of image formation and acquisition. Demonstrative examples and applications will mostly concern neuroscientific issues. In fact, the explosive progress in neuro-imaging provides an unparalleled demonstration of how imaging methods have become indispensable to modern biological sciences. A very illustrative overview of this recent progress is given in A.W. Toga's edition on neuro-imaging.[1] An additional reason to concentrate on neurobiological scenes arises from the abundance of place-coded information that is intrinsic to the organization of neural elements at various levels of spatial resolution.[2] Such a situation challenges the technological capability of computer imaging to handle an ever increasing amount of available data. In concert with increasingly powerful data acquisition techniques, there is an increasing demand for sophisticated data management techniques. Confirmation or re-interpretation of morphological data requires the introduction of graphical databases. Recent remarkable projects include "data-basing the brain" that will eventually allow a graphical excursion through the brain, from molecular to gross anatomical levels based on masses of collected image data.[3]

Yet another reason that digital imaging has received so much attention in recent years is the visualization of more than two dimensions by the addition of a third perspective spatial dimension[4] and color-coded fourth dimension.[5] The fourth dimension can possibly represent temporal events and be used to describe dynamic changes of objects. Together with recent advances in high-speed, low light imaging,[6] using gateable image intensifiers,[7] time resolved imaging holds great promise in localizing functional activity within images. Two essential features emerge from time-resolved imaging: tracer or marker kinetics, in which labelled or unlabelled compounds become temporarily bound by various tissue compartments and the resulting optical changes reflect time dependent metabolic steps of the tracer. Visualization of synaptic vesicle recycling by HRP was probably the first marker successfully used in this category.[8] More recently, potential-dependent fluorescent dyes are frequently used to map local activity in non-toxic and low invasive fashion.[9] Metabolic mapping employing radio-labelled 2-deoxyglucose (2-DG), although lacking cellular resolution, has proven to be most useful in brain metabolic mapping.[10] The second issue relevant to time-resolved methods revitalizes observations made 30 years ago[11] and may demonstrate the most remarkable form of imaging: spatio-temporal recording of intrinsic optical activities in living tissues.[12]

The final aspect that will be covered in this chapter is low light emission imaging which first became available during the 1970's.[13] The introduction of channel electron multiplied (CEM) image intensification has facilitated intensity gains up to 10^8 which allows the detection of one photon per sec and mm^2.[14]

Concordant with electrotechnical progress, the development of new generation chemoluminescent substrates to enzyme labelled molecules has been established. The substrates now seriously challenge the use of radio-isotopic based systems. In contrast to light absorption photometry, the signal intensity in light emission processes from most chemoluminescent reactions reaches steady state levels due to a degradation of reactive compounds. This principally permits the unambiguous association of signal intensity with unknown target concentration.

GENERAL NOTES ON MACHINE VISION OF BIOLOGICAL ATTRIBUTES

Although theoretical considerations often run into limitations imposed by physical reality, an overview of assorted techniques may be available from a short theoretical excursion. In Figure 1, the general setup of an imaging situation is outlined. With basically four types of energy sources (photons, electrons, magnetic fields and ultrasound), images are built up by primary detection and signal transfer. These images are subject to (human) observation and interpretation (inference to unknown energy sources).

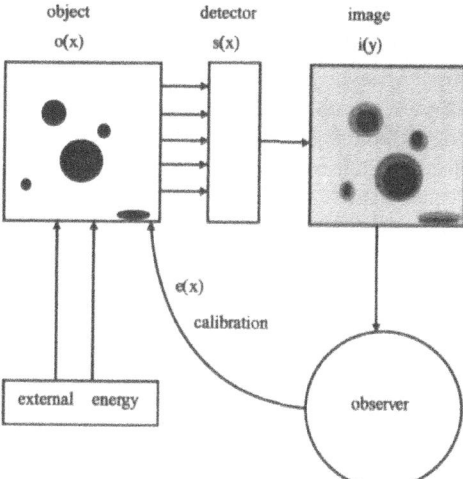

Figure 1. Design of imaging systems: the two-step process involves a transfer of object locations o(x) to an image i(y) by single or array detection s(x) and inference back to unknown object attributes by calibration e(x). There are two main types of imaging strategies: mapping intrinsic energy sources (emission tomography) and employing external energy (transmission tomography).

The overall process can be conceptualized by using a set of *transfer functions* that eventually relate object sources o(**x**) to images i(**y**) through the convolution

$$i(\mathbf{y}) = o(\mathbf{x}) \otimes s(\mathbf{x}) \tag{1}$$

and images to a biological attribute distribution $e(x)$ by calibration

$$e(x) = k^{-1} \cdot i(y) \tag{2}$$

where $\otimes$ denotes "convolution", $s(x)$ the *convolution kernel* representing the point spread function (PSF) of the image transfer and -k- the *calibration factor* that associates an external standard to image characteristics. Equation (1) covers technical aspects of image formation including signal intensity, detector characteristics, and image presentation. The most essential features of this transfer seem to be those contained in the term, *resolution*. For example, if photons are used to generate an image, *intensity resolution* and *range* are critical to signal source and object rendering. High light level signals (10^8 photons/mm^2 sec) usually emerge from extrinsic sources (transmitted light microscopy). Their transfer characteristics may be seen from the band width of the Fournier transform of $s(x)$, sometimes called its *modulation transfer function* (MTF).[15] As opposed to "strong" signals, the transfer of low level signals ($< 10^2$ photons/mm^2 sec) turns $s(x)$ in equation (1) into a probabilistic form, i.e. $s(y;x)$ representing the conditional probability that a photon arrives at location y and originates from an object source x. This situation applies to most intrinsic source emission signals (e.g. luminescence imaging).[16] The above examples demonstrate the fairly general relevance of the imaging equation (1) to machine vision studies of biological function. Different categories of sensing and acquisition approaches are modelled by a different form of transfer functions (*Table 1* is a summary of available sensing methods in terms of signal attributes).

Equation (2) implicitly contains the procedures required to finally arrive at the main target: the spatio-temporal distribution of biological signals. In its simplest form, the calibration factor -k- allows the conversion of image distances into physical dimensions such as absolute metric size and optical density. However, under most circumstances, the assessment of physiological variables entails more complex "calibration" or multiple transfer functions. This may involve the extraction of features from single images (e.g. morphometric features such as volume, surface and number)[17] or arithmetic operations among multiple images.[18] Thus, the estimation of biological attributes usually requires the calculation of more than one transfer function. As an example, transmission imaging of receptor binding by quantitative autoradiography requires at least two consecutive transformations where k_1 relates optical densities (OD) to tissue-radioactivity:

$$k_1 \cdot OD = f(dpm/mm^2) \tag{3}$$

by a function f that is found from least square approximations of empirical data collected from co-exposure of samples and standards.[19] Radioactive measurements resulting in counts per min, corrected for efficiency and radioactive decay are normalized by the area of the section to yield dpm/mm^2.

The second step relates the image (in terms of measured ODs) to biological attributes such as maximal binding sites B_{max} and to the concentration of radioligands that results in half-maximal specific binding k_d. This task can be achieved by pixel to pixel operations with multiple images that represent total binding (TB) and unspecific binding (UB) (sections incubated in excess of unlabelled ligand). Specific binding values are obtained from SpB = TB-UB and the concentration of free ligand FL from FL = total added ligand - SpB. Finally, the desired parameters can be obtained from a Rosenthal linear transformation[20]

$$B/FL = B_{max}/K_d - B/K_d \tag{4}$$

that allows linear extrapolation to B_{max} and estimation of k_d from $k_d = -1/slope$.

We should note that all the above transformations entail pixel by pixel operations in

2-dimensional images with each pixel of one image (e.g. TB image) corresponding to the coordinates of another image (e.g. UB image). Within reason, this makes *image alignment* very critical in order to avoid location correspondence errors. Alignment techniques are an important prerequisite for many synergistic approaches in image analysis and reconstructions of objects from sections. Some specific applications will be mentioned later.

Table I. Modes of imaging, separated in terms of energy source location and signal strength. Some images are only available from algebraic computation of coinciding signal locations (PET, CAT).

Energy source location	Energy Type and Strength		
	light low	light high	particles/else
tomography			
intrinsic	luminescence	PET,	scintillation images
extrinsic	fluorescence	CAT	electron microscopy
		light transmission microscopy	
surfaces			
intrinsic	luminescence		scanning tunneling
extrinsic		reflection light microscopy	scanning microscopy (SEM)

BIOLOGICAL ATTRIBUTES I. CALIBRATION OF SIGNAL SOURCES

The major principles of biological imaging techniques outlined in the present context are given by equations (1) and (2). It is important to note that physically dissimilar techniques (transmission vs emission tomography, see Table I) turn out as special cases of one general object to image relation. The nature of the imaging model (1) imposes constraints on subsequent calibration steps (2). In examining the latter claim, a comparison between transmission and emission imaging of receptor-ligand studies may be useful. As mentioned above, optical transmission imaging from receptor autoradiography requires a conversion of radioactive exposure to optical density (3). For exposure to electrons, this relation involves several parameters such as electron density, time, and film characteristics. It usually follows a sigmoid shaped negative Gombertz function

$$OD = a.\exp[-b.\exp(-kx)] \tag{5}$$

(*Fig. 2*) with x measured in terms of dpm/mm^2. To turn x into a measure of initial concentration, it has to be corrected for the radioactive decay that occurred during exposure time t, $x \rightarrow x/\exp(-l.t)$, with decay constant $l = -\ln 0.5/t_{1/2}$ related to the isotopes half-life. All the above calculations are subsumed by the calibration factor k_1. If we now replace optical transmission of autoradiographic films with a low level *intrinsic emission source* typical of luminescent imaging, the necessary steps for "biological calibration" change remarkably. In short, the labelled ligand in a luminescent assay generates light by the oxidation of a specific substrate that is externally applied in excess.[21] In case of the recently developed dioxetane substrates for alkaline phosphatase,[22] the time resolved light signal kinetics follows

$$LEU = 1/(a+b/t) \qquad a,b > 0 \tag{6}$$

where LEU is an acronym for light emission units, and -a- represents intensity at time t_o. The above curve resembles a saturation experiment and levels off to a maximum signal after 20 to 40 minutes (using Adamantyl-1,2-dioxetane phosphate -AMPPD- as a substrate, see *Fig. 2*). A linear version of equation (6) can be obtained from

$$1/\text{LEU} = a + b \cdot 1/t$$

which is equivalent to the Lineweaver-Burk transform in enzyme-kinetics.

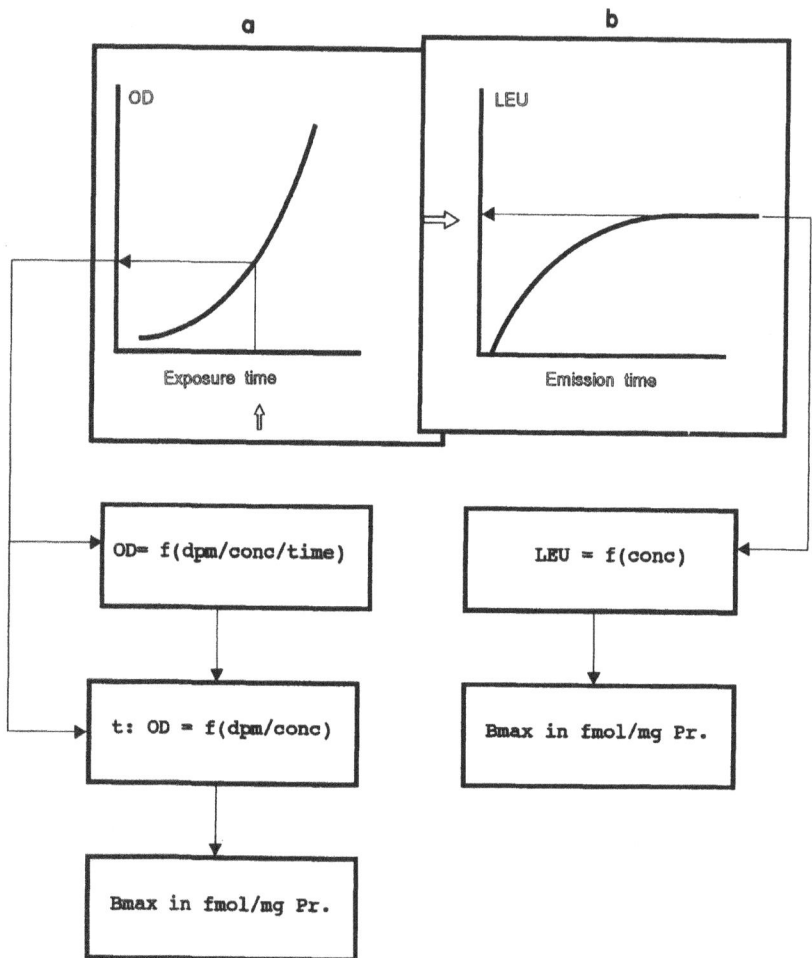

Figure 2. Calibration of signal sources in transmission receptor autoradiography (a) as compared with emission imaging of enzyme labelled receptor ligands (b). The conversion of radioactive exposure to biological attributes such as the maximum number of binding sites B_{max} entails at least three calibration steps: a time dependent process that provides optical density (OD) as a function of radioactivity (dpm), film characteristics, receptor concentration and exposure time. For a given exposure time, -t- tissue radioactivity (dpm/conc) must then be calibrated to OD (k2) and finally calculated into B_{max} (k3) (see text). Whereas in (b) intrinsic light emission units (LEU) are directly proportional to receptor concentration (k1) and specific binding is available to yield B_{max} by k2. The saturation curve (b) resembles the Michaelis-Menten model of enzyme reaction kinetics and its linearization allows direct e estimation of Kd (vs Km) and B_{max} (vs Vm) in enzyme limited emission imaging.

From this straight line, the parameters $a = 1/B_{max}$ and $b = kd/B_{max}$ can be found. Notice, that this replaces Rosenthal or Scatchard transformations of classical binding studies, demonstrating a remarkable "short-cut" for biological calibration with intrinsic and enzyme-limited emission imaging.

With specific luminogenic substrates, the kinetics of light emissions reaches steady state levels within minutes.[22,23] The maximum light emission (LEU in *Fig. 2*) represents a unique value that can be gained from a time resolved emission experiment. Enzyme ligand (probe or antibody) conjugates induce an enzyme catalyzed decomposition of a luminogenic substrate (luminol for horseradish peroxidase[24] and dioxetane phosphates for alkaline phosphatase[25]). Because 1, 2-dioxetanes carry their own oxidizing potential by intrinsic weak oxygen-oxygen bonds that facilitate exothermic light emissions upon decomposition, substrate dephosphorylation becomes associated with intrinsic light emissions without additional co-reactants. At maximal light emission (LEU_{max} in *Fig. 2*), the rate of enzymatic dephosphorylation balances the rate of luminogenic decomposition. Assuming that the amount of available substrate does not limit this reaction, LEU_{max} becomes a direct measure of enzyme-conjugate concentration and becomes remarkably independent of other confounding cofactors. Compared to optical transmission imaging of radiographic exposures, non-radiographic emission imaging allows a more direct calibration to the desired attributes such as specific and maximal binding.

Thus, the main step in calibrating intrinsic light emission sources in the present context involves the association of light emission units with the concentration of enzyme-ligand (probe or antibody) conjugates. In this lab, multi-labelled opioid ligand complexes that allow simultaneous signal visualization by both types of light emission imaging: fluorescent (external energy source) and luminescent (internal energy source) reactions have been recently designed.[26] Figure 3 shows calibration of light emission units to the concentration of a Naltrexone-FITC-anti-FITC-AP complex. The same complex was also tested for its ability to inhibit specific binding of the opioid-receptor radioligand ^{3}H-DAGO to quail brainstem homogenates. The resulting IC_{50} values (concentration of inhibitor that blocks 50% of binding in the absence of inhibition) were 8.41 nmol for Nal-FITC and 8.47 for Nal-FITC/anti-FITC (insert *Fig. 3*) indicating that labelling does not prevent competitive binding to opioid receptors in brain tissue homogenates. The opioid complex was visualized by a photon counting camera (BIQ, intensified CCD, Cambridge Imaging) that detects signals as low as 1 photon/sec and detector element.[27] Detection limits for alkaline phosphatase labelled ligands and AMPPD (Adamantyl-1, 2-dioxetane phosphate) as the lumigen substrate were $3-6 \times 10^{-15}$ mol/l with signal/noise (S/N) >3 and 100 fmol with S/N >8 (*Fig. 3*).

Further improvements of low light emission techniques, together with new luminescent chemicals, may soon seriously challenge traditional autoradiography. Light emitting sources that decompose as conjugates of target molecules, supply signals that are jointly related to the target biological attribute. It appears that we are directly visualizing a chemical process. This makes intrinsic emission imaging superior to high energy transmission techniques. Although recent technological advances inspire considerable enthusiasm, some natural constraints of low light emission imaging accompany the above advice with a warning. Single photon images turn the imaging equation (1) into a probabilistic form, i.e. registered photon location becomes inevitably associated with uncertainty or spread (see chapter 4). This may lead to a considerable loss of resolution and seriously limit quantification. Further, relatively stable "intermediates" enhance time resolved emission (such as the AMP-D ion that results from enzymatic decomposition of AMPPD), but suffer from considerable diffusion during light emission. More recently, membrane replicas of tissue sections (see below) or frozen section-cocktail sandwiches[28] were successfully used to minimize diffusion processes.

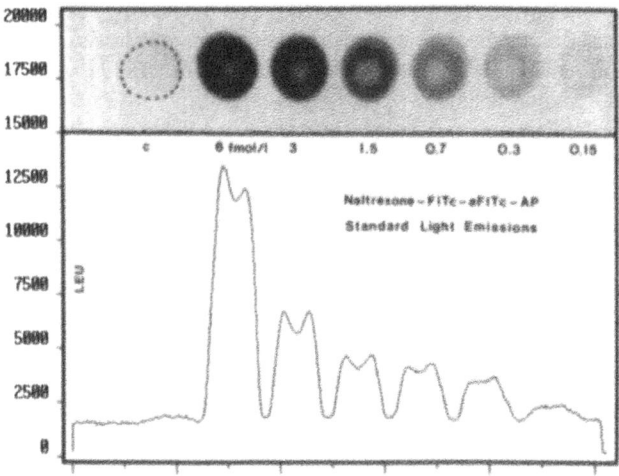

Figure 3. Calibration signals that associate light-emission units (LEU) with absolute values of an alkaline phosphatase labelled opioid-ligand (Naltrexone-FITC-antiFITC-AP) complex. 10^{-15} mol/l ligand are detected with S/N > 8. Images of 20 ul complex in TRIS with 10 ul LumiphosTM were obtained after 5 min signal integration using a MCP-intensified CCD camera that detects 1 photon/sec and detector element.

BIOLOGICAL ATTRIBUTES II. TOMOGRAPHIC FEATURES

Tomographic imaging monitors physical or optical sections along a given orientation. With the advent of a great variety of modern imaging techniques during the last decade, tomographic pictures have made an indispensable contribution to biomedical research. The ever increasing number of methodological and technical terms deserves an attempt at categorization. *Table I* outlines some major issues that emerge from present approaches. Different modes of tomography can be separated in terms of energy source location, signal strength and image formation.

Transmission tomography refers to imaging strategies in which energy sources and detectors are located on opposite sides of probe location (*Fig. 1*). The imaging particle beam may be a stream of photons (light microscopy, x-rays), electrons (electron microscopy), ions, neutrons or protons. In terms of signal strength, transmission light images emerge primarily from high light level sources (> 10^8 photons/mm^2 sec). The usual strategy in machine vision, registers imaging beams from transmission techniques through cathode tube or array cameras (CCDs). This may involve multiple beam conversion. For example, in transmission electron microscopy (TEM), primary electrons that have interacted with the probe are converted to light by an electron sensitive phosphor coated screen. A TV camera that is focussed on this screen subsequently converts light back into electrons to produce a video image (this second step is normally "intensified," e.g. by cooled long exposure CCDs or electron channel plate intensified CCDs). So far, the vast majority of quantitative histochemical information has been supplied by *absorbance photometry* based on tomic transmission images. Yet, other transmitted beam methods can provide images that encode additional features of interest. For example, matching the sampling frequency of a camera to the maximum usable magnification can significantly enhance the contrast based on a gradient of light-path differences. This is the mode of Video-Enhanced Differential Interference Contrast

(VEC-DIC) imaging that can detect time-resolved structures with less than 0.025 μm in diameter.[29]

Transmission images are usually obtained from physical sections produced with microtomes or vibratomes. This makes section thickness critical for beam intensity and object resolution. It also becomes evident that all information we intuitively ascribe to images are actually a matter of a three dimensional complexity. Properties of this complexity cannot be inferred simply on the basis of properties obtained from 2-dimensional images. Special procedures are required to obtain unbiased estimates of spatial entities.

Emission tomography generates images from either intrinsic energy sources (chemoluminescent images) or extrinsic excitation (fluorescence). During the middle to late 1980s, tomographic images emerged from *confocal optical slices* with the development of laser scanning confocal microscopy (LSCM).[30] This opened 3-dimensional confocal fluorescence imaging to biologists. Emission tomography also encompasses several *non-invasive techniques* of particular significance to neuro-imaging applications. These strategies revolve around efforts to compose 2-dimensional images from algebraic computation based on coinciding locations of emission (PET) or transmission sources (CAT) on one hand, and arrays of detectors on the other. Methods that "compose" images from corresponding locations of different physical signals become increasingly important in histochemical applications. Finally, it should be mentioned that numerous useful signals arise from a combination of transmission and emission processes. For example, a transmitting electron beam produces emission signals when it impinges on the specimen. Such "secondary" signals include X-rays, cathodoluminescence, secondary, backscattered and Auger electrons.[31] All of these signals can be used to form images and can be computationally combined with the "primary" image.

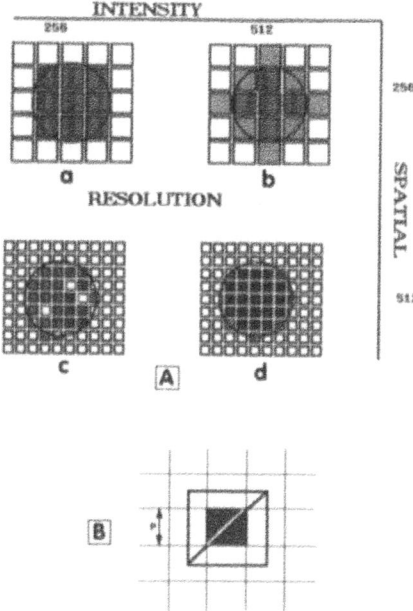

Figure 4. The relation of intensity (row) to spatial (column) resolution in sensing a circular feature (A). Spatial resolution (1/number of equally shaded squares) increases with intensity resolution (a - b, and c - d) for a given number of available pixels. Only maximum intensity and spatial resolution (d) preserves the shape of a circle. B: It needs at least 3 pixels to linearly cover an object. Thus the minimum detectable size of an object is 2 x pixel x SQR(2).

Consistent with the two-step process that associates objects' sources to biological attributes (equations 1 and 2), tomographic feature extractions revolve around two central issues: *resolution* and *morphometry*. The quality of the primary image depends on resolution, whereas inference to biological attributes entails the extraction of features from an image provided at a given resolution. In digital images, each point or pixel represents a value of relative intensity. Thus, the number of pixels required to code an object and the number of intensity levels assigned to pixels are mutually dependent. The factor of resolution may, therefore, be seen from two related aspects.

Intensity resolution is expressed as intensity range and number of levels within this range. Intensity range (or dynamic range) for light imaging extends from 10^4 LUX (lumens/square meter), e.g. day light, down to almost no light conditions of 10^{-12} LUX (equivalent to approx. 1 photon/cm^2/sec) which is still accessible to single photon imaging devices. No detecting device as yet can cover the full range of light intensity levels. The magnitude of typical dynamic ranges in video systems extends from 10^4 (wide range vidicons) down to 10^2 LUX (intensified cameras). Digitalization transforms the continuous intensity range into a limited, discrete number of intensity levels, typically 8 bit or 256 levels. The number of intensity levels emerging from digitalization and the dynamic range provided by sensing instruments need to work in unison. Naturally, the available bit level of intensities (e.g. 8 bit) is associated with the minimum number of pixels required to discriminate neighboring objects. This number determines the second issue of resolution.

Spatial resolution is defined as the smallest attribute of an image that can still be resolved in terms of differences in intensity levels of pixels. Figure 4 demonstrates the relation of intensity (row) to spatial (column) resolution in sensing a circular feature within an image.

It is evident that at least three pixels are required to discriminate an object from background along a single line of pixels. At least 1 bit gray levels are also required for the same task. To delineate or identify a circular shape (e.g. cell nuclei in light microscopes) from background, at least 64 pixels are needed. A table that compares various resolution limits in mouse and human brains is given by Hillman *et al*, 1990.[32] Figure 5 addresses the significant impact of resolution on all subsequent operational success. It demonstrates an extensive mid-sagittal map of catecholaminergic topography in the quail brain. Immunogold-silver enhanced cytochemistry was used to detect the presence of tyrosine hydroxylase (TH), the rate limiting enzyme during biosynthesis of catecholamines.[33] Paraformaldehyde perfused total brains were sliced on a vibratome to 70 μm thick sagittal sections with roughly 15 mm (longitudinal) x 8-10 mm (dorso-ventral). Almost all (95%) TH-L immuno-reactivity was observed within a 10 mm x 5 mm brainstem preparation extending from the Lobus parolfactorius (LPO) rostral to the Nucleus tractus solitarii (NS) caudal. One image (*Fig. 5A*) was digitized from a low magnification light optical video signal to 512^2 pixel x 8 bit intensity image. The same section was digitized into a $4x4x512^2$ pp x 8 bit image which is equivalent to a 2048^2 x 8 bit resolution (*Fig. 5B*). The first image provides a minimum resolvable linear extension of 56 μm (2 x SQR (2) x pixel-pixel distance, see *Fig. 4B*), whereas the second image can resolve 14 μm objects.

Average TH positive cell size for catecholaminergic brainstem neurons in the quail can be expected to range between 10 μm and 20 μm. Only the second image can provide spatial resolution necessary to locate single cell bodies within the image. At this point, a note of caution should be made about attempts to increase digitizing resolution without coherent integration of sensing resolution. That is, picture pixels must be matched to sensor pixels (e.g. number of pixels and read out performance of CCD cameras).[34] If the tomographic scene involves additional perspectives, such as color or motion, the minimum required resolution will increase. For example, RGB channels introduce a multiplication of three to intensity levels (8 bit grays versus 24 bit RGB).

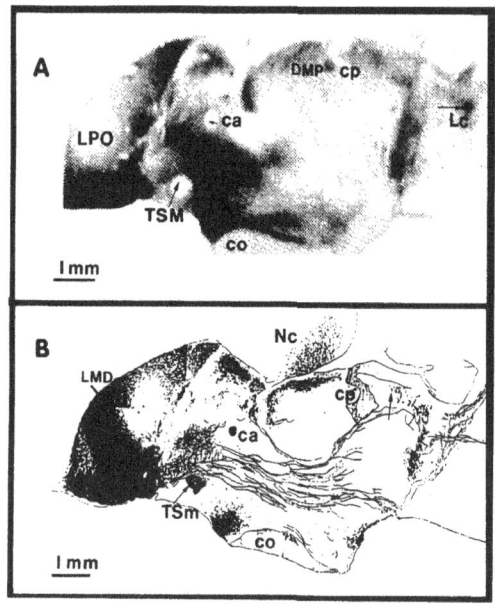

Figure 5. Effect of resolution in imaging tyrosine-hydroxylase like immunoreactivity in extensive (15 mm) paramedian, sagittal quail brain sections. (A) was digitized into 512^2 x 8 bit pixels and resolves approx. 56 μm objects, whereas (B) uses $2,048^2$ x 8 bit and can resolve 14 μm objects. Only the image in (B) provides cellular resolution (single cells indicated by arrows) and also detects single extensive fibers projecting from dorso-ventral brainstem sources (e.g. LC) to septum (SM), habenula (H) and LPO. CA: Commissura anterior, CP: Commissura posterior, Co: Chiasma opticum, LC: Locus coeruleus, LPO: Lobus parolfactorius, LMD: Lamina medullaris dorsalis, TSM: Tractus septomesencephalicus.

Morphometry

In order to relate images to biological attributes, it is often necessary to extract metric or topological properties from a discrete array of numbers.[35] *Computational geometry* established the body of methods that deals with approximation of "smooth" geometrical properties from discrete, step coded values available from digital images.[35-37] Intuition suffices to realize that the degree of resolution in the above sense critically confines the quality of approximation to object specific, continuous figures. Objects contained in spatial dimensions give rise to basically four geometrical functionals: volume, surface, mean curvature and connectivity (or simply number) that are all potentially significant for biological attributes. These continuous, position invariant and additive measures are usually inferred from their tomographic, i.e. two-dimensional projective or sectional appearance.[35] Further, geometrical functionals refer to binary images where the region of interest (ROI) becomes identical to the object from which the functionals should be measured. This makes *image segmentation or figure ground separation* a prerequisite for morphometry in image analysis. Of the many methods that are available today to separate objects from background, *thresholding*, the gray-scale distribution using *look-up table* transformation, is the most frequently used strategy. In most instances, volume and number are of major interest because they relate to mass-concentration of objects. The extensive list of stereological techniques and numerical procedures[38,39] that deal with volume and number was considerably simplified by the introduction of unbiased estimation methods during the mid 1980s.[40-42] The main steps involved in estimating volume and number of arbitrary shaped, isolated items (particles) follow.

Volume: the biological attribute contained in the number of particles (e.g. immuno-positive cells, synapses, vesicles, silver-grains) emerges from its reference to the region under study (e.g. cells within cortex, synapses per ganglion, silver grains per cell body). The reference volume can be reconstructed from the areas of delineated objects provided by optical or physical sections according the formula of Cavalieri (Italian mathematician, 1599-1647):

$$V = t . \sum_n a(prof) = \tilde{a} (prof) . t . n \qquad (7)$$

with area of profiles a(prof), from n serial sections each with section thickness t. Due to finite section thicknesses, transmission tomography of opaque objects embedded in transparent background may introduce an "error of projection" that depends on orientation and shape of an object (*Fig. 6*).

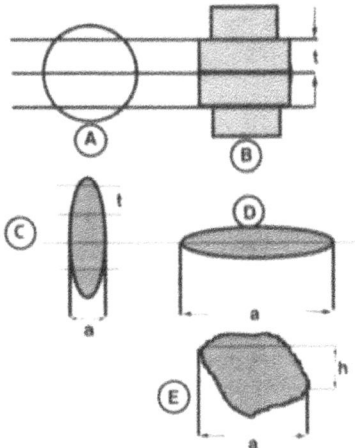

Figure 6. Transmission tomographs of opaque objects suffer from projection errors due to finite section thickness -t-. (A) and (B) demonstrate the effect of overprojection in slices through a sphere. "3-cuts" result in 4 sections, each with a maximum profile, the biggest section in the middle of (B) occurs in duplex. Errors due to finite section thickness depend on orientation (C and D) and shape of a figure (E). The error is proportional to (t x a) and h (an orthogonal distance between the two extreme points of a convex body; notice that this distance approaches zero in spherical objects). For corrections due to section thickness, see text.

A reasonable correction for section thickness effects in the (usual) case of opaque specimen (overprojection) subtracts the n-weighted fraction of the biggest section from the exhaustive stack of serial sections:

$$V = t [\sum_n a(prof) - 1/n . max \, a(prof)] \qquad (8)$$

Particle number is estimated from the number of profiles in sections that are contained within the area of reference; however, the number of profiles from tomographic probes not only depends on number but also on volume, linear extensions, shape and orientation of their "parent" particles. Within reason, an unbiased counting strategy for particles will have to exclude more than one profile count from anyone object. As an obvious consequence, the probe must be of the same dimension as the object (e.g. plane particles from 2-dimensional probes, spatial aggregates of particles from 3-dimensional probes).

430

This rationale suggests the use of two parallel optical or physical sections (thus a truly 3-dimensional probe) that can identify "tops" or "bottoms" of spatial particles. Gundersen discerned the rule to count profiles that are contained in one section but not in the adjacent or look-up plane.[40-42] The probe or dissector volume is available from

$$V \text{ dis} = a(\text{dis}) \cdot h \tag{9}$$

where a(dis) marks the area of the probe and h the height (section thickness for adjacent physical sections or distance of two confocal optical layers within one physical section). The number of particles contained in one section but not in the look up section C is summed over all probes and divided by the appropriate sum of dissector volumes, to yield

$$N_v = \quad C / \Sigma V \text{ dis} \tag{10}$$

This number is finally multiplied by the overall volume of reference (7) to highlight the desired number of items contained within the region of study

$$N = V \text{ ref} \cdot N_v \tag{11}$$

A quick verification of the result (11) shows that

$$N = C \cdot (\text{Vref}/\Sigma \text{ Vdis}) \tag{12}$$

which states that the "true number" of particles within a given volume is equal to the number of correctly counted profiles (one per particle) times the fraction of sampled volume. If the region of interest (ROI) is totally contained within the dissectors area, then Vref = Σ Vdis and N = C.

Figure 7 outlines the main steps involved in unbiased counting particles within a sphere.

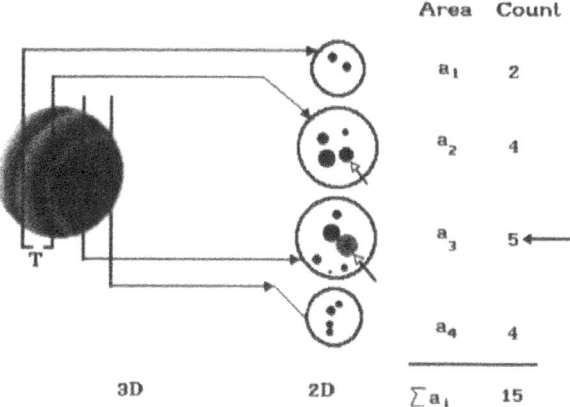

Figure 7. Unbiased counting of particles from their sectioned profiles. The sum of areas of all sections is determined according the formula of Cavalieri (see text) to yield the volume of reference Vref. Particles are counted only if they are not visible within their top look up section ('tops' of particles). In section 3 only 5 particles are counted, because one profile (arrow) was observed within two sections (section 2 and 3). There is no other way to determine number without bias for arbitrary shaped objects.

431

It should be noted that the above counting rule can also be justified by "reducing" the extension of particles to unique counting features (e.g. cell nuclei) that are not contained in more than one section (look ups are zero).

As we appreciate the simplicity behind the above counting paradigm, we also note the basic constraints imposed by tomographic sections. The optical or physical invasion of space elucidates its interior only at the sacrifice of connectivity.

BIOLOGICAL ATTRIBUTES III: COINCIDENCE IMAGING

Composing images from coincident locations of signal sources can provide information that would not be conceivable otherwise. For some images, the coincidence of physical events such as positron-electron annihilation into paired gamma photons, is used to compute the primary image. This is the case with PET images. Many additional perspectives arise from mapping the locations of qualitatively different, but spatially coinciding energy sources. The pixel-to pixel correlation between images of absolute ATP concentrations and blood flow in tumors may serve as an example.[43] In this case, absolute ATP concentrations were obtained from photon emission images using firefly luciferase as the lumigen. Iodo (^{14}C) antipyrine autoradiography on adjacent cryosections supplied information on regional blood flow. A pixel to pixel coincidence image revealed reduced flow and ATP values in necrotic areas. The combined image provides correlative information arising from two different signal emissions at quasi-identical locations that are obtained from two adjacent physical sections.

Reflecting on truly coincident images, we would expect signals from physically identical locations registered by different sensing equipment. Figure 8 demonstrates a location coincidence image (D) from emission (A) and transmission (C) tomographs that have been generated from the same physical section.

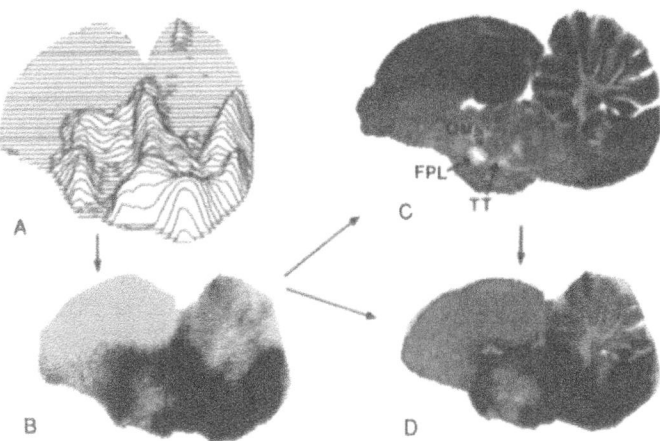

Figure 8. Multiple image operations for coincidence imaging. The 2-dimensional recording of intrinsic light emissions from alkaline phosphatases obtained from a nylon-membrane replica (A) is transformed into a gray-level coded image (B). A light-transmission image from a section taken after replication (C) is combined by pixel to pixel averaging with its light emission image to yield the image given in (D). Thus image -D- is composed from algebraic computation and correlates optical absorbance and intensity of light emissions at physically identical locations. FPL: Fasciculus proencephali lateralis, OM: Tractus occipitomesencephalicus, TT: Tractus tectothalamicus.

Protocol for Figure 8

The images were obtained from mid-sagittal quail brain vibratome sections. Emission tomographs were generated from a novel *membrane-replica technique* that allows the production of multiple images from one physical section. Positively charged nylon membranes (Hybond-N+, Amersham) were soaked for 5 min in Lumi-Phos™ (Lumigen), which provides the alkaline phosphatase substrate AMPPD together with a fluorescein derivative co-surfactant that enhances chemoluminescent efficiency up to 300-fold.[44] The membranes were then transferred to contact the section surface of brain blocks fixed in the vibratome bath (0.1M TRIS, pH 7.6, 4°C). Contact incubation lasted 5 min. The replicas were assessed by photon-imaging using an MCP-intensified CCD camera (BIQ-Image Res., Cambridge) and integrating the emission for 2 min. The resulting images were stored for further transformation on disk. From the remaining block of tissue, the vibratome section that exposed its surface to the replica (thickness 50 μm) was then mounted on poly-L-lysine coated glass slides, stained (methylene-blue, 0.5 g in 0.9% NaCl, 5 min), dehydrated and DPX mounted. Under the same magnification as the emission replica was recorded (*Fig. 8*), a digital light transmission image from a standard CCD camera was taken. The final step in producing a coincidence image impressively demonstrates the remarkable dynamics available from digital image transformations. There is usually a discrepancy between the size and shape of physically invaded material and its original parent structure. The difference may emerge from expansion due to physical force during sectioning or swelling in liquid environments. For example, digital images can easily be subjected to *uniform scaling* procedures and *geometrical alignment*. Thereby, correspondence of images can be achieved either by interactive, that is operator controlled, image manipulations or algorithmic approaches employing different measures of similarity among images.[4]

The coincidence image demonstrated in Figure 8D was obtained from a two step process: 1) uniform scaling to compensate tissue expansion of the section (C) as compared with its replica image (B) and 2) averaging every pixel of B with the location corresponding pixel in C. The result demonstrates a mixture of gray-value coded light emission activity obtained from AMPPD dephosphorylation by endogenous phosphatases and the gray-value coded transmission absorbance representing the cytoarchitecture of the same tissue probe. It appears that the limited resolution of the low-level signal becomes topographically confined by the much higher resolution contributed by the high-energy transmission image (approx. 162 μm). Although not available from the light emission image, the final image suggests strong light emissions originating from main brainstem areas except ventral mesencephalic fiber tracts (FPL, OM, TT) and optic chiasm. To further appreciate the coincidence image obtained by the above procedure, it is necessary to recall that the transfer function (equation 1), although different for both images A and C, maps the same spatial domain to the final image 8D. Multiple image operations establish a promising approach to synthesizing the ever increasing and chemically selective information from biological probes to integrative and spatially selective views. This may compensate for some global information that inevitably becomes lost during chemical targeting.

FINAL REMARKS

The methods discussed in this chapter indicate that computer-based image analysis can provide hitherto unobtainable information about biological attributes spatially coded by tissue and cells. Besides the ability of image analysis to associate numbers to physical items, perhaps the most promising issue emerges from its power to dispose spatial information by algebraic calculation. Several difficulties that are associated with the preparation of tissue for visualization and technical restrictions imposed on spatial and temporal resolution still remain. Also, many aspects of image interpretation and computational problems await resolution. Most problems arise from our intention of combining global, low scale images with single unit cell resolution. Francis Crick and Edward Jones have recently called for radically new techniques that are urgently needed in brain sciences where function is emergent from spatially coded signals.[45] Despite present problems, we hope that computer-assisted methods will contribute to providing physically sufficient, chemically specific but biologically low invasive images.

REFERENCES

1. A.W. Toga, "Three-Dimensional Neuroimaging," Raven Press, New York (1990).
2. G. M. Shepherd, "The Synaptic Organization of the Brain" (3rd Ed.), Oxford Univ Press (1990).
3. A. Gibbons, Data basing the brain, *Science* **258**:1872 (1992).
4. A.W. Toga, Three-dimensional reconstruction, *in:* "Three-Dimensional Neuroimaging", A.W. Toga, ed., Raven Press (1990).
5. R.C. Collins, E.M. Santori, and A.W. Toga, A three-dimensional image of brain function, *in:* "Three-Dimensional Neuroimaging", A.W. Toga, ed., Raven Press, 149-164 (1990).
6. G. Bernroider and A. Pinz (eds), "Image Acquisition and Real-Time Visualization," OCG Publ 56, R. Oldenburg, Vienna (1990).
7. H. Kume, H. Nakamura, and M. Suzuki, Gatable photonic detectors and it's image processing, *in:* "Proc 19th Int Congress on High Speed Photography & Photonics", 16-21 Sept. Cambridge, U.K. (1990).
8. R.C. Graham and M.J. Karnovsky, The early stages of absorption of injected horseradish peroxidase in the proximal tubule of the mouse kidney; ultrastructural cytochemistry by a new technique, *J Histochem Cytochem* **14**:291 (1966).
9. A. Grinvald, R.D. Frostig, E. Lieke, and R. Hildesheim, Optical imaging of neuronal activity, *Physiol Rev* **68**:1285 (1988).
10. G.E. Duncan and W. Stumpf, Brain activity patterns: assessment by high resolution autoradiographic imaging of radiolabeled 2-deoxyglucose and glucose uptake, *Prog Neurobiol* **34**:365 (1991).
11. B. Chance, P. Cohen, F. Jobsis, and B. Schoener, Intracellular oxidation-reduction states *in vivo*, *Science* (Washington), **137**:499 (1962).
12. M.M. Haglund, G.A. Ojemann, and D.W. Hochman, Optical imaging of epileptiform and functional activity in human cerebral cortex, *Nature* **358**:668 (1992).
13. G.T. Reynolds, Application of photosensitive devices to bioluminescence studies, *Photochem Photobiol* **27**:405 (1978).
14. C.E. Hooper and R.E. Ansorge, Quantitative photon images in the life sciences using intensified CCD cameras, *in:* "Bioluminescence and Chemoluminescence, Current Status", P.E. Stanley and L.J. Kricka, eds. 337-344, John Wiley & Sons, Chichester (1991).
15. J.W. Goodman, "Introduction to Fourier Optics," McGraw Hill, New York (1968).
16. G. Bernroider and K. Überriegler, The simulation of photon images from objects with very low light emissions, *in:* "Image Acquisition and Real Time Visualization", G. Bernroider and A. Pinz, eds. OCG Publ. 56, Oldenbourg, Vienna (1990).
17. G. Bernroider, The foundation of computational geometry, theory and application of the point-lattice concept within modern structure analysis. Lecture Notes in Biomathematics, **23**:153 (1978).
18. G. Mahoney, D.E. Hillman, and M. Canaday, High-resolution, large area image recording and analysis, *in:* "Three-Dimensional Neuroimaging" A.W. Toga, ed., Raven Press, New York (1990).
19. M.J. Kuhar and J.R. Unnerstall, Receptor autoradiography, *in:* "Methods in Neurotransmitter Receptor Analysis", H.I. Yamamura, S.J. Enna and M. J. Kuhar, eds., 177-218, Raven Press, New York (1990).
20. D.B. Bylund and H.I. Yamamura, Methods for receptor binding, *in:* "Methods in Neurotransmitter Receptor Analysis", H.I. Yamamura, S.J. Enna and M.J. Kuhar, eds., Raven Press, New york (1990).
21. E. Schramm, Evolution of bioluminescence ATP assays, *in:* "Bioluminescence and Chemoluminescence, Current Status", P. E. Stanley and L.J. Kricka, eds., John Wiley & Sons, Chichester (1991).
22. I. Bronstein and P. McGrath, Chemo uminsecence light up, *Nature* **338**:595 (1989).
23. I. Bronstein and L.J. Kricka, Improved chemoluminescent detection of alkaline phosphatase, *BioTechniques* **9**:No.2 (1990).

24. G.H.G. Thorpe and L.J. Kricka, Enhanced chemoluminescent reactions catalyzed by horse radish peroxide, *Methods Enzymol* **133**:331 (1986).
25. I. Bronstein and L.J. Kricka, Instrumentation for luminescent assays, *Am Clin Lab* **Jan** (1990).
26. G. Bemroider, M. Holztrattner and P. Hammerl, Imaging receptor binding by luminescence, *in*: "Proc. VII Int. Symp. on Bioluminescence and Chemoluminescence", Banff, Alberta (1993).
27. R.E. Ansorge, C.E. Hooper, W.W. Neale and J.G. Rushbrooke, Recent developments in low-light imaging systems, *in*: "Bioluminescence and Chemoluminescene, Current Status", P.E. Stanley and L.J. Kricka, eds., J. Wiley & Sons, Chichester (1991).
28. S. Walenta, M. Dellian, A.E. Goetz, G.E.H. Kuhnle,and W. Mueller-Klieser, Pixel to pixel correlation between images of absolute ATP concentrations and blood flow in tum ours, *Br J Cancer* **66**:1099 (1992).
29. R.D. Allen and N.S. Allen, Video-enhanced microscopy with a computer frame memory, *J Microsc* **129**:3 (1983).
30. V. Wielke, Laser scanning in microscopy, *SPIE* **396**:164 (1983).
31. M.M. Kersker, A. Buonaquisti, B.A. Weavers, and J.T. Mastovich, Imaging with electron microscopes, *Advanced Imaging* **Feb**:46 1990.
32. D.E. Hillman, R.R. Llinás, M. Canaday, and G. Mahoney, Concepts and methods of image acquisition, frame processing and image data representations, *in*: "Three-Dimensional Neuroimaging", A.W. Toga, ed., Raven Press, New York (1990).
33. L. Levitt, S. Spector, A. Sjoerdsma and S. Udenfriend, Elucidation of the rate limiting step in norepinephrine biosysnthesis in the perfused guinea-pig heart, *J Pharmacol Exp Ther* **148**:1 (1965).
34. D. Lake, Electronic cameras, sensors and electronic exposure control, *Advanced Imaging* **Nov**:24 (1990).
35. G. Bemroider, Recognition and classification of structure by means of stereological methods in neurobiology, *J Microsc* **107**:287 (1976).
36. G. Bemroider, Lattice, points and probability - the foundation of optical computing and stereology, *in*: "Special Issues of Pract. Metallography", J.L. Chermant, ed., Vol. 8 pp 51-60, Dr. Riederer Verl., Stuttgart (1978).
37. G. Bemroider and H. Adam, Synthetized point information sampling, *Mikroskopie (Vienna) Spec Suppl* **37**:119 (1980).
38. E.R. Weibel, "Practical Methods for Biological Morphometry," Academic Press (1980).
39. E.E. Underwood, "Quantitative Stereology," Addison Wesley, Reading, MA (1970).
40. D.C. Sterio, The unbiased estimation of number and sizes of arbitrary particles using the disector, *J Microsc* **134**:127 (1984).
41. L.M. Cruz-Orive, Particle number can be estimated using a disector of unknown thickness: the selector, *J Microsc* **145**:121 (1987).
42. H. Braendgaard and H.J. Gundersen, The impact of recent stereological advances on quantitative studies of the nervous system, *Jn Neurosci Meth* **18**:39 (1986).
43. S. Walenta, M. Dellian, A.E. Goetz, G.E.H. Kuhnle, and W. Mueller-Klieser, Pixel to pixel correlation between images of absolute ATP concentrations and blood flow in tumours, *Br J Cancer* **66**:1099 (1992).
44. A.P. Schaap, H. Akhavan, L.J. Romano, Chemoluminescent substrates for alkaline phosphatase: applications to ultrasensitive enzyme-linked immunoassays and DNA probes, *Clin Chem* **35**:1863 (1989).
45. F. Crick and E. Jones, Backwardness of human neuroanatomy, *Nature* **361**:109 (1993).

CONTRIBUTORS

ERICH ARRER, Central Laboratory, Immunohistochemistry and Biochemistry Unit, General Hospital, A-5020 Salzburg, Austria

SEBASTIAN BACHMANN, Institute for Anatomy and Cell Biology Ruprecht-Karls-University, Heidelberg, FRG

GUSTAV BERNROIDER, Institute for Zoology, University of Salzburg, A-5020 Salzburg, Austria

GORM DANSCHER, Institute of Anatomy, Department of Neurobiology, The Steno Institute, University, DK-8000 Aarhus, Denmark

OTTO DIETZE, Institute of Pathological Anatomy, Immunohistochemistry and Biochemistry Unit, General Hospital, Salzburg, Austria

PETER M. ECKL, Institute of Genetics and General Biology, University of Salzburg, Austria

GIAN-LUCA FERRI, Department of Cytomorphology, University of Cagliari, 09124 Cagliari, Italy

AXEL FISCHER, Institute for Anatomy and Cell Biology Ruprecht-Karls-University, Heidelberg, FRG

ROSA MARIA GAUDIO, Department of Cytomorphology, University of Cagliari, 09124 Cagliari, Italy

FRANK GIRARDI, University Clinic of Obstetrics and Gynecology, Landeskrankenhaus Graz, Austria

ANTON-HELMUT GRAF, Department of Gynecology and Obstetrics and the Institute of Pathological Anatomy, Immunohistochemistry and Biochemistry Unit, General Hospital, A-5020 Salzburg, Austria

LARS GRIMELIUS, Institute of Pathology, University of Uppsala, Uppsala, Sweden

JIANG GU, Deborah Research Institute, Browns Mills, NJ 08015, USA

GERHARD W. HACKER, Institute of Pathological Anatomy, Immunohistochemistry and Biochemistry Unit, General Hospital, A-5020 Salzburg, Austria

CORNELIA HAUSER-KRONBERGER, Institute of Pathological Anatomy, Immunohistochemistry and Biochemistry Unit, General Hospital, A-5020 Salzburg, Austria

ANTON HERMANN, University of Salzburg, Department of Animal-Physiology, Institute of Zoology, A-5020 Salzburg, Austria

VERENA HERMES, University Clinic of Obstetrics and Gynecology, Landeskrankenhaus Graz, Austria

HUBERT H. KERSCHBAUM, University of Salzburg, Department of Animal-Physiology, Institute of Zoology, A-5020 Salzburg, Austria

WOLFGANG KRAAZ, Department of Pathology, University Hospital, S-751 85 Uppsala, Sweden

LÄSZLÁ KRENÁCS, Department of Pathology, Albert Szent-Györgyi University of Medicine, Szeged, Hungary

TIBOR KRENÁCS, Department of Pathology, Albert Szent-Györgyi University of Medicine, Szeged, Hungary

WOLFGANG KUMMER, Institute for Anatomy and Cell Biology, Philipps-University, Marburg, FRG

PETER M. LACKIE, Southampton University Medicine, Southampton General Hospital, Southampton SO9 4XY, England

ANTHONY S.Y. LEONG, University of Adelaide, and Division of Tissue Pathology, Institute of Medical & Veterinary Science, Adelaide, South Australia, Australia

DORIS MACK, Department of Urology, Immunohistochemistry and Biochemistry Unit, Salzburg General Hospital, A-5020 Salzburg, Austria

WOLFGANG H. MUSS, Institute of Pathological Anatomy, Immunohistochemistry and Biochemistry Unit, General Hospital, A-5020 Salzburg, Austria

GERHARD NÖHAMMER, Institute of Biochemistry, Karl-Franzens-University Graz, Austria

ALDO PAOLICCHI, Institute of General Pathology, University of Pisa, Italy

HELLMUTH PICKEL, University Clinic of Obstetrics and Gynecology, Landeskrankenhaus Graz, Austria

JULIA M. POLAK, Department of Histochemistry, Royal Postgraduate Medical School, Hammersmith Hospital, London W12 ONN, UK

WANNEE ROJANAPO, National Cancer Institute, Bangkok, Thailand

JÜRGEN ROTH, Division of Cell and Molecular Pathology, Department of Pathology, University of Zürich, Switzerland

EVA RYLANDER, Department of Gynecology and Obstetrics, University Hospital, S-751 85 Uppsala, Sweden

JAN SÄLLSTRÖM, Department of Pathology, University Hospital, S-751 85 Uppsala, Sweden

LENA SCHEIBENPFLUG, Department of Pathology, University Hospital, S-751 85 Uppsala, Sweden

ANGELIKA SCHIECHL, Institute of Pathological Anatomy, General Hospital, A-5020 Salzburg, Austria

GERHARD SKOFITSCH, Histopharmacology Unit, Department of Zoology, University of Graz, A-8010 Graz, Austria

ANDERS STRAND, Department of Venereology and Dermatology, University Hospital, S-751 85 Uppsala, Sweden

HUICI SU, Department of Histology and Embryology, Fourth Military Medical University, Xián, Shoanxi, P.R. China

SIEW-KHIN TANG, Department of Anatomical Pathology, Alfred Hospital, Melbourne, Victoria, Australia

GIORGIO TERENGHI, Department of Histochemistry, Royal Postgraduate Medical School, Hammersmith Hospital, London W12 ONN, UK

ROBERTO TONGIANI, Institute of General Pathology, University of Pisa, Italy

BELA VIGH, 2nd Department of Anatomy, Semmelweis Medical University, H-1094 Budapest IX, Hungary

INGEBORG VIGH-TEICHMANN, Neuroendocrine Section of the Hungarian Academy of Sciences, Semmelweis Medical University, H-1094 Budapest IX, Hungary

ERIK WILANDER, Department of Pathology, University Hospital, Uppsala, Sweden

GERHARD WIRNSBERGER, Medical University Clinic, Department of Nephrology, Landeskrankenhaus Graz, Austria

INGEBORG ZEHBE, Department of Pathology, University Hospital, S-751 85 Uppsala, Sweden

INDEX